Lecture Notes in Artificial Intelligence 7416

Subseries of Lecture Notes in Computer Science

Thomas Röfer   N. Michael Mayer
Jesus Savage   Uluç Saranlı (Eds.)

RoboCup 2011:
Robot Soccer
World Cup XV

Springer

Series Editors

Randy Goebel, University of Alberta, Edmonton, Canada
Jörg Siekmann, University of Saarland, Saarbrücken, Germany
Wolfgang Wahlster, DFKI and University of Saarland, Saarbrücken, Germany

Volume Editors

Thomas Röfer
Deutsches Forschungszentrum für Künstliche Intelligenz, Cyber-Physical Systems
Enrique-Schmidt-Str. 5
28359 Bremen, Germany
E-mail: thomas.roefer@dfki.de

Norbert Michael Mayer
National Chung Cheng University, Department of Electrical Engineering
168 University Road
62102 Min-Hsiung, Chia-Yi, Taiwan
E-mail: mikemayer@ccu.edu.tw

Jesus Savage
Universidad Nacional Autonoma de Mexico, Bio-Robotics Laboratory
CP 04250 Mexico, D.F., Mexico
E-mail: robotssavage@gmail.com

Uluç Saranlı
Middle East Technical University Ankara, Department of Computer Engineering
06531 Ankara, Turkey
E-mail: saranli@ceng.metu.edu.tr

ISSN 0302-9743   e-ISSN 1611-3349
ISBN 978-3-642-32059-0   e-ISBN 978-3-642-32060-6
DOI 10.1007/978-3-642-32060-6
Springer Heidelberg Dordrecht London New York

Library of Congress Control Number: 2012942791

CR Subject Classification (1998): I.2, C.2.4, D.2.7, H.5, I.4, J.4

LNCS Sublibrary: SL 7 – Artificial Intelligence]

Typesetting: Camera-ready by author, data conversion by Scientific Publishing Services, Chennai, India

Printed on acid-free paper

Springer is part of Springer Science+Business Media (www.springer.com)

Preface

Fifteen years ago, RoboCup started with the vision "By the mid-21st century, a team of fully autonomous humanoid soccer players shall win the soccer game [..] against the winner of the most recent World Cup," a landmark project that has attracted a large number of international researchers over the years, the RoboCup community. Since the beginning, enormous progress has been made. RoboCup has also broadened its focus by addressing education (RoboCupJunior) as well as research fields closer to applications (RoboCupRescue and RoboCup@Home).

The RoboCup Symposium is the scientific core meeting of the RoboCup community. While the RoboCup competitions and demonstrations showcase the effectiveness of the systems in practice, the RoboCup Symposium answers the question of how they work. Each year the RoboCup Symposium is held on a different continent. In 2011, with Istanbul playing host, the 15th RoboCup Symposium became the first to take place on two continents at once.

We decided to organize the Symposium as a single track conference in 2011. The great advantage of this is that it allowed all participants to attend all sessions of the symposium. Thus, we blurred the classic separation of RoboCup into the three major areas of soccer, rescue, and service robotics. To further blur these distinctions, oral presentations in the three sessions were not divided by particular topic, but jumped between different areas and leagues. This approach demonstrated the breadth and depth of research within the RoboCup community and is reflected by the sequence of the papers in this book.

The consequence of a single-track schedule is that it results in a smaller number of talks. This was offset by presenting the majority of papers during two poster sessions, which allowed for greater interactions between presenters and their audience. To underline the importance of the poster presentations and to make the participants' choice about which posters to visit more informed, each poster session was preceded by a plenary presentation of so-called poster teasers, i.e., micro-talks of one minute and one slide per poster.

For the 15th RoboCup Symposium, we received 97 paper submissions, covering all areas of RoboCup research. All papers were carefully reviewed by the International Program Committee. Each submission was examined by three members of the Program Committee and the final decision was made by the Co-chairs. Twelve papers were selected for oral presentation and 32 for poster presentation. The authors Jörg Stückler and Sven Behnke received the best paper award for their contribution on "Compliant Task-Space Control with Back-Drivable Servo Actuators."

The RoboCup Symposium has a long tradition of inviting keynote speakers from outside the RoboCup community who have made outstanding contributions in the field of artificial intelligence. The speakers at the RoboCup 2011 Symposium, Luc Steels and Dieter Fox, talked about their ongoing research.

Luc works on language processing and humanoid robots, and asked the question "When Will They Start to Speak?". Dieter discussed the most talked-about sensor of the previous year in his keynote entitled "RGB-D Cameras: Challenges and Opportunities".

Another symposium tradition is the panel discussion. In contrast to previous years, where the panelists were usually RoboCup officials, in 2011 the leaders of some of the winning teams discussed the question "What Does It Really Take to Win?". The goal will be to develop a general perspective on the challenges of RoboCup built from the insights of the competition winners and their methods and approaches. They were also invited to contribute to this book with one paper per league. These "Champion Papers" were reviewed by members of the technical committees of the respective leagues and those accepted start the main part of this book.

We want to thank the RoboCup Organizing Committee for making the local arrangements, which worked out very well. In particular we want to thank the General Chair for RoboCup 2011, H. Levent Akin, and his Co-chairs, Çetin Meriçli and Tekin Meriçli. We would also like to thank the organizers of the previous RoboCup Symposium, who were very helpful and let us benefit from their experience. Further, we would like to thank all of the Program Committee members for their hard work and the great job of reviewing they did. Last but not least, we want to thank all of the authors for their contributions. The next RoboCup competition will run during June 18–24, 2012, in Mexico City with the conjoint symposium taking place on June 24.

March 2012   Thomas Röfer
                                                                     Norbert Michael Mayer
                                                                            Jesus Savage
                                                                           Uluç Saranlı

RoboCup Leagues in 2011

RoboCup consists of the four main areas RoboCupSoccer, RoboCupRescue, RoboCup@Home, and RoboCupJunior. Most of these areas are divided into a number of leagues that address different aspects of their overall research goals. While RoboCupJunior is presented on pages 63ff., the major leagues are described by their RoboCup Executive Committee members on the following pages.

Soccer Simulation League

Akiyama, Hideisa	AIST, Japan
Dorer, Klaus	Hochschule Offenburg, Germany
Talay, Sanem Sariel	Istanbul Technical University, Turkey

Soccer Small Size League

Akar, Mehmet	Boğaziçi University, Turkey
Sukvichai, Kanjanapan	Kasetsart University, Thailand
Weitzenfeld, Alfredo	USF Polytechnic, USA

Soccer Middle Size League

Cunha, Bernardo	University of Aveiro, Portugal
Merry, Roel	Eindhoven University of Technology, The Netherlands
Takemura, Yasunori	Nippon Bunri University, Japan

Soccer Standard Platform League

Lagoudakis, Michail	Technical University of Crete, Greece
Lee, Daniel	University of Pennsylvania, USA
Hall, Brad	University of New South Wales, Australia
Kaminka, Gal	Bar Ilan University, Israel

Soccer Humanoid League

Baltes, Jacky	University of Manitoba, Canada
Lupian, Luis F.	Universidad La Salle, Mexico
Behnke, Sven	University of Bonn, Germany

Rescue Robot League

Pellenz, Johannes	University of Koblenz-Landau, Germany
Kimura, Tetsuya	Nagaoka University of Technology, Japan
Mihankhah, Ehsan	K. N. Toosi University of Technology, Iran
Suthakorn, Jackrit	Mahidol University, Thailand

Rescue Simulation League

Ito, Nobuhiro   Aichi Institute of Technology, Japan
Balakirsky, Stephen   NIST, USA
Kleiner, Alexander   Freiburg University, Germany

@Home League

Iocchi, Luca   Sapienza University of Rome, Italy
Ruiz-de-Solar, Javier   Universidad de Chile, Chile
van der Zant, Tijn   University of Groningen, The Netherlands
Sugiura, Komei   NICT, Japan

Soccer Simulation League

Soccer Simulation is split into two subleagues: 2D simulation and 3D simulation.

2D Simulation

In 2D simulation, two teams of 11 simulated autonomous player agents and a coach agent play a game of soccer with very realistic rules and game play. In 2011, two major changes were introduced: the red card rule and the improved ball catch model for goalie agents. When a player performs an intentional foul command, an automatic referee might detect the red card rule. In that case, the player is sent off the field and the team has to reconsider its strategy.

The current major research topics in 2D simulation are flexibility and an online adaptation. Some teams introduced new approaches such as an online tactics planning method and a strategy management using a coach agent. In future competitions, online coaching techniques will become more important in order to adapt to the environment.

In all, 20 teams from 11 countries passed the qualification and finally 17 teams participated in the 2D competition. The winner was WrightEagle from the University of Science and Technology of China (China), followed by HELIOS from Fukuoka University, Osaka Prefecture University and AIST (Japan), and MarliK from University of Guilan (Iran).

3D Simulation

In 3D simulation, simulated Nao robots play in a physically realistic 3D environment. It is therefore more similar to real robot leagues than 2D simulation yet has the advantages of simulation that allow one to play in big teams, simplify learning approaches, and simplify robot model changes.

The 2011 competition introduced a couple of new features. The games were run in 9 vs. 9 player mode (was 6 vs. 6 in 2010). The field size was increased to keep the space per player roughly constant. New crowding rules were introduced to keep the number of players approaching the ball at a reasonable level. If two players of the same team are closer than 1 m to the ball, the farther of the two is beamed outside the field by the automated referee. Also, the number of defenders inside the goal area was limited to two. Both changes improved game quality considerably. Moreover, a new visualization software, roboViz, was used to show the games. Appart from high-end 3D graphics, it was used for the first time to allow spectators to watch the games in 3D using a specialized beamer and 3D glasses.

Technically, teams showed improved walking skills, a new skill to kick the ball high, a skill to kick the ball while running, backward pull kicks, and improved get-up behaviors.

In all, 22 teams from 12 countries participated in the 3D simulation competition. The winner was UTAustinVilla from the University of Texas (USA) followed by CIT3D from Changzhou Institute of Technology (China) and Apollo3D from Nanjing University (China).

Soccer Small Size League

The Small Size League (SSL) is one of the more exciting leagues of RoboCup, as two teams each consisting of 5 robots with limited size (max height 15 cm, max diameter 18 cm) play soccer at a high pace with an orange golf ball on a 6.05 m-by-4.0 5m green carpeted field. All objects on the field are tracked by a standardized vision system and off-field computers are used to communicate referee commands and position information to the robots. In SSL, building a successful team requires clever design, implementation, and integration of many hardware and software sub-components into a robustly functioning whole; hence, SSL soccer remains a very interesting and challenging domain for research and education.

The journey to SSL 2011 started when 28 teams from 12 different countries declared their interest in participation by submitting a pre-registration note by January 31, 2011. The deadline for submission of team description papers and videos for all teams was the end of February, after which 21 teams were choosen to participate in the tournament while two teams were conditionally qualified, with the requirement that they would demonstrate sufficient game play before the games started. Later, two teams dropped out owing to financial problems; however, the remaining 20 teams from 11 countries participated in RoboCup 2011. These 20 teams listed in Table 1 played each other first in four groups and then the games continued based on elimination starting from the second round, at the end of which team Skuba was the winner, Immortals the runner up, and MRL got the third place.

Two technical challenges (*Dynamic Navigation* and *Mixed Team Match*) and a (*Large Field*) demo was held in Istanbul 2011. The aim of the *Dynamic Navigation* technical challenge, was to examine the ability of robots to safely navigate in a dynamic environment (Winner: Skuba, 2nd place: MRL, 3rd place: Thunderbots). The *Mixed Team Match* technical challenge, which was also won by team Skuba, was a full match between mixed teams, each composed of five robots from two different SSL teams. The *Large Field* demo was carried out on a larger field twice as big as the regular one, with the objective of exploring the possibility of carrying out future SSL games on larger fields.

Table 1. Teams that participated in SSI 2011

Country of Origin	Teams
Brazil	RoboFEI
Canada	Thunderbots
China	ZjuNlict
Colombia	STOxs, Bochica
Germany	ER-Force, Tigers Mannheim
Iran	Immortals, MRL, Parsian, Cyrus
Japan	RoboDragons, KIKS, ODENS
Mexico	Eagle Knights
Thailand	Skuba
Turkey	BRocks, RoboTurk
USA	RoboJackets, RFC Cambridge

Soccer Middle Size League

The Middle Size League is, for the time being, the best interface league between Simulated 2D Soccer and real, fully autonomous, nonhumanoid robots. In this league robots of up to 50 cm in diameter and 80 cm in height play soccer in teams of up to five robots, using a regular size FIFA soccer ball on a field similar to a scaled human soccer field (currently 18 m × 12 m). Robots are fully autonomous with all sensors on-board. Cooperation is established by means of limited bandwidth wireless communication between team robots, and between robots and a nonoperated computer base station that can act as a coach. No human interaction with the robots is allowed. The current research focus is on cooperation at all levels, from team coordination using dynamic role assignment to attack and defense planning, going through active in game ball passing. Cooperation between heterogeneous robots, 3D efficient fast robotic vision, immunity to environment variable conditions (such as illumination changes), and perception levels are among other lines of research in this league.

The 2011 event presented some significant achievements mostly in the cooperation area, with a growing number of teams performing active in game passing, and in robustness and reliability of hardware and software (the first game ever played without the need to remove any robot from the field during the overall game). A demonstration on how to incorporate software from 2D simulation in robot team coordination using dynamic role and positioning assignment and role-based set-plays won the Scientific Challenge, while Hibikino-Musashi won the Technical Challenge involving playing with arbitrary colored balls, dribbling while navigating on a cluttered area, and team play with passing and scoring.

A total of 15 teams qualified for the 2011 RoboCup event. From those, 12 actually participated in the competition, representing teams from Europe, Asia, Middle East, and Australia. The competition was organized in three round robins followed by Mid-Finals and the Final. Table 2 shows the top eight final classification.

Table 2. Final classification

Rank	Team	Origin	Country
1	Water	Beijing I. S. T. University	China
2	Tech United Eindhoven	Eindhoven University of Technology	The Netherlands
3	CAMBADA	IEETA/DETI, University of Aveiro	Portugal
4	RFC Stuttgart	University of Stuttgart	Germany
5	MRL	Qazvin Islamic Azad University	Iran
6	Hibikino-Musashi	University of Kitakyushu	Japan
7	Carpe Noctem	University of Kassel	Germany
8	ISePorto	LSA-ISEP	Portugal

Soccer Standard Platform League

The Standard Platform League (SPL) [www.tzi.de/spl] of the RoboCup competition is unique among the robot soccer leagues, in that all participating teams compete with identical robot platforms, thereby accentuating advances in algorithmic development for fully autonomous robots. The current standard platform is the humanoid NAO robot, a 21 degrees-of-freedom bipedal robot measuring 58 cm in height and weighing 4.3 kg, built by Aldebaran Robotics in Paris, France.

Fig. 1. The SPL championship match held at RoboCup 2011 in Istanbul, Turkey.

The 2011 SPL soccer competition attracted 27 teams representing universities from across Europe, Asia, Africa, North and South America, and Australia. The soccer tournament featured two separate round robin qualifying rounds, with the ensuing final eight teams playing in the elimination round. The games consisted of autonomous teams (of four robots each) playing against each other with an orange hockey ball on a 4 × 6-m green soccer field with yellow and blue goalposts, solely using visual, ultrasonic, and proprioceptive sensors. The matches consisted of two 10-minute halves, with ties broken via penalty kick shootouts in the elimination round. The winners of the 2011 SPL soccer competition were:

1. *B-Human*, Universität Bremen and DFKI Bremen, Germany
2. *Nao Devils Dortmund*, TU Dortmund, Germany
3. *NTU RobotPAL*, National Taiwan University, Taiwan

In a separate open technical challenge competition, teams displayed interesting research in the field of autonomous robots. The winners were determined by votes from all participating teams, with the following results:

1. *RoboEireann*, National University of Ireland, Maynooth, Ireland
2. *Noxious-Kouretes*, University of Wales, UK, Oxford University, UK, TU Crete, Greece
3. *rUNSWift*, The University of New South Wales, Australia

The league also held a fun, unofficial challenge of a robot speed race from the corners to the center of the field. Discussion on future SPL competitions focused on less structured fields, uniformly colored goals, and teams with more robots.

Soccer Humanoid League

In the Humanoid League, mostly self-constructed robots with a human-like body plan compete in three size classes: KidSize (<60 cm), TeenSize (100–120 cm), and AdultSize (>130 cm). While the KidSize and Teensize robots are playing soccer games with three or two players per team, respectively, the AdultSize robots engage in 1 vs. 1 Dribble & Kick competitions. In addition, all three classes face technical challenges, such as dribbling the ball through an obstacle course, double-passing, and throw-in of the ball. In the RoboCup 2011 competitions, 22 KidSize teams, three TeenSize teams, and seven AdultSize teams participated.

The challenges of playing soccer with humanoid robots include fast and flexible bipedal locomotion, controlling dynamic full-body motions, maintaining balance in the presence of disturbances, robust visual perception of the game situation, individual soccer skills, and team coordination. The 2011 competition showed notable progress in both individual robot skills and team play.

One of the highlights of the KidSize class was shown by the goalie of team Darmstadt Dribblers: During a match, the robot picked up the ball and threw it across the field. Another notable development was the introduction of DARwIn-OP, an open humanoid robot, developed by Virginia Tech and University of Pennsylvania and commercialized by the Korean company Robotis. This robot performed very well and consequently Team DARwIn (USA) won the KidSize soccer tournament. Runner-up was CIT Brains (Japan) and the third place went to Darmstadt Dribblers (Germany), who also won the KidSize technical challenges. In the TeenSize class, the playing field was enlarged to 9×6 m. Team NimbRo of the University of Bonn (Germany), played with two robots: Bodo as goalie and Dynaped as field player. Dynaped scored very reliably and avoided obstacles carefully, such that it never lost balance in the final, which was played vs. team KMUTT (Thailand). NimbRo won the TeenSize competition for the third time in a row. In the AdultSize class, team CHARLIE of Virginia Tech

(USA), competed with their new robot Charlie 2. It reliably approached the ball, dribbled it, and kicked it toward the goal. CHARLE won the Dribble & Kick final vs. Robo Erectus (Singapore). Tsinghua Hephaestus (China) reached the third place. From the winners of the three size classes, the largest robot CHARLIE was elected Best Humanoid and received the Louis Vuitton award (see Fig. 2).

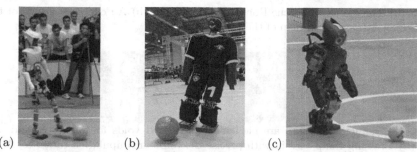

(a)   (b)   (c)

Fig. 2. Best humanoids: (a) Charlie (1st); (b) NimbRo (2nd); (c) DARwIn (3rd).

Rescue Robot League

The objective of the RoboCupRescue Robot League is to promote the development of intelligent, highly mobile, dexterous robots that can improve the safety and effectiveness of emergency responders performing hazardous operational tasks. We demonstrate and compare advances in robot mobility, perception, localization and mapping, mobile manipulation, practical operator interfaces, and assistive autonomous behaviors. We use annual competitions and subsequent field exercises with emergency responders to accomplish the following:

- Increase awareness about the challenges involved in deploying robots for emergency response applications
- Provide objective performance evaluations based on *DHS-NIST-ASTM International Standard Test Methods for Response Robots*
- Introduce Best-In-Class implementations to emergency responders within their own training facilities
- Support ASTM International Standards Committee on Homeland Security Applications (E54.08.01)

Participating robot teams search for simulated victims emitting several signs of life within a maze of terrains, obstacles, and manipulation tasks based on emerging standard test methods for response robots. As robots demonstrate successes against the obstacles posed in the arenas, the level of difficulty continually increases so the arenas always serve as a stepping-stone from the laboratory to the real world. Additional awards are given for Best-in-Class Autonomy, Best-in-Class Mobility, and Best-in-Class Manipulation. (More details at: http://wiki.robocup.org/wiki/Robot_League).

(a)   (b)   (c)

Fig. 3. (a) The RoboCupRescue Robot League arena. (b) An overview design of the arena. (c) A laser-scanned map of the as-built arena.

(a)   (d)

Fig. 4. (a) Simulated victims are hidden in fabricated voids for robots to find and map. (b) Crossing ramp terrain with precise dexterous manipulation tasks. (c) More advanced mobility obstacles. (d) The stepfield terrain provides the most comprehensive test of advanced mobility.

Rescue Simulation League

The Rescue Simulation League consists of three distinct competitions. These competitions are the agent competition, the virtual robots competition, and the infrastructure competition. Twenty-two teams from eight different countries participated in the 2011 competitions of this league. The main goal of the infrastructure competition is to foster the development of new simulator components, while the agent and virtual robot competition are focused on developing intelligent agents that make the right decisions in simulated disasters as described below.

The Agent Competition consists of a simulation platform that resembles a city after an earthquake. In this environment intelligent agents can be spawned, which influence the cause of events in the simulation. The agents have the role of police forces, fire brigades, and ambulance teams.

Recently, the simulation league has initiated a new type of challenge having the goal to significantly simplify the entry of newcomer teams. The idea is to extract from the entire problem addressed by the agents certain aspects such as task allocation, team formation, and route planning, and to present these sub-problems in an isolated manner as stand-alone problem scenarios with an abstract interface. As a consequence, participating teams can focus on their research on an aspect of the game, without having to solve all low-level issues.

The Virtual Robot Competition has as its goal the study of how a team of robots can work together to get a situation assessment of a devastated area as fast as possible. This will allow first responders to be well informed when they enter the danger zone.

In the previous years, different aspects of this task were studied as separate problems. This meant that a team could get points in a teleoperation test, a mapping test, and a deployment test. For the 2011 competition, all aspects were combined in one comprehensive mission; robots have to explore the environment, find victims, and stay with the victims (working as communication relay).

Fig. 5. Victim detection as presented by one of the Virtual Robot teams.

At the end of the Istanbul competition a workshop was held. Several teams reported that progress had been made on automatic victim detection, which allows the operator to concentrate on the coordination of the team. In the ideal case, the robot detects a victim on visual clues, which are confirmed by the human operator.

@Home League

The RoboCup@Home league aims to develop service and assistive robot technology with high relevance for future personal domestic applications. A set of benchmark tests is used to evaluate the robot's abilities and performance in a realistic nonstandardized home environment setting. The focus is on the following domains but is not limited to: Human–Robot–Interaction and Cooperation, Navigation and Mapping in Dynamic Environments, Computer Vision and Object Recognition Under Natural Light Conditions, Object Manipulation, Adaptive Behaviors, Behavior Integration, Ambient Intelligence, Standardization and System Integration. Using statistical benchmarking we came to the conclusion that the progress in @Home is very rapid. Tasks that were impossible to solve 5 years ago are now performed routinely by many robotic systems. About 40 teams are active in the world and the scientific problems in the @Home league resulted in at least 140 papers in 5 years, as far as we can track.

An interesting aspect of the @Home competition is that in the last few years the domain areas have begun to expand. Because themselves the robots have to behave in uncertain real-world situations, we also went into the real world. Since 2010 we have been shopping with the robots in real stores, showing high reliability, adaptiveness, and safety. This is exactly what we are looking for: robots that can cooperate with people in the real world. Also, the test where the robot has to follow a human using limited unsupervised training of maximally 60s was done outside the apartment, with great success.

Another important new feature in @Home is that, after 2 years of international discussions, we introduced the first cognitive benchmark, where the actual task to be performed is not known beforehand by the robot. As far as we know this is a first both in and outside RoboCup. During the test we put the robots in difficult situations, give them very complex spoken commands and also under-specified commands. For example, we might say to the robot: "Go to the kitchen and find John," while John is not there, or "I'm hungry" and the robot must figure out what to do.

For the past few years we have only allowed what we call "natural interaction". This has resulted in robots that we can talk to and we can also use gestures to control them. It is important that the interaction is natural, since we want everyday people to be able to use them. In the future we want to be able to state: RoboCup@Home has been an important factor in the creation of social robots that are commercially viable.

Organization

Symposium Co-chairs

Röfer, Thomas   German Research Center for Artificial
                                   Intelligence (DFKI), Germany
Mayer, Norbert Michael   National Chung Cheng University, Taiwan
Savage, Jesus   UNAM, Mexico
Saranlı, Uluç   Bilkent University, Turkey

International Program Committee

Acosta, Carlos   Singapore Polytechnic, Singapore
Akar, Mehmet   Boğaziçi University, Turkey
Akin, H. Levent   Boğaziçi University, Turkey
Almeida, Luis   University of Porto, Portugal
Amigoni, Francesco   Politecnico di Milano, Italy
Anderson, John   University of Manitoba, Canada
Baltes, Jacky   University of Manitoba, Canada
Behnke, Sven   University of Bonn, Germany
Birk, Andreas   Jacobs University, Germany
Bonarini, Andrea   Politecnico di Milano, Italy
Bredenfeld, Ansgar   Fraunhofer IAIS, Germany
Brena, Ramon   Tecnologico de Monterrey, Mexico
Burkhard, Hans-Dieter   Humboldt University Berlin, Germany
Caglioti, Vincenzo   Politecnico of Milano, Italy
Carpin, Stefano   University of California, Merced, USA
Cassinis, Riccardo   University of Brescia, Italy
Chalup, Stephan   The University of Newcastle, Australia
Chen, Weidong   Shanghai Jiao Tong University, China
Chernova, Sonia   Worcester Polytechnic Institute, USA
Chown, Eric   Bowdoin College, USA
Costa, Paulo   University of Porto, Portugal
Dias, M. Bernardine   Carnegie Mellon University, USA
Eguchi, Amy   Bloomfield College, USA
Farinelli, Alessandro   University of Southampton, UK
Frese, Udo   Universität Bremen, Germany
Frontoni, Emanuele   Università Politecnica delle Marche, Italy
Ghaderi, Ahmad   Iran
Gini, Giuseppina   Politecnico di Milano, Italy
Hengst, Bernhard   University of New South Wales, Australia
Hermans, Tucker   Georgia Institute of Technology, USA

Hester, Todd	University of Texas, Austin, USA
Hofmann, Alexander	FH Technikum Wien, Austria
Hong, Dennis	Virginia Tech, USA
Hugel, Vincent	Versailles University, France
Indiveri, Giovanni	Università del Salento, Italy
Iocchi, Luca	Sapienza University of Rome, Italy
Jahshan, David	University of Melbourne, Australia
Jamzad, Mansour	Sharif University of Technology, Iran
Kalyanakrishnan, Shivaram	University of Texas, Austin, USA
Kenn, Holger	European Microsoft Innovation Center, Germany
Kimura, Tetsuya	Nagaoka University of Technology, Japan
Kleiner, Alexander	Freiburg University, Germany
Kraetzschmar, Gerhard	Bonn Rhein Sieg University, Germany
Lagoudakis, Michail	Technical University of Crete, Greece
Laue, Tim	German Research Center for Artificial Intelligence (DFKI), Germany
Levy, Simon	Washington and Lee University, USA
Lima, Pedro U.	Instituto Superior Técnico, TU Lisbon, Portugal
Lopes, Gil	Minho University, Portugal
Matsumoto, Akihiro	Toyo University, Japan
Matteucci, Matteo	Politecnico di Milano, Italy
Menegatti, Emanuele	The University of Padua, Italy
Meriçli, Çetin	Boğaziçi University, Turkey
Meriçli, Tekin	Boğaziçi University, Turkey
Middleton, Rick	The National University of Ireland Maynooth, Ireland
Mingguo, Zhao	Tsinghua University, China
Monekosso, Dorothy	University of Ulster, UK
Nakashima, Tomoharu	Osaka Prefecture University, Japan
Nardi, Daniele	Sapienza University of Rome, Italy
Naruse, Tadashi	Aichi Prefectural University, Japan
Nazemi, Eslam	Shahid Beheshti University, Iran
Noda, Itsuki	National Institute of Advanced Industrial Science and Technology, Japan
Obst, Oliver	CSIRO ICT Centre, Australia
Ohashi, Takeshi	Kyushu Institute of Technology, Japan
Pagello, Enrico	University of Padua, Italy
Pirri, Fiora	Sapienza University of Rome, Italy
Plöger, Paul G.	Bonn Rhein Sieg University of Applied Sciences, Germany
Polani, Daniel	University of Hertfordshire, UK
Quinlan, Michael	University of Texas, Austin, USA

Reis, Luis Paulo	University of Porto, Portugal
Ribeiro, A. Fernando	Minho University, Portugal
Rojas, Raúl	Free University of Berlin, Germany
Ruiz-del-Solar, Javier	Universidad de Chile, Chile
Rybski, Paul	Carnegie Mellon University, USA
Sammut, Claude	University of New South Wales, Australia
Santos, Vitor	University of Aveiro, Portugal
Schiffer, Stefan	RWTH Aachen University, Germany
Shiry, Saeed	Amirkabir University of Technology, Iran
Sridharan, Mohan	Texas Tech University, USA
Steinbauer, Gerald	Graz University of Technology, Austria
Takahashi, Tomoichi	Meijo University, Japan
Takahashi, Yasutake	Fukui University, Japan
Tawfik, Ahmed	University of Windsor, Canada
Tu, Kuo-Yang	National Kaohsiung First University of Science and Technology, Taiwan
van der Zant, Tijn	University of Groningen, The Netherlands
Velastin, Sergio	University of Kingston, UK
Visser, Ubbo	University of Miami, USA
von Stryk, Oskar	Technische Universität Darmstadt, Germany
Weitzenfeld, Alfredo	USF Polytechnic, USA
Williams, Mary-Anne	University of Technology, Sydney, Australia
Wotawa, Franz	Technische Universität Graz, Austria
Xie, Guangming	Peking University, China
Zickler, Stefan	Carnegie Mellon University, USA

Additional Reviewers

Abeyruwan, Saminda	University of Miami, USA
Aboul-Ela, Magdy	The French University, Egypt
Abreu, Pedro	University of Porto, Portugal
Almeida, Fernando	University of Aveiro, Portugal
Basaran, Ersin	Boğaziçi University, Turkey
Birbach, Oliver	German Research Center for Artificial Intelligence (DFKI), Germany
Browning, Brett	Carnegie Mellon University, USA
Calisi, Daniele	Sapienza University of Rome, Italy
Dallalibera, Fabio	The University of Padua, Italy
Danis, F. Serhan	Boğaziçi University, Turkey
Friedmann, Martin	Technische Universität Darmstadt, Germany
Gianni, Mario	Sapienza University of Rome, Italy
Haider, Sajjad	University of Technology, Sydney, Australia
Hung, Emmet	National Chung Cheng University, Taiwan
Kohlbrecher, Stefan	Technische Universität Darmstadt, Germany
Lau, Nuno	University of Aveiro, Portugal

Marchetti, Luca	INRIA Sophia Antipolis, France
Munaro, Matteo	The University of Padua, Italy
Nicklin, Steven	The University of Newcastle, Australia
Ozkucur, Nezih Ergin	Boğaziçi University, Turkey
Papadakis, Panagiotis	Sapienza University of Rome, Italy
Pierris, Georgios	University of Wales, Newport, UK
Pretto, Alberto	The University of Padua, Italy
Previtali, Fabio	Sapienza University of Rome, Italy
Randelli, Gabriele	Sapienza University of Rome, Italy
Sun, Dali	University of Freiburg, Germany
Van Dijk, Sander	University of Hertfordshire, UK

Table of Contents

Champion Papers

Best Paper

Papers with Oral Presentation

Papers with Poster Presentation

WrightEagle and UT Austin Villa: RoboCup 2011 Simulation League Champions

Aijun Bai[1], Xiaoping Chen[1],
Patrick MacAlpine[2], Daniel Urieli[2], Samuel Barrett[2], and Peter Stone[2]

[1] Department of Computer Science, University of Science and Technology of China
xpchen@ustc.edu.cn
[2] Department of Computer Science, The University of Texas at Austin
pstone@cs.utexas.edu

Abstract. The RoboCup simulation league is traditionally the league with the largest number of teams participating, both at the international competitions and worldwide. 2011 was no exception, with a total of 39 teams entering the 2D and 3D simulation competitions. This paper presents the champions of the competitions, WrightEagle from the University of Science and Technology of China in the 2D competition, and UT Austin Villa from the University of Texas at Austin in the 3D competition.

1 Introduction

The RoboCup simulation league has always been an important part of the RoboCup initiative. The distinguishing feature of the league is that the soccer matches are run in software, with no physical robot involved. As such, some of the real-world challenges that arise in the physical robot leagues can be abstracted away, such as image processing and wear and tear on physical gears. In exchange, it becomes possible in the simulation league to explore strategies with larger teams or robots (up to full 11 vs. 11 games), and to leverage the possibility of automating large numbers of games, for example for the purpose of machine learning or to establish statistically significant effects of strategy changes.

In 2011, as in recent past years, there were two separate simulation competitions held at RoboCup. In both cases, a *server* simulates the world including the dynamics and kinematics of the players and ball. Participants develop a fully autonomous team of client *agents* that each interact separately with the server by i) receiving sensations representing the view from its current location; ii) deciding what actions to execute; and iii) sending the actions back to the server for execution in the simulated world. The sensations are abstract and noisy in that they indicate the approximate distance and angle to objects (players, ball, and field markings) that are in the direction that the agent is currently looking. The server proceeds in real time, without waiting for agent processes to send their actions: it is up to each agent to manage its deliberation time so as to keep up with the server's pace. Furthermore, each agent must be controlled by a completely separate process, with no file sharing or inter-process communication (simulated low-bandwidth verbal communication is available via the server).

T. Röfer et al. (Eds.): RoboCup 2011, LNCS 7416, pp. 1–12, 2012.

Though similar in all of the above respects, the 2D and 3D simulators also differ in some important ways. As their names suggest, the 2D simulator models only the (x, y) positions of objects, while the 3D simulator includes the third dimension. In the 2D simulator, the players and the ball are modeled as circles. In addition to its (x, y) location, each player has a direction that its body is facing, which affects the directions it can move; and a separate direction in which it is looking, which affects its sensations. Actions are abstract commands such as turning the body or neck by a specified angle, dashing forwards or backwards with a specified power, kicking at a specified angle with a specified power (when the ball is near), or slide tackling in a given direction. Teams consist of 11 players, including a goalie with special capabilities such as catching the ball when it is near. In particular, the 2D simulator does not model the motion of any particular physical robot, but does capture realistic team-level strategic interactions.

In contrast, the 3D simulator implements a physically realistic world model and an action interface that is reflective of that experienced by real robots. The simulator uses the Open Dynamics Engine[1] (ODE) library for its realistic simulation of rigid body dynamics with collision detection and friction. ODE also provides support for the modeling of advanced motorized hinge joints used in the humanoid agents. The agents are modeled after the Aldebaran Nao robot,[2] which has a height of about 57 cm, and a mass of 4.5 kg. The agents interact with the simulator by sending actuation commands to each of the robot's joints. Each robot has 22 degrees of freedom: six in each leg, four in each arm, and two in the neck. In order to monitor and control its hinge joints, an agent is equipped with joint perceptors and effectors. Joint perceptors provide the agent with noise-free angular measurements every simulation cycle (20 ms), while joint effectors allow the agent to specify the direction and speed (torque) in which to move a joint. Although there is no intentional noise in actuation, there is slight actuation noise that results from approximations in the physics engine and the need to constrain computations to be performed in real-time. The 3D simulator thus presents motion challenges similar to that of the humanoid soccer leagues, especially the standard platform league (SPL) which uses physical Nao robots. In particular, it is a non-trivial challenge to enable the robots to walk and kick without falling over. The 3D simulation league teams consist of 9 (homogeneous) players, rather than 4 as in the SPL.

In 2011, the 2D and 3D simulation competitions included 17 and 22 teams, respectively, from around the world. This paper briefly describes and compares the two champion teams, WrightEagle from USTC in the 2D simulation league, and UT Austin Villa from UT Austin in the 3D simulation league.

The remainder of the paper is organized as follows. Section 2 introduces the WrightEagle team, particularly emphasizing its heuristic approximate on-line planning for large-scale and sparse-reward MDPs. Section 3 introduces the UT Austin Villa Team, focussing especially on its learning-based walk. Section 4 concludes.

---

[1] http://www.ode.org/
[2] http://www.aldebaran-robotics.com/eng/

2　WrightEagle: 2D Simulation League Champions

This section describes the RoboCup 2011 2D competition champion team, WrightEagle. First the overall system structure, based on hierarchical MDPs, is introduced, which then leads into an overview of the team's main research focus: heuristic approximate planning in such MDPs.

2.1　System Structure of WrightEagle

The team WrightEagle, including the latest version, has been developed based on the Markov decision processes (MDPs) framework [9], with the MAXQ hierarchical structure [3] and heuristic approximate online planning techniques strengthened particularly in the past year.

MDP Framework. Formally, an MDP is defined as a 4-tuple $\langle S, A, T, R \rangle$, where

- S is the set of possible states of the environment,
- A is the set of available actions of the agent,
- T is the transition function with $T(s'|s, a)$ denoting the next state probability distribution by performing action a in state s,
- R is the reward function with $R(s, a)$ denoting the immediate reward received by the agent after performing action a in state s.

A set of standard algorithms exist for solving MDP problems, including linear programming, value iteration, and policy iteration [9]. However, in large-scale and sparse-reward domains such as the 2D simulator, solving the MDP problem directly is to some degree intractable. In WrightEagle, some techniques including the MAXQ hierarchy and heuristic approximate online planning were applied to overcome this difficulty.

MAXQ Hierarchical Decomposition. The MAXQ framework decomposes a given MDP into a hierarchy of sub MDPs (known as subtasks or behaviors) $\{M_0, M_1, \cdots, M_n\}$, where M_0 is the root subtask which means solving M_0 solves the entire original MDP [3]. Each of the subtasks is defined with a subgoal, and it terminates when its subgoal is achieved.

Over the hierarchical structure, a *hierarchical policy* π is defined as a set of policies for each of the subtasks $\pi = \{\pi_0, \pi_1, \cdots, \pi_n\}$. Each subtask policy π_i is a mapping from states to actions $\pi_i : S_i \to A_i$, where S_i is the set of relevant states of subtask M_i and A_i is the set of available primitive actions or composite actions (i.e. its subtasks) of that subtask.

A hierarchical policy is *recursively optimal* if the local policy for each subtask is optimal given that all its subtasks are in turn recursively optimal [11].

Dieterich [2] has shown that a recursively optimal policy can be found by computing its recursively optimal V function, which satisfies:

$$Q(i, s, a) = V(a, s) + C(i, s, a) , \tag{1}$$

$$V(i,s) = \begin{cases} \max_a Q(i,s,a) & \text{if } i \text{ is composite} \\ R(s,i) & \text{otherwise} \end{cases}, \tag{2}$$

and

$$C(i,s,a) = \sum_{s',N} \gamma^N T(s',N|s,a)V(i,s'), \tag{3}$$

where $T(s',N|s,a)$ is the probability that the system terminates in state s' with a number of steps N after action a is invoked.

To solve $V(i,s)$, a complete search of all paths through the MAXQ hierarchy starting from subtask M_i and ending at primitive actions should be performed, in a depth-first search way. If $V(i,s)$ for subtask M_i is known, then its recursively optimal policy can be generated from $\pi_i(s) = \text{argmax}_a Q(i,s,a)$.

Hierarchical Structure of WrightEagle. The main hierarchical structure of WrightEagle is shown in Fig. 1. The two subtasks of root task WrightEagle are Attack and Defense. Attack has its subtasks including Shoot,

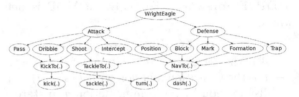

Fig. 1. MAXQ task graph for WrightEagle

Dribble, Pass, Position and Intercept, while Defense has its subtasks including Block, Trap, Mark and Formation. In WrightEagle, these subtasks are called *behaviors*.

Behaviors share their same subtasks, including KickTo, TackleTo and NavTo. Note that, the parentheses after these subtasks in Fig. 1 indicate that they are parameterized. In turn, these subtasks also share their same subtasks consisting of kick, turn, tackle and dash, which are parameterized primitive actions originally defined by the 2D domain.

2.2 Heuristic Approximate Online Planning

Theoretically, the full recursively optimal V function can be solved by some standard algorithms. But in practice this is intractable in 2D domain, because:

1. the state and action space is huge, even if discretization methods are used;
2. the reward function is very sparse, as the ball is usually running for thousands of cycles without any goals being scored;
3. the environment is unpredictable as teammates and opponents are all autonomous agents having the ability to make their own decisions, which prevents the original MDP problem to be solved completely *offline*.

For these reasons, the WrightEagle team focuses on approximate but not exact solutions. Our method is to compute the recursively optimal V function by heuristic *online* planning techniques under the consideration of simplification.

Current State Estimation. To model the RoboCup 2D domain as an MDP which assumes that the environment's state is fully observable, the agent must overcome the difficulty that it can only receive local and noisy observations, to obtain a precise enough estimation of the environment's current state. In our team, the agent estimates the current state from its *belief* [7]. A belief b is a probability distribution over state space, with $b(s)$ denoting the probability that the environment is actually in state s. We assume conditional independence between individual objects, then the belief $b(s)$ can be expressed as

$$b(s) = \prod_{0 \leq i \leq 22} b_i(s[i]), \tag{4}$$

where s is the full state vector, $s[i]$ is the partial state vector for object i, and $b_i(s[i])$ is the marginal distribution for $s[i]$. A set of m_i weighted samples (also known as particles) are then used to approximate b_i as:

$$b_i(s[i]) \approx \{x_{ij}, w_{ij}\}_{j=1...m_i}, \tag{5}$$

where x_{ij} is a sampled state for object i, and w_{ij} represents the approximated probability that object i is in state x_{ij} (obviously $\sum_{1 \leq j \leq m_i} w_{ij} = 1$).

In the beginning of each step, these samples are updated by Monte Carlo procedures using the domain's *motion model* and *sensor model* [1]. It is worth noting that, the agent can not observe the actions performed by other players, so it always assume that they will do a random kick if the ball is kickable for them, or a random walk otherwise. Finally, the environment's current state s is estimated as:

$$s[i] = \sum_{1 \leq j \leq m_i} w_{ij} x_{ij}. \tag{6}$$

Transition Model Simplification. Recall that, computing Equations 3, 1, and 2 recursively can find the recursive optimal policy over the MAXQ hierarchy. However, to completely represent either the transition function T or the completion function C in the 2D domain is intractable. Some approximate methods are applied to overcome this difficulty, as described next.

In WrightEagle, based on some pre-defined rules, the entire space of possible state-steps pairs (s', N) for each subtask is dynamically split into two classes: the success class and the failure class, denoted as $sucess(s, a)$ and $failure(s, a)$ respectively.

After splitting, the expected state-steps pair (called pre-state in WrightEagle) of each class is used to represent it, which can be calculated either by Monte Carlo methods or approximately theoretical analysis, denoted as s_s and s_f respectively. Then the completion function can be approximately represented as

$$C(i, s, a) \approx pV(i, s_s) + (1 - p)V(i, s_f), \tag{7}$$

where $p = \sum_{(s', N) \in sucess(s,a)} T(s', N|s, a)$.

To efficiently calculate p in Equation 7, a series of approximate methods were developed in WrightEagle. They are classified into two groups: subjective probability based on some heuristic functions, and objective probability based on some statistical models.

Heuristic Approximation of Value Function. Some heuristic evaluation functions (denoted as $H(i, s_s|s, a)$ and $H(i, s_f|s, a)$ respectively) were developed to approximate $V(i, s_s)$ and $V(i, s_f)$, because:

1. the pre-states s_s and s_f are hard to estimate due to the unpredictable property of the environment, especially when they are far in the future;
2. completely recursively computing costs too much to satisfy the real-time constraints in the 2D domain.

For each subtask, different H functions were developed, according to different subgoals. For low level subtasks and primitive actions, the reward functions $R(s, a)$ are too sparse to be used directly. As a substitute, some pseudo-reward functions are developed in WrightEagle. Take the KickTo subtask for an example, the maximum speed that the ball can be kicked in a range of given cycles plays a key role in the local heuristic function.

Heuristic Search in Action Space. By now, we have almost solved the MAXQ hierarchy used in WrightEagle by heuristic approximate online planning techniques, but there's still one difficulty remaining in Equation 2: as for the parameterized actions (including KickTo, kick, etc.), the action space is too huge to be searched directly by some brute force algorithms.

Some heuristic search methods are introduced to deal with this issue, e.g. an A* search algorithm with some special pruning strategy is used in the NavTo subtask when searching the action space of dash and turn, a hill-climbing method is used in the Pass subtask when searching the action space of KickTo, etc [4,10].

Particularly, for the Defense subtask and its subtasks, which are more involved with cooperation between agents, theoretical analysis is more difficult. Some work has been done on this based on the decentralized partially observable Markov decision processes (DEC-POMDPs) [13,14].

2.3   RoboCup 2011 Soccer 2D Simulation League Results

In RoboCup 2011, the WrightEagle team won the champion with no lost games, and achieved an average goal difference of 12.33 in total 12 games.[3] The Helios team, a united team from Fukuoka University, Osaka Prefecture University, and the National Institute of Advanced Industrial Science and Technology of Japan, won second place, and the MarliK team from University of Guilan of Iran won third place.

---

[3] The detailed competition results can be found at:
http://sourceforge.net/apps/mediawiki/sserver/index.php?
title=RoboCup2011/Competition

3   UT Austin Villa: 3D Simulation League Champions

This section describes the RoboCup 2011 3D competition champion team, UT Austin Villa, with particular emphasis on the main key to the team's success: an optimized omnidirectional walk engine. Further details about the team, including an inverse kinematics based kicking architecture and a dynamic role assignment and positioning system, can be found in [8].

3.1   Agent Architecture

The UT Austin Villa agent receives visual sensory information from the environment which provides distances and angles to different objects on the field. It is relatively straightforward to build a world model by converting this information about the objects into Cartesian coordinates. This of course requires the robot to be able to localize itself for which the agent uses a particle filter. In addition to the vision perceptor, the agent also uses its accelerometer readings to determine if it has fallen and employs its auditory channels for communication.

Once a world model is built, the agent's control module is invoked. Figure 2 provides a schematic view of the control architecture of the UT Austin Villa humanoid soccer agent.

At the lowest level, the humanoid is controlled by specifying torques to each of its joints. This is implemented through PID controllers for each joint, which take as input the desired angle of the joint and compute the appropriate torque. Further, the agent uses routines describing inverse kinematics for the arms and legs. Given a target

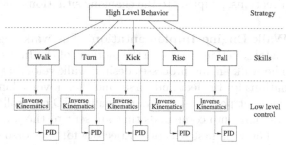

Fig. 2. Schematic view of UT Austin Villa agent control architecture

position and pose for the foot or the hand, the inverse kinematics routine uses trigonometry to calculate the angles for the different joints along the arm or the leg to achieve the specified target, if at all possible.

The PID control and inverse kinematics routines are used as primitives to describe the agent's skills. In order to determine the appropriate joint angle sequences for walking and turning, the agent utilizes an omnidirectional walk engine which is described in subsection 3.2. When invoking the kicking skill, the agent uses inverse kinematics to control the kicking foot such that it follows an appropriate trajectory through the ball. This trajectory is defined by set waypoints, ascertained through machine learning, relative to the ball along a cubic Hermite spline. Two other useful skills for the robot are falling (for instance, by the goalie to block a ball) and rising from a fallen position. Both falling and rising are accomplished through a programmed sequence of poses and specified joint angles.

Because the team's emphasis was mainly on learning robust and stable low-level skills, the high-level strategy to coordinate the skills of the individual agents is relatively straightforward. The player closest to the ball is instructed to go to it while other field player agents dynamically choose target positions on the field based on predefined formations that are dependent on the current state of the game. For example, if a teammate is dribbling the ball, one agent positions itself slightly behind the dribbler so that it is ready to continue with the ball if its teammate falls over. The goalie is instructed to stand a little in front of its goal and, using a Kalman filter to track the ball, attempts to dive and stop the ball if it comes near.

3.2   Omnidirectional Walk Engine and Optimization

The primary key to Austin Villa's success in the 2011 RoboCup 3D simulation competition was its development and optimization of a stable and robust fully omnidirectional walk. The team used an omnidirectional walk engine based on the research performed by Graf et al. [5]. The main advantage of an omnidirectional walk is that it allows the robot to request continuous velocities in the forward, side, and turn directions, permitting it to approach its destination more quickly. In addition, the robustness of this engine allowed the robots to quickly change directions, adapting to the changing situations encountered during soccer games.

Walk Engine Implementation. The walk engine uses a simple set of sinusoidal functions to create the motions of the limbs with limited feedback control. The walk engine processes desired walk velocities given as input, chooses destinations for the feet and torso, and then inverse kinematics are used to determine the joint positions required. Finally, PID controllers for each joint convert these positions into commands that are sent to the joints.

The walk first selects a trajectory for the torso to follow, and then determines where the feet should be with respect to the torso location. The trajectory is chosen using a double linear inverted pendulum, where the center of mass is swinging over the stance foot. In addition, as in Graf et al.'s work [5], the simplifying assumption that there is no double support phase is used, so that the velocities and positions of the center of mass must match when switching between the inverted pendulums formed by the respective stance feet.

The walk engine is parameterized using more than 40 parameters, ranging from intuitive quantities, like the step size and height, to less intuitive quantities like the maximum acceptable center of mass error. These parameters are initialized based on an understanding of the system and also testing them out on an actual Nao robot. This initialization resulted in a stable walk. However, the walk was extremely slow compared to speeds required during a competition. We refer to the agent that uses this walk as the *Initial* agent.

Walk Engine Parameter Optimization. The slow speed of the *Initial* agent calls for using machine learning to obtain better walk parameter values. Parameters are optimized using the CMA-ES algorithm [6], which has been successfully applied in [12]. CMA-ES is a policy search algorithm that successively generates and evaluates sets of candidates. Once CMA-ES generates a group of candidates, each candidate is evaluated with respect to a *fitness* measure. When all the candidates in the group are evaluated, the next set of candidates is generated by sampling with probability that is biased towards directions of previously successful search steps.

As optimizing 40 real-valued parameters, can be impractical, a carefully chosen subset of 14 parameters was selected for optimization while keeping all the other parameters fixed. The chosen parameters are those that have the highest potential impact on the speed and stability of the robot, for instance: the maximum step sizes, rotation, and height; the robot's center of mass height, shift amount, and default position; the fraction of time a leg is on the ground and the time allocated for one step phase; the step size PID controller; center of mass normal error and maximum acceptable errors; and the robot's forward offset.

Similarly to a conclusion from [12], Austin Villa has found that optimization works better when the robot's fitness measure is its performance on tasks that are *executed during a real game*. This stands in contrast to evaluating it on a general task such as the speed walking straight. Therefore, the robot's in-game behavior is broken down into a set of smaller tasks, and the parameters for each one of these tasks is sequentially optimized. When optimizing for a specific task, the performance of the robot on the task is used as CMA-ES's fitness value for the current candidate parameter set values.[4]

In order to simulate common situations encountered in gameplay, the walk engine parameters for a goToTarget subtask are optimized. This consists of an obstacle course in which the agent tries to navigate to a variety of target positions on the field. The goToTarget optimization includes quick changes of target/direction for focusing on the reaction speed of the agent as well as holding targets for longer durations to improve the straight line speed of the agent. Additionally the agent is instructed to stop at different times during the optimization to ensure that is stable and doesn't fall over when doing so. In order to encourage both quick turning behavior and a fast forward walk, the agent always walks and turns toward its designated target at the same time. This allows for the agent to swiftly adjust and switch its orientation to face its target, thereby emphasizing the amount of time during the optimization that it is walking forward. Optimizing the walk engine parameters in this way resulted in a significant improvement in performance with the *GoToTarget* agent able to quickly turn and walk in any direction without falling over. This improvement also showed itself in actual game performance as when the *GoToTarget* agent played 100 games against the *Initial* agent, the *GoToTarget* agent won on average by 8.82 goals with a standard error of .11.

To further improve the forward speed of the agent, a second walk engine parameter set was optimized for walking straight forward. This was accomplished

---

[4] Videos of the agent performing optimization tasks can be found online at
http://www.cs.utexas.edu/ AustinVilla/sim/3dsimulation/
AustinVilla3DSimulationFiles/2011/html/walk.html

by running the goToTarget subtask optimization again, but this time the *goToTarget* parameter set was fixed while a new parameter set, called the *sprint* parameter set, was learned. The *sprint* parameter set is used when the agent's orientation is within 15° of its target. By learning the *sprint* parameter set in conjunction with the *goToTarget* parameter set, the new *Sprint* agent was stable switching between the two parameter sets and also increased the agent's speed from .64 m/s to .71 m/s as timed when walking forward for ten seconds after starting from a standstill.

Although adding the *goToTarget* and *sprint* walk engine parameter sets improved the stability, speed, and game performance of the agent, the agent was still a bit slow when positioning to dribble the ball. This slowness makes sense because the goToTarget subtask optimization emphasizes quick turns and forward walking speed while positioning around the ball involves

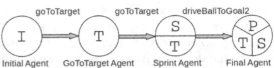

Fig. 3. UT Austin Villa walk parameter optimization progression. Circles represent the set(s) of parameters used by each agent during the optimization progression while the arrows and associated labels above them indicate the optimization tasks used in learning. Parameter sets are the following: I = *initial*, T = *goToTarget*, S = *sprint*, P = *positioning*.

more side-stepping to circle the ball. To account for this discrepancy, the agent learned a third parameter set called the *positioning* parameter set. To learn this new parameter set a driveBallToGoal2 optimization was created in which the agent is evaluated on how far it is able to dribble the ball over 15 seconds when starting from a variety of positions and orientations from the ball. The *positioning* parameter set is used when the agent is within .8 meters of the ball. Both *goToTarget* and *sprint* parameter sets are fixed and the optimization naturally includes transitions between all three parameter sets, which constrains them to be compatible with each other. Adding both the *positioning* and *sprint* parameter sets further improved the agent's performance such that it, the *Final* agent, was able to beat the *GoToTarget* agent by an average of .24 goals with a standard error of .08 across 100 games. A summary of the progression in optimizing the three different walk parameter sets can be seen in Figure 3.

3.3   RoboCup 2011 Soccer 3D Simulation League Results

The UT Austin Villa team won the 2011 RoboCup 3D simulation competition in convincing fashion by winning all 24 matches it played, scoring 136 goals and conceding none. The CIT3D team, from Changzhou Institute of Technology of China, came in second place while the Apollo3D team, from Nanjing University of Posts and Telecommunications of China, finished third. The success UT Austin Villa experienced during the competition was no fluke as when playing 100 games against each of the other 21 teams' released binaries from the competition, the UT Austin Villa team won by at least an average goal difference of 1.45 against every team. Furthermore, of these 2100 games played, UT Austin Villa won all but 21 of them which ended in ties (no losses). The few ties were all against three of the better teams: Apollo3D, Bold Hearts, and RoboCanes. We can therefore conclude that UT Austin Villa was the rightful champion of the competition.

While there were multiple factors and components that contributed to the success of the UT Austin Villa team in winning the competition, its omnidirectional walk was the one which proved to be the most crucial. When switching out the omnidirectional walk developed for the 2011 competition for a fixed skill based walk used in the 2010 competition, and described in [12], the team did not fare nearly as well. The agent with the previous year's walk had a negative average goal differential against nine of the teams from the 2011 competition, suggesting a probable tenth place finish. Also this agent lost to our *Final* agent by an average of 6.32 goals across 100 games with a standard error of .13. One factor that did not come into play in UT Austin Villa winning the competition, however, was its goalie. The team's walk allowed it to dominate possession of the ball and keep it away from the opposing team such that the opponent never had a chance to shoot on the goal, and thus the goalie never touched the ball during the course of gameplay.

4   Conclusion

This paper introduced the champions of the RoboCup 2011 simulation leagues.

First, we described the MAXQ hierarchical structure of WrightEagle, and the online planning method combining with heuristic and appropriate techniques over this hierarchy. Based on this effort, The WrightEagle team has a very flexible and adaptive strategy, particularly for unfamiliar teams. It has won 3 championships and 4 runner-ups in the past 7 years of RoboCup competitions.[5]

Second, we described the learning-based omnidirectional walk of UT Austin Villa, and the series of fitness functions that were employed during learning. The resulting walk was quick, agile, and stable enough to dribble around most of the other teams in the competition. 2011 was the first victory for UT Austin Villa in the 3D simulation league.[6]

Acknowledgments. The authors would like to thank the additional contributing members of WrightEagle (Feng Wu, Zongzhang Zhang, Haochong Zhang and Guanghui Lu) and UT Austin Villa (Shivaram Kalyanakrishnan, Frank Barrera, Nick Collins, Adrian Lopez-Mobilia, Art Richards, Nicu Stiurca, and Victor Vu).

The work of WrightEagle is supported by the Natural Science Foundations of China under Grant No. 60745002, and No. 61175057, as well as the National Hi-Tech Project of China under Grant No. 2008AA01Z150.

The work of UT Austin villa took place in the Learning Agents Research Group (LARG) at the Artificial Intelligence Laboratory, The University of Texas at Austin.

---

[5] More information about the WrightEagle team can be found at the team's website: http://www.wrighteagle.org/2d

[6] More information about the UT Austin Villa team, as well as video highlights from the competition, can be found at the team's website: http://www.cs.utexas.edu/~AustinVilla/sim/3dsimulation/

LARG research is supported in part by grants from the National Science Foundation (IIS-0917122), ONR (N00014-09-1-0658), and the Federal Highway Administration (DTFH61-07-H-00030).

References

1. Dellaert, F., Fox, D., Burgard, W., Thrun, S.: Monte carlo localization for mobile robots. In: Proceedings 1999 IEEE International Conference on Robotics and Automation (Cat. No.99CH36288C), vol. 2, pp. 1322–1328. IEEE (2001)
2. Dietterich, T.G.: Hierarchical reinforcement learning with the MAXQ value function decomposition. Journal of Machine Learning Research 13(1), 63 (1999)
3. Dietterich, T.G.: The MAXQ method for hierarchical reinforcement learning. In: Proceedings of the Fifteenth International Conference on Machine Learning, vol. 8(c), pp. 118–126. Morgan Kaufmann (1999)
4. Fan, C., Chen, X.: Bounded incremental real-time dynamic programming. In: Frontiers in the Convergence of Bioscience and Information Technologies, pp. 637–644 (2007)
5. Graf, C., Härtl, A., Röfer, T., Laue, T.: A robust closed-loop gait for the standard platform league humanoid. In: Proc. of the 4th Workshop on Humanoid Soccer Robots in Conjunction with the 2009 IEEE-RAS Int. Conf. on Humanoid Robots, pp. 30–37 (2009)
6. Hansen, N.: The CMA Evolution Strategy: A Tutorial (January 2009), http://www.lri.fr/~hansen/cmatutorial.pdf
7. Kaelbling, L.P., Littman, M.L., Cassandra, A.R.: Planning and acting in partially observable stochastic domains. Artificial Intelligence 101(1-2), 99–134 (1998)
8. MacAlpine, P., Urieli, D., Barrett, S., Kalyanakrishnan, S., Barrera, F., Lopez-Mobilia, A., Ştiurcă, N., Vu, V., Stone, P.: UT Austin Villa 2011: A champion agent in the RoboCup 3D soccer simulation competition. In: Proc. of 11th Int. Conf. on Autonomous Agents and Multiagent Systems (AAMAS 2012) (June 2012)
9. Puterman, M.L.: Markov Decision Processes: Discrete Stochastic Dynamic Programming. John Wiley & Sons, Inc. (1994)
10. Shi, K., Chen, X.: Action-driven markov decision process and the application in robocup. Journal of Chinese Computer Systems 32, 511–515 (2011)
11. Sutton, R.S., Precup, D., Singh, S.: Between mdps and semi-mdps: A framework for temporal abstraction in reinforcement learning. Artificial Intelligence 112(1-2), 181–211 (1999)
12. Urieli, D., MacAlpine, P., Kalyanakrishnan, S., Bentor, Y., Stone, P.: On optimizing interdependent skills: A case study in simulated 3D humanoid robot soccer. In: Proc. of the Tenth Int. Conf. on Autonomous Agents and Multiagent Systems (AAMAS 2011), May 2011, pp. 769–776 (2011)
13. Wu, F., Chen, X.: Solving Large-Scale and Sparse-Reward DEC-POMDPs with Correlation-MDPs. In: Visser, U., Ribeiro, F., Ohashi, T., Dellaert, F. (eds.) RoboCup 2007. LNCS (LNAI), vol. 5001, pp. 208–219. Springer, Heidelberg (2008)
14. Wu, F., Zilberstein, S., Chen, X.: Online planning for multi-agent systems with bounded communication. Artificial Intelligence 175(2), 487–511 (2011)

Robot Hardware, Software, and Technologies behind the SKUBA Robot Team

Kanjanapan Sukvichai, Teeratath Ariyachartphadungkit, and Krit Chaiso

Faculty of Engineering, Kasetsart University, 50 Phaholyothin Road, Thailand
fengkpsc@ku.ac.th,
{teeratath,nuopolok}@hotmail.com

Abstract. SKUBA is a winner from RoboCup 2011, Turkey. The aim of this paper is to explain the basic concepts of the design of the robot hardware and the AI system. The robot mechanics are explained in detail along with the electronic boards. Protection circuits are added to make the robot more robust. A torque controller is implemented as low-level controller to reduce the effect of surfaces. The high level AI system is separated into three major modules: predictor/tracker, strategy, and control module. The tracking algorithm and the Kalman Observer are implemented in the predictor/tracker module. The strategy module takes care of playing and evaluates the probability of each play against the opponent. Finally, the control module is explained. Path planning algorithms and a modified kinematics equation are implemented in this module in order to make the robot move along the desired trajectory with less velocity error.

Keywords: Small Size League, mobile robot, torque controller, modified kinematics.

1 Introduction

The Small Size League is a part of the RoboCup robot competitions, which take place every year. The aim of RoboCup is to promote robotics and AI research by offering a publicly appealing, but formidable challenge. The Small Size League is a robot soccer competition that is designed based on the FIFA rules. The main interesting and strong characteristic of this league are the complex AI algorithms, fast speed game play, and multi-agents cooperation. Each team has five robots on the field and plays against the opponent. The overhead wide-angle cameras are mounted over each side of the field and send their images to the SSL-Vision system [1]. The SSL-Vision will broadcast the positions of the ball and all robots to each team via a network cable.

Skuba is a Small Size League soccer robot team from Kasetsart University, which has participated in the RoboCup competition since 2006. We won the championships in the last three years, i.e. the RoboCup 2009 in Graz, Austria and 2010 in Singapore and RoboCup 2011 in Istanbul, Turkey. We also won the RoboCup Iran Open in Tehran in April 2011. In this paper, the SKUBA robot structure and SKUBA software are revealed and explained.

T. Röfer et al. (Eds.): RoboCup 2011, LNCS 7416, pp. 13–24, 2012.
© Springer-Verlag Berlin Heidelberg 2012

2   SKUBA Robot

This section describes the robot electronics system that is used in the driving system including its designs and components. Details about operations and algorithms are presented in the firmware section. The robot consists of two electronics boards: the main board and the kicker/chipper board. The main board handles all of the robot tasks except kicking. The kicker board controls the entire kicker system. The robot mechanic is described in the last section of this topic.

2.1   Robot Electrical Circuit

The board consists of a Xilinx Spartan-3 XC3S400 FPGA, motor driver, user interface, some add-on modules and a debugging port. The microprocessor core (RISC-32 architecture) and interfacing logic for external peripherals are implemented using FPGA in order to handle the low-level control of the brushless motors such as velocity and position control. The main electronics board receives commands from the main software running on an off-board computer. The board integrates the processing components together with the power components to keep the board compact and to minimize wiring. With limited space, almost all components are implemented as small SMD packages. There are two types of brushless motors in the robot: the driving motor and the dribbling motor. Each driving motor is a 30 W Maxon EC45 flat motor with a custom back-extended shaft for attaching an encoder wheel. The motor itself can produce a feedback signal from Hall sensors for measuring wheel velocity. However, this multi-pole motor sends only roughly 48 pulses per revolution; therefore, this motor is equipped with a US Digital E4P encoder, which has a higher resolution due to its 1440 pulses per revolution. The dribbling motor is a high speed 15 W Maxon EC16 motor connected with the encoder in the same fashion as the EC45. Figure 1 shows the SKUBA main PCB board.

Fig. 1. SKUBA main PCB board

In order to drive the brushless motor, the driven PWM signal must be generated in particular sequences through several groups of MosFETs. The digital sequencer component is implemented in an FPGA via the VHDL language in order to control the driving sequence perfectly. There are two critical issues of the driving system that can damage the robot: motor dead-time and over-current. The dead time of the motor is one of the most critical issues for the robot, because it can damage the driver

MosFET, the battery, and the power supply of the robot; hence the dead-time protection is also implemented in the FPGA. The dead-time value can be found in the motor datasheet. The motor sequence diagram and the VHDL code are shown in Fig. 2.

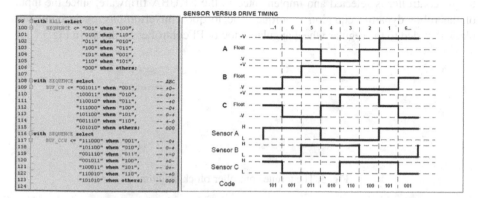

Fig. 2. The VHDL code of the motor sequencer (left) and the motor trimming diagram (right)

The second problem is the over-current problem. This problem occurs when the load of the motor is larger than during normal operation such as when two robots collide, or push against each other. The motor driving current is measured by an ACS712 and the measured data is used in the over-current protection circuit inside the FPGA. If the current is above the set boundary, the firmware in the microprocessor will reduce the PWM duty cycle to keep the current always below the motor limit.

Other circuits, which are needed in the small size robot, are the kicker, the chipper, the ball IR sensor and the wireless circuit. Kicker and chipper circuits are similar since they are the boost convertors, which pump the voltage from 18.5 volts to 200 volts and store it in two capacitors. The IGBT will discharge the stored energy to the kicker and chipper solenoids when needed. The ball infrared sensor is placed at the front of the robot to check for the presence of a ball. Checking for the ball is important, because if the robot kicks or chips without a ball, the kicker and chipper mechanisms can be damaged by the reaction force. A Nordic wireless module (nRF24L01+), which operates at 2.4 GHz, is used as the robot's wireless communication module. In each SKUBA robot, two wireless modules are used for bi-directional communication since the robot's current information is important for the AI, for example the battery level, whether a ball is present, some error codes, etc.

2.2   Robot Firmware

The robot firmware is the low-level motion control software. The robot controller is the most important part of the SKUBA system, because high-level programs such as the AI and the strategy module cannot work properly if the robot cannot respond to the command as it should do. The command sent by the AI consists only of the desired translational and angular velocities of the robot, not of the wheels. This command must be converted into wheel angular velocities before it can be used. The challenge of the low-level controller design is to make the robot move with good accuracy. The basic control scheme is a wheel angular velocity PD or PI controller. This controller is easy to implement, but it does not guarantee consistency, when the floor friction is changed. A torque controller is selected and implemented in the SKUBA firmware, since the input of the robot dynamic equation is torque. The torque convertor/duty cycle convertor block is added to the regular wheel angular velocity PI controller as shown in Fig 3.

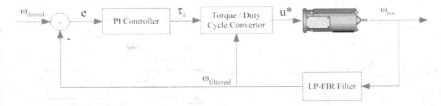

Fig. 3. The torque controller block diagram

The torque convertor equation and its parameters can be found by using the data-sheet or experiments. In this paper, the Maxon brushless motor datasheet is used to derive the torque convertor equation as shown in Fig 4. The conservation of energy shown in equation (1) and the motor torque-angular velocity conversion parameters are used to solve the torque convertor equation. The motor torque is the function of the applied voltage and current angular velocity as shown in equation (2).

Fig. 4. Model of the Maxon brushless motor

$$U \cdot I = \frac{\pi}{30,000} \cdot \dot{\phi} \cdot \tau_m + R \cdot I^2 \qquad (1)$$

$$\tau_m = \left(\frac{k_m}{R}\right) \cdot U - \left(\frac{k_m}{k_n \cdot R}\right) \cdot \dot{\phi} \qquad (2)$$

where

τ_m is the output torque of the motor

I, U, R are input current, voltage, and armature resistance, respectively

k_m, k_m are the electro-mechanic and angular velocity constant, respectively

$\dot{\phi}$ is the current angular velocity of the motor

Equation (2) cannot be used directly, since the input of the motor is a PWM signal and not a constant voltage. Thus the ratio between the maximum output and the desired torque at a particular speed is used to construct the PWM duty cycle equation. The desired torque is found by a PI controller with the error between the filtered desired velocity and the actual angular velocity of the motor used as input. The complete controller rule can be written in discrete form as equation (3):

$$DutyCycle = u^*[n] = \frac{k_p e[n] + k_i \sum_{1}^{n} e[i]}{\left(\dfrac{k_m}{R}\right) \cdot V\max - \left(\dfrac{k_m}{k_n \cdot R}\right) \cdot \omega_{filtered}} \tag{3}$$

where

    $\omega_{filtered}$ is the currently measured angular velocity after filtering
    e is the error between filtered desired velocity and actual angular velocity
    $V\max$ is the power supply voltage that is applied to the motor driver
    u^* is the duty cycle of the PWM signal
    k_p, k_i are the proportional and integral PI controller gains, respectively

    The optimal PI gains are found by manually tuning the robot on three different surfaces. The benefit of this controller is the robustness of the robot motion when the robot moves on surfaces with different frictions [2] such as the competition carpets. Fig. 5 shows the robot tuning on different surfaces. Fig 6 shows the trajectories of the robot, comparing the PI controller and the torque controller on three surfaces.

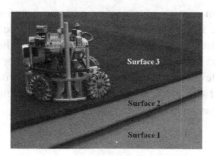

Fig. 5. Robot tuning on three different surfaces

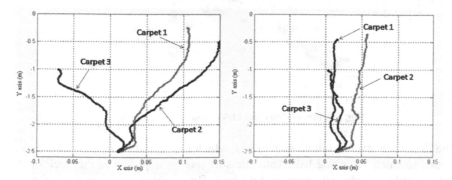

Fig. 6. Robot trajectories on 3 different carpets

2.3   Robot Hardware

The SKUBA robot consists of 4 omni-direction wheels, a kicker, a chipper, and a dribbler mechanism. Most parts of the robot are made from aluminum number 7075 except for the kicker and chipper solenoid, which are made from hard plastic (the previous version used glass epoxy). The wheels are located at 33, 147, 225, and 315 degrees. The idea of the symmetry design of the robot wheels is that the robot usually moves forward and backward more often than left and right, thus this design allows higher speeds for moving forward and backward than it does for moving left and right. The robot chassis and wheel location are shown in Fig. 7.

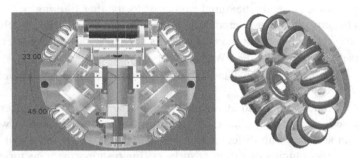

Fig. 7. Robot chassis (left) and robot wheel (right)

The kicker mechanism consists of a moving part and a stationary part. The solenoid, which is fixed to the robot chassis, is made from multiple layers of copper wire number 22 wrapped around a cylindrical and rectangular parallelepiped shape of hard plastic for the kicker and the chipper, respectively. The kicker plunger is made from iron and aluminum while the chipper plunger is only made from iron. An iron rod is in the middle between two aluminum rods in the kicker plunger. Fig. 8 shows the kicker and chipper mechanisms.

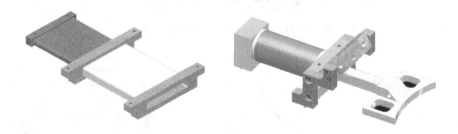

Fig. 8. Chipper (left) and kicker (right) mechanism

The dribbler is made from aluminum rod, which is covered by rubber. This part is also used to catch a moving ball. Thus the soft sponge is placed at the back of the dribbler in order to reduce the energy of the ball. The dribbler part is shown in Fig 9.

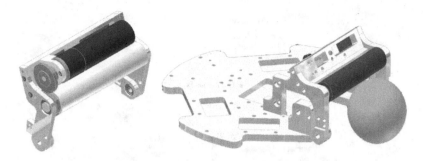

Fig. 9. Dribbler mechanism (left) and assembled to chassis (right)

3   SKUBA Software

In this topic, the software modules of the SKUBA team are explained. The SKUBA software system can be separated into three major parts, which are predictor/tracker, strategy planner, and control module.

Although the information about the robots and the ball are sent by the SSL-vision system frequently, they are not fast enough for a very fast ball or game. Moreover, the latency of the communication between the SSL-vision server and the team AI, the processing time, and the communication from the team computer to the robots can make the system a 'non-real-time system'. The predictor/tracker module is used to correct this information and to predict the future states of the game before sending it to the next modules. Basically, the predictor can be implemented by using a Kalman filter (observer). The ball and the robots can be modeled as particles running in the free space (field) in a continuous time domain without the control input. There is no control input in this case because it is not possible to know the commands that are sent to the other robots. The measurement noise and state error will be added to the Kalman filter. The optimal state transition matrix for the observer can be found by the algebraic Riccati equation. Finally, the continuous time result will be converted by using bilateral conversion that converts the continuous time system to a discrete time system. In the SKUBA system, the ball predictor and the robots predictor are separated, because some constraints are not the same. The tracker is also important, since the SSL-vision can send out 'more than one' ball and robots. The tracker is a state machine program that considers the current position of the ball/robots state compared with the previous state and makes the decision by probability. The maximum speed of the ball and the robots are accounted for in the equation of possibility.

The tracking of the chipped ball is one of the most challenging topics. The ball is captured from the real world to the image frame axis by the SSL-vision and sent to the AI software. The three-dimensional information is lost. In order to reconstruct the ball height, a coordinate transformation is applied. This process can be considered as the reverse process of the SSL-vision. The ball position from the SSL-vision is represented as $\{x', y', 0\}$ while the real position should be $\{x, y, z\}$ as shown in Fig. 10.

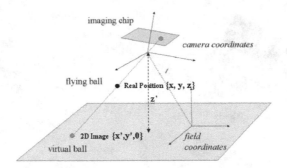

Fig. 10. Three-dimensional ball trajectory

The equation of the motion of a ball in three dimensions is

$$\{x, y, z\} = \left\{x_0 + v_x t + \tfrac{1}{2} g_x t^2, y_0 + v_y t + \tfrac{1}{2} g_y t^2, z_0 + v_z t + \tfrac{1}{2} g_z t^2\right\} \tag{4}$$

The values of $\{v_x, v_y, v_z, x_0, y_0, z_0\}$ are needed to predict the trajectory of the ball. If the camera is stationary above the field, the height z' is a constant. By using the ratio of two similar triangles, the ratio of frame i can be expressed as:

$$\left\{\begin{matrix} \alpha \\ \beta \end{matrix}\right\}\Bigg|_{Frame\ i} = \left\{\frac{x'}{z'}, \frac{y'}{z'}\right\}\Bigg|_{Frame\ i} = \left\{\frac{x_0 + v_x t + \tfrac{1}{2} g_x t^2}{z_0 + v_z t + \tfrac{1}{2} g_z t^2}, \frac{y_0 + v_y t + \tfrac{1}{2} g_y t^2}{z_0 + v_z t + \tfrac{1}{2} g_z t^2}\right\}\Bigg|_{Frame\ i} \tag{5}$$

By using the rotation matrix that converts from earth axes to image axes and considering the trajectory as a parabolic curve, the final equation can be written in the following form when $\{g_x, g_y, g_z\} = \{0, 0, -9.81\}$

$$\begin{bmatrix} z_0 \alpha_i + v_z \alpha_i \cdot t_i - x_0 - v_x t_i \\ z_0 \beta_i + v_z \beta_i \cdot t_i - y_0 - v_y t_i \end{bmatrix} = \begin{bmatrix} -\tfrac{1}{2} g_z \alpha_i t_i^2 \\ -\tfrac{1}{2} g_z \beta_i t_i^2 \end{bmatrix} \tag{6}$$

There are 3 unknown variables, therefore it needs at least 3 frames to compute the real three-dimensional trajectory. The result of the reconstruction is shown in Fig. 11. Fig. 12 shows the trajectory of a curveball in SKUBA system.

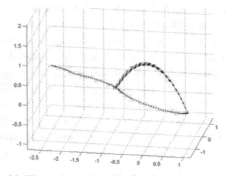

Fig. 11. Three-dimensional ball trajectory reconstruction

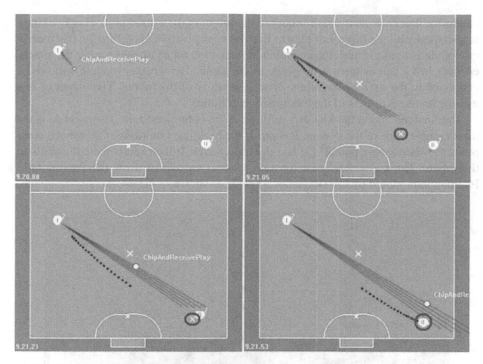

Fig. 12. A curveball in the SKUBA system. The circle shows the predicted drop point.

Fig. 13. Strategy module and its sub systems

After the position of the robots and the ball is corrected by the predictor/tracker module, this data will be sent to the strategy module shown in Fig. 13. The strategy module consists of four subsystems, which are PlayBook, PlayTransition, SkillSet, and Weather. SkillSet is a collection of robot individual skills (basic operation or behavior that each robot can do) such as move forward, open dribbler, and a kick skill. The SkillSet is the lowest level of the strategy module. PlayBook is the collection of the play. Play is the sequence of skills that robots have to perform such as free kick etc. In this subsystem, the history of play is recorded, because this information will be used to evaluate the quality of that play and if the quality of the play is high, that play will be used more in this particular game. PlayTransition acts like the conductor that controls the skill sequence rhythm to make a good play. PlayTransition is also used with the history of play to find the next possible play. Finally, the Weather module is used to check the condition of the game, for example whether the attacker robot lost a ball or there is no goalie in the game. The weather module will send a command to the strategy module to change the duty of the robots. The output of this module are the positions of the robots and their duties.

The final module in the SKUBA software is the control module. This module is the most complex one of the system. It controls the high-level motions of all robots, computes the "reasonable points" [3] and the trajectories. It then sends the commands to the robots. The reasonable point for each robot is calculated using the Passing Grid Evaluation proposed by CMDragon in 2009 [3] as shown in Fig 14.

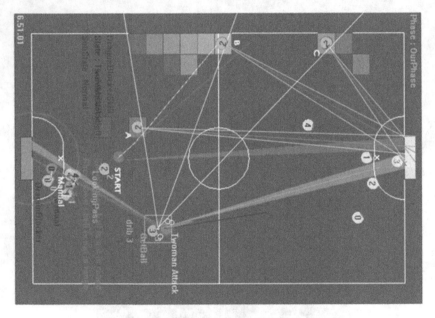

Fig. 14. Grid Evaluation in the SKUBA system

The path planning generation techniques that are used in the SKUBA system are Rapidly-exploring Random Tree (RRT), Sub Goal, and Open loop motion. The RRT is used to generate the trajectory at the regular motion. Sub Goal is used when there are many obstacles along the robot path. Finally, the Open loop motion is the special motion that is used when the robot needs to play a special skill, for example, the robot runs and hits the ball while the ball is still moving in the free space etc. The Open loop path planning uses a geometry calculation and a physical speed calculation to achieve such a movement.

The control algorithm in the SKUBA system that differs most from other teams is the robot velocity control, instead of a robot position control [4]. In the SKUBA system, we definitely control the velocity along the trajectory that is generated by

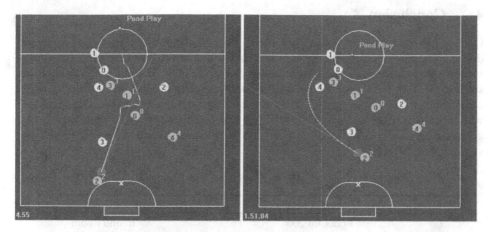

Fig. 15. RRT path (left) and Sub Goal path (right)

modified trapezoid path profiles and not the position. Velocity is more important, because the robot must arrive at a point at a deadline. This enables the robot to play in a very fast game.

The position is cared less about in this algorithm, therefore its overshoot response can lead to a major problem, which is the 'two defenders' penalty. But on the other hand, this approach has advantages when attacking, because position overshoots will help the robot to prevent opponent robots from getting a ball. In order to make a precise velocity control, the velocity PID control loop is added to the control module along with modified kinematics.

The modified kinematics is fully provided in the SKUBA ETDP 2011 [5]. The idea is to merge the friction effect with the regular mobile robot kinematics. The friction matrix is found in experiments. This matrix can be time-varying or constant, in SKUBA's case, this matrix is constant. Because of the complexity, the modified kinematics is implemented in the AI and not in the low-level robot controller. Thus the final equation will be reform to the function of the old command that generates the friction effect matrix as shown below from the trapezoid path profiles generator pulse:

$$\zeta_{send} = \zeta_{send_{old}} + \vartheta \tag{7}$$

where,

$\zeta_{send} = \begin{bmatrix} v_x & v_y & \omega \end{bmatrix}^T$, and ϑ is the friction effect matrix

There are so many things needed to be tuned and recorded during the competition. There is an auto calibration system implemented in the SKUBA system. This system helps to reduce the large amounts of time when manually tuning, and it allows team members to focusing more on the human strategic planning and fixing the robots. The friction effect matrix and the ball/robots predictor are automatically calibrated on the competition field. The Open Loop control also needs to be tuned.

4   Conclusion

Our system has been continuously improved since the beginning. This year, we introduced some improvements in the high-level motion controller and the necessity of the velocity controller to achieve a higher quality of game play. The automatic calibration software is fully used since RoboCup 2010 and it greatly reduces the amount of team setup time, which allowed us to focus more on the strategic planning. The software, which runs the robot team, was built in 2006 and has been improved each year. It has given us very successful competition results for the last several years; the results are summarized in Table 1.

Table 1. Competition results for SKUBA SSL RoboCup team

Competition	Result
RoboCup 2006	Round Robin
RoboCup Thailand Championship 2007	3rd Place
RoboCup Thailand Championship 2008	2nd Place
RoboCup 2008	3rd Place
RoboCup 2009	1st Place
RoboCup China Open 2009	1st Place
RoboCup 2010	1st Place
RoboCup Iran Open 2011	1st Place
RoboCup 2011	1st Place

There are two main focuses for RoboCup 2012, the flux residue in the chip solenoid and game play data mining. This flux will generate the unpredictable chipping behavior. This information will be also be used to improve the trajectory model of the flying ball using an energy equation. The game play data mining is already implemented this year and the result can be used for a defender, but next year we will develop a new algorithm that can mine the useful information for an attacker. We hope that our robot team will perform better in this year and we are looking forward to sharing experiences with other great teams around the world.

References

1. Zickler, S., Laue, T., Birbach, O., Wongphati, M., Veloso, M.: SSL-Vision: The Shared Vision System for the RoboCup Small Size League. In: Baltes, J., Lagoudakis, M.G., Naruse, T., Ghidary, S.S. (eds.) RoboCup 2009. LNCS (LNAI), vol. 5949, pp. 425–436. Springer, Heidelberg (2010)
2. Sukvichai, K., Wasuntapichaikul, P., Tipsuwan, Y.: Implementation of Torque controller for brushless motor on the omni-directional wheeled mobile robot. In: ITC-CSCC 2010, Pattaya, Thailand, pp. 19–22 (2010)
3. Zickler, S., James Bruce, J., Biswas, J., Licitra, M., Veloso, M.: CMDragon Extended Team Description Paper. In: RoboCup 2009 (2009)
4. Sukvichai, K., Wechsuwanmanee, P.: Development of the modified kinematics for a wheeled mobile robot. In: ITC-CSCC 2010, Pattaya, Thailand, pp. 88–90 (2010)
5. Chaiso, K., Sukvichai, K.: SKUBA Extended Team Description Paper. In: RoboCup 2011 (2011)

B-Human 2011 – Eliminating Game Delays*

Tim Laue[1], Thomas Röfer[1], Katharina Gillmann[2],
Felix Wenk[2], Colin Graf[2], and Tobias Kastner[2]

[1] Deutsches Forschungszentrum für Künstliche Intelligenz,
Sichere Kognitive Systeme, Enrique-Schmidt-Str. 5, 28359 Bremen, Germany
{Tim.Laue,Thomas.Roefer}@dfki.de
[2] Universität Bremen, Fachbereich 3 – Mathematik und Informatik,
Postfach 330 440, 28334 Bremen, Germany
{kathy,fwenk,cgraf,dyeah}@informatik.uni-bremen.de

Abstract. After having won the Standard Platform League competitions in 2009 and 2010, the B-Human software already included sophisticated solutions for most relevant subtasks, such as vision, state estimation, and walking. Therefore, the development towards RoboCup 2011 did not focus on replacing specific low-quality components, but was guided by an overall goal: eliminating game delays by more efficient actions and faster reactions to game state changes. This required several changes all over the system. In this paper, we present some of the developments that had the most impact regarding our goal: different ball models and corresponding cooperative ball tracking and retrieval strategies, a path planner as well as new approaches for tackling situations.

1 Introduction

B-Human is a joint RoboCup team of the Universität Bremen and the German Research Center for Artificial Intelligence (DFKI). The team consists of numerous undergraduate students as well as three researchers. The students participate in the team in the form of a two-year project course. Afterwards, some of them also write their thesis in the team's context. The researchers have already been active in a number of other RoboCup teams, such as the GermanTeam and the Bremen Byters (both Four-Legged League), B-Human and the BreDoBrothers (Humanoid Kid-Size League), and B-Smart (Small-Size League). Due to this particular continuity, the team always has a significant number of experienced members and we have been able to incrementally improve the overall software performance without major breaks during the past years.

Since its start in the Standard Platform League in 2009, the team B-Human has won every tournament it participated in. The status of the team's software after RoboCup 2010 can be considered as complete regarding solutions for most of the league's major challenges, such as a robust vision system, precise self-localization, and fast and stable walking. However, the overall game performance

---

* The authors would like to thank all B-Human team members for providing the software base for this work.

indicated a significant lack of reactivity in some regularly occurring situations, such as retrieving a lost ball, winning a tackle against a dribbling opponent, or avoiding a walking obstacle.

This situation has led to the overall goal for 2011: eliminating game delays. In this paper, we describe some of the most significant developments that contributed to achieve this goal. To overcome any delays resulting from failed ball tracking, a cooperative ball model as well as corresponding cooperative ball tracking and retrieval strategies have been developed. Obstacle avoidance has become more efficient by the implementation of an RRT-based path planner. Finally, tackling situations can now be handled more successfully due to a new perception of the opponent's feet and the ability to carry out kicking motions within the walking pattern.

This paper is organized as follows: Section 2 briefly summarizes the Standard Platform League's state of the art, focusing on previous works of the B-Human team. The different ball models are presented together with the cooperative tracking and retrieval strategies in Sect. 3. The path planner is described in Sect. 4, followed by the developments regarding tacklings in Sect. 5.

2   State of the Art

As aforementioned, the current major challenges of the Standard Platform League can be considered as more or less solved, not only by B-Human, but also by a number of other teams. In this section, we briefly describe the currently used approaches, with a focus on developments related to the works presented in this paper.

Vision is confined mostly by NAO's limited computing resources. Therefore, most teams rely on manual color classification and detect cues by a combination of heuristics and grid-based or blob-based clustering approaches. However, some teams such as HTWK [16] already have systems that perform an automatic color classification. Whereas it is obviously common to reliably detect major objects such as the ball and the goals, only a few teams perform visual obstacle detection. Current solutions for this task include a detection based on the robots' waistbands [4] as well as the usage of color histogram features [12]. In Sect. 5.1, we present a new approach for detecting nearby robots in tackle situations.

For self-localization, probabilistic state estimation approaches such as (different variations of) Kalman filters [8], Monte-Carlo localization [5], or a combination of both are successfully applied by almost all teams. Several teams also estimate the ball's velocity in a sophisticated manner, as indicated by a number of effectively jumping goal keepers. Furthermore, it can be assumed that most teams communicate the ball state among their robots and possibly also perform a fusion of the different measurements. In Sect. 3, we present the cooperative ball tracking approach employed by our team.

Probably due to the limited computational resources of the NAO platform, elaborated planning approaches for action selection or motion planning are currently not common in the Standard Platform League. Another reason might be the fact that most problems in this domain can still be handled by simpler approaches, such as finite state machines. However, Steffens *et al.* [15] have already presented an A* path planner for the SPL. In Sect. 4, we describe the advantages of using an RRT-based path planning approach.

Although Aldebaran Robotics already provides a robust walking implementation for the NAO robot, all successful teams rely on their own walking approaches which are able to reach higher speeds. Recent works have been published by Czarnetzki *et al.* [3] and Graf and Röfer [6]. Kicking implementations are in many cases based on static sequences of key frames. However, in recent years, some teams introduced dynamic kicking motions, such as HTWK [17], NaoDevils [2], and B-Human [13]. An approach for combining walking and kicking is presented in Sect. 5.2.

3 Ball Models

The ball is the most important object in a soccer game as its state determines the behaviors of all robots at any moment during a game. Therefore, it is highly advantageous for a team if all robots know its correct position as often as possible. In addition, knowing the velocity of the ball is important, because this allows predicting future ball positions. For instance, for the goal keeper to decide when to dive, it must know when the ball would cross the goal line. Together with a friction model, it can also be predicted, where a ball will come to a halt, allowing robots to directly head to that position instead of chasing a moving target.

Keeping track of the ball was one of our major goals for 2011. The *local ball model* estimates the ball's position and velocity for each individual robot. The *global ball model* fuses the local ball models of all players to a team-wide estimate of the ball's position and velocity. The *synchronized head control* tries to make sure that the team does not miss unexpected ball movements. Finally, the *field coverage model* is used to coordinate the search for the ball in case it has been lost.

3.1 Local Ball Model

The local ball model uses Kalman filters [8] to derive the actual ball motion from the perceptions of the ball delivered by the vision system. Since ball motion on a RoboCup soccer field has its own peculiarities, as for instance instantaneous speed changes due to kicks and ball repositioning due to referee interventions, the belief about the ball state is inherently multimodal. Since a single Kalman filter cannot represent a multimodal belief state, we use multiple multivariate Gaussian probability distributions (currently twelve) to represent the belief concerning the ball. Each of these distributions is used independently for the prediction step and the correction step of the filter. Only one of those distributions is used to generate the actual ball model. That distribution is chosen depending on how well the current measurement, i. e. the position the ball is currently seen at, fits to the mean of the distribution and how small the variance of that distribution is. That way we get a pretty accurate estimate of the ball motion while being able to quickly react on displacements of the ball, for example when the ball is moved by the referee after being kicked off the field.

To further improve the accuracy of the estimation, the distributions are equally divided into two sets, one for rolling balls and one for balls that do not move. Both sets are maintained at the same time and get the same measurements for the correction steps. Since the perceived position and motion of the ball can change rapidly at almost any time, the worst distribution of each set in each frame gets reset to effectively throw one filter away and replace it with a newly initialized one.

There are some situations in which a robot changes the motion of the ball. After all, we filter the ball position to finally get to the ball and kick it. The robot influences the motion of the ball either by kicking it or by just standing in the way of a rolling ball. To incorporate these influences into the ball model, the mean value of the best probability distribution from the last frame gets clipped against the robot's feet. In such a case, the probability distribution is reset, so that the position and the velocity of the ball get overwritten with new values depending on the motion of the foot the ball is clipped against. Since the vector of position and velocity is the mean value of a probability distribution, a new covariance matrix is calculated as well. The covariance matrix determining the process noise for the prediction step is fixed over the whole process. Contrary to that, the covariance for the correction step is derived from the actual measurement; it depends on the distance between robot and ball.

3.2 Global Ball Model

Unlike some other domains, such as the Small Size League, the robots in the SPL do not have a common and consistent model of the world, but each of them has an individual world model, estimated on the basis of its own limited perception. However, the rules allow the robots to communicate with each other over WLAN, using a limited bandwidth of 500 kbit/s per team. Since a shared model is a necessity for creating cooperative behavior, we implemented a combined world model that lets all robots of a team have an estimate of the current state of the world, even if parts of it were not seen by each individual robot. This estimate is consistent among the team of robots (aside from delays in the WLAN communication) and consists of three different parts (*global ball model*, *positions of the teammates*, and *positions of opponent players*).

In this paper, we focus on the global ball model. It is calculated locally by each robot, but takes the ball models of all teammates into account. This means that the robot first collects the last valid ball model of each teammate, which is in general the last received, except for the case that the teammate is not able to play, for instance because it is penalized or fallen down. In this case, the last ball model computed before that incident is used. Since a ball model might already be impeded by misreadings when a robot detects, for instance, that it is falling, the last valid ball model is the one that was received 500 ms before the robot reports that it is incapacitated. To still be able to access this model, each robot buffers the ball models from its teammates for the last 500 ms in a ring buffer.

The only situation in which a teammate's ball model is completely ignored is when the ball was seen outside the field, which is considered a false perception. After the collection of the ball models, they are combined in a weighted sum calculation to get the global ball model. The ball model of each individual robot is weighted by the product of different factors. The first factor is the validity of the self-localization estimate (pose). The validity reflects how unimodal the belief about the robot's pose is. We use a combination of particle filter and Kalman filter for self-localization. Basic self-localization is provided by an Augmented Monte-Carlo localization [7]. However, when the belief state about the pose seems to be unimodal, the Kalman filter takes over and tracks the pose provided by the particle filter, until the particle filter suggests a completely different pose, for instance because the robot was kidnapped. When the Kalman filter is in charge, the validity of the pose is set to 1. Otherwise, it is the ratio between the number of particles in the largest cluster of the particle filter, i. e. the one from which the pose is calculated, and the overall number of particles. The second factor is the period of time Δn in seconds since the ball was last seen by the robot, i. e. how long it has not been seen. The third factor is the period of time Δm in seconds for which the ball is missing, i. e. how long the ball was not seen although it should have appeared in the robot's camera image according to the ball model. The final factor is the approximate deviation of the distance to the ball based on the bearing. Technically, σ is computed as the expected change of the distance measurement if the vertical bearing of the ball would be wrong by $1°$. From these factors, a weight w_r is calculated for each robot r. While a higher validity results in a bigger weight, larger values for the other three factors reduce that weight:

$$w_r = validity_{pose_r} \left(1 - \frac{1}{1 + e^{\alpha - \beta \Delta n_r}}\right) \left(1 - \frac{1}{1 + e^{\gamma - \delta \Delta m_r}}\right) \frac{1}{\sigma_r} \qquad (1)$$

The values used for the constants in the equation are $\alpha = 5, \beta = 1, \gamma = 4, \delta = 4$, which means that missing a ball that should have appeared in the camera image is significantly worse than not having seen the ball because the robot looked somewhere else. Based on the weights for all N robots considered, a common ball model is calculated that contains an approximate position and velocity of the ball.

3.3 Synchronized Head Control

Although the ball is the most important object in a soccer game, even the robot approaching or having possession of the ball has to regularly look away from it to perceive field lines and goal posts to update its self-localization. In particular, this is necessary when the robot is close to the ball, since in this case, camera images showing the ball will usually not contain any objects that support self-localization. While the robot is looking away from the ball, the risk of missing unexpected ball movements is rather high. On the one hand, an opponent robot might kick the ball away, and on the other hand, the robot itself might inadvertently touch the ball and thereby move it away. Since searching for the ball delays the game, the teammates are responsible for keeping track of the ball when the robot closest to the ball (i. e. the *striker*) looks away.

To accomplish this, all robots continuously broadcast whether their current head motion will show them the ball in their camera image according to their ball model. If it will not and if the sending robot is currently the striker, its teammates will look at the ball instead, independently of what their original intention for a head motion was. It is not guaranteed that each of them will actually be able to see the ball, because other robots could be blocking the view. However, the fact that all of them are looking at the ball increases the chance that at least one of them will actually be able to see it. But even if none of them does, it is still advantageous to observe the region where the ball would appear eventually when the striker has lost it. Note that there is no negotiation between the robots involved. In particular, the robot having possession of the ball is not limited in any way by this behavior.

Experiments have shown that as a result of this synchronization, the periods of time in which no robot of the team sees the ball are below 100 ms most of the time and very rarely climb above one second.

3.4   Field Coverage Model

All the methods described above cannot completely prevent the team from losing track of the ball, for instance when it is hidden by an opponent robot or it has been moved by a referee. Therefore, it is also important to optimize the process of finding the ball again after it has been lost. In general, searching for the ball is a three-step process. First, a robot just moves its head and sweeps the area it can see without moving its body. Then, it starts turning on the spot. Finally, it starts walking around the field to get a different perspective. Where the robot looks at during the search can be optimized by managing both the areas it has already covered and the areas that have been covered by its teammates.

To keep track of which parts of the field are visible to a robot, the field is divided into a very coarse grid of cells, each cell being a square that has a size of $0.5m \times 0.5m$ (cf. Fig. 1a, b). To determine which of the cells are currently visible, the current image is projected onto the field. Then all cells the centers of which lie within the projected image are candidates for being marked as visible, unless either robots are obstructing the view to that cell or the cell is so far away (2 m) that other robots would not be recognized safely. Having determined the set of visible cells, each of those cells is timestamped. These timestamps are later used to build the global field coverage model and to determine the least-recently-seen cell that can be used to generate the head motion to scan the field while searching for the ball.

A special situation arises when the ball goes out. If this happened, the cell in which the ball most likely has been put back is determined by the last intersection of the trajectory of the ball with an outer field line before the referee

a)

b)

c)

d)

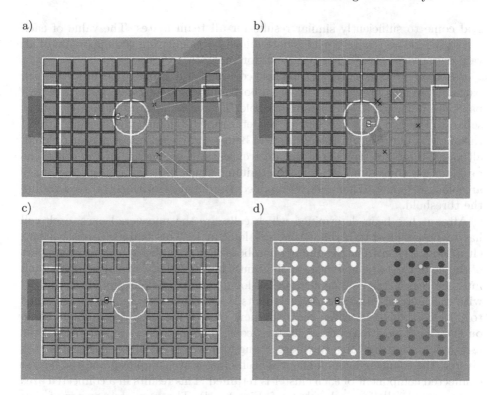

Fig. 1. Local and global field coverage models. a) The visible area of a single robot projected on the field. Uncovered cells are black. Other robots (each denoted by a black X) cast shadows over the visible area. b) The model generates the best camera targets to search for the ball. The white X marks the best target that is reachable without turning. The gray X marks the overall best target. c) The global grid obtained by merging the communicated local grids. d) The largest connected components of each cluster of uncovered cells. Each cell of such a component is marked by a circle in a different color to indicate the cluster assignment. The three extra circles mark the resulting search targets.

computer sent the signal that indicates that the ball is out. Knowing this cell, the timestamps of the entire grid are reset such that this cell, in which the ball most likely is, appears to be the most outdated one and the cells at the left and right field borders appear to be more outdated than the rest of the grid. Of course, this grid resetting can only work well if the ball motion was estimated accurately and if the referees put the ball on the correct position on the field. However, without resetting, the information stored in the grid would not be useful anyway.

In addition to its own local field coverage grid, each robot maintains the field coverage grids of its teammates, which are incrementally updated in every team communication cycle. All these grids have to be merged into a single global grid that looks roughly the same for all teammates so that calculations based on the grid come to sufficiently similar results for all team mates. The value of each cell of the global coverage grid is determined by calculating the maximum of all values stored by each individual robot for that cell.

Based on the values in the global field coverage grid, it has to be decided which parts of the field are covered by the robots and which parts are not, i.e. which parts are unknown to the team as a whole. Therefore, a threshold is required to separate the two classes. It has to be determined dynamically, because a fixed threshold could result in the entire field being considered uncovered or covered, although there are still significant differences in the coverage of the cells. The problem has some similarities to determining which parts of a gray scale image are black or white. Therefore, we applied the Otsu algorithm [14] to compute the threshold.

After it has been determined which cells are the uncovered ones, each cell has to be assigned to a robot that will look at it. This is done using k-means clustering. k is set to be the number of robots that are able to cover a certain part of the field, i.e. to be included, a robot must not be fallen down or penalized and must be reasonably confident in its self-localization. The clusters are initialized with the current positions of the robots and each uncovered cell is assigned to its closest cluster. After that, the new cluster means are computed based on the center positions of the cluster's cells. This process is repeated until the assignments do not change anymore. Using four-way flood-fill on each cell of each cluster, the connected components of each cluster are computed and the largest connected component of each cluster is retained. This results in a connected area of uncovered cells for each robot (cf. Fig. 1c, d). The geometric center of that area is calculated and it is used as a target position for that robot from where it will search for the ball.

4   Path Planning

In recent years, we have used a reactive obstacle avoidance approach based on ultrasonic measurements. This approach has not been effective as the robot only reacted on recently measured obstacles in its immediate vicinity, a behavior that caused multiple problems (cf. Fig. 2a). First and foremost, avoiding an obstacle whilst keeping the original walking target in focus requires walking sidewards, which is exceedingly slow on the NAO. Additionally, reactive obstacle avoidance recurrently guides the robot into situations that thwart fluent play. The most prominent example for such a situation are local minima, which occur often in presence of multiple robots (especially after the last increase of the number of players), but also overtaking an opponent robot during a footrace towards the ball is a hard task.

To overcome these problems, we developed a path planner that is based on the established *Rapidly-Exploring Random Tree* algorithm [10]. This non-optimal algorithm is based on the random exploration of the search space and works on continuous values. It builds up a tree that quickly expands in few directions of the search space, as in each step of the algorithm the tree is enlarged by one

a)   b)

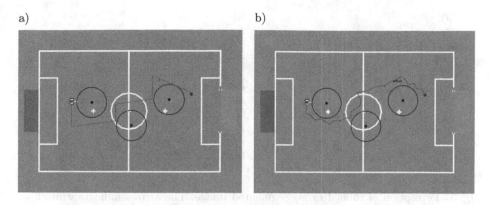

Fig. 2. Paths resulting from different approaches for obstacle avoidance: a) The previously used reactive behavior and b) the RRT planning algorithm

edge in a random direction (cf. Fig. 2b). For this general algorithm, different variants such as *RRT-Extend* [10], *RRT-Connect* [9] and *RRT-Bidirectional* [11] exist. We use *Extend* and a slightly modified variant of *Bidirectional*. Using the former variant restricts the expansion towards the random position to a given distance that is the same for each expansion, which has a direct influence on the expansion of the tree, whereas the latter variant has no influence on the tree itself but decreases the runtime. This is achieved by creating two separate trees, one beginning from the start point and one from the end point. Another modification is to replace the random position by the target position or a waypoint of the last found path with a given probability. Using this modifications helps to avoid oscillations of the path, for instance if there is an obstacle on the direct way to the target.

One advantage of using a path planner is its ability to consider a number of distant obstacles to prevent the aforementioned situations. For this purpose, the path planner not only uses the local ultrasonic measurements, but also integrates a combined world model, which is similar to the global ball model as described in Sect. 3.2, including the positions of all teammates and of the robots tracked by them. Moreover, having a planned path, the time for avoiding an obstacle can be decreased since the robot walks on a circular path around the obstacles, preferring a fast forward walk and rotations over slow sidewards motions (cf. Fig. 2b).

5   Efficient Tacklings and In-walk Kicks

Even a fast reaction regarding ball state changes and an efficient path planning approach cannot always prevent a situation that regularly occurs in each game: the presence of an opponent robot near the ball. Experiences from previous competitions showed that the resulting situations require fast actions as the robot that loses the tackling has only a minor chance of regaining ball possession.

a)   b)

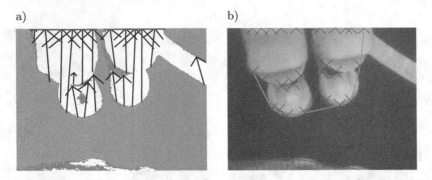

Fig. 3. Perception of robot feet. a) Segmented image. The arrows indicate white line segments detected in a preprocessing step but rejected by the field line detection algorithm. For being considered as robot parts, at least some of these segments must start at the upper image border. Therefore the single arrow on the right side of the image will be ignored. b) Raw image with the convex hull around the start and end points of the accepted segments is computed. The closest (thick circle), the leftmost, and the rightmost point (marked by the smaller circles) are determined to describe the feet.

Therefore, we spent much efforts not only in behavior tuning but also in the development of two approaches that strongly contributed to winning a majority of all tacklings: the perception of a nearby robot's feet as well as the ability to carry out kicks within the normal walking pattern.

5.1 Foot Perception

To successfully solve tackle situations by dribbling around an opponent, knowledge about the position of the opponent is necessary. For short ranges, the NAO's ultrasonic sensors provide reasonable distance measurements (and thus allow the detection of the presence of an obstacle) but quite poor angular information, including many false positives in case of centered obstacles.

To overcome these problems, a simple but yet effective solution has been found: the visual detection of the opponent's feet. This approach has two advantages: Firstly, when looking at a nearby ball, a blocking robot's feet are, in general, in the field of view and no additional search motions of the head are needed. Secondly, to dribble a ball around a robot, its feet are the only body parts of interest as other parts, such as upper body and arms, are obviously not able to block the ball. In addition, the perception of these parts is probably less precise and might distort the overall position estimate.

However, the position of nearby feet can be determined quite reliably and straightforward by clustering white image segments that have previously been rejected by the line detection algorithm, similar to the input used by Metzler *et al.* [12]. An example is shown in Fig. 3. The resulting perceptions are only used in tackle situations, where they override all other, comparably imprecise obstacle information. For path planning, the position of feet is not useful, as for this task, the upper bodies of other robots are the main obstacles.

5.2   In-walk Kicks

As already mentioned in Sect. 2, the tasks of walking and kicking are often treated separately, each solved by different approaches. In presence of opponent robots, such a composition might waste precious time as certain transition phases between walking and kicking are necessary to ensure stability. Direct transitions between walking and kicking are likely to let the robot stumble or, in the worst case, to fall over. Therefore, the B-Human walking implementation [6] is able to carry out sidewards and forward kicks within the walk cycle.

Such an in-walk-kick is described by a number of parameters. On the one hand, the sizes and speeds of the step before the kick and the step during the kick are defined. On the other hand, a 6-D trajectory (three degrees for translation and three degrees for rotation) relative to the original trajectory of the swinging foot is defined that overlays the original trajectory and thereby describes the actual kicking motion. The kick retains the start and end positions and speeds of a normal step. The instability resulting from the higher momentum of the kick is compensated by the walk during the steps following the kick.

6   Conclusions and Future Works

In 2011, B-Human showed again a strong overall performance and won every match at the German Open as well as at RoboCup 2011 in Istanbul. The developments presented in this paper significantly contributed to these achievements such that almost all tackles have been won and all robots were able to quickly react on any ball state changes. The latter, in combination with the new path planner, strongly decreased the number of situations in which an opponent robot was alone at the ball, having the chance to score.

Several components of B-Human's current system require a proper calibration to perform as desired. Therefore, the focus of future work is probably not the development of new soccer features – except for necessary adaptations regarding major rule changes – but the integration of approaches and tools that reduce calibration efforts and enable a more efficient testing. Currently ongoing theses works already include a new vision approach that does not rely on manual color calibration as well as two new approaches for robot behavior specification, a scripting language based on the programming concept of generators and a strategy definition based on the playbook approach [1] respectively, both leading towards more compact and faster ways to adapt behavior definitions.

References

1. Bowling, M., Browning, B., Veloso, M.: Plays as effective multiagent plans enabling opponent-adaptive play selection. In: Proc. of the 14th Int. Conf. on Automated Planning and Scheduling (ICAPS), Whistler, BC, Canada, pp. 376–383 (2004)
2. Czarnetzki, S., Kerner, S., Klagges, D.: Combining Key Frame Based Motion Design with Controlled Movement Execution. In: Baltes, J., Lagoudakis, M.G., Naruse, T., Ghidary, S.S. (eds.) RoboCup 2009. LNCS (LNAI), vol. 5949, pp. 58–68. Springer, Heidelberg (2010)

3. Czarnetzki, S., Kerner, S., Urbann, O.: Applying Dynamic Walking Control for Biped Robots. In: Baltes, J., Lagoudakis, M.G., Naruse, T., Ghidary, S.S. (eds.) RoboCup 2009. LNCS (LNAI), vol. 5949, pp. 69–80. Springer, Heidelberg (2010)

4. Fabisch, A., Laue, T., Röfer, T.: Robot recognition and modeling in the RoboCup Standard Platform League. In: Pagello, E., Zhou, C., Behnke, S., Menegatti, E., Röfer, T., Stone, P. (eds.) Proc. of the Fifth Workshop on Humanoid Soccer Robots at the 2010 IEEE-RAS Int. Conf. on Humanoid Robots, Nashville, TN, USA (2010)

5. Fox, D., Burgard, W., Dellaert, F., Thrun, S.: Monte-Carlo localization: Efficient position estimation for mobile robots. In: Proc. of the Sixteenth National Conf. on Artificial Intelligence, Orlando, FL, USA, pp. 343–349 (1999)

6. Graf, C., Röfer, T.: A center of mass observing 3D-LIPM gait for the RoboCup Standard Platform League humanoid. In: Röfer, T., Mayer, N.M., Savage, J., Saranli, U. (eds.) RoboCup 2011. LNCS, vol. 7416, pp. 102–113. Springer, Heidelberg (2012)

7. Gutmann, J.S., Fox, D.: An Experimental Comparison of Localization Methods Continued. In: Proc. of the 2002 IEEE/RSJ Int. Conf. on Intelligent Robots and Systems (IROS 2002), Lausanne, Switzerland, vol. 1, pp. 454–459 (2002)

8. Kalman, R.E.: A new approach to linear filtering and prediction problems. Transactions of the ASME-Journal of Basic Engineering 82(Series D), 35–45 (1960)

9. Kuffner, J.J., LaValle, S.M.: RRT-connect: An efficient approach to single-query path planning. In: Proc. of the 2000 IEEE Int. Conf. on Robotics and Automation (ICRA 2000), San Francisco, CA, USA, vol. 2, pp. 995–1001 (2000)

10. LaValle, S.M.: Rapidly-exploring random trees: A new tool for path planning. Tech. Rep. TR 98-11, Computer Science Dept., Iowa State University (1998)

11. LaValle, S.M., Kuffner, J.J.: Randomized kinodynamic planning. In: Proc. of the 1999 IEEE Int. Conf. on Robotics and Automation (ICRA 1999), Detroit, MI, USA, vol. 1, pp. 473–479 (1999)

12. Metzler, S., Nieuwenhuisen, M., Behnke, S.: Learning Visual Obstacle Detection Using Color Histogram Features. In: Röfer, T., Mayer, N.M., Savage, J., Saranli, U. (eds.) RoboCup 2011. LNCS, vol. 7416, pp. 149–161. Springer, Heidelberg (2012)

13. Müller, J., Laue, T., Röfer, T.: Kicking a Ball – Modeling Complex Dynamic Motions for Humanoid Robots. In: Ruiz-del-Solar, J., Chown, E., Ploeger, P.G. (eds.) RoboCup 2010. LNCS (LNAI), vol. 6556, pp. 109–120. Springer, Heidelberg (2010)

14. Otsu, N.: A threshold selection method from grey level histograms. IEEE Transactions on Systems, Man, and Cybernetics 9(1), 62–66 (1979)

15. Steffens, R., Nieuwenhuisen, M., Behnke, S.: Multiresolution path planning in dynamic environments for the standard platform league. In: Pagello, E., Zhou, C., Behnke, S., Menegatti, E., Röfer, T., Stone, P. (eds.) Proc. of the Fifth Workshop on Humanoid Soccer Robots in conjunction with the 2010 IEEE-RAS Int. Conf. on Humanoid Robots, Nashville, TN, USA, pp. 59–64 (2010)

16. Tilgner, R., Reinhardt, T., Borkmann, D., Seering, S., Kalbitz, T., Fritzsche, R.: Nao-Team HTWK – team description paper. In: Röfer, T., Mayer, N.M., Savage, J., Saranli, U. (eds.) RoboCup 2011: Robot Soccer World Cup XV Preproceedings, RoboCup Federation (2011)

17. Xu, Y., Mellmann, H.: Adaptive Motion Control: Dynamic Kick for a Humanoid Robot. In: Dillmann, R., Beyerer, J., Hanebeck, U.D., Schultz, T. (eds.) KI 2010. LNCS, vol. 6359, pp. 392–399. Springer, Heidelberg (2010)

RoboCup 2011 Humanoid League Winners

Daniel D. Lee[1], Seung-Joon Yi[1], Stephen McGill[1], Yida Zhang[1],
Sven Behnke[2], Marcell Missura[2], Hannes Schulz[2], Dennis Hong[3],
Jeakweon Han[3], and Michael Hopkins[3]

[1] GRASP Lab, Engineering and Applied Science, Univ. of Pennsylvania, USA
{ddlee,yiseung,smcgill3,yida}@seas.upenn.edu
http://www.seas.upenn.edu/~robocup
[2] Autonomous Intelligent Systems, Computer Science, Univ. of Bonn, Germany
{schulz,behnke,missura}@cs.uni-bonn.de
http://ais.uni-bonn.de
[3] RoMeLa, Mechanical Engineering, Virginia Tech, USA
{dhong,jk4robot,hopkns}@vt.edu
http://www.romela.org/robocup

Abstract. Over the past few years, soccer-playing humanoid robots advanced significantly. Elementary skills, such as bipedal walking, visual perception, and collision avoidance have matured enough to allow for dynamic and exciting games. In this paper, the three winning Humanoid League teams from the KidSize, TeenSize, and AdultSize class present their soccer systems. The KidSize winner team DARwIn used the recently introduced DARwIn-OP robot. The TeenSize winner NimbRo used their self-constructed robots Dynaped and Bodo. The AdultSize Louis Vuitton Best Humanoid Award winner CHARLI detail the technology behind the outstanding performance of its robot CHARLI-2.

1 Introduction

In the RoboCup Humanoid League, mostly self-constructed robots with a human-like body plan compete with each other on the soccer field. The league comprises three size classes: KidSize (<60 cm), TeenSize (100-120 cm) and AdultSize (>130 cm). While the KidSize robots are playing 3 vs. 3 soccer games, the TeenSize robots started to play 2 vs. 2 soccer games in 2010, and the AdultSize robots engage in 1 vs. 1 Dribble-and-Kick competitions. In addition, all three classes face technical challenges, like dribbling the ball through an obstacle course, double-passing, and throw-in of the ball. In this paper, the three winning teams of the RoboCup 2011 championship in Istanbul — DARwIn, NimbRo, and CHARLI — detail their hard- and software approaches to solve the problems of playing humanoid soccer. These include fast and flexible bipedal locomotion, controlling dynamic full-body motions, maintaining balance in the presence of disturbances, robust visual perception of the game situation, individual soccer skills, and team coordination.

T. Röfer et al. (Eds.): RoboCup 2011, LNCS 7416, pp. 37–50, 2012.

2   KidSize Winner Team DARwIn

Team DARwIn is a joint team of the University of Pennsylvania's GRASP lab and Virginia Tech's RoMeLa lab. The DARwIn-OP proved to be a reliable design and became a commercialized product for general robotics research. Complementing Virginia Tech's tradition of humanoid hardware development, the University of Pennsylvania utilizes its long-time experience of RoboCup Standard Platform League participation. With Penn's open source release of its source code [6] and the open source DARwIn-OP, Team DARwIn based its performance on fully open source engineering.

2.1   DARwIn-OP Robot Hardware

We used the DARwIn-OP robot designed by RoMeLa lab as the robotic platform for RoboCup KidSize competition. It is 45 cm tall, weighs 2.8 kg, and has 20 degrees of freedom. It has a web camera for visual feedback, a 3-axis accelerometer and 3-axis gyroscope for inertial sensing. Position controlled Dynamixel servos are used for actuators, which are controlled by a custom microcontroller connected by an Intel Atom based embedded PC at a control frequency of 100Hz. One noticeable feature of our robotic platform is that after years of joint development, it has become reliable enough to be a commercial product produced by Robotis, co., Ltd. This gave us a big logistic advantage as we could work with a significant number of standardized robots.

2.2   Unified Humanoid Robotics Platform

We have been working with a number of the DARwIns for research-based tasks, in addition to robotic soccer. To exploit the commonalities of different platforms and tasks, and to reduce development time, we developed a flexible cross-robot software architecture, of which the main goals are modularity and portability. Every component of this architecture remains individually interchangeable, which ensures that we can easily port code between robots.For the robotic soccer task, we can use basically the same behavioral logic -the high level controller for interacting with the environment - regardless of which humanoid the platform interacts with. For this year's RoboCup, we used the same basic code for five different humanoid platforms: DARwIn-HP and DARwIn-OP for KidSize class, CHARLI for AdultSize class, Nao for standard league, and a Webots model of Nao for the Robostadium simulation league. The overall structure of our software architecture is shown in Fig. 1. It consists of two subsystems: A motion subsystem and a vision subsystem, as well as a behavioral logic module which governs high-level behavior. The vision subsystem processes the video stream and extracts vision cues such as balls, goalposts and lines and passes them to the behavioral logic, which controls the high level behavior such as setting walk velocity or initiating special actions such as kicking or diving. Finally, the motion subsystem communicates with robot-specific actuators and sensors, and generates joint trajectories for numerous motions according to behavior commands

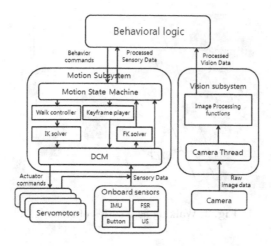

Fig. 1. Block Diagram of the Software Architecture

from the behavioral logic. In addition to robotic soccer, we use this software platform for other research projects. An open source version of our platform can be downloaded at the UPennalizers website[1].

2.3 Walking

We use a zero moment point (ZMP) [13] based walk controller which uses the 3D linear inverted pendulum model (LIPM) to calculate the torso trajectory so that the actual ZMP lies inside the support foot. However, our walk engine has two notable features. It is not periodic; instead we allow each step to have an arbitrary support foot, walk velocity and step duration. Additionally, we calculate the torso trajectory for each step period using an analytic solution of the ZMP equation assuming a piecewise linear ZMP trajectory. This induces a discontinuous torso velocity at each step transition, but as it occurs during the (most stable) double support phase, it does not hamper the stability much in practice. On the other hand, our approach enables high maneuverability and flexibility. We have achieved a maximum walk speed of 36 cm/s which is very high considering the relatively small size of our robot.

Foot Trajectory Generation: We divide the walking into a series of steps which can be generally defined as

$$STEP_i = \{SF, t_{STEP}, L_i, T_i, R_i, L_{i+1}, T_{i+1}, R_{i+1}\} \tag{1}$$

As illustrated in Fig. 2.3, SF denotes the support foot, t_{STEP} is the duration of the step, and L_i, T_i, R_i, and $L_{i+1}, T_{i+1}, R_{i+1}$ are the initial and final 2D poses of left foot, torso and right foot in (x, y, θ) coordinates. The L_i, T_i, R_i poses are

---

[1] https://fling.seas.upenn.edu/~robocup/wiki/

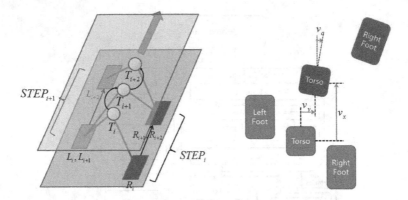

Fig. 2. Walking as a series of steps

determined by the final feet and torso poses from the last step, and L_{i+1}, R_{i+1} are calculated using the commanded walk velocity and current foot configuration to enable omnidirectional walking. Foot reachability and self-collision constraints are also taken into account when calculating target foot poses. We use a FIFO queue structure to handle special sequences of steps, such as dynamic kick. When the current step $STEP_k$ is over, a new step $STEP_{k+1}$ is determined, and foot trajectories for the new step are generated accordingly.

Torso Trajectory Generation: After generating foot trajectories, the torso trajectory should be generated such that the resulting ZMP lies inside the support polygon during the single support phase. In general, we need to solve an optimization problem. However, we opt to use an analytic torso trajectory solution with zero ZMP error assuming the following piecewise linear reference zero moment point (ZMP) trajectory $p_i(\phi)$ for the left support case

$$p_i(\phi) = \begin{cases} T_i(1 - \frac{\phi}{\phi_1}) + L_i\frac{\phi}{\phi_1} & 0 \leq \phi < \phi_1 \\ L_i & \phi_1 \leq \phi < \phi_2 \\ T_{i+1}(1 - \frac{1-\phi}{1-\phi_2}) + L_i\frac{1-\phi}{1-\phi_2} & \phi_2 \leq \phi < 1 \end{cases} \tag{2}$$

where ϕ is the walk phase and ϕ_1, ϕ_2 are the timing parameters determining the transition between the single support and double support phases. This ZMP trajectory yields for following $x_i(\phi)$ solution with zero ZMP error during the step period:

$$x_i(\phi) = \begin{cases} p_i(\phi) + a_i^p e^{\phi/\phi_{ZMP}} + a_i^n e^{-\phi/\phi_{ZMP}} \\ \quad -\phi_{ZMP} m_i \sinh\frac{\phi-\phi_1}{\phi_{ZMP}} & 0 \leq \phi < \phi_1 \\ p_i(\phi) + a_i^p e^{\phi/\phi_{ZMP}} + a_i^n e^{-\phi/\phi_{ZMP}} \\ & \phi_1 \leq \phi < \phi_2 \\ p_i(\phi) + a_i^p e^{\phi/\phi_{ZMP}} + a_i^n e^{-\phi/\phi_{ZMP}} \\ \quad -\phi_{ZMP} n_i \sinh\frac{\phi-\phi_2}{\phi_{ZMP}} & \phi_2 \leq \phi < 1 \end{cases} \tag{3}$$

where $\phi_{ZMP} = t_{ZMP}/t_{STEP}$ and m_i, n_i are ZMP slopes which are defined as follows for the left support case

$$m_i = (L_i - T_i)/\phi_1 \qquad (4)$$

$$n_i = -(L_i - T_{i+1})/(1 - \phi_2) \qquad (5)$$

The parameters a_i^p and a_i^n can be uniquely determined by the boundary conditions $x_i(0) = T_i$ and $x_i(1) = T_{i+1}$.

Active Stabilization: The physical robot differs much from the ideal LIPM model, and external perturbations from various sources can make the open loop walk unstable. With the DARwIn-OP robot, we have 3 sources of sensory feedback: Filtered IMU angles, gyro rate readings and proprioception information based on joint encoders. We use this information to apply stabilizing torques at the ankle joints, called "ankle strategy." In addition, we also implemented other human-inspired push recovery behaviors, hip and step strategies, and used machine learning to find an appropriate controller to reject disturbances using those strategies [14]. In spite of some success in controlled situations, we found the learned controller was not reliable enough for competition and we only used the ankle strategy controller. Overall, our robots were very stable during fast walking, but they fell down when colliding into other robots. Implementing a reliable push recovery controller that can prevent falling from collision remains a big challenge.

2.4   Kicking

Instead of using the typical key frame method for kicking, we use the walk engine to generate a set of parameterized kick motions. There are many advantages to this approach. Designing and tuning a new kick is much easier than making a key frame kick in joint space, the active stabilization can be used during kicking, and kicking can be seamlessly integrated with walking. Utilizing the walk engine also allows us to perform a dynamic kick. Typically, the robot would put its center of mass (COM) within its support polygon during kicking to be statically stable. Instead, the robot puts its ZMP within the support polygon to be dynamically stable during kicking.

Static Kick: To allow our step-based walk engine to execute static kicks, we first define a kick as a sequence of kick steps $KICK_i$

$$KICK_i = \{SF, t_{STEP}, L_i, T_i, R_i, L_{i+1}, T_{i+1}, R_{i+1}\} \qquad (6)$$

which has the same format as $STEP_i$ but has 6D coordinate$(x, y, z, \psi, \theta, \phi)$ for L, T, R. Each kick step corresponds to an elementary action during the kick such as lifting, kicking and landing. With the help of this simple operation space definition for kicking, we can easily make and test a number of different kicks in a short time. The frontal kick we used for our matches consists of 7 kick steps and takes approximately 4 seconds to complete, and can kick balls up to 5 meters.

Dynamic Kick: Static kicks can be powerful, but the main disadvantage is that it requires a longer time to stabilize before beginning to walk again. However, when opponents are nearby, kicking fast is much more important than kicking strong. Thus, we need a dynamic kick, which can be much faster as it does not require a complete stop and static balancing. Dynamic front kick is implemented by putting two steps in the step queue, support step and kick step. For kick step we use longer step period and special foot trajectory so that it can maximize the foot velocity at hitting the ball. After the robot kicks the ball, the step queue is emptied and it resumes walking according to its commanded walk velocity without stopping. Similarly, the dynamic side kick consists of three steps including two normal steps and one special step.

We have found that as the body is also moving forward, the dynamic front kick has more range than its static kick counterpart. It can be executed very fast, too – it takes 3 steps in worst case which takes 0.75 sec. The main disadvantage is the weak kick strength: in most cases, the dynamic kick cannot shoot more than 1.5 meters. However, as the robot completes its kick way faster and it is moving forward during kicking, it can quickly catch up to the ball and kick again. We have tested several different tactics for choosing between dynamic and static kicks, and we have found fast dynamic kicks are much more effective against good teams. There is less probability of kicking out of bounds, and there is the unexpected side effect of deceiving enemy goalies.

2.5   Optimal Approaching

One challenge for robotic soccer is to arrive at the target pose in the shortest time possible while satisfying all the locomotion constraints. The basic strategy for arriving at the target pose is the rotate-chase-and-orbit strategy, which approaches the ball in straight path and orbits around the ball until it reaches the target pose. However, this is actually a motion planning problem, and can be formalized as a reinforcement learning problem with state S, action A and reward R:

$$
\begin{aligned}
S &= \{r, \theta_{ball}, \theta_{goal}\} \\
A &= \{v_x, v_y, v_a\} \\
R &= 100 \; at \; target \; pose, \; -1 \; otherwise
\end{aligned}
\tag{7}
$$

We approximate the policy function by a heuristic controller with 5 continuous parameters and train it using a policy gradient RL algorithm. After 150 episodes of training, average steps to reach the target pose decreased by about 25% compared to baseline rotate-chase-orbit strategy. During the match, we found an unexpected side effect of this strategy – it tends to make the robot reach the kicking position first and then turn to face the goal, which effectively blocks opponents' kick towards our goal.

3 TeenSize Winner Team NimbRo

Team NimbRo has a long and successful history in RoboCup with overall nine wins in international Humanoid League competitions since 2005. In 2011, our team won the TeenSize class for the third time in a row. Our robots also won the 2011 technical challenges with a good performance in the Double Pass and the Obstacle Dribbling challenges. This year, the main rule change in the TeenSize class was an increase in size of the field to 9×6 m. We successfully adapted our system to the larger field size and repeated last year's reliable performance in the finals without a single fall and without the need for human intervention. In the remainder of this section, we describe the mechanical and electrical design of our robots, the visual perception of the game situation, and the generation of soccer behaviors in a hierarchical framework.

3.1 Mechatronic Design of NimbRo TeenSize Robots

Fig. 3 shows our two TeenSize robots: Dynaped and Bodo. Their mechanical design focused on simplicity, robustness, and weight reduction.

Dynaped is 105 cm tall and weighs 7.5 kg. The robot has 13 DOF: 5 DOF per leg, 1 DOF per arm, and one joint in the neck that pans the head. Its legs include parallel kinematics that prevents the robot from tilting in sagittal direction. Dynaped's leg joints are driven by master-slave pairs of Robotis Dynamixel EX-106 actuators. Bodo is 103 cm tall and has a weight of about 5 kg. The robot is driven by 14 Dynamixel actuators: six per leg and one in each arm.

The robot skeletons are constructed from rectangular milled aluminum tubes. The feet are made from sheets of composite carbon and heads are produced by 3D printing of polymer material. Both robots are protected against mechanical stress by a 'mechanical fuse' between the hip and the spine. This mechanism includes a pre-loaded spring that yields to large external forces. Together with foam protectors, it allows the robots to dive quickly to the ground as a goalie [8].

Fig. 3. RoboCup 2011 TeenSize finals: NimbRo vs KMUTT Kickers. Our team played with the robots Dynaped (striker) and Bodo (goalie).

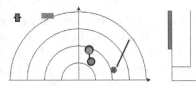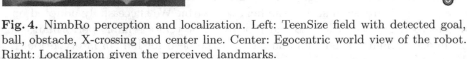

Fig. 4. NimbRo perception and localization. Left: TeenSize field with detected goal, ball, obstacle, X-crossing and center line. Center: Egocentric world view of the robot. Right: Localization given the perceived landmarks.

The robots are controlled by a pocket size PC, a Sony Vaio UX, which features an Intel 1.33 GHz ULV Core Solo Processor. This PC runs computer vision, behavior control, motion generation, and WLAN communication. The robots are also equipped with a HCS12X microcontroller board, which manages the detailed communication with all joints via an RS-485 bus. The microcontroller also reads in a dual-axis accelerometer and two gyroscopes. The robots are powered by high-current Lithium-polymer rechargeable batteries, which last for about 20 minutes of operation.

3.2   Proprioception, Visual Perception and Self-localization

The perception of humanoid soccer robots can be divided into two categories: proprioception and computer vision. For proprioception, we use the joint angle feedback of the servos and apply it to the kinematic robot model using forward kinematics. Additionally, we fuse accelerometer and gyroscope measurements to estimate the tilt of the trunk in roll and pitch direction. Knowing the attitude of the trunk and the configuration of the kinematic chain, we rotate the entire model around the current support foot such that the attitude of the trunk matches the angle we measured with the IMU. This way, we obtain a robot pose approximation that can be used to extract the location and the velocity of the center of mass. Temperatures and voltages are also monitored for notification of overheating or low batteries, respectively.

For visual perception of the game situation, we capture and process 752×480 YUV images from a IDS uEye camera with fish eye lens (Fig. 4 left). Pixels are color-classified using a look-up table. In down-sampled images of the individual colors, we detect the ball, goal-posts, poles, penalty markers, field lines, corners, T-junctions, X-crossings, obstacles, team mates, and opponents using size and shape information. We estimate distance and angle to each detected object by removing radial lens distortion and by inverting the projective mapping from field to image plane (Fig. 4 center). To account for camera pose changes during walking, we learned a direct mapping from the IMU readings to offsets in the image. We also determine the orientation of lines, corners and T-junctions relative to the robot.

We track a three-dimensional robot pose (x, y, θ) on the field using a particle filter [12] (Fig. 4 right). The particles are updated using a motion model which is a simple linear function of the gait velocity commanded to the robot. Its parameters are learned from motion capture data [9]. The weights of the particles are updated according to a probabilistic model of landmark observations (distance and angle) that accounts for measurement noise. To handle unknown data association of ambiguous landmarks, we sample the data association on a per-particle basis. The association of field line corner and T-junction observations is simplified using the orientation of these landmarks. Further details can be found in [10].

3.3   Hierarchical Reactive Behavior Control

We control our robots using a framework that supports a hierarchy of reactive behaviors [2]. This framework allows for structured behavior engineering. Multiple layers that run on different time scales contain behaviors at different abstraction levels. When moving up the hierarchy, the update frequency of sensors, behaviors, and actuators decreases. At the same time, they become more abstract. Raw sensor input from the lower layers is aggregated to slower, abstract sensors in the higher layers. Abstract actuators enable higher-level behaviors to configure lower layers in order to eventually influence the state of the world.

Currently, our implementation consists of three layers. The lowest, fastest layer is responsible for generating motions, such as walking, kicking and the goalie dive. Our omnidirectional gait [1] is based on rhythmic lateral weight shifting and coordinated swinging of the non-supporting leg in walking direction. This open-loop gait is self-stable when undisturbed. In order to reject larger disturbances, we recently extended our gait engine with a lateral capture step controller [7] that modifies the timing and the lateral location of the footsteps to maintain balance. This controller uses a linear inverted pendulum model to predict the motion of the robot's center of mass. For the goalie, we designed a motion sequence that accelerates the diving motion compared to passive sideways falling from an upright standing posture [8]. The goalie jump decision is based on a support vector machine that was trained with real ball observations.

At the next higher layer, we abstract from the complex kinematic chain and model the robot as a simple holonomic point mass that is controlled with a desired velocity in sagittal, lateral and rotational directions. We are using a cascade of simple reactive behaviors based on the force field method to generate ball approach trajectories, ball dribbling sequences, and to implement obstacle avoidance.

The topmost layer of our framework takes care of team behavior, game tactics and the implementation of the game states as commanded by the referee box.

4   AdultSize Winner Team CHARLI

Stemming from the success of team DARwIn in the KidSize class, team CHARLI has participated in the AdultSize class since its beginning at RoboCup 2010.

Because the AdultSize class is still relatively new, the main focus of most teams in this class, including team CHARLI, has been on the development of a stable bipedal walking platform to serve as a basis for autonomy. At RoboCup 2010, we introduced our first adult-size platform, CHARLI-L (Cognitive Humanoid Autonomous Robot with Learning Intelligence — Lightweight), and at RoboCup 2011 we introduced a second version named CHARLI-2. Both robots share an emphasis on lightweight design in order to reduce costs, increase safety during use, and improve ease of handling. These characteristics are especially important for robotics competitions such as RoboCup. This section presents some of the innovative mechanical design features of CHARLI-L and CHARLI-2, as well as an overview of the system architecture followed by our vision for future generations of the platform.

4.1 Mechanical System

CHARLI-L: One of the main goals of the mechanical design of CHARLI-L was to minimize the weight of the overall system by utilizing lightweight off-the-shelf actuators in the lower body. The challenge is that, when used in a conventional configuration, these actuators cannot produce enough torque for the application. Three features that enabled CHARLI-L to use such actuators were: a parallel four-bar linkage design for the legs, a synchronized actuator configuration, and tension springs to reduce the required actuator torque at the joints.

Fig. 6 shows the parallel four-bar linkage and orientation of the actuators. Unlike conventional adult-size humanoids, CHARLI-L does not use gear reduction mechanisms such as harmonic drives. Instead, multiple EX-106+ Dynamixel actuators are used in tandem to actuate each joint. Assuming the two actuators move simultaneously with identical torque, the overall torque is doubled. The parallel four-bar configuration makes the packaging of such a configuration easy to implement as two actuators can actuate a revolute joint each in the single kinematic chain. The design of the CHARLI-L's parallel four-bar linkage is such that each foot is constrained to be parallel to the ground at all times, enabling a walking gait with only 5 DOF [4] [3]. Eliminating one of the degrees of freedom

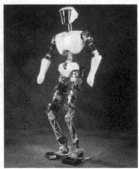

Fig. 5. CHARLI-L (left) and CHARLI-2 (middle and right, on a soccer field)

of the leg resulted in further weight reduction, as fewer actuators were needed. Another advantage of the parallel four-bar approach was the ease of implementation of a tension spring to provide additional torque. The configurations of these springs are such that, when CHARLI-L's leg supports the upper body during walking, the resulting tensile force reduces the required torque of the actuators as shown in Fig. 6 (right). Using this approach, CHARLI-L was able to achieve stable walking using off-the-shelf components, while reaching an overall weight of only 12.7 kg.

CHARLI-2: Although the innovative mechanical design of CHARLI-L proved successful, a new design was chosen for the following version in order to investigate the benefits of a different approach. Thus, CHARLI-2 utilizes a more conventional serial chain configuration instead of the previous four-bar configuration. There were four main reasons behind this decision. First, although the four-bar configuration was thought to reduce the overall weight by eliminating a set of actuators, it turned out that the additional linkages outweighed the extra actuators. Second, the elimination of an active DOF limited the motion of the foot, constraining the possible walking strategies. Third, although the tension springs did indeed help reduce the required torque at the joints, the nonlinear behavior of springs and hysteresis caused problems for the control algorithms. Lastly, a new walking approach under development required more torque than the previous design was able to provide [5] [11].

To address these issues, CHARLI-2 was designed using a serial configuration with 6 active DOF per leg; furthermore, an additional gear train was added between the output of each pair of actuators and the output stage replacing the four-bar configuration. Fig. 7 shows the gear train of CHARLI-2's knee. The reduction ratio of this gear train is 3:1. Two identical EX-106+ actuators rotate in tandem to actuate the joint. Using this configuration, the maximum holding torque of the joint can reach 60 Nm — twice the required torque for normal walking. The hip roll and pitch joints are implemented using a similar gear system. For the other joints (i.e. the hip yaw, ankle roll and ankle pitch), a

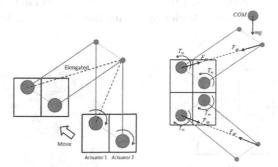

Fig. 6. Left: Concept of the spring assisted parallel four-bar linkage with synchronized actuation. Right: Diagram of CHARLI-L's leg.

single EX-106+ actuator was used, as these joints require less torque (under 20 Nm). The result of this design change was a reduction in overall weight to 12.1 kg, and in increase in torque from 30 Nm to 60 Nm (for the knee). Additionally, the new design eliminated the control difficulties associated with the use of springs in the previous version. This led to a significant increase in performance and stability. Currently, CHARLI-2 can walk at a speed of up to 0.4 m/s, and with further optimization of the walking algorithm, we believe that this can be increased by 20%.

4.2   Electronic System

CHARLI-2 shares a common system architecture with our KidSize humanoid robot platform, DARwIn-OP. All high-level processing and control is performed on an Intel-based PC running GNU/Linux. A ROBOTIS Co. CM-730 sub-controller board acts as the communication relay between the Dynamixel actuators and PC, providing services for both sensor acquisition and actuator control. CHARLI-2's software architecture is based on the humanoid robotics framework employed by team DARwIn for the KidSize competition. (Please refer to the section regarding DARwIn-OP's software design.) Due to the differences in scale between the two robots, a specialized motion module was developed for CHARLI to allow stable ZMP-based walking.

4.3   Future of CHARLI

The research of Virginia Tech's team CHARLI (and team DARwIn) is founded on innovative platform development. Against conventional wisdom, we have developed a brand new platform from scratch every year we have participated in RoboCup. This process produced two high performance platforms, leading to first place wins in both the AdultSize class (CHARLI-2) and KidSize class (DARwIn-OP). As a contribution to the robotics community, we have made the successful DARwIn-OP platform completely open source, both in hardware design and software, through sponsorship by the National Science Foundation and through collaboration with Purdue University, University of Pennsylvania, and ROBOTIS Co. We are considering doing the same for a future version of

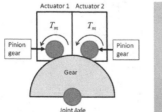

Fig. 7. Gear train of CHARLI-2's knee. Left: Design concept. Right: Real part.

CHARLI, as we believe the open source approach is the quickest and most effective way to accelerate the development of robotics technology.

5   Conclusions

Robotic soccer, like human soccer, relies on a combination of the overall team strategy and individual skill sets. The 2011 competition showed notable progress in both areas.

In KidSize, for example, team DARwIn mainly focused on the individual skill of a striker, such as the ability to take possession of the ball and to move it as fast as possible. This strategy required developing algorithms such as quick omnidirectional walking, dynamic kicking and optimal approaching. As the KidSize league is inherently the most dynamic league among RoboCup leagues, DARwIn will keep focusing on extending the dynamic behavior of their robots.

From their experience with a broad range of humanoid robots, DARwIn developed an open-source, common software structure for humanoid robots. With its help, the team was able to save time preparing for different RoboCup leagues – and to win two championships in KidSize and Adult Size. The open-source release of their code base, in addition to the commercial release of the DARwIn-OP robot, will certainly help other RoboCup teams and stimulate general robotic research using small humanoid robots. A similar open platform is desirable for TeenSize and AdultSize.

In TeenSize, team NimbRo showed a very stable robot that played several matches without falling. For 2012, the team is developing a new robot that will be able to get up after a fall, and perform throw-ins. The team is working to improve robot balance after pushes and collisions and to integrate footstep planning into their behavior architecture.

In AdultSize, team CHARLI demonstrated reliable dribbling and kicking. Their performance was acknowledged by the team leaders with the Louis Vuitton Best Humanoid Award. Team CHARLI (and team DARwIn) are committed to continue the introduction of brand new platforms every year with innovative features and approaches which depart from the conventional. The team plans to introduce new effective walking strategies using compliant linear actuators with force control, and to demonstrate safe falling and recovery in the near future.

In the future, the Humanoid League will continue to raise the bar. Possible rule changes include enlarging the field, increasing the number of players, reducing color-coding of objects, soccer games for AdultSize robots, and new technical challenges, such as kicking the ball high. This will keep the competition challenging and will contribute towards the ultimate goal of humanoid robots playing soccer with humans.

Acknowledgements. The authors would like to recognize the National Science Foundation and the Office of Naval Research for partially supporting this work through grants CNS 0958406 and ONR 45006.

References

1. Behnke, S.: Online trajectory generation for omnidirectional biped walking. In: Proceedings of International Conference on Robotics and Automation (ICRA) (2006)
2. Behnke, S., Stückler, J.: Hierarchical reactive control for humanoid soccer robots. International Journal of Humanoid Robots (IJHR) 5, 375–396 (2008)
3. Choi, H., Park, Y.: Development of a biped walking robot actuated by a closed-chain mechanism. Robotica 24(1), 31–37 (2006)
4. Hamon, A., Aoustin, Y.: Cross four–bar linkage for the knees of a planar bipedal robot. In: Proc. of 10th Int. Conf. on Humanoid Robots, Humanoids (2010)
5. Han, J., Hong, D.: Development of a full–sized bipedal humanoid robot utilizing spring assisted parallel four–bar linkages with synchronized actuation. In: IDETC/CIE Conference (2011)
6. McGill, S.G., Brindza, J., Yi, S.-J., Lee, D.D.: Unified humanoid robotics software platform. In: The 5th Workshop on Humanoid Soccer Robots (2010)
7. Missura, M., Behnke, S.: Lateral capture steps for bipedal walking. In: Proceedings of IEEE-RAS International Conference on Humanoid Robots, Humanoids (2011)
8. Missura, M., Wilken, T., Behnke, S.: Designing Effective Humanoid Soccer Goalies. In: Ruiz-del-Solar, J., Chown, E., Ploeger, P.G. (eds.) RoboCup 2010. LNCS (LNAI), vol. 6556, pp. 374–385. Springer, Heidelberg (2010)
9. Schmitz, A., Missura, M., Behnke, S.: Learning Footstep Prediction from Motion Capture. In: Ruiz-del-Solar, J., Chown, E., Ploeger, P.G. (eds.) RoboCup 2010. LNCS (LNAI), vol. 6556, pp. 97–108. Springer, Heidelberg (2010)
10. Schulz, H., Liu, W., Stückler, J., Behnke, S.: Utilizing the Structure of Field Lines for Efficient Soccer Robot Localization. In: Ruiz-del-Solar, J., Chown, E., Ploeger, P.G. (eds.) RoboCup 2010. LNCS (LNAI), vol. 6556, pp. 397–408. Springer, Heidelberg (2010)
11. Song, S., Ryoo, Y., Hong, D.: Development of an omni–directional walking engine for full–sized lightweight humanoid robots. In: IDETC/CIE Conference (2011)
12. Thrun, S., Burgard, W., Fox, D.: Probabilistic Robotics. MIT Press (2001)
13. Vukobratovic, M., Borovac, B.: Zero-moment point - thirty five years of its life. International Journal of Humanoid Robotics (IJHR), 157–173 (2004)
14. Yi, S.-J., Zhang, B.-T., Hong, D., Lee, D.D.: Learning full body push recovery control for small humanoid robots. In: Proceeding of IEEE International Conference on Robotics and Automation (ICRA), pp. 2047–2052 (2011)

Towards Robust Mobility, Flexible Object Manipulation, and Intuitive Multimodal Interaction for Domestic Service Robots

Jörg Stückler, David Droeschel, Kathrin Gräve, Dirk Holz,
Jochen Kläß, Michael Schreiber, Ricarda Steffens, and Sven Behnke

Rheinische Friedrich-Wilhelms-Universität Bonn
Computer Science Institute VI: Autonomous Intelligent Systems
Friedrich-Ebert-Allee 144, 53113 Bonn, Germany
{stueckler,droeschel,graeve,holz,schreiber,behnke}@ais.uni-bonn.de,
{klaehs,steffens}@cs.uni-bonn.de
http://www.NimbRo.net/@Home

Abstract. In this paper, we detail the contributions of our team NimbRo to the RoboCup @Home league in 2011. We explain design and rationale of our domestic service robot Cosero that we used for the first time in a competition in 2011. We demonstrated novel capabilities in the league such as real-time table-top segmentation, flexible grasp planning, and real-time tracking of objects. We also describe our approaches to human-robot cooperative manipulation and 3D navigation. Finally, we report on the use of our approaches and the performance of our robots at RoboCup 2011.

1 Introduction

The RoboCup @Home league has been established in 2006 to research and benchmark autonomous service robots in everyday scenarios. In the first years of the league, basic capabilities of the robots have been tested. The robots had to show object recognition and grasping, safe indoor navigation, and basic human-robot interaction (HRI) skills such as speech recognition and synthesis. Progress in the league allowed to introduce more complex test procedures in 2010.

Our team NimbRo participates in the @Home league since 2009. In the first year, we could demonstrate basic mobile manipulation and HRI skills with our self-constructed robots Dynamaid and Robotinho. We could reach the third place in the competition and received the Innovation Award for innovative robot body design, empathic behaviors, and robot-robot cooperation. In the second year, we participated with a mechanically improved version of Dynamaid. We tackled the more complex tests and showed many new mobile manipulation and human-robot interaction skills such as gesture recognition. We also demonstrated as the first team the opening and closing of a refrigerator at RoboCup. Overall, we could reach the second place in 2010.

T. Röfer et al. (Eds.): RoboCup 2011, LNCS 7416, pp. 51–62, 2012.

Fig. 1. The cognitive service robot *Cosero* cooperatively carries a table with a user and bakes omelett at the RoboCup@Home finals 2011 in Istanbul

In this year's competition, we participated with Dynamaid and its successor, Cosero. While Cosero still retains the light-weight design of Dynamaid, we improved its construction and appearance significantly and made it more precise and stronger actuated. In the tests, the robots showed their human-robot interaction and mobile manipulation capabilities. We introduced many new developments, like grasp planning to extend the range of graspable objects, real-time scene segmentation and object tracking, and human-robot cooperative manipulation of a table. Our performance was well received and has been awarded the first place in 2011. In this paper, we summarize our main novel contributions to the RoboCup@Home league. We detail the construction of Cosero, and the algorithms we developed in the context of the league.

2   Design of Cognitive Service Robot Cosero

Our everyday environments are adapted to the specific capabilities and constraints of the human body. When robots perform similar tasks like humans in such environments, it is a natural design rationale to equip the robots with human-like motion and perception abilities. A further advantage of a human-like body is that the robot's behavior is predictable and can easily be interpreted by humans. In everyday scenarios, robots also may interact physically with humans. This imposes requirements on the safety of such a robot. A light-weight design makes a household robot inherently less dangerous than a heavy-weight industrial-grade robot.

We focused the design of our robots Dynamaid and Cosero (s. Fig. 1) on such requirements. Cosero's mobile base has a small footprint of 59×44 cm and drives omnidirectionally. This allows Cosero to maneuver through the narrow passages found in household environments. Its two anthropomorphic arms resemble average human body proportions and reaching capabilities. A yaw joint in the torso enlarges the workspace of the arms. In order to compensate for the missing torso pitch joint and legs, a linear actuator in the trunk can move the upper body vertically by approx. 0.9 m. This allows the robot to manipulate on similar heights like humans – even on the floor.

Cosero has been constructed from light-weight aluminum parts. All joints in the robot are driven by Robotis Dynamixel actuators. These design choices allow for a light-weight and inexpensive construction, compared to other domestic service robots. While each arm has a maximum payload of 1.5 kg and the drive has a maximum speed of 0.6 m/sec, the low weight (in total ca. 32 kg) requires only moderate actuator power. The robot's main computer is a HP Pavillion dv6 notebook with an Intel i7-Q720 processor.

Cosero perceives its environment with a variety of complementary sensors. The robot senses the volume in front of it in 3D with a Microsoft Kinect RGB-D camera in its head that is attached to the torso with a pan-tilt unit in the neck. For obstacle avoidance and tracking in farther ranges and larger field-of-views than the Kinect, the robot is equipped with multiple laser-range scanners. One Hokuyo URG-04LX is placed on a roll actuator in the lower torso. Aligned horizontally, it detects objects on horizontal surfaces such as tables or shelves. In vertical alignment, it measures distance and height of these objects. A further Hokuyo URG-04LX measures objects in a height of ca. 4 cm above the floor. The main sensor for 2D localization and mapping is a SICK S300 sensor in the mobile base which perceives the environment in the horizontal plane at a height of approx. 27 cm. In the upper torso, we mounted a Hokuyo UTM-30LX laser-range scanner on a tilt actuator to acquire precise range measurements in up to 30 m for 3D obstacle avoidance, mapping, and localization. The laser-range scanners are also useful to track persons in the robot's surroundings.

To improve the robustness of manipulation, the robot can measure the distance to obstacles directly from the grippers. We attached infrared distance sensors to each gripper that point downward and forward in the finger tips. Another sensor in the palm measures distance to objects within the gripper.

Finally, the sensor head also contains a shotgun microphone for speech recognition. By this, the robot points the microphone towards human users and at the same time directs its visual attention to the user. We attached a human face mask to support the interpretation of the robot's gaze by the human.

3   Mobile Manipulation

One significant part of the competition in the @Home league tests the mobile manipulation capabilities of the robots. The robots shall be able to fetch objects from various locations in the environment. To this end, the robot must navigate through the environment, recognize objects, and grasp them.

3.1   Perception

Real-Time Table-Top Segmentation: In household environments, objects are usually located on planar surfaces such as tables. We therefore base our object detection pipeline on fast horizontal plane segmentation of the depth images of the Kinect [14]. Fig. 2 shows an exemplary result of our approach in a table-top scene. Our method processes depth images with a resolution of

Fig. 2. Object detection. Left: example table top setting. Center: raw point cloud from the Kinect with RGB information. Right: each detected object is marked with a distinct color.

Fig. 3. Grasp planning. Left: object shape properties. The arrows mark the principal axes of the object. Center: we rank feasible, collision-free grasps (red, size prop. to score) and select the most appropriate one (large, RGB-coded). Right: example grasps.

160×120 at frame rates of approx. 16 Hz on the robot's main computer. This enables our system to extract information about the objects in a scene with a very low latency for further decision-making and planning stages. For object identification, we utilize texture and color information [9].

In order to process the depth images efficiently, we combine rapid normal estimation [7] with fast segmentation techniques. The normal estimation method exploits the principle of integral images to estimate surface normals in a fixed image neighborhood in constant time. Overall, the runtime complexity is linear in the number of pixels for which normals are calculated. Since we search for horizontal support planes, we find all points with vertical normals. We segment these points into planes using RANSAC [5] and find the objects by clustering the points above the convex hull of the support plane.

Grasp Planning: We investigate grasp planning to enlarge the set of graspable objects and to allow for obstructions by obstacles [14]. In our approach, we assume that the object is rigid and symmetric along the planes spanned by the principal axes of the object, e.g., cylindrical or box-shaped. We found that our approach also may yield stable grasps when an object violates these assumptions.

Fig. 4. Learning of object models. Left: during training the user selects points (red dots) to form a convex hull around the object. Center: color and shape distribution modeled at 5 cm resolution. Lines indicate surface normals (color-coded by orientation). Right: color and shape distribution modeled at 2.5 cm resolution.

Fig. 3 illustrates the main steps in our grasp planning pipeline and shows example grasps.

We consider two kinds of grasps: A side-grasp that approaches the object horizontally and grasps the object along the vertical axis in a power grip. The complementary top-grasp approaches the object from the top and grasps it with the finger tips along horizontal orientations. Our approach extracts the object's principle axes in the horizontal plane and its height. We sample pre-grasp postures for top- and side-grasps and evaluate the grasps for feasibility under kinematic and collision constraints. The remaining grasps are ranked according to efficiency and robustness criteria. The best grasp is selected and finally executed with a parametrized motion primitive. For collision detection, we take a conservative but efficient approach that checks simplified geometric constraints.

Real-Time Object Tracking: When a robot interacts with objects, it has to estimate its pose with respect to the objects. Frequently, localization in a map is not precise enough for this purpose. For example, the place of many household objects such as tables or chairs is subject to change. The robot must then be able to detect the object in its current sensor view and estimate the relative pose of the object.

We develop methods for real-time tracking of objects with RGB-D cameras [12]. In our approach, we train a multi-resolution surfel map of the object (s. Fig. 4). The map is represented in an octree where each node stores a normal distribution of the volume it represents. In addition to shape information, we also model the color distribution in each node.

For fast object teach-in, we use checkerboard patterns laid out around the object. In the images, the user selects points on a convex hull around the object in the common plane of the checkerboards. The visual markers yield a precise map reference frame, in which various views on the object can be merged.

Once the map has been obtained, we build multi-resolution surfel maps with color information from new RGB-D images [13]. Then, we register this map to

Fig. 5. Left: panorama image of an office. Center: 3D surfel map learned with our approach (surfel orientation coded by color). Right: 2D navigation map extracted from the 3D surfel map (gray: unknown, white: traversable, black: untraversable).

the object map with an efficient multi-resolution strategy. We associate each node in the image map to its corresponding node in the object map using fast nearest-neighbor look-ups. We optimize the matching likelihood to find the most likely pose. Our efficient implementation supports real-time registration of RGB-D images on the robot's main computer.

3D Perception for Navigation: Reliable obstacle avoidance is a prerequisite for safe navigation of a robot in indoor environments. We implemented means to incorporate 3D measurements from laser scanners and depth cameras [3]. We maintain the measurements in a 3D point cloud in the close surrounding of up to 10 m. In order to handle dynamic objects, a temporal unlearning strategy discards measurements that are older than a fixed amount of time. When a sensor sweeps a map region with its field-of-view, the points in this volume are removed after shorter duration.

Many approaches to indoor localization and mapping use 2D laser scanners to acquire 2D footprints of the environment. Occupancy grid maps are used to represent the map, because they provide dense information about free and occupied space for localization and path planning. One problem of such 2D maps occurs in path planning, if untraversable obstacles cannot be perceived on the laser scanners height. Localization with 2D lasers imposes further restrictions if dynamic objects occur, or the environment changes in the scan plane of the laser. Then, localization may fail since large parts of the measurements are not explained by the map.

We address these problems by building 3D maps of the environment [8]. Fig. 5 demonstrates example maps generated with our approach. We choose to represent the map in a 3D surfel grid. The robot acquires full 3D scans at several locations in the environment. Given the location w.r.t. the environment, we perform mapping in known poses to generate the surfel map from several scans. We obtain the trajectory estimate using a 2D SLAM approach (GMapping, [6]) and refine it by registering the 3D scans with ICP [1].

Once the map has been obtained, we extract a 2D navigation map. First, we find the traversable cells in the 2D map by region growing in the 3D map with the robot's scan poses as seed points. The region growing algorithm expands to

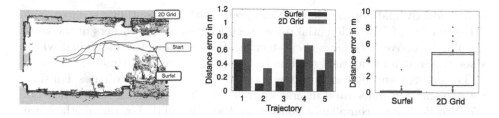

Fig. 6. Localization in dynamic environments. Left: examplary result of the tracking performance of 2D localization (2D Grid) and our 3D localization approach (Surfel). Center: localization accuracy for pose tracking. Right: global localization accuracy.

cells when the gap between the surfels at the cell borders is small enough to traverse it. Additionally, we check if all cells are free within the robot's height range.

For localization, we developed a Monte Carlo method that can incorporate full 3D scans as well as 2D scans. When used with 3D scans, we extract surfels from the scans and evaluate the observation likelihood. From 2D scans, we extract line segments and associate them with surfels in the map. In both cases, we use a nearest-neighbor look-up grid to efficiently associate measurements with surfels.

Localization in 3D maps is specifically useful in crowded environments. The robot can then leverage measurements above the height of people to localize at the static parts of the environment. More general, by representing planar surface elements in the map, we can also concentrate the localization on planar structures, as they more likely occur in static environment parts.

Fig. 6 shows experimental results for pose tracking and global localization. In the experiments, we compare the reliability of standard Monte Carlo localization in 2D occupancy maps and our localization method which prefers measurements above the persons' heights. Eight persons were randomly walking in the test environment. We quantify, how often and how accurate the localization methods estimate the final position of the trajectory. We use the SICK S300 laser scanner on the mobile base for 2D localization. For 3D localization, the laser scanner in the chest is continuously tilted. When the robot stands during a full sweep, the complete 3D scan is integrated. Otherwise, we use the immediate 2D scans. In pose tracking, we initialize the localization at the correct position. It can be seen that our approach localizes the robot more accurately. For global localization, we initialize the methods with a uniform distribution of 5000 particles. We evaluated global localization at 45 starting points in various trajectories. Global localization in the 2D map only succeeds in ca. 30% of the runs, whereas our approach achieves 97.5% success rate at a distance threshold of 0.5 m. While our approach yields superior results, it still retains the efficiency of 2D localization.

3.2 Behavior Control

Motion Control: The design of the mobile bases of our robots supports omnidirectional driving [9]. The linear and angular driving velocities can be set independently and can be changed to speeds within a continuous range. The drive consists of four differential drives each located at the corners of the rectangular base. We determine their steering direction and the individual wheel velocities from an analytical solution to the drive's inverse kinematics.

The anthropomorphic arms support control in Cartesian coordinates. For this, we implemented differential inverse kinematics with redundancy resolution [9]. We also developed compliance control for the arms [11]. For our method, we exploit that the servo actuators are back-drivable and that the torque which the servo applies for position control can be limited. Compliance can be set for each direction in task-space separately. For example, the end-effector can be kept loose in both lateral directions while it keeps the other directions at their targets.

We implemented several motion primitives like grasping with one or two arms, pointing or waving gestures, and object placement. Motion primitives such as side- and top-grasps or pointing gestures can be parametrized in the target.

Mobile Manipulation Control: For mobile manipulation, we developed controllers that combine control of the drive and the arms with perception capabilities. Cosero can grasp objects on horizontal surfaces on tables and in shelves in a height range from ca. 0.3 m to 1 m [9]. It also carries the object and hands it to human users. We further developed solutions to pour-out containers, to place objects on horizontal surfaces, to dispose objects in containers, to grasp objects from the floor, and to receive objects from users. When handing an object over, the arms are compliant in upward direction so that the human can pull the object, the arm complies, and the object is released.

The robots can also open and close doors, if the door leaf can be moved without the handling of an unlocking mechanism. For example, fridges or cabinets are commonly equipped with magnetically locked doors that can be pulled open without special manipulation of the handle. To open a door, the robot drives in front of the door, detects the door handle with its torso laser, approaches the handle, and grasps it. The drive moves backward while the gripper moves to a position to the side of the robot in which the opening angle of the door is sufficiently large to approach the open fridge or cabinet. The gripper follows the motion of the door handle through compliance in the lateral and the yaw directions. The robot moves backward until the gripper reaches its target position. For closing a door, the robot has to grasp the handle and moves forward while it holds the handle at its relative initial grasping pose. The arm deviates from this pose by the constraining motion of the door leaf, and the robot drives to keep the handle at its initial pose, relative to the robot. The closing of the door can be detected when the arm is pushed back towards the robot.

4   Human-Robot Interaction

A service robot in everyday environments not only needs mobile manipulation abilities. It closely interacts with humans, even physically. This interaction should be natural and intuitive such that laymen can operate the robot and understand its actions.

4.1 Person Awareness

A key prerequisite for human-robot interaction is the robot's awareness of the persons that sourround it. Our robots maintain a belief on the location and identity of persons [10]. We implemented a multi-hypothesis tracker that initializes new person believes, when faces or upper bodies are detected at reasonable locations in the environment. Using laser scanners, the position and moving speed of the persons is then tracked at high frame rates.

4.2 Speech Recognition and Synthesis

Speech is the primary modality for the communication of complex statements between humans. We therefore support speech in our robots employing the Loquendo SDK. Its speech recognition is speaker-independent and uses a small-vocabulary grammar which we change with the dialog state. The grammar definition of the Loquendo speech recognition system allows to tag rules with semantic attributes. When speech is recognized, a semantic parse tree is provided that we process further. We use the parsed semantics to interpret sentences for complex commands and to generate appropriate behavior.

4.3 Gesture Recognition and Synthesis

An important non-verbal communication cue is the recognition and performance of gestures. We equipped our robots with several gestures. For example, the robot can draw a user's attention to certain locations in the environment by simply pointing at it. Our robots can also perceive gestures such as pointing, showing of objects, or stop gestures [4]. The robots sense these gestures using the RGB-D camera. For pointing gestures, we accurately estimate the pointing direction from body features such as the position of the head, hand, shoulder, and elbow. We also investigated the use of Gaussian Process regression to learn an interpretation of the pointing direction [2] using the body features.

4.4 Human-Robot Cooperative Manipulation

We study physical interaction between a human user and a robot in a cooperative manipulation task [12]. In our scenario, the human and the robot cooperatively carry a large object, i.e., a table. For the successful performance of this task, the robot must keep track of the human actions.

In order to approach the table, the robot tracks its 6D pose with our real-time object tracking approach. The robot waits then until the human lifts the table. As soon as the human lifts the table, the robot measures a significant pitch of the table. Then, the robot also lifts the table and begins to hold the table compliant in the horizontal plane. The human user pulls and pushes the table into the desired directions, and the robot compensates for the displacement of its end-effectors by driving accordingly with its mobile base. When the target location of the table is reached, the user can simply put the table down which is detected by the robot.

5   Experiments at RoboCup 2011

With Dynamaid and Cosero, we competed in the RoboCup@Home 2011 competition in Istanbul. Our robots participated in all tests of stage I and II, and performed very well. We accumulated the highest score of all 19 teams in both stages. Our final demonstration was also awarded the best score such that we achieved the first place in the competition.

In the *Robot Inspection and Poster Session* test in Stage I, Cosero and Dynamaid registered themselves. Meanwhile, we presented our work to leaders of other teams in a poster session. Overall, we have been awarded the highest score in this test. In the *Follow Me* test, Cosero met a previously unknown person and followed him reliably through an unknown environment. Cosero showed that it can distinguish the person from others and that it recognizes stop gestures. In *Who Is Who*, two previously unknown persons introduced themselves to Cosero. Later in the test, our robot found one of the previously unknown persons, two team members, and one unknown person and recognized their identity correctly. The *Open Challenge* allows the teams to show their research in self-defined demonstrations. In this challenge, Cosero fetched a bottle of milk, opened it, and poured it into a cereal bowl. Then, Cosero grasped a spoon using our approach to grasp planning and placed it next to the bowl. Cosero understood a complex command partially and went to a correct place in the *General Purpose Service Robot I* test. In the *Go Get It!* test, Cosero found a correct object and delivered it. After stage I, we were leading the competition.

In stage II, Cosero participated in the *Shopping Mall* test. It learned a map of a previously unknown shopping mall and navigated to a shown location. In the *General Purpose Service Robot II* test, Cosero first understood a partially specified command and asked questions to obtain missing information about the object to grasp and about the location of the object. After successful execution, it worked on a task with erroneous information. It detected that the ordered object is not at the specified location, and went back to the user to report the error. In the *Demo Challenge*, we demonstrated pointing gestures by showing the robot in which baskets to put colored and white laundry. The robot then cleaned the appartment, picked white laundry from the floor, and put it into the correct basket. It then picked carrots and teaboxes from a table. The objects could be chosen and placed by a jury member. The technical commitee awarded us the highest score. We reached the finals with 8,462 points followed by WrightEagles from China with 6,625 points.

In the finals, we demonstrated the cooperative carrying of a table by Cosero and a human user. Then, a user showed Cosero where it finds a bottle of dough to make an omelett. Our robot then went to the cooking plate to switch it on. It succeeded partially in turning the plate on. Then, Cosero drove to the location of the dough and grasped it. At the cooking plate, it opened the bottle and poured it into the pan. We applied our real-time object tracking method in order to approach the cooking plate. Meanwhile, Dynamaid opened a refrigerator, grasped a bottle of orange juice out of it, and placed the bottle on the breakfast

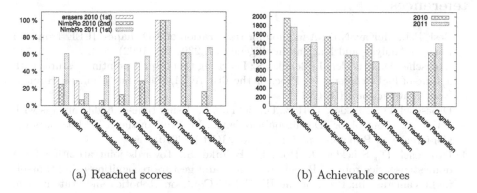

(a) Reached scores   (b) Achievable scores

Fig. 7. Reached (a) and achievable (b) scores in the predefined test procedures per functionality

table. Our performance received the best score by the jury consisting of members of the executive committee and external judges from science and the media.

Fig. 7 summarizes the scores achieved for individual functionalities as proposed in [15]. Note that due to the sequential nature of the predefined test procedures, in some tests our robots did not reach specific sub-tasks. For instance, in *Enhanced Who Is Who* or *Shopping Mall*, our system had difficulties to understand the orders by the human user and, hence, did not have the chance to gain score for object manipulation. The results demonstrate that we improved most functionalities compared to 2010 and achieved well in developing a balanced domestic service robot system.

6   Conclusion

In this paper, we presented our developments for the RoboCup@Home league in 2011. We detailed our approaches to real-time scene segmentation, object tracking, 3D navigation, and human-robot cooperative manipulation. We use the RoboCup@Home competitions to evaluate our methods in a realistic setting. With our domestic service robots, we won the competitions in 2011.

In future work, we aim to further advance the versatility of the mobile manipulation and human-robot interaction skills of our robots. The learning of models of arbitrary objects and the real-time tracking of these models is one step in this direction. Equally important, we are working to improve the perception of persons and the interpretation of their actions. We also plan to remove the necessities to adapt the tools of the robot to its current end-effectors. In order to improve the manipulation skills of our robots, we will improve the design of the grippers. We plan to construct thinner fingers with touch sensors. Then, we could devise new methods to grasp smaller objects or to use smaller tools.

References

1. Besl, P.J., McKay, N.D.: A method for registration of 3-D shapes. IEEE Trans. on Pattern Analysis and Machine Intelligence 14, 239–256 (1992)
2. Droeschel, D., Stückler, J., Behnke, S.: Learning to interpret pointing gestures with a time-of-flight camera. In: Proc. of the 6th ACM/IEEE Int. Conf. on Human-Robot Interaction, HRI (2011)
3. Droeschel, D., Stückler, J., Holz, D., Behnke, S.: Using time-of-flight cameras with active gaze control for 3D collision avoidance. In: Proc. of the IEEE International Conference on Robotics and Automation, ICRA (2010)
4. Droeschel, D., Stückler, J., Holz, D., Behnke, S.: Towards joint attention for a domestic service robot – Person awareness and gesture recognition using time-of-flight cameras. In: Proc. of the IEEE Int. Conf. on Robotics and Automation, ICRA (2011)
5. Fischler, M.A., Bolles, R.C.: Random sample consensus: a paradigm for model fitting with applications to image analysis and automated cartography. Communications of the ACM 24(6), 381–395 (1981)
6. Grisetti, G., Stachniss, C., Burgard, W.: Improved techniques for grid mapping with Rao-Blackwellized particle filters. IEEE Trans. on Robotics 23(1) (2007)
7. Holz, D., Holzer, S., Rusu, R.B., Behnke, S.: Real-Time Plane Segmentation using RGB-D Cameras. In: Röfer, T., Mayer, N.M., Savage, J., Saranli, U. (eds.) RoboCup 2011. LNCS, vol. 7416, pp. 306–317. Springer, Heidelberg (2012)
8. Kläß, J., Stückler, J., Behnke, S.: Efficient mobile robot navigation using 3D surfel grid maps. In: Proc. of 7th German Conf. on Robotics, ROBOTIK (2012)
9. Stückler, J., Behnke, S.: Integrating indoor mobility, object manipulation, and intuitive interaction for domestic service tasks. In: Proc. of the IEEE Int. Conf. on Humanoid Robots, Humanoids (2009)
10. Stückler, J., Behnke, S.: Improving People Awareness of Service Robots by Semantic Scene Knowledge. In: Ruiz-del-Solar, J., Chown, E., Ploeger, P.G. (eds.) RoboCup 2010. LNCS (LNAI), vol. 6556, pp. 157–168. Springer, Heidelberg (2010)
11. Stückler, J., Behnke, S.: Compliant Task-Space Control with Back-Drivable Servo Actuators. In: Röfer, T., Mayer, N.M., Savage, J., Saranli, U. (eds.) RoboCup 2011. LNCS, vol. 7416, pp. 78–89. Springer, Heidelberg (2012)
12. Stückler, J., Behnke, S.: Following human guidance to cooperatively carry a large object. In: Proc. of the IEEE-RAS Int. Conf. on Humanoid Robots (2011)
13. Stückler, J., Behnke, S.: Robust real-time registration of RGB-D images using multi-resolution surfel maps. In: Proc. of 7th German Conf. on Robotics, ROBOTIK (2012)
14. Stückler, J., Steffens, R., Holz, D., Behnke, S.: Real-time 3D perception and efficient grasp planning for everyday manipulation tasks. In: Proc. of the 5th European Conference on Mobile Robots, EMCR (2011)
15. Wisspeintner, T., van der Zant, T., Iocchi, L., Schiffer, S.: RoboCup@Home: Scientific competition and benchmarking for domestic service robots. Interaction Studies 10(3), 392–426 (2009)

RoboCupJunior – A Decade Later

Amy Eguchi[1], Nicky Hughes[2], Matthias Stocker[3], Jiayao Shen[4], and Naomi Chikuma[5]

[1] Bloomfield College, New Jersey, USA
amy_eguchi@bloomfield.edu
[2] Open University, UK
n.a.hughes@btinternet.com
[3] Universität Ulm, Germany
stocker.matthias@googlemail.com
[4] Singapore Polytechnic, Singapore
jyshen@sp.edu.sg
[5] RoboCupJunior Japan Committee, Japan
nchiku@comet.ocn.ne.jp

Abstract. As RoboCupJunior reached a decade mark in 2011, we feel the need for examining the current situation after 12 revisions and modifications to the league rules and structures since its launch in 2000. RoboCupJunior International is now attracting over 250 teams involving approximately 1,000 students originating from more than 30 countries. This paper aims to report on the progress achieved thus far, both technologically and educationally, and the issues currently addressed, together with suggestions for the future of RoboCupJunior.

1 Introduction

Educational robotics tournaments have had the greatest impact on the growing popularity of educational robotics in K-12 setting around the world. There is a growing number of robotics competitions and events that are available both at national and international levels. Some of the most popular robotics competitions include the FIRST Robotics Competition, the FIRST Tech Challenge, FIRST LEGO League, and Junior FIRST LEGO League organized by The FIRST organization (http://www.usfirst.org/); BotBall (http://www.botball.org/); World Robot Olympiad (http://www.wroboto.org/); and RoboCupJunior (http://www.robocupjunior.org).

Those tournaments employ the goal-oriented approach to teaching, which is a popular approach in the fields of engineering, computer science, and artificial intelligence. Each educational robotics tournament sets a goal for teams to achieve, which leads to their learning. For example, each year, all FIRST competitions have new goals and/or themes that teams work on to solve. Those educational robotics events have reported positive impacts on the students and teachers/mentors who participated [1-6]. For example, from the Botball survey, 89% of students surveyed felt more confident with technology and 100% of these students planed to pursue a degree in a technical or math-related field after their participation [2]. FIRST reported that 89% of the participants in FIRST Robotics

T. Röfer et al. (Eds.): RoboCup 2011, LNCS 7416, pp. 63–77, 2012.

Competition from 2002 to 2005 indicated an increased understanding of the role of science and technology in solving real-world problems, and 69% indicated increased interest in science and technology careers. Also, 95% indicated an increased understanding of the value of working on a team, and 89% indicated increased self-confidence [1]. The studies of the impact of the RoboCupJunior also suggest that the experiences of the participating students are positive. More than 50 % of participants surveyed indicated that the participation of the event made positive impacts on their learning on physics, programming, mechanical engineering, electronics, science, as well as communication skills, teamwork, and personal development [6].

Those survey results show the benefits that educational robotics competitions can bring to participating students. However, RoboCupJunior (RCJ) stands apart from other educational robotics programs for several reasons. First, it focuses more on education than competition. Second, its challenges remain the same from one year to the next, providing a scaffolded learning environment in which students can develop more sophisticated solutions as they grow and expand their knowledge. Also, since RCJ sits at the entry-level of the international RoboCup (RC) initiative, which is strongly committed to research, education and involvement of young people in technology, students can continue to develop their skills and knowledge in more advanced research programs. Third, its challenges, called leagues, use topics – soccer, rescue and dance – that are familiar to a broad range of our societies to attract and motivate students into educational robotics. All three Junior leagues emphasize both cooperative and collaborative nature of design, programming and building in a team setting [4].

Fig. 1-3. Soccer, Rescue and Dance

The idea to create a league for young robotics participants at RoboCup was first introduced in 1998 in a demonstration by Henrik Hautop Lund and Luigi Pagliarini [4]. In the demonstration the LEGO Mindstorm kits were used to create soccer playing robots. After further discussion, a pilot project was implemented at RoboCup Euro, 2000. Twelve teams with a total of 50 students, ages 13 to 16 from eight schools, developed soccer robots to play one-on-one robot soccer games [4]. Also in 2000, the first RCJ competition was organized during RoboCup, Melbourne, 2000, with 25 teams from three countries. At the first RCJ competition, three challenges (leagues) were introduced – Dance, Sumo and Soccer. Initially, there were specific age restrictions for each challenge. For example, the Dance challenge focused on students up to 12 years old (primary school age). Sumo was for students from 12 to 15 years old, and Soccer was for children between the ages of 12 to 18 years (secondary school). The success of the first RCJ competition led to creation of subsequent annual RCJ International competitions. In 2001, RCJ International was held in Seattle, USA, where 25 teams with 104 participants, including both students and mentors, from four countries participated: Australia, Germany, the UK and the USA. The most significant differences in the league rules were that in 2001 the age restrictions were taken away from all challenges, and the Sumo challenge was replaced by the Rescue challenge, which has been one of the iconic challenges of RCJ since its launch.

Since 2000, RCJ has grown to be a very popular educational activity for school age children in many countries from around the world. In 2011, a mere decade later, the RCJ International competition was held in Istanbul, Turkey, with a total of 251 teams comprised of 955 students from 30 countries. RCJ has also grown to align its structure with the RC Major leagues. Each RCJ league consists of its own Technical Committee (OC) and Organizing Committee (OC). The TC is composed of six members elected every year at the international competition by RCJ National Representatives from participating countries and RC Trustees and Executive Committee members representing RCJ. The TC is in charge of making rule changes by closely examining and discussing the learning needs of participating teams. The OC is composed of six members assigned by the RCJ General Chair and RC Trustees representing RCJ. Their role is to plan and organize the annual RCJ International competition in coordination with the Local Organizing Committee.

This paper presents and discusses the development of RCJ by each league, focusing on the issues and challenges that we face, and the directions that we would like to take the leagues in the future. The paper reports on observational accounts of each league organizers and technical committee members rather than scientific and/or statistical analysis of data from RCJ competitions. The following sections introduce and explain the organization, progress observed and issues presented in 2011 concerning the three RCJ leagues and two demonstrations, ending with suggestions for the future.

2   RCJ Soccer

The Soccer league was inspired by Lund's demonstration and the major Soccer leagues. Two teams with two soccer robots on each team (2-on-2) play on a special field. Though initially a greyscale mat, the floor of the field is now the same green carpet that is used by the major Soccer league. During the game, the robots are programmed to detect and maneuver a soccer ball emitting infrared light. Early in the history of Junior Soccer, there were games with one robot on one team competing against another team's robot (1-on-1). However, as time progressed, 1-on-1 competitions were discontinued at the international level due to the advanced skills of teams participating. For the 2011 competition, there were two Junior Soccer sub-leagues – Open League and Light Weight league. With the Open league, the maximum weight of a robot is 2,500g, whereas with the Light Weight league, the maximum weight of a robot is 1,250g. The Light Weight league was created to avoid heavy robots crashing into and damaging light-weight robots during game play. The sub-leagues are further divided by the type of a field and/or age of team members.

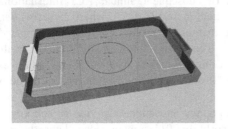 

Fig. 4. Soccer A Field Diagram   **Fig. 5.** Soccer B Field Diagram

There are two types of fields used in the Soccer league. The size of Soccer A field is 122cm x 183cm with walls surrounding the boarder of the field (Fig.4). The size of Soccer B field itself is the same as the Soccer A field, however; it has an additional 30cm outer area, which is surrounded by walls (Fig.5).

Age categories are set by the RCJ general rules. Primary category is for students up to 14 years old who can construct and program a robot on their own (without adult assistance). Secondary category is for students ranging in age from 15 to 19 years. However, not all of the sub-leagues specify the same age categories. The technical committee of each league decides the skills and knowledge required by each sub-league and determines the age categories appropriate for the sub-league. For example, the Light Weight Soccer A league has primary and secondary age categories as their sub-leagues. However, the Open Soccer A league permits any age up to 19 years old (the maximum age for participating in RCJ). Table 1 shows all Junior Soccer sub-leagues at RCJ 2011[1].

Table 1. Soccer Sub-Leagues

Sub-Leagues	Field Type	Age Categories
Light Weight Soccer A Primary	Field A	Primary Age
Light Weight Soccer A Secondary	Field A	Secondary Age
Light Weight Open Soccer A	Field A	Open up to 19 yr old
Open Soccer B	Field B	Open up to 19 yr old

In total there were 109 teams participating in Junior Soccer in 2011, as shown in Table 2.

Table 2. Breakdown of Junior Soccer 2011 participating teams

	LW A Prim	LW A Sec	Open A	Open B	total
Students	62	63	102	163	390
Teams	22	17	24	46	109

---

[1] For more detailed soccer rules, visit
`http://rcj.robocup.org/rcj2011/soccer_2011.pdf`

2.1   Progress

IR Soccer Ball. In 2009, a new IR soccer ball (RCJ-05) was introduced to the Junior Soccer community. Similar to previous version of the IR soccer ball, RCJ-05 emits infrared light. However, its mode A emits IR pulse light, where previous soccer ball uses un-modulated IR light. Before we made it as a requirement, we needed to assess feasibility of mode A with teams participating in international games. In 2010, mode A was implemented for the firs time with Soccer B games as a pilot because mode A requires teams to either develop or purchase a sensor that can detect the pulsed infrared light. Since Soccer B in general requires advanced skills and knowledge to compete, our assumption was that the teams were capable of switching to the newly introduced mode. The report from 2010 indicated positive results. In 2011, mode A was implemented with all of the sub-leagues. After our close observation of the games, we came to conclude that teams have adapted to RCJ-05 pulse mode. It was reported that the ball could be detected from a longer distance than was possible with mode B. The ball also gives the added benefit of not requiring as frequent battery changes. From the progress demonstrated in 2011, it is conceivable that in a few years Junior teams may be able to use cameras to detect a passive ball's shape or color, no longer needing the active infrared light emission.

2.2   Issues

Open Soccer B. Field B, with its additional 30 cm outer area around the playing field, was introduced a few years ago to simulate out-of-bounds area in human soccer games. It was hoped that teams would be able to develop robots that could intercept a ball and keep it in play on the field most of the time. This may become true a few years down the road; however, it has not been the case with the current players. On Field B, the ball tends to spend too much time outside the playing field. Robots are expected to be constructed and programmed to locate a ball, quickly move to its location, and then direct the ball toward the opponent's goal for a goal. While most robots handle these objectives well, some robots are designed to use excessive brute force to attack and kick the ball, resulting in damage to the ball, other robots and/or the field. Although there are rules penalizing powerful robots that cause damage, some incidents are difficult to judge due to the high-speed nature of the event. The popular strategy of building and programming aggressive robots places too much focus on winning of the game by any means, rather than accomplishing the better goal of playing a successful game by keeping the ball inside of the play field. Open Soccer B needs careful reexamination each year by its technical committee to help teams accomplish its learning objectives – collaboration, cooperation and advancement of their skills to accomplish the goal set by the game.

Team Interview. As part of the participation, each team is required to be interviewed by a set of interviewers including technical and organizing committee members. The interviewers reported that a good variety of programming languages, skills and sophistication were observed. Existing rules do emphasize that team members must take an ownership of the construction and the programming of their robots. However, some students struggled with the thorough explanation of their program. The organizing committee members and judges have reported that they encountered incidents where team mentors or/and parents provided substantial help which, as a result, disabling students from taking an ownership of the work. In addition to the regular practice of requiring team member(s) to be able to explain particularities about their robot, its construction and programming during an interview, if in doubt, the interviewer should ask a team member responsible for a particular robot skill or attribute to demonstrate his capability. The interviewer could also ask the designated programmer to write a new simple code on the spot (example: write a program that will make the robot find a light or a dark wall, stop for a second, back up a bit and turn 180 degrees).

SuperTeam. In 2005, a SuperTeam scheme was introduced to Soccer League in order to facilitate and encourage collaboration between participating teams. In case of the Soccer league, a SuperTeam consists of three individual teams, possibly from different countries, working together to play several games against other SuperTeam teams. The organizing committee upon scheduling of each year's event randomly assigns the teams on a SuperTeam. Each SuperTeam plays one match against another SuperTeam. A match consists of three games. In the first game each individual team plays against one individual team from the opponent SuperTeam. After that game each SuperTeam can choose the pairing for one of the following two games.

Although the initial purpose was to encourage teams from different countries to work together, share their expertise and experience, we have seen more complaints than evidence of success from teams. Many teams have complained that it is unfair to have weak SuperTeam partners assigned to them, making it difficult to win games. It has been reported that very limited cooperation has been observed between SuperTeam partner teams. Although more structured feedback from participating teams should be collected for further examination of the issue, there is the need for future technical and organizing committee to reexamine the existing SuperTeam scheme to reinforce the collaboration among all teams involved. A new SuperTeam scheme has been proposed by some of the technical committee members that utilizes Small Size League field or a similar sized field for SuperTeams to hold a 5-on-5 soccer game in which five robots from five different teams per SuperTeam to play a game. For 2012, we will try to run demonstration games to examine the possibility of making the new SuperTeam scheme into an official SuperTeam event for Junior Soccer.

3   RCJ Rescue

Inspired by the major Rescue league, Junior Rescue was implemented in 2001. A rescue team is required to develop a rescue robot that can navigate through the rescue arena, which represents a scaled-down, simulated disaster scenario, and find a victim(s). The Rescue league has two sub-leagues – Rescue A and Rescue B. With Rescue A, teams use line-following strategies to navigate through the rescue arena where debris and obstacles are scattered, possibly blocking the line. The robot needs to climb up a ramp to the second floor to rescue a victim by pushing or pick up and move the victim into the evacuation zone. Rescue A has games for primary and secondary age groups. In 2011, Rescue B sub-league was officially added to provide challenges for more advanced teams. Rescue B is open to any age up to age 19. With Rescue B, a robot needs to navigate through a maze using wall following algorithms, while moving over debris and avoiding obstacles. Victims emitting heat are scattered across the arena, which a robot needs to rescue by finding their locations by stopping in front of each victim.

There were 93 teams in total participating in Junior Rescue in 2011, as Table 3 shows.

Table 3. Breakdown of Junior Rescue 2011 participating teams

	Res A Pri	Res A Sec	Res B	CoSpace	total
Students	93	112	72	33	310
Teams	32	33	21	7	93

3.1   Progress

Rescue A - Locating the Victim. Before 2010, the Rescue A victims were located on the floor (2D) with color coded sticker/tapes (Silver and Green). Most of teams used light sensors facing the floor to detect the color of the surface to locate the victims. Since 2010, the victim has been changed to 3D object with silver surface (a soda can wrapped by aluminum foil). With this victim, a light sensor cannot provide accurate information to determine the location of the victim since there might be other light source around the area. To receive more accurate information about the location of the victim, some teams used both light and distance sensor for finding the victim (if the light source is close, i.e. within 30cm range, it is the victim). Also there are teams using an infrared sensor which can also provide the direction of the victim, followed by using compass sensor to determine its own location to define the direction that it needs to move the victim to rescue. The observation of teams' strategies suggests that the change has challenged many teams to try out different strategies that require more than one sensor to locate the victim. The technical and organizing committees need to continue monitoring the advancements of participating teams so that the game can provide best challenges to further their learning.

SuperTeam. Rescue SuperTeam is organized differently from other Junior leagues. In Rescue, after the regular competition, the top 12 teams participate in the final SuperTeam competition. Rescue SuperTeam games present different challenges from the regular Rescue games. Each SuperTeam consists of two teams. The purpose of the challenge is to test, not only the robotics skills and knowledge, but also the collaboration between teams. In previous years, a SuperTeam pair was decided by a draw. However, in 2011, it was for teams to choose their pair. The suggestion was made for teams to find another team that employed a similar strategy. The result was very encouraging. The teams successfully selected their partner team based on their observation of other teams. Although structured feedback from teams should be collected to further examine the effect, it is reported that teams were more excited and engaged in their SuperTeam activity than previous years. Since this was a success, it might be beneficial to develop a scheme to provide the SuperTeam experience to all of the participating Rescue teams for the upcoming years.

3.2   Issues

Rescue B - Algorithm Strategies. Although Rescue B is a new addition to Junior Rescue, there are some teams that could employ advanced strategies. With Rescue B, the main tasks are to 1) navigate through the maze and 2) find all victims. Most of the teams participating in 2011 used one side of the wall to navigate through the maze. This strategy sounds promising; however, it does not guarantee that the robot can find all victims since it misses the ones on the other side of the wall. A couple of teams employed Simultaneous Localization And Mapping (SLAM). Using this strategy, a robot tries to build up a map of the unknown area, while at the same time it keeps track of its current location. Although their strategy was more advanced and the robot could find all victims in the arena, the result was not encouraging because they were not the top teams. Sometimes mapping the whole area would take more time than using one wall to go through the maze. It can provide more accuracy but might require more time. The technical committee needs to reexamine its assessment system and consider revising it to encourage teams to use advanced strategies, like SLAM.

4   RCJ Dance

Dance league, one of the original Junior leagues since 2000, attracts many more girls than Rescue or Soccer because of its focus on arts and technology. Each dance team is required to build a robot or multiple robots that move to music for up to two minutes. The creative and innovative presentation of robot(s) is emphasized in Dance league. There is no size or number limit for dance robots as long as they stay on the 6m x 4m stage performance area. Team members are also encouraged to perform on the stage with their robots. Although it was introduced as an entry-level event that focused on primary school children in 2000, over years, the dance performances have gained in complexity and now require advanced construction and programming of the performing robots. Originally, the robot performances were dances – a robotic dance performance to music with some synchronization to the rhythm. In recent years, we noticed a different type of performance similar to a theater performance with a story or theme. In 2009, we started to examine the trend and later decided to use two distinctively different performance score sheets – dance and theatre - to make sure that the score sheets equally benefit both types of performances. Score sheets are used as rubrics for teams to understand how their performance will be assessed. The score sheets emphasize the demonstration of creativity, innovation, taking risks with complicated or advanced programming and construction, and creative use of different sensors. All teams are required to be interviewed by a set of technical judges including technical and organizational committee members as well as performance judges. Sub-leagues of Dance league are defined by the age of team members (primary and secondary) [2].

There were 49 teams in total participating in Junior Dance in 2011, as shown in Table 4. In 2011, it was reported that the overall team performances were of good quality; however, risk-taking and innovative use of technology to enhance their performance, which generally require advanced skills and knowledge, were observed in very few performances.

Table 4. Breakdown of Junior Dance 2011 participating teams

	Dance Pri	Dance Sec	CoSpace	total
Students	99	111	45	255
Teams	19	23	7	49

4.1   Progress

SuperTeam. scheme was introduced to the Dance league in 2007. In the case of Dance SuperTeam, two to three teams from different countries/regions/continents form a SuperTeam, and recreate their SuperTeam performance within a day or less. A SuperTeam requires teams to collaborate by sharing and discussing as they prepare their new performance. Since the first implementation of SuperTeam in 2007, we have seen many very successful instances of collaboration among teams and received positive feedbacks from teams and mentors/parents. Dance SuperTeams are also required to produce documentation of their collaborative work in a visual presentation (i.e. PowerPoint, Video) and/or verbal presentation on stage. The presentation makes it possible for judges, organizers and audience to "experience" their collaboration. Every year, it was reported that teams enjoyed their experience with SuperTeam as they share and learn each other's culture and language, as well as robotics skills. Since there are almost always language barriers between teams, mentors/parents are encouraged to become a medium of communication and collaboration as translators. The SuperTeam experience has an added benefit of promoting collaboration among those adults as well.

In 2011, the teamwork demonstrated by the SuperTeams was exceptional. The SuperTeams were actively reprogramming their robots, and in some cases each other's robots. Their performances clearly displayed the fun and enjoyment gained from their collaborative work. Some teams initially complained as they did not speak a common language; however, the students overcame the language barrier and learned to communicate using other means such as translation software found on the Internet. For many teams, participating on a SuperTeam was the most rewarding and memorable experience of the competition.

Visual Presentation. Many teams use visual presentation as part of their performances, even adding some scenery through a video/slide show to make the performance look

---

[2] For more detailed Dance rules:
http://rcj.robocup.org/rcj2011/dance_2011.pdf

real or to tell the story of the performance. Some teams use advanced programs to create those visual presentations. The technical committee needs to evaluate the effect of the visual presentation and their educational value to team members in the future. Consideration needs to be given to the possibility of adding evaluation marks for the visual presentation as part of the overall performance value.

4.2 Issues

Technical Interview. Higher achieving teams are careful to study the rules and score sheets in order to gain high marks in the technical interview. To obtain a high score, the students need to demonstrate that their robots use many different sensors and effective and efficient codes, as well as incorporate interesting mechanical parts in their construction. One problem is that in some cases technology shown to the judges in the technical interviews appeared briefly if at all in the performance. In other instances the students were not clear in communicating all their innovative and complex work. As a result, their innovative and complex work could not be reflected in the interview scores. In general, although their performances were still of a high quality, the advancement in technology over the past few years have not been reflected in the robotic performances in 2011. To encourage more innovative use of technology, a change in interview scoring system to fairly grant points to innovative use of technology in their performance.

Use of Predesigned Robot Kits. Several teams included off the shelf humanoid robotic kits that were placed center stage in their performances, with often the more interesting home-/hand-made robotic achievements being used as the scenery. Usually, those off the shelf humanoid robotic kits provide almost no space for modification of its construction or program, hence no educational value. Teams using those kits tend to get fewer marks on construction, creativity, innovation or programming. However, some teams still prefer to use those kit-robots as the center of their performance since they look good on stage for the audience and are enticing to the students. Although it is clear from the score sheets that teams do not gain high marks by using kit-robots, the rules need to be clearer in directing teams away from using such kits. Also, we need to find ways to better communicate to the mentors/parents of teams that the focus of RCJ activities is "education" so that their suggestions to the teams will benefit the students.

Use of Mains Electricity. In 2011, several teams assumed that they could use mains electricity during their performance, for example for high-powered motors used to lift up heavy backdrops. The use of mains electricity, as well as the use of massive water or explosives on the stage, has been banned since 2007. It is important to work diligently to avoid potentially dangerous situations. Unfortunately, in the 2011 competitions, mains electricity was used for tasks that could have been achieved with the use of batteries. Restricting the teams to using only batteries encourages students to use innovative methods to overcome potential problems.

5   RCJ CoSpace Demonstration

In CoSpace Robotics world, the technology of co-existing Space is applied to the interoperability between physical and virtual worlds of robotics to produce an amalgamation of experiences that synthesize the benefits of both physical and virtual worlds. CoSpace Robotics combines and connects robots in a real, physical space with a 3D virtual-reality world. It allows students to experience and interact with robots not only in the real world but also in virtual reality that is based on the physical model in the real world.

In a CoSpace environment, physical robots and virtual robots can interact simultaneously in real-time. Locations and events in the physical world can be captured through the use of sensors and mobile devices, which can be materialized in a virtual world. Correspondingly, certain actions or events within the virtual domain can affect those in the physical world.

CoSpace Demonstration was introduced to RCJ in 2010 where several teams from Singapore participated, using the CoSpace platform – CoSpace Robot (CsBot) developed by the Advanced Robotics and Intelligent Control Centre (ARICC)[3]. In 2011, there were two CoSpace Demonstrations – CoSpace Rescue and CoSpace Dance, which involved more teams from different countries.

5.1   CoSpace Rescue

The theme of the CoSpace Rescue for RCJ 2011 was Treasure Hunt. With Treasure Hunt challenge, first in the virtual environment, a treasure map with a list of treasures was provided to teams. Each team had to develop appropriate AI strategies for a virtual robot to navigate through the treacherous terrain by avoiding obstacles and collect treasures in the 3D virtual environment while competing against an opponent robot performing the same mission. Next, the teams applied the same AI strategies to the identical real robot to search the treasures in the real world with the same set-up of the virtual arena. In RCJ 2011, there were seven teams participating in the CoSpace Rescue.

5.2   CoSpace Dance

The CoSpace Dance requires team members and robots (both real & virtual) come together to create a performance in a co-existing space with real-time communication. CoSpace dance teams need to build real robot(s), set-up real environment and props, and design virtual robot(s) and environment using 3D objects. It is a requirement for teams to establish a communication between real and virtual robots. Multimedia, such as music and video, can be integrated to both real and virtual environment to enrich their performances. In 2011, seven teams participated in the CoSpace Dance. These teams were also paired with Secondary Dance teams in SuperTeam performances.

---

[3] For more information, visit http://cospacerobot.org/

Fig. 7 & 8. RCJ CoSpace Rescue and Dance Demonstration

5.3　Issues

Teaching Materials and the Revision of the Platform. Although many team members enjoyed the new experience of navigating a robot in the virtual environment as well as in the real world, teams struggled with the steep learning curve required. The team members expressed lack of resources available for them to fully learn the platform. With this, the technical committee along with the development group at ARICC will develop more teaching materials including video and lessons in the future.

Also, the platform was not reliable enough to avoid situations when teams had to restart the platform several times while preparing for their games. Such technical issues will be examined by the development group at ARICC.

Virtual-Real Communication. Another issue raised was the lack of communication between virtual and real robots. The distinctive feature of CoSpace that differentiates it from other robotics activities is its ability to bridge the real and virtual worlds. However, CoSpace Rescue has no emphasis on this feature. With CoSpace Dance, although it was stated as one of the requirements in its rules, very few teams had successfully employed its CoSpace interaction. For CoSpace Rescue 2012, we are planning to require the communication between real and virtual robots as one of its game play. With CoSpace Dance, the 2012 rules will be revised to mandate the virtual-real communication as part of robotics performance.

6　Future Challenges

This section addresses the issues overarching all Junior leagues that we have been continuously facing in past several years and the challenges that they entail. Those issues include 1) the overall number of participating Junior teams and the selection strategy, 2) mentor/teacher involvements, and 3) how to involve populations which are in some sense marginalized from RCJ activities.

6.1　Selection of Participating Teams

Since 2000 the RCJ community has been growing around the world from three countries to over 30 countries participating in the annual international competition.

The largest number of teams and participants reached was in RCJ 2010, where 289 teams with 1,004 team members participated. The number of countries with active RCJ communities is 33 as of July 2011. We are expecting more Latin American countries to be involved in RCJ activities in coming years since RCJ 2012 will be held in Mexico City, Mexico. The challenge in each year is how to accommodate teams willing to participate in the annual competition from all active countries. The number of teams we can accept is determined by the size of the venue each year. For example, in 2006, 240 teams with 885 students from a total of 23 countries participated. However, in following years, the number of participating teams went down due to the size of the Junior venue. The more countries willing to participate in the RCJ annual competition, the less the number of teams we can accept from each country. Some of the countries with a large number of Junior teams, including China, Japan and Australia, with more than 1,000 active teams, cannot send a number of teams that represents the size of their activity, while countries with small number of active teams can send one team per league, which frequently represents almost their whole national RCJ activity. This asymmetry is becoming a problem that we can perhaps alleviate organizing regional selection events where teams from neighboring countries participate to select teams to represent their region in the international competition.

6.2   Dealing with Mentors/Parents of Teams

As stated at the beginning of the paper, the focus of RCJ is "education". Our mission is to provide the best opportunity possible for the participating students to learn from their experience during the preparation as well as at the competition where they interact with other teams. However, often times, the intense competitiveness at the venue leads to negative involvements of mentors and teachers by helping their teams too much with physically writing codes for and/or constructing the robots. In general, the team set-up area is strictly for team members only and mentors/parents are not allowed to stay inside. However, this is not always respected. To reduce this negative involvement we have been working to provide ways for mentors/teachers to also have positive learning experiences. One of the examples is the Dance SuperTeam with which, mentors/parents are expected to actively help teams by providing communication support. Another example is the RCJ RobotDemo workshop where the adults and team members can share their expertise and/or experiences and learn from each other during the competition. It is usually offered in the evening so that team members interested in can also participate. In 2010, the RoboCup Symposium also included a workshop focusing on Educational Robotics. We believe that this kind of venues for mentors/teachers to have professional development experience and to exchange and share their expertise are extremely valuable and should be encouraged in the future.

6.3   Possibility to Open Doors to Robotics for New Populations by Reducing the Competitiveness

Reducing the intense competitiveness between teams is always one of our challenges every year. Since the event itself is a competition, some degree of competitiveness is obviously unavoidable. However, it should not get intensified to the point where mentors/parents do the work for the students, or teams intentionally destroy or damage the robotics of opponent teams by using devices forbidden in the rules. The nature of intense competitiveness might exclude some population of students who might not bear it. Despite being commonly accepted that educational robotics competitions attract and inspire the participants who might not often be motivated through regular school curricula, some researchers argue that they might limit participant diversity [7, 8]. For students who do not necessarily favor the competitive nature, robotics competitions might not be a comfortable event to participate [7, 8]. Among the Junior leagues, we conjecture that Dance attracts girls the most because of its less competitive nature and focus on artistic expression. Still on gender, Rescue attracts more girls than Soccer, in spite the majority of participants being boys.

Digital Robotic Exhibitions. Rather than competitions, art exhibitions, where robots can interact with people or with another robot, or exhibitions of innovative robotics projects might reduce the competitiveness and promote creativity using the technology with different disciplines, such as art. Rusk, Resnick, Berg and Pezalla-Granlund suggest that combining art and engineering for artistic and self-expression can inspire girls and boys to think more creatively [8]. They also suggest that robotics projects can be demonstrated through a style of exhibitions. Some future possibilities should be discussed among interested parties, including Junior technical committees as well as interested mentors/parents.

7   Conclusion

After a decade since RCJ international competition launched in 2000, there are several thousands of teams participating in RCJ activities in more than 30 countries. RCJ and the participating teams have progressed tremendously to be able to present complex robotics knowledge and skills in their performances at the competition, which now gathers around 1000 participants. Since the focus of RCJ is "education," we strive to embrace the learning experience of participating students. For this reason, SuperTeam scheme has been adapted to all of Junior leagues to promote collaboration among teams and team interview has received substantial weight emphasizing the learning experience of participating students to RCJ activities. These, however, had different impact across different leagues and each one is facing different issues and exhibiting different stages of progress, which we discussed along the paper.

Finally, we also addressed a few RCJ transversal issues, namely the challenges of dimension and team selection, involving mentors/teachers in constructive ways, and keeping competitiveness under healthy levels.

RCJ has made a long way with a growing success through careful steps, despite being a purely volunteer-based organization. As such, it relies on substantial effort after hours of many people around the world that carefully prepare and then run each event, motivated by the satisfaction of the participants. We believe these are enough ingredients to a successful continuation.

References

1. FIRST: About FIRST - Impact (2006),
   http://www.usfirst.org/about/impact.htm
2. KISS Institute for Practical Robotics: Botball Educational Robotics (2009),
   http://botball.org/about (retrieved December 11, 2010)
3. Sklar, E., Eguchi, A.: RoboCupJunior — Four Years Later. In: Nardi, D., Riedmiller, M., Sammut, C., Santos-Victor, J. (eds.) RoboCup 2004. LNCS (LNAI), vol. 3276, pp. 172–183. Springer, Heidelberg (2005)
4. Sklar, E.: RoboCupJunior 2002: The State of the League. In: Kaminka, G.A., Lima, P.U., Rojas, R. (eds.) RoboCup 2002. LNCS (LNAI), vol. 2752, pp. 489–495. Springer, Heidelberg (2003)
5. Sklar, E., Eguchi, A., Johnson, J.: Examining the Team Robotics through RoboCupJunior. In: proceedings of the Annual Conference of Japan Society for Educational Technology, Nagaoka, Japan (2002)
6. Sklar, E., Eguchi, A., Johnson, J.: Scientific Challenge Award: RoboCupJunior - Learning with Educational Robotics. AI Magazine 24(2), 43–46 (2003)
7. Hamner, E., Lauwers, T., Bernstein, D., Nourbakhsh, I., DiSalvo, C.F.: Robot Diaries: Broadening Participation in the Computer Science Pipeline through Social Technical Exploration. In: Proceedings of the AAAI Spring Symposium on Using AI to Motivate Greater Participation in Computer Science, Palo Alto, CA (2008)
8. Rusk, N., Resnick, M., Berg, R., Pezalla-Granlund, M.: New Pathways into Robotics: Strategies for Broadening Participation. Journal of Science Education and Technology (February 2008)

Compliant Task-Space Control
with Back-Drivable Servo Actuators

Jörg Stückler and Sven Behnke

University of Bonn
Computer Science Institute VI, Autonomous Intelligent Systems
Friedrich-Ebert-Allee 144, 53113 Bonn, Germany
stueckler@ais.uni-bonn.de, behnke@cs.uni-bonn.de

Abstract. In this paper, we propose a new approach to compliant task-space control for high degree-of-freedom manipulators driven by position-controlled actuators. The actuators in our approach are back-drivable and allow to limit the torque used for position control. Traditional approaches frequently achieve compliance through redundancy resolution. Our approach not only allows to adjust compliance in the null-space of the motion but also in the individual dimensions in task-space. From differential inverse kinematics we derive torque limits for each joint by examining the contribution of the joints to the task-space motion. We evaluate our approach in experiments with specific motions. We also report on the application of our approach at RoboCup 2010, where we successfully opened and closed the fridge in the RoboCup@Home finals.

Keywords: compliance control, task-space control, domestic service robots, servo actuators.

1 Introduction

In todays industrial settings, robots require fast, precise, and reliable execution of motions. The use of high-stiffness motion control can guarantee robust operation in this domain, but it also demands precise models of the dynamics of the robot mechanism and manipulated objects. Furthermore, precautions need to be taken to prevent physical interaction with humans under any circumstances. However, this approach is not applicable to domains, such as service robotics, in which the environment is less structured, i.e., uncertainty is involved in the models, or physical interaction with humans can not be avoided.

Compliance in motion control opens up new application domains for manipulation robots. Since small errors in model acquisition and estimation can be compensated through compliant control, the robot is able to operate despite measurement errors. It is also possible to use compliant motion for explorative manipulation of objects, especially of articulated objects. Finally, compliant motion allows for direct but safe physical interaction with humans, for example, for teaching by guidance or for physical intervention by humans.

T. Röfer et al. (Eds.): RoboCup 2011, LNCS 7416, pp. 78–89, 2012.
© Springer-Verlag Berlin Heidelberg 2012

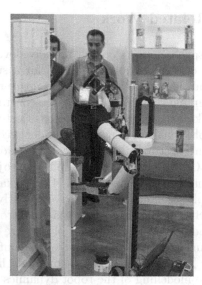

Fig. 1. Our domestic service robot Dynamaid opens and closes the fridge at RoboCup 2010 in Singapore

For compliant motion control, only the torque necessary to achieve a position, velocity, or force is exerted through the robot actuators. In combination with a light-weight robot construction, this control approach can achieve inherently safe motion, since small forces and torques are required for control. When the desired motion of the robot to achieve a task does not constrain all degrees of freedom of a robot mechanism, redundancy needs to be resolved. The robot may be controlled fully compliant in the null-space of the task-constrained motion.

In this paper, we propose compliant task-space control for redundant manipulators driven by servo actuators. The actuators in our approach are back-drivable and allow to configure the maximum torque used for position control. From differential inverse kinematics we derive a method to limit the torque of the joints depending on how much they contribute to the achievement of the motion in task-space. Furthermore, our approach not only allows to adjust compliance in the null-space of the motion but also in the individual dimensions in task-space. This is very useful when only specific dimensions in task-space shall be controlled in a compliant way. We evaluate our approach quantitatively in experiments for specific task-space motions. We also report on the use of compliant task-space motion control for manipulating articulated objects at RoboCup, where we successfully applied our approach to open and close the fridge in the RoboCup@Home finals 2010 in Singapore (s. Fig. 1).

The remainder of the paper develops as follows. After a brief overview of related work in the fields of compliant and task-space motion control in the next section, we will state our method in Sec. 3. In Sec. 4 we give further insights into our method with an example application. We finally report on the experimental evaluation of our approach in Sec. 5.

2   Related Work

Task-space motion control, initially developed by Liegeois [5], is a well established concept in robotics (s. [6] for a survey). Common to task-space control methods is to transfer motion in a space relevant to a task to joint-space motion. One simple example of task-space control is the control of the end-effector of a serial kinematic chain along pose trajectories in Cartesian space. The task-space control formalism allows to consider secondary objectives, when the task-space motion constrains less degrees of freedom than available in the robot mechanism. Optimization criteria are then projected into the null-space of the motion in joint-space. De Lasa et al. [4], for example, demonstrate how to incorporate multiple secondary objectives consistently into the task-space control framework.

Early approaches to task-space control have been velocity-based [5]. In these methods, velocity-based control laws are derived by differentiation and inversion of a function that maps joint-space configurations to task-space. Acceleration-based [1] and force-based [3] methods have also been proposed. They require precise modelling of the robot dynamics and have been shown to be difficult to implement [6].

For compliant motion control in task-space, acceleration- and force-based methods are naturally suited. Velocity-based methods have been reported to be ill-suited for compliant control, when compliance is established with redundant degrees of freedom of the robot kinematics [6]. Instead of relying on redundancy resolution for compliant control, we propose to adjust compliance for each dimension and direction in task-space as well as in the null-space of the motion when the robot kinematics is redundant for the task.

3   Compliant Task-Space Control with Position-Controlled Actuators

In our method, we employ velocity-based task-space control and derive a control law for compliant motion. We assume that the robot actuators follow position trajectories with servo control loops and that the torque used for control is limitable.

3.1   Velocity-Based Task-Space Controller

Central to task-space controllers is a mapping f from joint states $q \in \mathbb{R}^m$ to states $x \in \mathbb{R}^n$ in task-space, i.e., the forward kinematics

$$x = f(q). \tag{1}$$

By linearization, one obtains the differential relationship

$$\dot{x} \approx J(q)\dot{q} \tag{2}$$

between velocities in joint-space and task-space. The inversion of this relation-ship yields a mapping from task-space velocities to joint-space velocities,

$$\dot{q} \approx J^{\dagger}\dot{x}, \tag{3}$$

where $J^{\dagger} := J^{\dagger}(q)$ is the pseudo-inverse of $J(q)$. When the task-space has less dimensions than degrees of freedom are available in the robot kinematics, the inverse mapping has a null-space,

$$\dot{q} \approx J^{\dagger}\dot{x} + \left(I - J^{\dagger}J\right)\dot{q}^{0}, \tag{4}$$

in which joint motion $\dot{q}^{0}$ can be projected such that the tracking behavior in task-space is not altered. In this case, we call the robot kinematics redundant for the task.

Given a desired trajectory in task-space $x_d(t)$, we derive a control scheme to follow the trajectory with a position-controlled servo actuator,

$$\begin{aligned}
\dot{x}(t) &= K_x\left(x_d(t) - x(t)\right), \text{ and} \\
\dot{q}(t) &= K_q\left(J^{\dagger}\dot{x}(t) + \alpha\left(I - J^{\dagger}J\right)\nabla g(q(t))\right),
\end{aligned} \tag{5}$$

where K_x and K_q are gain matrices which can be adjusted in each time step to limit velocities in task- and joint-space, respectively. The cost function $g(q(t))$ optimizes secondary criteria in the null-space of the motion, and α is a step-size parameter. Cost criteria typically cover joint limit avoidance or the preference of a convenient joint state.

3.2   Compliant Task-Space Control

The control loop in each servo actuator implements torque control to achieve a target position in joint-space. In our approach, we assume that the torque applied by the actuator can be limited. We derive the responsibility of each joint for the motion in task-space, and distribute a desired maximum torque onto the involved joints according to their responsibility.

We measure the responsibility of each joint for the task-space motion through the inverse of the Jacobian

$$R_{task}(t) := \text{abs}\left[J^{\dagger}(q(t))\begin{pmatrix} \Delta x_1 & 0 & \cdots & 0 \\ 0 & \Delta x_2 & \ddots & \vdots \\ \vdots & \ddots & \ddots & 0 \\ 0 & \cdots & 0 & \Delta x_n \end{pmatrix}\right], \tag{6}$$

where $\Delta x := K_x\left(x_d(t) - x(t)\right)$ is the target motion in task-space and abs de-termines absolute values of a matrix element-wise. Each entry (i, j) of the ma-trix $R_{task}(t)$ measures the contribution of the velocity of the j-th task component to the velocity of joint i.

In addition, we also define the responsibility of each joint for the null-space motion

$$R_0(t) := \text{abs} \left[\alpha \left(I - J^\dagger J \right) \nabla g(q(t)) \right].$$ (7)

Finally, we obtain the responsibility $R(t) := (R_{task}(t), R_0(t))$ of task-space and null-space motion through concatenation of the individual responsibilities.

We determine the compliance $c \in [0,1]^n$ in dependency of the deviation $d_i := x_d(t) - x(t)$ of the actual state $x(t)$ from the target state $x_d(t)$ in task-space, i.e.,

$$c_i := \begin{cases} 1 - \frac{d_i - d^-}{d^+ - d^-} , & \text{if } d^- \leq d_i \leq d^+ \\ 1 & , \text{ if } d_i < d^- \\ 0 & , \text{ if } d_i > d^+ \end{cases}$$ (8)

such that the compliance is one for $d_i \leq d^-$ and d^+ and zero for $d_i \geq d^+$.

For each task dimension i the motion can be set compliant in the positive and the negative direction seperately. The direction of motion is given by the deviation of the actual state in task-space from the target. When the task dimension is not set compliant, we choose high holding torque τ_i^x in this dimension. If it is set compliant, the holding torque

$$\tau_i^x = c_i \tau_i^{x-} + (1 - c_i) \tau_i^{x+}$$ (9)

interpolates between a minimal holding torque τ_i^{x-} for full compliance $c_i = 1$ and a maximal holding torque τ_i^{x+} for minimal compliance $c_i = 0$. The minimal and maximal holding torques should be chosen for the task at hand. For example, when the motion is set compliant along the vertical axis, gravity can be compensated by sufficient minimal holding torque. The holding torque for the null-space motion can also be set to the desired compliance.

We distribute the torques for the individual task dimensions on the joints responsible for the motion in these dimensions. First, we determine the activation matrix

$$A(t) := R(t) \begin{pmatrix} \left(\sum_j R_{j,1} \right)^{-1} & 0 & \cdots & 0 \\ 0 & \left(\sum_j R_{j,2} \right)^{-1} & \ddots & \vdots \\ \vdots & \ddots & \ddots & 0 \\ 0 & \cdots & 0 & \left(\sum_j R_{j,n} \right)^{-1} \end{pmatrix},$$ (10)

by normalizing the responsibility of the joints to sum to one along each task dimension. The task component torques are then distributed according to the activation of each joint

$$\tau^q = A(t) \tau^x$$ (11)

to the individual joint torque limits τ^q.

In order to obtain the responsibility matrix, we linearize the relationship between task- and joint-space at the actual joint positions and incorporate the

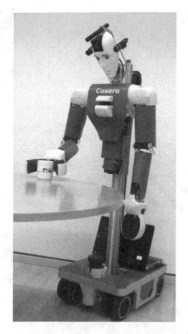

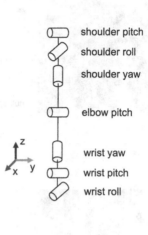

Fig. 2. Left: Cognitive Service Robot Cosero. Right: Schematics of the seven degrees of freedom in the anthropomorphic arms.

deviation of the actual state from the target state in task-space. When large deviations shall be allowed, the linear approximation is coarse and leads to large deviations in uncompliant task dimensions. Instead, we propose to adapt an intermediate target $\tilde{x}_d(t)$ in task-space which complies towards the actual state $x(t)$

$$\tilde{x}_d(t) := x_d(t) + \eta\,(x(t) - x_d(t)), \tag{12}$$

where $\eta \in [0,1)$ is adjustable. For $\eta \to 1$ the intermediate target fully follows external influences in the compliant task dimensions. Intermediate values of η also control how fast the robot returns to the actual target.

4   Example Application

We exemplify our approach with the mobile manipulation of door leaves by our domestic service robot Cosero. The robot is shown in Fig. 2 together with a schematics of the kinematic model of the 7-DOF anthropomorphic arms of the robot. Cosero's joints are mainly driven by Robotis Dynamixel EX-106+ (10.7 Nm holding torque, 154 g) and RX-64 (6.4 Nm holding torque, 116 g) actuators.

Several approaches exist to manipulate doors when no precise articulation model is known. For instance, Niemeyer and Slotine [7] propose to follow the motion of the door handle using force control. Jain and Kemp [2] use compliant

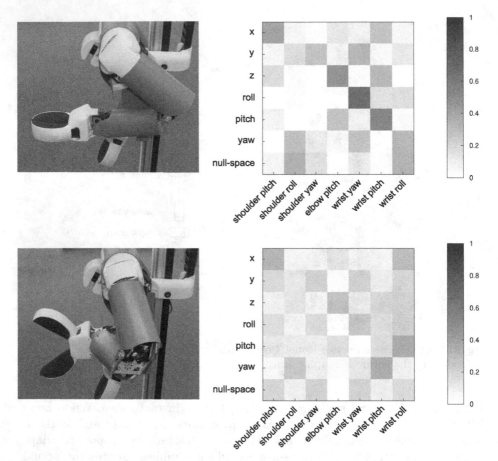

Fig. 3. Examples of activation matrices for two arm poses. The task-space dimensions correspond to forward/backward (x), lateral (y), vertical (z), and rotations around the x-axis (roll), y-axis (pitch), and z-axis (yaw).

equilibrium point control to push a door open. They use force sensors to decide, when the controller fails to grasp the handle or when the door is blocked. Schmid et al. [8] design a controller to adjust the gripper position and orientation in a fixed pose towards the door leaf. They measure deviations from the intended end-effector pose using force and tactile feedback. Our approach does not require feedback from force or tactile sensors.

In order to open the door, the robot grasps the door handle and exerts a force in backward direction, orthogonal to the closed door leaf. We set the motion of the end-effector compliant in the lateral direction and in rotation around the vertical axis, such that the end-effector may comply to the motion of the door handle, which is constrained on a circle by the rotational joint in the door hinge. The robot moves back, until the door handle has reached its maximal position in backward direction and resists further motion. The robot closes the

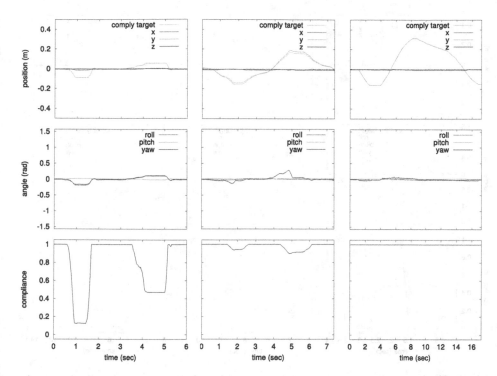

Fig. 4. Deviations from targets in task-space and compliance c for compliant control in y-direction in the vertical grasp pose without target adaptation ($\eta = 0$, left), with intermediate target adaptation ($\eta = 0.9$, center), and with strong target adaptation ($\eta = 0.999$, right)

door by grasping the door handle and exerting a force in forward direction, while following and turning to keep the door handle at the initial grasp position relative to the robot. It pushes the door until the door is closed and resists further forward motion of the end-effector.

In both control applications, the end-effector must comply to the constrained motion of the door handle in specific directions in the task-space, while keeping other directions at their targets. In Fig. 3, we give examples for the activation matrix $A(t)$ in two grasping poses for a door handle. The upper row shows the configuration of the joints and the activation matrix, when the end-effector grasps a vertically aligned door-handle. The grasp in the lower row is aligned to horizontal door handles. The task-space dimensions correspond to forward/backward (x), lateral (y), vertical (z), and rotations around the x-axis (roll), y-axis (pitch), and z-axis (yaw).

It can be seen from the activation matrices, that for both poses, motion in x-direction in task-space involves the elbow and shoulder pitch actuators. For the vertical grasp (upper row), the wrist pitch joint also contributes to the motion in x-direction. In constrast the wrist roll joint contributes to the x-direction for the

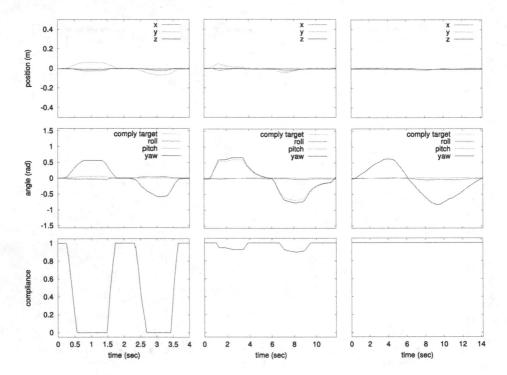

Fig. 5. Deviations from targets in task-space and compliance c for compliant control in yaw-direction in the vertical grasp pose without target adaptation ($\eta = 0$, left), with intermediate target adaptation ($\eta = 0.9$, center), and with strong target adaptation ($\eta = 0.999$, right)

horizontal grasp (lower row). Also, other shoulder and wrist joints add smaller contributions to the motion.

In task-space y-direction, the shoulder roll and yaw joints and the wrist yaw joints are primarily responsible for the motion when grasping vertically. For the horizontal grasp, all joints but the elbow pitch joint contribute significantly to the motion.

For the rotational motion in task-space, we observe, that yaw-rotation strongly involves the shoulder roll, wrist yaw, and wrist roll joints. This is due to the fact, that for a yaw rotation of the end-effector, the shoulder roll joint has to move the elbow in- and outwards, which also induces a roll rotation of the end-effector that is compensated by the wrist roll joint. In vertical grasping alignment, the yaw-rotation is achieved primarily with the shoulder roll and the wrist pitch joint.

5   Experiments

We evaluate the performance of our approach for compliance control with our domestic service robot Cosero (s. Fig. 2). Throughout the experiments we use

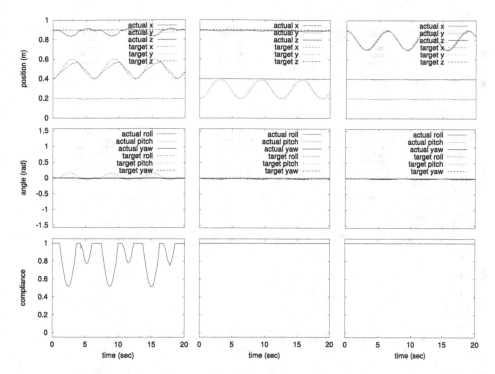

Fig. 6. Tracking performance for sinusoid motion in compliant task-space dimensions (left: x, center: y, right: z)

the settings $d^- = 0.01\,$m and $d^+ = 0.1\,$m in x-, y-, and z-direction. In the roll-, pitch-, and yaw-directions we choose $d^- = 0.04\,$rad and $d^+ = 0.4\,$rad.

5.1 Compliance Control in Static Poses

In a first set of experiments, we set the motion compliant in y-direction, while the target pose is kept constant at the vertical grasp pose (s. Fig. 3, top left). Fig. 4 shows the reaction of our controller on applied forces in lateral directions on the wrist for different settings of the target adaptation rate η. When the adaptation rate is set to $\eta = 0$, the end-effector complies only by a fraction of d^+. Compliance decreases with deviation and the limit torques increase until the arm resists further motion in y-direction. Since the target is not adapted to the actual pose of the end-effector, other task dimensions also deviate from their target position. As soon as the applied force ceases, the end-effector moves back to its target pose. At an adaptation rate of $\eta = 0.9$, the end-effector may comply farer to the external force, while motion in other task dimensions is reduced. When the applied force is suddenly reduced to zero, the end-effector slowly moves back to its target pose. For $\eta = 0.999$, the end-effector fully complies to the applied force, while other task dimensions can be controlled close at their

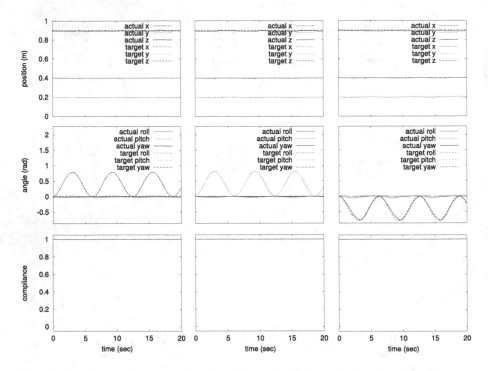

Fig. 7. Tracking performance for sinusoid motion in compliant task-space dimensions (left: roll, center: pitch, right: yaw)

target positions. Since gravity always acts on the robot arm, the end-effector moves to an equilibrium pose when the applied force is removed.

We also evaluate our controller for different settings of the target adaptation rate η in the yaw-direction. On the end-effector we apply torques in yaw direction. From Fig. 5 it can be seen that the controller behaves similarly to applied torques in yaw-directions like in the experiments with forces in the compliant y-direction.

5.2   Tracking with Compliance Control

We further evaluate the tracking behavior of our approach without target adaptation, i.e. $\eta = 0$. We set each each task dimension compliant individually and examine the tracking of a sinusoid target motion in the compliant task dimension. For the y-, z-, roll-, pitch-, and yaw-direction, the controller follows the target motion very well.

In the x-direction, the controller temporarily looses track in the z-dimension in downwards direction when the arm extends far forward. This is due to the fact that the pitch joints are concurrently involved in the motion in x-direction as well as in z-direction. Since the deviation in z-direction also leads to a tracking error in x-direction, compliance in x decreases. By this, the available torque in the pitch joints increases until the deviation in z-direction can be corrected.

6   Conclusions

In this paper we proposed an approach to redundant task-space control that allows for compliance control in individual directions of the task-space. Our approach extends velocity-based task-space control and is suited for back-drivable position-controlled actuators for which the available torque can be limited.

From differential inverse kinematics, we derive a controller that distributes torque limits for compliant directions in task-space onto the joints that are responsible for the motion. The controller allows to adjust the compliance range in task-space and the rate with which it adapts its target to follow external forces.

In experiments we show that our controller achieves the desired compliance behavior when it holds a static pose. We evaluated the controller for different target adaptation rates. We also demonstrated that the controller is capable to track motion in compliant task-space dimensions. Finally, at RoboCup 2010 in Singapore, we successfully applied our approach to open and close the fridge in the finals of the RoboCup@Home league. The performance of the robot was well received by the jury.

Our approach is easy to implement when a velocity-based controller is available. In future work, we could improve the distribution of forces and torques in task-space onto the joints by considering the dynamics of the robot mechanism.

References

1. Hsu, P., Mauser, J., Sastry, S.: Dynamic control of redundant manipulators. Journal of Robotic Systems 6(2), 133–148 (1989)
2. Jain, A., Kemp, C.C.: Pulling open novel doors and drawers with equilibrium point control. In: Proc. of the 9th IEEE-RAS Int. Conf.on Humanoid Robots (2009)
3. Khatib, O.: A unified approach for motion and force control of robot manipulators: The operational space formulation. IEEE Journal of Robotics and Automation 3(1), 43–53 (1987)
4. de Lasa, M., Hertzmann, A.: Prioritized optimization for task-space control. In: Proceedings of the IEEE/RSJ International Conference on Intelligent Robots and Systems (IROS), pp. 5755–5762 (2009)
5. Liegeois, A.: Automatic supervisory control of the configuration and behavior of multibody mechanisms. IEEE Transactions on Systems, Man and Cybernetics 7(12), 868–871 (1977)
6. Nakanishi, J., Cory, R., Mistry, M., Peters, J., Schaal, S.: Operational space control: a theoretical and empirical comparison. International Journal of Robotics Research 27(6), 737–757 (2008)
7. Niemeyer, G., Slotine, J.-J.E.: A simple strategy for opening an unknown door. In: Proc. of the IEEE Int. Conf. on Robotics and Automation (1997)
8. Schmid, A.J., Gorges, N., Goger, D., Worn, H.: Opening a door with a humanoid robot using multi-sensory tactile feedback. In: Proc. of the IEEE Int. Conf. on Robotics and Automation, ICRA (2008)

Planning Stable Paths for Urban Search and Rescue Robots

Mohammad Norouzi, Freek De Bruijn, and Jaime Valls Miró

Faculty of Engineering and IT, University of Technology, Sydney (UTS)
Sydney NSW 2007, Australia
mohammad.norouzi@student.uts.edu.au, {Freek.DeBruijn,javalls}@uts.edu.au
http://www.eng.uts.edu.au

Abstract. Rescue robots are platforms designed to operate in challenging and uneven surfaces. These robots are often equipped with manipulator arms and varying traction arrangements. As such, it is possible to reconfigure the kinematic of robot in order to reduce potential instabilities, such as those leading to vehicle tip-over. This paper proposes a methodology to plan feasible paths through uneven topographies by planning stable paths that account for the safe interaction between vehicle and terrain. The proposed technique, based on a gradient stability criterion, is validated with two of the best known path search strategies in 3D lattices, i.e. the A* and the Rapidly-Exploring Random Trees. Using real terrain data, simulation results obtained with the model of a real rescue robot demonstrate significant improvements in terms of paths that are able to automatically avoid regions of potential instabilities, to concentrate on those where the freedom of exploiting posture adaptation permits generation of optimally safe paths.

Keywords: Stability Analysis, Pose Reconfigurability, Path Planning, Rescue Robot.

1 Introduction and Motivation

This paper addresses the efficient deployment of field mobile robots in environments such as planetary exploration, mining or search and rescue, where efficient and safe locomotion over irregular terrain is paramount. These environments are often inaccessible or considered too dangerous for humans to operate in, thus naturally suited for the use of autonomous robotic platforms. For effective operation, these vehicles have to deal with rough, uneven surfaces and many uncertainties, and these characteristics have a strong influence on the robot's ability to perform as planned and need to be accounted for. In general, mobile robots operating under these scenarios operate at low speeds, thus the robot dynamics can be safely disregarded and a simpler quasi-static representation be adopted instead. For the specific case of reconfigurable robots operating under these conditions, as the one described in Section 2, their kinematical configuration plays a crucial role in defining the interaction forces between vehicle and terrain. Appropriately selecting this that can lead to improvements in robot locomotion.

T. Röfer et al. (Eds.): RoboCup 2011, LNCS 7416, pp. 90–101, 2012.

Fig. 1. The iRobot PackBot robot with pan and tilt sensor payload unit, in a mock-up Urban Search And Rescue (USAR) test arena

Providing the robot the ability to actively assume poses that reduce potential tip-over instabilities by repositioning their Center of Mass (CM) during planning (and/or control) is the strategy proposed in this paper to enable safer navigation over irregular terrains.

Various criteria have been proposed in the literature to analyze the qualitative performance of robot stability, mostly with the aim of real-time short-term tip-over monitoring, prediction and prevention [14,6], or off-line path optimization [1]. A fuzzy controller for autonomous negotiation of stairs was presented in [4]. In this work a stability measure is also employed to provide a reliable non-Euclidean metric that is derived from the vehicle stability constraints about each tip-over axis of the robot, as detailed in Section 3. This metric is then exploited in a variational formulation of cost-driven path planning in three-dimensional (3D) irregular lattices, and tested with tow of the most widely used graph-search methods for robot path planning: A* and Rapidly-Exploring Random Trees (although strictly speaking RRTs build the tree space where search then takes place, in this work the first path to reach the goal is returned to solve the path planning problem, hence RRTs are hereby treated as a graph-search method). The section 4 gave a brief overview of the methods for robot path Planning. The feasibility of the proposed approach in biasing the planning towards safe traversable paths over irregular terrain is demonstrated by the results given in Section 5.

2   Robot Model

The multi-tracked iRobot Packbot depicted in Figure 1 was the platform employed to validate the practical aspects of this research. It consists of a skid-steer vehicle base, equipped with two front flippers that enable the robot to traverse obstacles and rough terrain. A manipulator arm is attached to the vehicle base via a 1 Degree of Freedom (DoF) shoulder joint. It carries a 2-DoF pan-and-tilt unit equipped with several cameras and lights. An additional sensor head unit is also mounted on top of the arm head to enhance the Search and Rescue capabilities of the robot in its navigational and victim identification activities. It incorporates a laser scanner, a 3D time-of-flight camera and a thermal camera. The robot is battery powered and features two battery compartments on its left and right hand side.

The computation of the robot posture is based on the calculation of the position of CM. This, in turn, is significantly influenced by the kinematic configuration during robot motion. The position of the arm and flippers has a strong influence on the CM, as does the number and location of the batteries attached. All masses and dimensions necessary to reliably determine the CM were identified and validated through experimental measurements. The influence of head panning and tilting on the robot's CM is very small in comparison to the effects that arise from the position of the arm and flippers. These were therefore neglected. Thus, the arm and flipper's joints are the key reconfigurable degrees of freedom (DoFs) considered. While the formulation allows for full dynamic effects to be readily incorporated, as the rover is assumed to be operated at low speeds during search and rescue operations, gravity is the only force considered in this work.

3   Stability Metric and Terrain Modeling

There have been a number of propositions to address the issue of stability of mobile robots. Some research has focused on the analysis of the robot's CM to find suitable controls to cope with specific scenarios like overcoming obstacles and small ditches [21]or climbing stairs [2]. A real-time rollover protection strategy based on whole-body touch sensors that are embedded in the tracks of the robot and an energy stability margin that serves as an indicator for unsafe robot configurations have also been developed [8]. Stabilizing actions to protect the robot from roll-over based upon empirical flipper movements were also proposed.More general approaches for the stability control of reconfigurable mobile robots have also taken into account other constraints such as traction optimisation [17], [3]. Both these used the original Force-Angle stability margin (FA) originally reported in [13]. A multi tracked robot on a steep slope was examined in [18] to determine boundaries for the CM and came up with a strategy to traverse a given slope. The stability measure employed considered the angle between the vector through CM and tip-over axis and the vector of the resulting force through the CM. This may be sufficient for a robot with relatively high CM's that do not change significantly, but is not representative for the actual stability in many other cases, such as robots more strongly subjected to external forces and moments. A revised version to the original FA was proposed by Papadopoulos and Rey to allow for dynamical changes in the robot configuration [15]. This constitutes a more suitable stability measure for mobile robots/manipulators as it exhibits a more simplistic geometric interpretation and thus could be more easily computed. It is for this reason that this is the metric employed in this work.

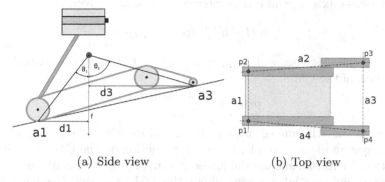

(a) Side view   (b) Top view

Fig. 2. Example FA measure

As explained, the metric was introduced in two different versions, and these are briefly reviewed in the next Section to better understand the influence of the CM's height for platforms that can significantly reposition their CM to improve stability in uneven terrains.

3.1 The Force-Angle Stability Margin (FA)

The FA measure β was first proposed in 1996 [13] as

$$\beta = min(\theta_i \cdot \|\mathbf{f_i}\|) \tag{1}$$

where $\mathbf{f_i}$ is the net force (including all static and dynamic forces, as well as moments) contributing to a potential roll-over about a particular tip-over axis $\mathbf{a_i}$. The tip-over axes $\mathbf{a_i}$ are given as the lines between m arbitrary supporting points $\mathbf{p_i}, i = \{1, .., m\}$

$$\mathbf{a_i} = \mathbf{p_{i+1}} - \mathbf{p_i}, i = \{1, .., m-1\} \tag{2}$$

$$\mathbf{a_m} = \mathbf{p_1} - \mathbf{p_m} \tag{3}$$

θ_i is the angle between $\mathbf{f_i}$ and the tip-over axis normal through the tip-over axis and the CM. Figure 2 illustrates these parameters in an example, where $\mathbf{a_1}$ and $\mathbf{a_3}$ are perpendicular to the paper representing the tip-over axes through $\mathbf{p_1}/\mathbf{p_2}$ and $\mathbf{p_3}/\mathbf{p_4}$ respectively.

The revised version of the FA measure was published in 2000 [15] and besides $\mathbf{f_i}$ and θ_i, it also included $\mathbf{d_i}$, the distance between $\mathbf{a_i}$ and $\mathbf{f_i}$ as

$$\beta = min(\theta_i \cdot \|\mathbf{d_i}\| \cdot \|\mathbf{f_i}\|) \tag{4}$$

This enables the metric to become sensitive to varying heights of the CM. The greater the value of the stability measure β_i, the more stable the vehicle becomes in terms of tipping over about the given axis. Negative values of the measure indicate an occurring tip-over instability.

The tip-over axis normal l_i that intersects the CM is given by

$$l_i = (I - \hat{a}_i\hat{a}_i^T)(p_{i+1} - p_{CM}) \tag{5}$$

where $\hat{a}_i$ is the unit vector along a_i, I is the 3×3 identity matrix and p_{CM} is the position of the CM.

$$p_{CM} = \frac{\sum_j p_{mass_j} m_j}{m_{tot}} \tag{6}$$

where m_j is the jth lumped mass with inertial location of p_{mass_j} and m_{tot} is the total system mass. Given f_r, the net force acting on the CM which includes gravitational, external and inertial forces, and n_r, the net moment encompassing all external and inertial moments about the CM axis, the effective net force f_i that contributes to a potential tip-over about one specific axis a_i can be determined by

$$f_i = (I - \hat{a}_i\hat{a}_i^T)f_r + \frac{\hat{l}_i \times ((\hat{a}_i\hat{a}_i^T)n_r)}{\|l_i\|} \tag{7}$$

The first term considers the part of the net force perpendicular to the tip-over axis. The second term considers the moment about the tip-over axis, converted into an equivalent force couple, where one member of the couple passes through the CM and thus can be added to the net force, whereas the other member passes through the tip-over axis. The latter clearly does not contribute to f_i.

The angle θ_i for each tip-over axis can then be computed by

$$\theta_i = \sigma_i \cos^{-1}(\hat{f}_i\hat{l}_i) \tag{8}$$

where

$$\sigma_i = \begin{cases} +1 & (\hat{f}_i \times \hat{l}_i)\hat{a}_i > 0 \\ -1 & \text{otherwise} \end{cases} \tag{9}$$

The revised FA also requires the shortest distance d_i between a_i and f_i, which can be obtained by adding the projection of l_i on f_i to negative l_i, i.e.

$$d_i = -l_i + (l_i^T \hat{f}_i)\hat{f}_i \tag{10}$$

The normalised measure is given by:

$$\hat{\beta}_i = \frac{\beta_i}{\beta_{norm}} \tag{11}$$

where β_{norm} is chosen to be the stability criterion for the vehicle in the "home" pose. For more details on these derivations, the reader is referred to [15].

3.2   Terrain Modeling and Stability Axes Evaluation

The environment perception and terrain modeling is a fundamental problem that must take place before any path planning or navigation algorithm occurs. The terrains have been modeled by regular 2.5D Delaunay triangulation applied to the 3D point cloud provide by RGB-D camera. Figure 3 shows the mock-up Urban Search And Rescue (USAR) test arena and the generated 2.5D model. The dimensions of the USAR arena is 6 by 8 meters and the grid resolution (GR) used in the model was chosen at 4cm to guarantee accurate detailing. In order to make a fair comparison between the original and stable path planners, a pre-processing traversability algorithm is applied to the model in order to find out the areas obviously are not traversable, e.g. walls and steep slopes.

The contact footprint between the platform and the surface is defined by at most four points. Assuming a rigid body, the calculation of the contact footprint between the platform and the surface is done by assuming the robot sitting on a fake plane at a given orientation, with the size of the plane varying depending on the configuration of the flippers. The projection of this plane on the grided terrain results in a set of candidate points to define the stability axes. The red points depicted in Figure 4a show an example. The outermost points which form a convex support polygon would represent the configuration of the vehicle sitting on the terrain at that location. A maximum of four possible contact points are assigned to form the vertices of the projected plane, while at least 3 contact points are needed for the pose to be regarded as stable. If the configuration proves to be unstable, the plane is rotated around in a manner that CM's height will decrease. At that point, a new projection of the plane is calculated as before to establish whether the configuration is stable or not. Figure 4b shows an example for a resulting contact area with three tip-over axes.

4 Graph Theory Methods for Robot Path Planning

The extraction of shortest paths has been extensively surveyed in the literature, and it is not the objective of this paper. However, some background is hereby provided, which is by no means comprehensive but is useful for contextualizing the proposed stability-driven optimization search. Probably the canonical method for computing shortest paths on graphs or discretized settings is Dijkstra's algorithm [5]. To speed up the computations, some heuristics have been proposed that reduced the search space, and the best-first A* search algorithm[7] is extensively used in many search applications [11], particularly when a path from a given start point to a known goal needs to be calculated in real-time. However, it should be noted that A* requires different searches for each pair of start and goal points, thus other more efficient tree-search strategies such as IDA* [10] have also been proposed. For the case of Euclidean metrics, the exploitation of specific data structures have given rise to faster algorithms, such as Visibility Graphs [16] or Rapidly-Exploring Random Trees (RRTs) [20]. Despite savings in processing time, in general terms a notable constraint of these discrete-computation methods is the need of "smoothing" operators to produce realistic paths, as the solutions need to follow existing griding connections. A* and RRT's have been employed in this work to benchmark the proposed metric for planning stable paths. The next Sections provide an overview of the two methods, and the mechanism to incorporate the stability metric into the planning process.

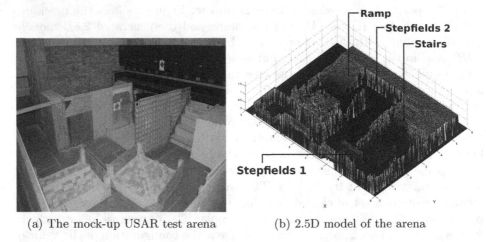

(a) The mock-up USAR test arena   (b) 2.5D model of the arena

Fig. 3. The mock-up USAR test arena and its 2.5D model

4.1 Rapidly-Exploring Random Trees Graph Generation

Rapidly-Exploring Random Trees (RRTs) [20] are a class of sampling based motion planners in that they take random samples from the configuration space to plan their paths. Under general conditions, they are regarded as probabilistically complete, i.e., given sufficient time, they will eventually find a solution. They were first proposed as efficient sampling based planners to solve path planning problems in high dimensional spaces, like in [12] where an RRT algorithm was able to plan paths for a 6-DOF manipulator. Further variations have addressed continuous cost functions such as those based on energy usage of a robot [9], or with a bias towards certain areas in both continuous and binary cost spaces [19]. The RRT method's key feature is that it will rapidly grow into unexplored areas within the configuration space of the robot, as it is heavily biased towards areas with large Voronoi regions. As RRTs keep the sampled nodes connected at all times, thereby forming a parent-child tree of expanding nodes in the environment, when the goal is reached, the algorithm can then quickly trace back the final path by going from the goal to the start point.

RRT with Stability Optimisation. Algorithm 1 summarises the pseudo code of adapted RRT algorithm used in this paper to plan a stable path from an initial position, Ni. The algorithm generates a new node (N_{new}) and checks if it is acceptable in terms of cost and collision. If N_{new} doesn't fulfill the conditions the generation and checking process will be repeated in a *while* loop to find an acceptable node. Then a search within the existing tree will be performed to find the nearest neighbors of the node. The minimum cost node among neighbor nodes will be selected and finally the new node is added to the tree. The nearest neighbor is assigned as the parent for the new node. This loop will run until either the goal is reached or until a predefined number of iterations k is reached.

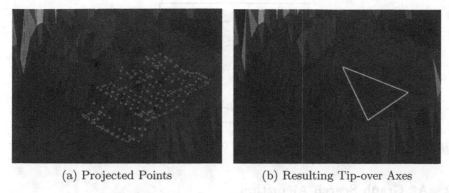

<table>
(a) Projected Points   (b) Resulting Tip-over Axes
</table>

Fig. 4. Calculation of the contact footprint between the platform and the terrain surface

---

Algorithm 1. The Proposed RRT algorithm
_____
1: create $RRT(N_i, k)$
2: $tree = N_i$
3: **for** $j = 1$ to k **do**
4:    N_{new} = create new random node
5:    **while** the cost of N_{new} is not acceptable **do**
6:       N_{new} = create new random node
7:    **end while**
8:    N_{near} = $nearest\_neighbours$
9:    find the minimum cost node(N_{near},N_{new})
10:   connect(N_{near},N_{new})
11:   add N_{new} to the $tree$
12: **end for**
13: return $tree$
_____

The implementation of the cost derived from the FA stability metric into the RRT algorithm is accomplished when considering the connect(N_{near},N_{new}) function. Instead of connecting the tree to the nearest neighbor based on Euclidean distance between the neighbor and a new node N_{new}, this function will now take the cost derived from the stability metric into consideration.

An example where three N_{near} nodes are considered at each iteration is given in Figure 5, where the grey area indicates uneven terrain (but not unstable, since unstable nodes will be immediately disregarded), and white represents relatively flatter areas. In this example, the three nearest Euclidean distance neighbors are shown within the dotted circle. However, as the two closest neighbor's added cost will be higher than that in the white area, the latter will be chosen as parent of the new node and added to the tree.

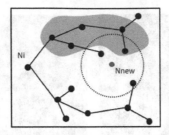

Fig. 5. Nearest Neighbour selection based on cost derived from the stability metric

4.2   A* Graph Search Algorithm

The A* algorithm [7] (and its many variants, D*, D* Lite etc) is a flexible heuristic algorithm, probably the most widely applied in graph searching applications to find minimum cost paths. In the A* searching mechanism, when planning begins the cells adjacent to the starting point are evaluated and ranked by a cost function $f(n) = g(n) + h(n)$ to select the cell with minimum cost that will be explored next. In this calculation, $g(n)$ is the accumulated cost of an optimal path from the start node to node n and $h(n)$ is an admissible (not overestimating) heuristic estimation of the remaining cost to get from node n to the goal. Thus, A* is a greedy best-first search algorithm which expands the most promising nodes first.

A* Search with Stability Optimization. The definition of the evaluation functions are critical to come up with effective paths. In the work proposed here, the heuristic function $h(n)$ considers the Euclidean distance from the node under consideration to the target. The $g(n)$ function has been weighted in favour of the FA stability criterion as

$$g(n) = \alpha * g_d(n) + (1 - \alpha) * g_s(n), \quad 0 \le \alpha \le 1 \tag{12}$$

where $g_d(n)$ is the distance term, $g_s(n)$ comes from the FA measure and α is a weighting factor. Note that by varraying α the emphasis can be set on stability or distance. The default configuration of the robot pose is to try and keep its arm orthogonal with respect to the global coordinate frame to provide the best possible field of view (for victim detection), while the flippers are prevented from touching the terrain unless it is necessary. If a stable solution can not be found along the path being considered, the algorithm will try to find a different path so that the stability of the robot can be guaranteed along the entire lenght of the path.

5   Results and Analysis

A distinctive example is provided to show the proposed algorithm in operation in a challenging and representative environment. The algorithm has been applied

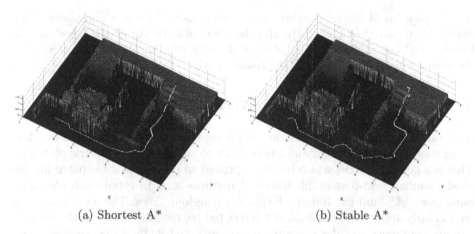

(a) Shortest A*   (b) Stable A*

Fig. 6. Resulted path from A* algorithm. Colour coding indicates height of the terrain from blue to red.

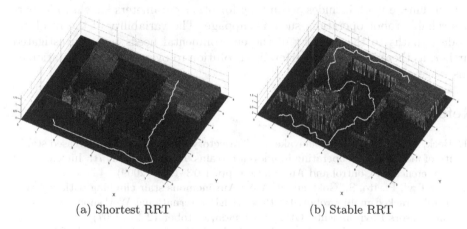

(a) Shortest RRT   (b) Stable RRT

Fig. 7. Resulted path from RRT algorithm. Colour coding indicates height of the terrain from blue to red.

to the modelled arena depicted in Figure 3. Results representative of A* and RRT are collected in Figures 6 and 7 respectively. For illustrative matters, the front walls of the arena have been removed, but they are considered during the planning procedure.

Figures 6a shows the output of the original A* algorithm, it can be seen stability is not guaranteed in the whole path especially where path is crossing over edges of step-fields and near the walls. The resulted path from the stable A* algorithm is depicted in Figures 6b, where it is shown how the proposed planner is able to come up with more stable geodesics than those obtained when stability is not considered. In particular higher stability costs result in paths that bypass on of the step fields. The path obtained from a non-stable RRT planner is depicted in Figures 7a and the result for the stable RRT is shown in Figures 7b. The stable path is longer and more winding, but it traverses over the ramp which is the most stable possible path in this arena.

6   Conclusions and Future Works

A variational computation based on the FA measure has been proposed as a formalization of using a stability metric in order to supplement geometric planning. This is a generic metric, which has been proved to provide feasible paths for the most common best-first graph search algorithms used in robot path planning, namely the A*, and the Rapidly-Exploring Random Trees. The methodology is particularly applicable, although not restricted to, reconfigurable platforms that can actively move to safer poses to reduce potential instabilities, such as those leading to vehicle tip-over when operating in uneven terrains. Simulations with real 3D data sets have been provided to demonstrate the performance of the algorithm with the quasi-static model of a real rescue vehicle operating in rough terrain. Future work includes accounting for other parameters known to jeopardies reliable robot operation, such as slippage. The variability of the resulting geodesic paths to the gridding of the environmental needs to be investigated further, as there is no known generalized solution to overcome this limitation at this stage.

References

1. Beck, C., Miro, J.V., Dissanayake, G.: Trajectory optimisation for increased stability of mobile robots operating in uneven terrains. In: Proc. IEEE 7th International Conference on Control and Automation, pp. 1913–1919 (2009)
2. Ben-Tzvi, P., Ito, S., Goldenberg, A.A.: Autonomous stair climbing with reconfigurable tracked mobile robot. In: Proc. IEEE International Workshop on Robotic and Sensors Environments, Ottawa, Canada, October 12-13 (2007)
3. Besseron, G., Grand, C., Amar, F.B., Bidaud, P.: Decoupled control of the high mobility robot hylos based on a dynamic stability margin. In: Proc. IEEE/RSJ International Conference on Intelligent Robots and Systems, Nice, France, September 22-23 (2008)
4. Chonnaparamutt, W., Birk, A.: A Fuzzy Controller for Autonomous Negotiation of Stairs by a Mobile Robot with Adjustable Tracks. In: Visser, U., Ribeiro, F., Ohashi, T., Dellaert, F. (eds.) RoboCup 2007. LNCS (LNAI), vol. 5001, pp. 196–207. Springer, Heidelberg (2008)
5. Dijkstra, E.W.: A note on two problems in connexion with graphs. Numerische Mathematik 1, 269–271 (1959)
6. Freitas, G., Gleizer, G., Lizarralde, F., Hsu, L.: Kinematic reconfigurability control for an environmental mobile robot operating in the amazon rain forest. Journal of Field Robotics 27, 197–216 (2010)
7. Hart, P.E., Nilsson, N.J., Raphael, B.: A formal basis for the heuristic determination of minimum cost paths. IEEE Transactions on Systems Science and Cybernetics 4, 100–107 (1968)

8. Inoue, D., Ohno, K., Nakamura, S., Tadokoro, S.: Whole-body touch sensors for tracked mobile robots using force-sensitive chain guides. In: Proc. IEEE International Workshop on Safety, Security and Rescue Robotics, Sendai, Japan (October 2008)
9. Jaillet, L., Cortés, J., Simoén, T.: Sampling-based path planning on configuration-space costmaps. IEEE Transactions on Robotics 26, 635–646 (2010)
10. Korf, R.: Depth-first iterative-deepening: An optimal admissible tree search. Artificial Intelligence 27(1), 97–119 (1985)
11. Nielsson, N.: Problem-Solving Methods in Artificial Intelligence. McGraw-Hill (1971)
12. Oh, K., Hwang, J.P., Kim, E., Lee, H.: Path planning of a robot manipulator using retrieval rrt strategy. World Academy of Science, Engineering and Technology 25, 171–175 (2007)
13. Papadopoulos, E.G., Rey, D.A.: A new measure of tipover stability margin for mobile manipulators. In: Proc. IEEE International Conference on Robotics and Automation, Minneapolis, USA, vol. 4, pp. 3111–3116 (1996)
14. Rey, D.A., Papadoupoulos, E.G.: Online automatic tipover prevention for mobile manipulators. In: Proc. IEEE/RSJ International Conference on Intelligent Robots and Systems, vol. 3, pp. 1273–1278 (1997)
15. Rey, D.A., Papadoupoulos, E.G.: The force angle measure of tipover stability margin for mobile manipulatiors. Vehicle System Dynamics 33, 29–48 (2000)
16. Rohnert, H.: Shortest path in the plane with convex polygonal obstacles. Information Processing Letters 23(2), 71–76 (1986)
17. Schenker, P., Huntsberger, T., Pirjanian, P., Dubowsky, S., Iagnemma, K., Sujan, V.: Rovers for agile, intelligent traverse of challenging terrain. In: Proceedings of the 7th International Symposium on Artificial Intelligence, Robotics and Automation in Space, i-SAIRAS, Nara, Japan (May 2003)
18. Shoval, S.: Stability of a multi tracked robot traveling over steep slopes. In: Proc. IEEE International Conference on Robotics and Automation, New Orleans, LA, USA (2004)
19. Urmson, C., Simmons, R.: Approaches for heuristically biasing rrt growth. In: Proc. IEEE/RSJ International Conference on Intelligent Robots and Systems, pp. 1178–1183 (2003)
20. Valle, S.L.: Rapidly-exploring random trees: A new tool for path planning. Tech. rep., Iowa State University, Dept. of Computer Science (1998)
21. Wang, W., Du, Z., Sun, L.: Dynamic load effect on tracked robot obstacle performance. In: Proc. of International Conference on Mechatronics WA1-B-4, Kumamoto, Japan (2007)

A Center of Mass Observing 3D-LIPM Gait for the RoboCup Standard Platform League Humanoid

Colin Graf[1] and Thomas Röfer[2]

[1] Universität Bremen, Fachbereich 3 – Mathematik und Informatik,
Postfach 330 440, 28334 Bremen, Germany
cgraf@informatik.uni-bremen.de
[2] Deutsches Forschungszentrum für Künstliche Intelligenz,
Sichere Kognitive Systeme, Enrique-Schmidt-Str. 5, 28359 Bremen, Germany
Thomas.Roefer@dfki.de

Abstract. In this paper, we present a walking approach for the Nao robot that improves the agility and stability of the robot when walking on a flat surface such as the soccer field used in the Standard Platform League. The gait uses the computationally inexpensive model of an inverted pendulum to generate a target trajectory for the center of mass of the robot. This trajectory is adapted using the observed real motion of the center of mass. This approach does not only allow compensating the inaccuracies in the model, but it also allows for reacting to external perturbations effectively. In addition, the method aims at facilitating a preferably fast walk while reducing the load on the joints.

1 Introduction

Since 2008, the humanoid robot *Nao* [4] that is manufactured by the French company Aldebaran Robotics is the robot used in the RoboCup Standard Platform League. The Nao has 21 degrees of freedom. It is equipped with a 500 MHz processor, two cameras, an inertial measuring unit, sonar sensors in its chest, and force-sensitive resistors under its feet. The camera takes 30 images per second while other sensor measurements are delivered at 100 Hz (50 Hz until 2009). The joints can be controlled at the same time resolution, i. e. walking means to generate 100 sets of 21 target joint angles per second.

Since the beginning of 2010, Aldebaran Robotics provides a gait for the Nao [4] that, although being a closed-loop walk, only takes the actual joint angles into account, not the measurements of the inertial measurement unit in Nao's chest. Thus the maximum speed reachable with the walk provided is still severely limited. As delivered by the manufacturer, it is approximately 10 cm/s. For RoboCup 2008, Kulk and Welsh designed an open-loop walk that keeps the stiffness of the joints as low as possible to both conserve energy and to increase the stability of the walk [9]. The gait reached 14 cm/s although it was based on the previous walking module provided by Aldebaran Robotics. Two groups

T. Röfer et al. (Eds.): RoboCup 2011, LNCS 7416, pp. 102–113, 2012.

worked on walks that keep the Zero Moment Point (ZMP) [13] above the support area using preview controllers. Both implement real omni-directional gaits. Czarnetzki *et al.* [1] reached speeds up to 20 cm/s with their approach. In their paper, this was only done in simulation. However, at RoboCup 2009 their robots reached similar speeds on the actual field, but they seemed to be hard to control and there was a certain lack in robustness, i.e., the robot fell down quite often. Strom *et. al* [12] modeled the robot as an inverted pendulum in their ZMP-based method. They reached speeds of around 10 cm/s.

In [6], [11], and [5], we already presented a robust closed-loop gait for the Nao. The active balancing used in the approach is based on the pose of the torso of the robot. In addition, we also presented an analytical solution to the inverse kinematics of the Nao, solving the problems introduced by the special hip joint of the Nao, i.e. dealing with the constraint that both legs share a degree of freedom in the hip. The gait presented in this paper is a continuation of this work.

The main contribution of this paper is presenting a computational inexpensive way for using the inverted pendulum model with dynamic phase duration for modeling a fast but omnidirectional and responsive walk and to use the same model to react to perturbations. Hence in addition to previous works, the focus is placed on using sensor feedback to observe the state of the robot and to adjust the inverted pendulum model to the observed state in order to improve the stability of the walk. The resulting walk is one of the fastest omnidirectional walks implemented on the Nao so far.

The structure of this paper is as follows: in the next section, modeling the walking robot as an inverted pendulum to control position and speed of its center of mass is discussed. In Section 3, the integration of sensor feedback is presented. Section 4 discusses the results achieved, followed by Section 5, which concludes the paper and gives an outlook on future work.

2   Using the Inverted Pendulum to Create Walking Motions

Generating a walking motion for humanoid robot basically means to create a sequence of joint angle sets, where each joint angle set will be executed successively. To be able to create a single set and a series of joint angles, a method is required to represent the state and the change of the state of the robot while it is walking. The approach presented in this paper reduces the model of the robot to its center of mass and uses the position and desired velocity of the center of mass to describe the state of the robot. The change of the state is described by determining a trajectory for the movement of the center of mass. From the position of the center of mass, the actual joint angles are determined by generating an additional trajectory for the position of the nonsupporting foot and by applying inverse kinematics to both legs (details are given in [5]).

To describe the movement of the center of mass, the 3-Dimensional Linear Inverted Pendulum Mode (3D-LIPM) [7] is used, which provides an approximation of a physically respectable model for the motion of the center of mass. In addition, changes in rotation, as they occur while rotating on the spot or while walking along a curve, are ignored. Hence, the position and velocity of the center of mass on a plane of height h in parallel to the ground relative to the origin of the inverted pendulum (see Fig. 1) are given by

$$x(t) = x_0 \cdot \cosh(k \cdot t) + \dot{x}_0 \cdot \frac{1}{k} \cdot \sinh(k \cdot t) \qquad (1)$$

$$\dot{x}(t) = x_0 \cdot k \cdot \sinh(k \cdot t) + \dot{x}_0 \cdot \cosh(k \cdot t) \qquad (2)$$

where $k = \sqrt{\frac{g}{h}}$, g is the gravitational acceleration ($\approx 9.81 \frac{m}{s^2}$), $x_0 \in \mathbb{R}^2$ is the position of the center of mass relative to the origin of the inverted pendulum at $t = 0$, and $\dot{x}_0 \in \mathbb{R}^2$ is the velocity of the center of mass at $t = 0$. A point under the currently supporting foot is used as origin of the inverted pendulum (see Fig. 4).

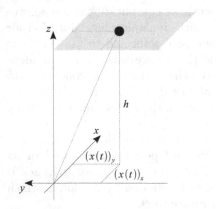

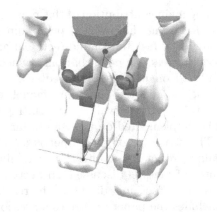

Fig. 1. An inverted pendulum in the three-dimensional space with fixed height h

Fig. 2. An inverted pendulum attached to an obliquely forwards walking simulated model of the Nao

In a single support phase, the inverted pendulum defines the motion of the center of mass according to its position and velocity relative to the origin of the inverted pendulum. Hence at the beginning of a single support phase, the position and velocity of the center of mass should be in a state that leads to the proper position and velocity for the next single support phase (of the other leg). The origin of the inverted pendulum should thereby be placed as close as possible to an optimal position under the foot (see Fig. 4). Since the step sizes to be performed can be chosen without severe constraints, the movement of the center of mass has to be adjusted for each step so the origins of the inverted pendulums used fit to the feet positions that are defined by the step sizes. Most walking approaches applied on the Nao [12,3,1] use a short double support phase for accelerating or decelerating the center of mass to achieve such an adjustment. To maximize the possible range that can be covered within a phase, the single support phase should make up as much as possible of the whole step phase to reduce the accelerations that are necessary for shifting the foot. Hence, the approach presented in this paper aims on eliminating the need of a double support phase, while keeping the origins of the inverted pendulums close to their optimal positions.

Even though no double support phase is used, a method to manipulate the movement of the center of mass is required. Therefore, the point in time for altering the support leg is used to control the velocity of the center of mass in the y (right→left) direction. To control the velocity in the x (back→front) direction, the origin of the inverted pendulum is shifted along the x-axis towards the elongated shape of the feet (see Fig. 4). This way the velocity of the center of mass can be manipulated enough to cover a specific distance (step size) while swinging from one leg to the other.

2.1   The System of Coordinates

Walking is a sequence of single support phases. In this paper, the symbols used to describe the current single support phase have no extra markings, while the symbols used to describe the following single support phase are dashed (e. g. x vs. $\bar{x}$). For each new single support phase, a new coordinate system Q is used to describe the set points of the center of mass and the feet positions. The origin of the inverted pendulum has the distance $r \in \mathbb{R}^2$ on the x-y-plane from the origin of $\bar{Q}$. This distance remains (almost) constant within a single support phase. The origin of Q is located between both feet so that a step size $\bar{s}$ describes the offset from the origin of Q to the origin of $\bar{Q}$ (see Fig. 3 and Fig. 4) where $\bar{Q}$ is the coordinate system Q of the upcoming single support phase. If the robot walks in place, the step size is 0 and Q is the same as $\bar{Q}$.

2.2   Computing Step Durations

To apply the functions (1) and (2) for generating walking motions, a definition of the point in time $t = 0$ is required to determine when to alter the support leg. $t = 0$ is defined as the inflection point of the pendulum motion where the y-component of the velocity is 0 ($(\dot{x}_0)_y = 0$). The position of the center of mass at this point $(x_0)_y$ is an arbitrary parameter and has a value of greater or lower than 0 depending on the active support leg. Since $(\dot{x}_0)_y = 0$, the function

$$x_y(t) = (x_0)_y \cdot \cosh(k \cdot t) \qquad (3)$$

in the range $t \geq t_b$ and $t \leq t_e$ can be used to compute the y-component of the center of mass position relative to the origin of the inverted pendulum. A single support phase starts at $t = t_b$ ($t_b < 0$) and ends at $t = t_e$ ($t_e > 0$).

If the nonsupporting foot should be placed with a distance of $\bar{r}_y + \bar{s}_y - r_y$ to the supporting foot at the end of the single support phase (see Fig. 5) and if $\bar{x}(\bar{t})$ and $\dot{\bar{x}}(\bar{t})$ are position and velocity of the center of mass relative to the next

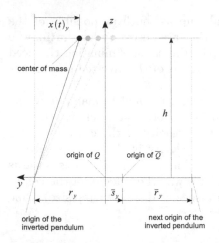

Fig. 3. y-z-cross section of the coordinate system used for the altering inverted pendulums

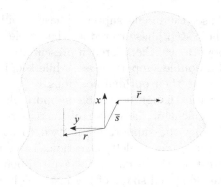

Fig. 4. x-y-cross section showing the step size $\bar{s}$ and the inverted pendulum origins r and $\bar{r}$. The small gray circle marks the foot position that is also referred as optimal inverted pendulum origin. The dotted line marks allowed inverted pendulum origins.

pendulum origin, the point in time to alter the support leg can be determined by finding the ending of a single support phase t_e where:

$$(x(t_e))_y - (\bar{x}(\bar{t}_b))_y = \bar{r}_y + \bar{s}_y - r_y \tag{4}$$

$$(\dot{x}(t_e))_y = (\dot{\bar{x}}(\bar{t}_b))_y \tag{5}$$

The ending of a single support phase t_e and a matching beginning of the next single support phase $\bar{t}_b$ cannot be found by simply solving equation (4) and (5) for t_e and $\bar{t}_b$. This is not possible since it cannot be assumed that the functions $(x(t))_y$ and $(\bar{x}(\bar{t}))_y$ are symmetric. To handle this problem an iterative method is used, in which t_e is initially guessed. The equation (5) can be transformed into

$$\bar{t}_b = \frac{1}{\bar{k}} \cdot \operatorname{arcsinh} \left(\frac{(x_0)_y \cdot k \cdot \sinh(k \cdot t_e)}{(\bar{x}_0)_y \cdot \bar{k}} \right) \tag{6}$$

to compute a value for $\bar{t}_b$ that matches to the guessed t_e. The guessed t_e can then be refined using the velocity of the center of mass at t_e and the length $(x(t_e))_y - (\bar{x}(\bar{t}_b))_y$.

2.3   Walking Forwards and Backwards

Up to now, the length of a single support phase and the y-component of the center of mass position at every point in time can be determined. To cover a step size $\bar{s}_x$, the origin of the inverted pendulum of the next single support phase should be placed in a distance of $\bar{r}_x - \bar{s}_x - r_x$ from the origin of the current inverted pendulum. In addition, the velocities of the center of mass relative to

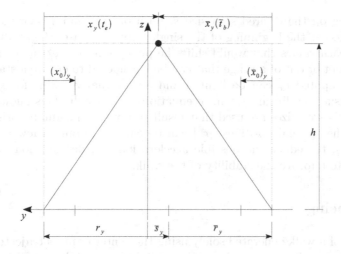

Fig. 5. Two facing inverted pendulum pendulums used to cover step size $\bar{s}_y$

each pendulum origin should be equal at the point in time when the support leg alternates. So, analogically to the equations (4) and (5) for the y-direction, the following equations should apply in the x-directions:

$$(x(t_e))_x - (\bar{x}(\bar{t}_b))_x = \bar{r}_x + \bar{s}_x - r_x \tag{7}$$

$$(\dot{x}(t_e))_x = (\dot{\bar{x}}(\bar{t}_b))_x \tag{8}$$

Two properties are planned ahead for the next single support phase. On the one hand, the position of the origin of the inverted pendulum should be optimal, so that $\bar{r}_x = 0$. On the other hand, the center of mass should be exactly over the next pendulum origin $(\bar{x}(0))_x = 0$ at $t = 0$. The latter is substantiated on simple forward walking ($\bar{s}_x = 0$) where $-t_b = t_e$, so that $(\bar{x}(0))_x = 0$ guarantees an evenly distributed center of mass motion that allows using optimal inverted pendulum origins (r_x and $\bar{r}_x = 0$) when walking with a constant step size. In order to cover a distance of $\bar{s}_x$, the origin of the inverted pendulum r_x of the current single support phase is chosen in a way that equation (7) applies.

When the center of mass has the position x_{t_b} and velocity $\dot{x}_{t_b}$ relative to the origin of Q at the beginning of a single support phase, r_x can be computed by using the following linear system of equations:

$$
\begin{aligned}
r + x_0, \cosh(k\,t) \;+\; \dot{x}_0\,\tfrac{\sinh(k\,t)}{k} &= x_t \\
x_0\,k\sinh(k\,t) \;+\; \dot{x}_0\cosh(k\,t) &= \dot{x}_t \\
x_0\,k\sinh(k\,t_e) + \dot{x}_0\cosh(k\,t_e) - \bar{x}_0\,\bar{k}\sinh(\bar{k}\,\bar{t}_b) - \dot{\bar{x}}_0\cosh(\bar{k}\,\bar{t}_b) &= [0,0]^T \\
r + x_0\cosh(k\,t_e) \;+\; \dot{x}_0\,\tfrac{\sinh(k\,t_e)}{k} \;-\; \bar{x}_0\cosh(\bar{k}\,\bar{t}_b) \;-\; \dot{\bar{x}}_0\,\tfrac{\sinh(\bar{k}\,\bar{t}_b)}{k} - \bar{s} &= \bar{r}
\end{aligned}
\tag{9}
$$

It is not only possible to compute r_x using the linear system of equations (9), but also to compute $(x_0)_x$ and $(\dot{x}_0)_x$, so that a complete set of pendulum parameters (r, x_0 and $\dot{x}_0$) can be determined.

Depending on the desired step size $\bar{s}$, position x_{t_b}, and velocity $\dot{x}_{t_b}$ of the center of mass at the beginning of the single support phase, the absolute value of r_x can reach values that would shift the pendulum origin out of the convex hull of the foot or out of a range that can be considered to ensure a stable walk. Hence, a computed r_x can be limited and an alternative value for $\bar{s}_x$ can be computed using the linear system of equations (9) as well. This allows making sure that only step sizes are used that result in inverted pendulum origins close enough to the optimal positions. $|r_x|$ can be capped to only a few millimeters (e. g. 4 mm), to reduce the possible acceleration and deceleration of walking speeds and to improve the stability of the walk.

3 Balancing

The results of a walk generated solely using the center of mass trajectory as described in Section 2 are not convincing. It might be possible to find parameters that keep the robot upright and that allow slow locomotion, but it is obvious that the model alone is not suitable to keep the walk permanently stable. Furthermore, the robot is not capable to react on perturbations of any kind.

Without balancing the trajectory of the center of mass is static and can only be executed as it was computed before. In the case that an external perturbation affects the robot or when the model used is simply not precise enough to represent the dynamics of the robot, the motion of the center of mass does not follow the trajectory as intended and the robot may fall in any direction. There are several approaches to handle this problem. As soon as a deviation in the motion of the center of mass is detected, a counteraction can be executed to bring the center of mass back to the desired position. Another approach is to compute a slightly modified trajectory for the center of mass to continue the deviated motion according to the inverted pendulum model and to adjust future steps to the deviated motion. The walk presented in this paper uses the second approach. Using the first approach, it is hardly possible to compensate a perturbation completely, but it is quite useful to prevent the error from increasing with further movement of the center of mass. The advantage of the second approach is that it acts in a more farsighted manner. If an error is in the system, it is absorbed to continue the intended step as stable as possible. But at first, both approaches require observing the actual position of the center of mass to detect an error.

3.1 Observing the Center of Mass

For determining an observed position of the center of mass, a four-dimensional (simple) Kalman filter [8] is used for the x and y-components of the position and the velocity of the center of mass. It is actually implemented using two independent two-dimensional Kalman filters. Estimating the z-component is left out, since the 3D-LIPM used can not handle any dynamic pendulum heights. Instead of computing the sensor readings for a predicted center of mass position, the error Δx_i between the expected center of mass position x_{e_i} relative to Q and the measured center of mass position x_{mi} at time t_i are computed and used as innovation. The expected center of mass position is computed using $x(t)$ with the inverted pendulum parameters of the current single support phase. The measured center of mass position x_{mi} is basically computed by using an estimated orientation of the robot torso and the kinematic chain to the supporting foot that is constructed using the joint angle sensor readings of the Nao.

$$x_{ei} = r + x(t_i) \tag{10}$$

$$\Delta x_i = x_{mi} - x_{ei} \tag{11}$$

The Kalman filter uses x_{ei} and the function $\dot{x}(t_i)$ with the inverted pendulum parameters of the current single support phase as the predicted state μ_i'. The covariance Σ_i' of the predicted state is computed as in an ordinary four-dimensional Kalman filter that estimates a position and a velocity of an object in two-dimensional space with a process noise $\Sigma_{\varepsilon i}$.

$$\mu_i' = \begin{bmatrix} (x_{ei})_x \\ (x_{ei})_y \\ (\dot{x}(t_i))_x \\ (\dot{x}(t_i))_y \end{bmatrix}, \Sigma_i' = A \cdot \Sigma_{i-1} \cdot A^T + \Sigma_{\varepsilon i}, A = \begin{bmatrix} 1 & 0 & \Delta t_i & 0 \\ 0 & 1 & 0 & \Delta t_i \\ 0 & 0 & 1 & 0 \\ 0 & 0 & 0 & 1 \end{bmatrix}, \Delta t_i = t_i - t_{i-1} \tag{12}$$

The Kalman gain K_i for the innovation Δx_i can be computed assuming covariance Σ_{mi} for the measurement x_{mi}.

$$K_i = \Sigma_i' \cdot C^T \cdot (C \cdot \Sigma_i' \cdot C^T + \Sigma_{mi})^{-1} \quad \text{with } C = \begin{bmatrix} 1 & 0 & 0 & 0 \\ 0 & 1 & 0 & 0 \end{bmatrix} \tag{13}$$

And finally, the filtered position x_{f_i}, the filtered velocity $\dot{x}_{f_i}$, and the updated covariance Σ_i can then be computed.

$$\mu_i = \begin{bmatrix} (x_{f_i})_x \\ (x_{f_i})_y \\ (\dot{x}_{f_i})_x \\ (\dot{x}_{f_i})_y \end{bmatrix} = \mu_i' + K_i \cdot \Delta x_i, \Sigma_i = \Sigma_i' - K_i \cdot C \cdot \Sigma_i' \tag{14}$$

The parameters of the Kalman filters, i.e. the assumed process noise $\Sigma_{\varepsilon i}$ and the deviation of the computed error Σ_{mi}, can control how much sensor feedback is used to correct the pendulum parameters (see Section 3.2). Using a small process noise and a large deviation of the computed error results in only little corrections according to the measured position of the center of mass.

3.2 Correcting the Inverted Pendulum Parameters

The filtered position x_{f_i} and the filtered velocity $\dot{x}_{f_i}$ of the Kalman filter that estimates the true position and velocity of the center of mass are used to re-determine the parameters of the inverted pendulum of the current single support phase. Since the prediction of the natural motion of the center of mass is based on the pendulum parameters, the main purpose of the correction is to improve the prediction in the next iteration. So the pendulum parameters r, x_0, $\dot{x}_0$, and t_i are re-determined to find a pendulum function that fits to the estimated position and the estimated velocity of the center of mass:

$$r' + x'(t_i') = x_{f_i} \tag{15}$$

$$\dot{x}'(t_i') = \dot{x}_{f_i} \tag{16}$$

The y-component $(x_0)_y'$ of the center of mass at the point in time $t = 0$ as well as the corrected current time t_i' can be computed using the estimated position and the estimated velocity. $(x_0)_y'$ is calculated by using equation (3) and its derivation:

$$(x_f)_y - r_y' = (x_0)_y' \cdot \cosh(k' \cdot t_i') \tag{17}$$

$$(\dot{x}_f)_y = (x_0)_y' \cdot k' \cdot \sinh(k' \cdot t_i') \tag{18}$$

Equation (17) can be solved for $k' \cdot t_i'$ and inserted into (18) to solve the resulting equation for $(x_0)_y'$:

$$(x_0)_y' = \sqrt{((x_{f_i})_y - r_y')^2 - \frac{(\dot{x}_{f_i})_y{}^2}{k'^2}} \tag{19}$$

t_i' can then be computed by solving equation (18) for t_i':

$$t_i' = \frac{1}{k'} \cdot \operatorname{arcsinh}\left(\frac{(\dot{x}_f)_y}{k' \cdot (x_0)_y}\right) \tag{20}$$

Given the corrected y-component of the pendulum position at $t' = 0$ (which is possibly shifted to $t = 0$) a corrected point in time t_e' for the ending of the current single support phase can be computed using the iterative method as described in Section 2.2 with the current step size $\bar{s}$. The x-components of the estimated position x_{f_i} and the estimated velocity $\dot{x}_{f_i}$ can be used for computing the corrected pendulum origin r_x' and the other pendulum parameters $(x_0)_x'$ and $(\dot{x}_0)_x'$ by using the linear system of equations (9).

3.3   Controlling the Predicted Positions of the Center of Mass

A general problem with balancing a walk is that the sensor readings are not in sync with the controlled joint angles. On the Nao, 40 ms or 4 cycles in the walk generation process elapse until a reaction from a joint angle request becomes recognizable. Hence, a pendulum motion in sync with the measurements is considered at first. At this frame, the estimated position and the estimated velocity of the center of mass are used to compute the corrected pendulum parameters. To control a position of the center of mass x_{p_i}, the position of the center of mass 40 ms in the future is predicted according the corrected pendulum parameters. If the 40 ms exceed the ending of the current single support phase, the pendulum parameters of the next single support phase are used instead:

$$x_{p_i} = \begin{cases} r' + x'(t_i' + 40\text{ms}) & \text{if } t_i' + 40\text{ms} < t_e' \\ \bar{s} + \bar{r} + \bar{x}(\bar{t}_b' + t_i' + 40\text{ms} - t_e') & \text{otherwise} \end{cases} \qquad (21)$$

4   Results

To find parameters for the gait, the robot was walking in a stationary position and some initial parameters were slightly changed to minimize the average error between expected and measured center of mass position and to result in a preferably constant and smooth motion from side to side (see Fig. 6). Some parameters define the stance and the motion with values such as the height of the center of mass above the ground, the offset between both feet (normally 10 cm because of the robot's leg design), the step height, and the magnitude of a step-bound rotation around the x-axis to simplify the center of mass shifting to a side. Furthermore, there are parameters that define the motion of the inverted pendulum such as the pendulum width $(x_0)_y$, the position of the optimal pendulum origin within a foot r_y, and the assumed height of the pendulum h, which can be chosen independently from the actual position of the center of mass to adjust the inverted pendulum model to the actual dynamics of the robot. The parameters can be chosen by hand or with an automatic method such as the Particle Swarm Optimization [2] that uses the averaged error between the expected and the measured position of the center of mass to rate the quality of a set of parameters.

The walking parameters used in the work presented aim at compromising between a maximum walking speed and a minimum load on joints. Hence, the center of mass is located quite high at 262 mm over ground, so that thigh and lower leg stand with an obtuse angle to each other to reduce the load on the knee joint. This can be a huge advantage when the Nao should stand or walk for a long period of time, since the knee joint can overheat quickly at a sharper angle due to the higher load (e. g. in less than 20 minutes, i. e. the duration of a game). Another advantage of the high center of mass is that—compared to a lowered stance—smaller changes in the joint angles are necessary to reach target foot positions that are far away. But a drawback of the high center of mass is that the maximum reachable target foot distance is more limited by the maximum length of a leg.

Using the best performing parameters found so far, the robot reaches 31 cm/s forwards with a step size of 7 cm, 12 cm/s sideways with a step size of 8 cm, 22 cm/s backwards, and 92°/s when rotating on the spot. At full speed, the average error of the position of the center of mass in the y-direction is 7.76 mm. The error is smaller with reduced walking speed (e. g. 5.88 mm at 10 cm/s forwards). With reduced walking speed, the gait is also substantially more robust against perturbations. The correction of the inverted pendulum parameters causes the motion of the center of mass to adjust quickly to the expected trajectory (see Fig. 7).

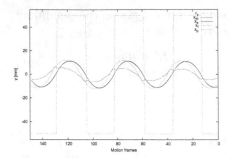 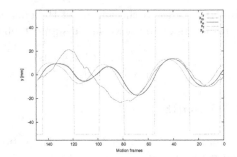

Fig. 6. A plot showing the y-component of the measured (red) and expected (black) center of mass position while the robot was walking on the spot. The motion of the robot is steady although the average error appears to be immense (4.26 mm).

Fig. 7. A plot showing the y-component of the measured (red) and expected (black) center of mass position when the robot was pushed from the side. The expected center of mass position chases the measured position and the walk stabilizes quickly.

5   Conclusions and Future Work

In this paper, we present a robust closed-loop gait for the Nao robot. The gait uses the center of mass as simplified representation of the walking robot and a model for the movement of this center of mass that is based on two alternating inverted pendulums. The model allows eliminating the need of a double-support phase by dynamically adjusting the point in time at which the support leg alternates. Thus, the load on the joints for bridging over larger distances can be reduced.

In addition, we have suggested a method for estimating the actual position and the actual velocity of the center of mass. We have explained how the estimate can be used for correcting the inverted pendulum model. The correction does not only allow using the inverted pendulum model on real hardware without perfect joint calibration, but it also adds robustness against perturbations such as forces exerted on the robot. The maximum speed achieved could be increased in comparison to the results of previous works.

Work that was already started for RoboCup 2010 and is still ongoing is to add a mechanism to learn the difference between the inverted pendulum model and the real dynamics of the robot. Furthermore, we plan to extend the walking engine to a general motion engine that, e. g., will also integrate dynamic kicks as presented by [10]. This will significantly improve the robot's ability to dribble the ball and speed up the transitions between walking and kicking.

Achnowledgement. The authors would like to thank all B-Human team members for providing the software base for this work.

References

1. Czarnetzki, S., Kerner, S., Urbann, O.: Observer-based dynamic walking control for biped robots. Robotics and Autonomous Systems 57(8), 839–845 (2009)
2. Eberhart, R.C., Kennedy, J.: A new optimizer using particles swarm theory. In: Sixth International Symposium on Micro Machine and Human Science, pp. 39–43 (1995)
3. Gouaillier, D., Collette, C., Kilner, C.: Omni-directional closed-loop walk for NAO. In: 2010 10th IEEE-RAS International Conference on Humanoid Robots (Humanoids), pp. 448–454 (December 2010)
4. Gouaillier, D., Hugel, V., Blazevic, P., Kilner, C., Monceaux, J., Lafourcade, P., Marnier, B., Serre, J., Maisonnier, B.: The NAO humanoid: a combination of performance and affordability. CoRR abs/0807.3223 (2008)
5. Graf, C., Röfer, T.: A closed-loop 3D-LIPM gait for the RoboCup Standard Platform League humanoid. In: Zhou, C., Pagello, E., Behnke, S., Menegatti, E., Röfer, T., Stone, P. (eds.) Proceedings of the Fourth Workshop on Humanoid Soccer Robots in Conjunction with the 2010 IEEE-RAS International Conference on Humanoid Robots (2010)
6. Graf, C., Härtl, A., Röfer, T., Laue, T.: A Robust Closed-Loop Gait for the Standard Platform League Humanoid. In: Zhou, C., Pagello, E., Menegatti, E., Behnke, S., Röfer, T. (eds.) Proceedings of the Fourth Workshop on Humanoid Soccer Robots in Conjunction with the 2009 IEEE-RAS International Conference on Humanoid Robots, Paris, France, pp. 30–37 (2009)
7. Kajita, S., Kanehiro, F., Kaneko, K., Fujiwara, K., Yokoi, K., Hirukawa, H.: A realtime pattern generator for biped walking. In: Proceedings of the 2002 IEEE International Conference on Robotics and Automation (ICRA 2002), Washington, D.C., USA, pp. 31–37 (2002)
8. Kalman, R.E.: A new approach to linear filtering and prediction problems. Transactions of the ASME–Journal of Basic Engineering 82(Series D), 35–45 (1960)
9. Kulk, J.A., Welsh, J.S.: A low power walk for the NAO robot. In: Kim, J., Mahony, R. (eds.) Proceedings of the 2008 Australasian Conference on Robotics & Automation, ACRA 2008 (2008)
10. Müller, J., Laue, T., Röfer, T.: Kicking a Ball – Modeling Complex Dynamic Motions for Humanoid Robots. In: Ruiz-del-Solar, J., Chown, E., Ploeger, P.G. (eds.) RoboCup 2010. LNCS (LNAI), vol. 6556, pp. 109–120. Springer, Heidelberg (2010)
11. Röfer, T., Laue, T., Müller, J., Burchardt, A., Damrose, E., Fabisch, A., Feldpausch, F., Gillmann, K., Graf, C., de Haas, T.J., Härtl, A., Honsel, D., Kastner, P., Kastner, T., Markowsky, B., Mester, M., Peter, J., Riemann, O.J.L., Ring, M., Sauerland, W., Schreck, A., Sieverdingbeck, I., Wenk, F., Worch, J.H.: B-Human team report and code release 2010 (2010),
http://www.b-human.de/file_download/33/bhuman10_coderelease.pdf
12. Strom, J., Slavov, G., Chown, E.: Omnidirectional Walking Using ZMP and Preview Control for the NAO Humanoid Robot. In: Baltes, J., Lagoudakis, M.G., Naruse, T., Ghidary, S.S. (eds.) RoboCup 2009. LNCS (LNAI), vol. 5949, pp. 378–389. Springer, Heidelberg (2010)
13. Vukobratovic, M., Borovac, B.: Zero-moment point – thirty five years of its life. International Journal of Humanoid Robotics 1(1), 157–173 (2004)

Ball Interception Behaviour in Robotic Soccer

João Cunha, Nuno Lau, and João Rodrigues

Universidade de Aveiro

Abstract. In robotic soccer the ball is the most crucial factor of the game. It is therefore extremely important for a robot to retrieve it as soon as possible. Thus ball interception is a key behaviour in robotic soccer. However, currently most MSL teams move to the ball position without considering the ball velocity. This often results in inefficient paths described by the robot. This paper presents the CAMBADA solution for a ball interception behaviour based on a uniformly accelerated robot model, where not only the ball velocity is taken into account but also the robot current velocity as well as the robot acceleration, maximum velocity and sensor-action delays are considered. The described work was introduced in the Portuguese robotics open Robótica2009 and RoboCup 2009 and improved the team performance contributing to the first and third places, respectively.

1 Introduction

Robotic soccer, much like human soccer, revolves around a key aspect of the game: the ball. To win, a team must make an efficient use of it, either by passing it to a team-mate, by dribbling it towards the goal, or by shooting it on goal in order to score. However even the most technically evolved robot cannot perform such moves if it doesn't regain possession of the ball as fast as possible. In order to obtain the ball, depending on the situation, a robot must catch a loose ball, receive a pass or tackle it from an opponent. However, most MSL teams, currently move to the ball position, without considering its velocity. Given the increasingly dynamic aspect of the game over recent years, the ball is constantly moving therefore this method doesn't provide the most efficient path to regain ball possession. This paper describes the developed solution implemented within the framework of the CAMBADA project at the University of Aveiro.

The CAMBADA is the University of Aveiro robotic soccer team competing in the RoboCup [4] Middle Size League. The project started in 2003 by researchers of IEETA[1] ATRI[2] research group and students from DETI[3] of the University of Aveiro. The multidisciplinary project includes diverse research areas such as

---

[1] Instituto de Engenharia Electrónica e Telemática de Aveiro - Aveiro's Institute of Electronic and Telematic Engineering.

[2] Actividade Transversal em Robótica Inteligente - Transverse Activity on Intelligent Robotics.

[3] Departamento de Electrónica, Telecomunicações e Informática - Electronics, Telecomunications and Informatics Department.

T. Röfer et al. (Eds.): RoboCup 2011, LNCS 7416, pp. 114–125, 2012.

image analysis and processing, control, artificial intelligence, multi-agent coordination and sensor fusion. Since its origin, the CAMBADA team has competed in several national and international competitions having won the last four national championships as well as the 2008 edition of RoboCup World Championship. More recently, the CAMBADA team placed in third in both RoboCup 2009 in Graz, Austria and RoboCup 2010 in Singapore.

The CAMBADA team is composed by six robots designed to play soccer. Competing in the RoboCup's Middle Size League, the CAMBADA robots must not exceed the maximum dimensions of $52cm \times 52cm \times 80cm$. The rules however don't impose a particular shape leaving that decision to each team. CAMBADA robots have a conical shape with a base radius of $24cm$ and are $71cm$ high as can be seen in Fig 1.

Fig. 1. A CAMBADA robot

The remainder of this paper is organized as follows: Section 2 describes the CAMBADA software architecture upon which the interception behaviour was developed. Section 3 presents the notion of a behaviour in the CAMBADA context. The interception behaviour implementation is detailed in Section 5. Section 6 discusses the obtained results. Finally, Section 7 presents the conclusions.

2   CAMBADA Architecture

The CAMBADA robots were designed and built at the University of Aveiro. The hardware is distributed in three layers which facilitate replacement and maintenance.

The top layer has the robot's vision system. The CAMBADA robots have an omni-directional vision obtained by means of a CCD camera pointed upwards towards an hyperbolic mirror which enables a robot to see in 360 degrees[10][3].

The middle layer houses the processing unit, currently a 12" laptop, which collects the data from the sensors and computes the commands provided to the actuators. The laptop executes the vision software along with all high level and decision software and can be seen as the brain of the robot. Given the positional advantage, a ball retention device is placed on this layer.

A network of micro-controllers is placed beneath the middle layer to control the low-level sensing/actuation system, or the nervous system of the robot. The sensing and actuation system is highly distributed, meaning that each node in the network controls different functions of the robot, such as, motion, odometry, kick, compass and system monitor.

The lower layer is composed by the robot motion system and kicking device. The robots move with the aid of a set of three omni-wheels, disposed at the periphery of the robot at angles that differ 120 degrees from each other, powered by three 24 V / 150 W Maxon motors. On this layer there is also an electromagnetic kicking device. Also, for ball handling purposes, a barrier sensor is installed underneath the robot's base, that signals the higher level that the ball is under control.

In the context of an interception, omni-directional vision offers great advantages over other kinds of vision since the robot doesn't need to reposition the camera or itself to see the ball.

The locomotion system is also a factor that greatly affects a wheeled robot ability to intercept the ball as an holonomic motion robot can move to the interception point with the front of the robot oriented towards the ball. It would be far more complex for a robot to intercept a ball using Ackerman steering or differential motion.

It is no surprise that the MSL has evolved towards these types of vision and motion systems, used by almost every team, as they offer significant advantages concerning ball detection and consequent interception.

Following the CAMBADA hardware approach, the software is also distributed. Therefore, five different processes are executed concurrently. All the processes run at the robot's processing unit in Linux.

All processes communicate by means of an RTDB[4] which is physically implemented in shared memory. The RTDB is a data structure which contains the essential state variables to control the robot. The RTDB is divided in two regions, the local and shared regions.

The local section holds the data needed by the local processes and is not to be broadcasted to the other robots. The shared section is divided between all running agents to contain the data of the world state as perceived by the team. Each sub-divided area is allocated to one robot where it stores the perceived state of the world. There is also one sub-divided area specific for the coach information. As the name implies the shared section is broadcasted through the

---

[4] Real-Time DataBase.

team, as each agent transmits the owned sub-divided shared section, achieving information sharing between the team.

The RTDB implementation guarantees the temporal validity of the data, with small tolerances [1].

The software architecture is depicted in Fig 2.

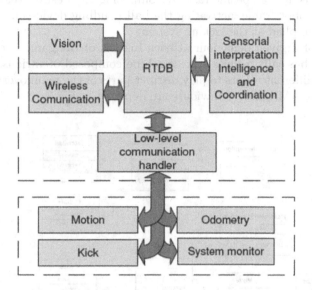

Fig. 2. The software architecture, adapted from [2]

The processes composing the CAMBADA software are:

Vision which is responsible for acquiring the visual data from the cameras in the vision system, processing and transmitting the relevant info to the CAMBADA agent. The transmitted data is the position of the ball, the lines detected for localization purposes and obstacles positions. Given the well structured environment the robots play in, all this data is currently acquired by color segmentation [10][9].

Agent is the process that integrates the sensor information and constructs the robot's worldstate. The agent then decides the command to be applied, based on the perception of the worldstate, accordingly to a pre-defined strategy [5].

Comm that handles the inter-robot communication, receiving the information shared by the team-mates and transmitting the data from the shared section of the RTDB to the team-mates [13][14].

HWcomm or hardware communication process is responsible for transmitting the data to and from the low-level sensing and actuation system.

Monitor that checks the state of the remaining processes relaunching them in case of abnormal termination.

Given the real-time constraints, all process scheduling is handled by a library specifically developed for the task, *pman*, process manager [11].

3 Behaviours

The different CAMBADA behaviours represent the basic tasks to be performed by the robot, such as move to a position in the field, dribble or kick the ball. A behaviour can then be seen as the basic block of a CAMBADA robot attitude. A behaviour executes a specific task by computing the desired velocities to be applied at the robot frame, activating the ball handling device and the desired strength to be applied at the kicking system.

The choice of a given behaviour at each instant of the game is executed by a role which is basically a finite-state machine composed of various behaviours that allow the different robots to play distinct parts of the team overall strategy.

The various CAMBADA behaviours are depicted in Fig 3.

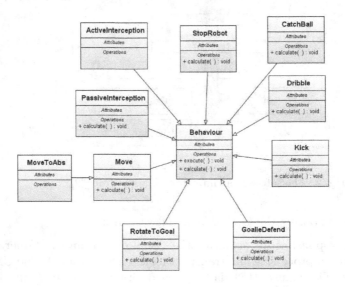

Fig. 3. Class diagram of all CAMBADA behaviours

All the CAMBADA behaviours derive from a generic behaviour class *Behaviour*. This class implements the method *execute* which inserts in the RTDB the different values that will later be translated to the powers to be applied at the various robot actuators.

Since the control carried out on the ball handling system is on/off, method *grabberControl* implemented on the *Behaviour* class activates the device based on the position of the ball and when the ball is engaged.

All the derived behaviours have the responsability of calculating the desired velocities to be applied at the robot frame and the behaviour *Kick* in particular calculates the strength to be applied at the kicking system.

4 Sensor Fusion and World Modelling

In order to efficiently intercept a rolling ball, a robot needs to have an accurate and robust perception of its surrounding environment with great emphasis on the ball position and velocity. Furthermore a precise self-localization and ego-motion estimates are also required.

In order to localize itself on the field, the robot uses information gathered from the field lines seen by the omni-vision camera. From this information, the robot applies an error minimization method described in [6]. The robot velocity is given by applying a linear regression over the robot recent past positions.

To estimate the ball position, a Kalman filter is applied to filter the noise observed in the information retrieved by the visual sensor. This yields smoother ball trajectories. Although the Kalman filter is able to provide the robot velocity, the resulting estimation has slow convergence when the ball movement changes direction. This is specially critical since during the highly dynamic environment of the game the ball is constantly changing direction. To address this issue a linear regression over the Kalman estimates with an adaptive buffer size has been implemented. The linear regression keeps track of the recent ball positions. However when the ball position becomes inconsistent with the ball velocity the older ball positions are discarded. Using this method faster convergence is achieved when the ball movement is changed [15][7].

5 Implementation

The ability to intercept the ball in its path is of major importance in the robotic soccer context. The alternative, moving to the current ball position, is by no means optimal, since a robot takes more time to catch the ball and in some cases it might not even catch it. Fig 4 shows a possible robot path when the it moves to the estimated ball position.

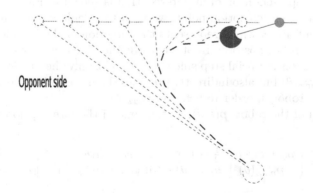

Fig. 4. Described path when the robot moves to the estimated ball position

When considering that the ball could be dribbled by an opponent robot in the goal's direction, not intercepting the ball could have severe negative consequences. The problem scales as the other teams strive to improve their robot's top speed.

5.1   Related Work

There are some solutions to the ball interception problem in the robotic soccer RoboCup scenarios, especially in the simulation league. In [12] this problem is solved by approximating the robot movement to an uniform movement with velocity v_r. A different solution is proposed by [16] abstracts the robot of the numerical values of the robot relative position to the ball and ball velocity using qualitative descriptions of the environment. Despite the existent solutions in simulated environments few teams in the Middle Size League exhibit ball interception skills. An exception is the Brainstormers Tribots team that was able to teach a robot to intercept a rolling ball using Reinforcement Learning [8]. The ISocRob team [17] has also proposed a solution similar to the one presented in [12]. While this solution is able to intercept a ball in some situations it assumes that the robot can instantaneously achieve v_r. As the velocity of robots in the MSL increases this approximation tends to provide worse results.

5.2   Proposed Solution

The proposed solution assumes an uniformly accelerated kinematic model instead of a uniform movement kinematic model. The value of the maximum acceleration imposed on the robot movement is known, $a_{max} = 3m/s^2$. Therefore the acceleration and current speed are taken into account in the interception point calculation, as well as the robot's maximum speed, $speed_{max}$. Since the maximum velocity depends on the direction of the movement and wheel slippage and slacks, its value was empirically obtained. The value used in the CAMBADA team is $1.8m/s$.

A geometric representation of an interception is shown in Fig 5.

To determine the interception point, we need to calculate t the time to intercept the ball, but also θ which represents the direction of the interception point in respect to the robot position. In other words the direction to where the robot will. Calculating θ is a crucial step since it gives, not only the maximum velocity of the robot $v_{max}(\theta)$ but also indirectly provides the direction of the acceleration imposed on the robot in order to achieve $v_{max}(\theta)$.

The position of the robot, $p(t, \theta)$, at the time of the interception is given by,

$$p(t,\theta) = \begin{cases} p_0 + v_0 \cdot t + \frac{1}{2} \cdot a(\theta) \cdot t^2 \text{ if } \|v_0 + a(\theta) \cdot t\| \leq \|v_{max}(\theta)\|; \\ p_{max}(\theta) + v_{max}(\theta) \cdot t' \text{ if } \|v_0 + a(\theta) \cdot t\| \geq \|v_{max}(\theta)\|. \end{cases} \quad (1)$$

where

$$v_{max}(\theta) = \{speed_{max} \cdot \cos(\theta), speed_{max} \cdot \sin(\theta)\} \quad (2)$$

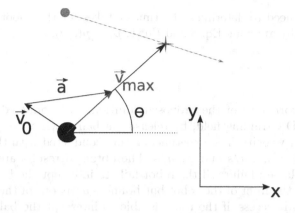

Fig. 5. Geometric representation of an interception

$$t_{max} = \frac{\|v_{max}(\theta) - v_0\|}{a_{max}} \tag{3}$$

$$a(\theta) = \{a_{max} \cdot \cos(\phi(\theta)), a_{max} \cdot \sin(\phi(\theta))\} \tag{4}$$

$$\phi(\theta) = \arctan(v_{0_y} - v_{max_y}(\theta), v_{0_x} - v_{max_x}(\theta)) \tag{5}$$

$$t' = t - t_{max} \tag{6}$$

$$p_{max}(\theta) = p_0 + v_0 \cdot t_{max} + \frac{1}{2} \cdot a(\theta) \cdot t_{max}^2 \tag{7}$$

This means that the robot will move according to an uniformly accelerated movement with aceleration $a(\theta)$ until its velocity $v_0 + a(\theta) \cdot t$ saturates, which happens at moment t_{max}. At this point the robot should be at $p_{max}(\theta)$. From this moment on the robot will move according to an uniform movement with its maximum velocity $v_{max}(\theta)$.

Since the Eq 1 is non-linear, a numerical method would be required to find an interception point $p(t, \theta)$. To simplify the calculations performed by the robot, an iterative (hill-climbing) solution was developed. The solution tests consecutive points in the ball path. A valid interception point is found when the robot reaches the considered point before the ball.

The consecutive points are generated using a time step of 0.1 seconds. Since the ball is assumed to move according to an uniform movement model with velocity v_b, the i-th iteration interception test point p_i is given by,

$$p_i = p_b + v_b \cdot i \cdot 0.1 \tag{8}$$

The time the ball takes to reach the considered is directly obtained by $0.1 \cdot i$. On the other hand by testing a specific point, we can obtain the direction the robot will move, θ, which is given by

$$\theta = \arctan(p_{i_y} - p_{o_y}, p_{i_x} - p_{0_x})) \tag{9}$$

Hence, we only need to determine the time it takes for the robot to reach p_i. This is possible by applying Eq. 8 and Eq. 1, $p_i = p(t, \theta)$.

6   Results

To test the performance of the proposed solution, we conducted experiments in the CAMBADA training field, by releasing a ball down a ramp in order to achieve a desired velocity. The experiments were conducted with three different ball velocities, $1m/s$, $2m/s$ and $2.5m/s$. The obtained results are classified in three different classes: failure, if the robot fails to intercept the ball, contact, if the ball touches the front of the robot but bounces away out of the range of the robot control and success, if the robot is able to intercept the ball and keep it under control.

Table 1 presents the obtained results in the tests performed on the real robots.

Table 1. The results obtained in the interception tests performed on the CAMBADA robots

Ball Velocity(m/s)	Failure	Contact	Success
1	0%	0%	100%
2	0%	20%	80%
2.5	0%	100%	0%

Furthermore, experiments were conducted to test the performance of the developed solution against the strategy of moving towards the ball without considering its velocity. Using the same initial conditions such as the robot position and velocity and ball position and velocity, a robot intercepts the ball in under 2 seconds while a robot moving towards the ball is not even able to come in contact with the ball. Figure 6 presents the paths performed by the robot using both strategies. For visualization purposes the evolution of the distance between the robot and the ball during the course of the experiment is presented. Keep in mind that the robot has an approximate radius of $25cm$. Thus the distance between the ball and the robot cannot be smaller than this value.

7   Discussion

In robotic soccer, ball interception skills are of major importance and provide serious advantages during a game. Although the RoboCup Simulation League has a variety of solutions that address this problem, the transition to real systems presents some issues that hinder such effort. This paper presented a solution that solves the ball interception problem considering an uniformly accelerated with saturation robot motion model. The obtained results in Table 1 show that the robot is able to consistently intercept the ball in its path. However the

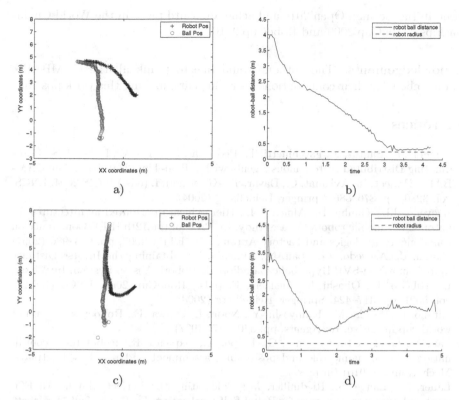

Fig. 6. Examples of interception and moving to the ball strategies. The circles represent the ball positions while the crosses represent the robot positions. a) the paths described by the robot and the ball using the developed interception algorithm. b) the corresponding distance between the ball and the robot throughout the experiment. c) the paths described by the ball and the robot when the latter is naively moving the the ball position. d) the corresponding distance between the ball and the robot.

performance decreases as the ball velocity increases. This is due to the fact that the obtained solution is the shortest time interception point. This causes the robot to intercept the ball as soon as possible without attempting to absorb the impact of the ball. Hence the ball can bounce away in such situations.

Although the interception behaviour is not ideal for receiving a pass from a team-mate, this skill is still very useful in defensive situations in order to stop an opponent dribbling the ball. This advantage is clearly depicted in Fig. 6 where the robot using the interception behaviour is able to intercept the ball. On the other hand a robot that moves towards the ball position is unable to catch the ball before it leaves the field.

The developed solution has been successfully integrated in the CAMBADA competition strategy helping the team to reach important results such as winning the National Championships, Robotica'2009 and Robotica'2010, placing second in the German Open'2010 and achieving third place in the World Championships, RoboCup'2009 and RoboCup'2010.

Acknowledgements. The authors would like to thank all the CAMBADA team members for their contributions over the years, making this work possible.

References

1. Almeida, L., Santos, F., Facchinetti, T., Pedreiras, P., Silva, V., Lopes, L.S.: Coordinating Distributed Autonomous Agents with a Real-Time Database: The CAMBADA Project. In: Aykanat, C., Dayar, T., Körpeoğlu, İ. (eds.) ISCIS 2004. LNCS, vol. 3280, pp. 876–886. Springer, Heidelberg (2004)
2. Azevedo, J.L., Cunha, B., Almeida, L.: Hierarchical distributed architectures for autonomous mobile robots: A case study. In: Proc. of the 12th IEEE Conference on Emerging Technologies and Factory Automation, ETFA 2007, pp. 973–980 (2007)
3. Cunha, B., Azevedo, J.L., Lau, N., Almeida, L.: Obtaining the Inverse Distance Map from a Non-SVP Hyperbolic Catadioptric Robotic Vision System. In: Visser, U., Ribeiro, F., Ohashi, T., Dellaert, F. (eds.) RoboCup 2007. LNCS (LNAI), vol. 5001, pp. 417–424. Springer, Heidelberg (2008)
4. Kitano, H., Asada, M., Kuniyoshi, Y., Noda, I., Osawa, E.: Robocup: The robot world cup initiative. In: Agents, pp. 340–347 (1997)
5. Lau, N., Lopes, L.S., Corrente, G., Filipe, N., Sequeira, R.: Robot team coordination using dynamic role and positioning assignment and role based setplays. Mechatronics (2010) (in press)
6. Lauer, M., Lange, S., Riedmiller, M.: Calculating the Perfect Match: An Efficient and Accurate Approach for Robot Self-localization. In: Bredenfeld, A., Jacoff, A., Noda, I., Takahashi, Y. (eds.) RoboCup 2005. LNCS (LNAI), vol. 4020, pp. 142–153. Springer, Heidelberg (2006)
7. Lauer, M., Lange, S., Riedmiller, M.A.: Modeling Moving Objects in a Dynamically Changing Robot Application. In: Furbach, U. (ed.) KI 2005. LNCS (LNAI), vol. 3698, pp. 291–303. Springer, Heidelberg (2005)
8. Müller, H., Lauer, M., Hafner, R., Lange, S., Merke, A., Riedmiller, M.: Making a Robot Learn to Play Soccer Using Reward and Punishment. In: Hertzberg, J., Beetz, M., Englert, R. (eds.) KI 2007. LNCS (LNAI), vol. 4667, pp. 220–234. Springer, Heidelberg (2007)
9. Neves, A.J.R., Martins, D.A., Pinho, A.J.: A hybrid vision system for soccer robots using radial search lines. In: Proc. of the 8th Conference on Autonomous Robot Systems and Competitions, Portuguese Robotics Open, ROBOTICA 2008, Aveiro, Portugal, April 2008, pp. 51–55 (2008)
10. Neves, A.J.R., Corrente, G.A., Pinho, A.J.: An Omnidirectional Vision System for Soccer Robots. In: Neves, J., Santos, M.F., Machado, J.M. (eds.) EPIA 2007. LNCS (LNAI), vol. 4874, pp. 499–507. Springer, Heidelberg (2007)
11. Paulo Pedreiras, L.A.: Task Management for Soft Real-Time Applications Based on General Purpose Operating Systems. In: Robotic Soccer, pp. 243–252. I-Tech Education and Publishing (December 2007)
12. Riley, P., Stone, P., McAllester, D.A., Veloso, M.M.: ATT-CMUnited-2000: Third Place Finisher in the Robocup-2000 Simulator League. In: Stone, P., Balch, T., Kraetzschmar, G.K. (eds.) RoboCup 2000. LNCS (LNAI), vol. 2019, pp. 489–492. Springer, Heidelberg (2001)

13. Santos, F., Almeida, L., Pedreiras, P., Lopes, L.S., Facchinetti, T.: An Adaptive TDMA Protocol for Soft Real-Time Wireless Communication among Mobile Autonomous Agents. In: Proc. of the Int. Workshop on Architecture for Cooperative Embedded Real-Time Systems, WACERTS 2004 (2004)
14. Santos, F., Corrente, G., Almeida, L., Lau, N., Lopes, L.: Selfconfiguration of an Adaptive TDMA wireless communication protocol for teams of mobile robots. In: Proc. of the 13th Portuguese Conference on Artificial Intelligence, EPIA 2007 (2007)
15. Silva, J., Lau, N., Neves, A.J.R., Rodrigues, J.M.O.S., Azevedo, J.L.: World modeling on an MSL robotic soccer team. Mechatronics 21(2), 411–422 (2011)
16. Stolzenburg, F., Obst, O., Murray, J.: Qualitative Velocity and Ball Interception. In: Jarke, M., Koehler, J., Lakemeyer, G. (eds.) KI 2002. LNCS (LNAI), vol. 2479, pp. 283–298. Springer, Heidelberg (2002)
17. van der Vecht, B., Lima, P.U.: Formulation and Implementation of Relational Behaviours for Multi-robot Cooperative Systems. In: Nardi, D., Riedmiller, M., Sammut, C., Santos-Victor, J. (eds.) RoboCup 2004. LNCS (LNAI), vol. 3276, pp. 516–523. Springer, Heidelberg (2005)

Rigid and Soft Body Simulation Featuring Realistic Walk Behaviour

Oliver Urbann, Sören Kerner, and Stefan Tasse

Robotics Research Institute
Section Information Technology
TU Dortmund University
44221 Dortmund, Germany

Abstract. Using a simulation for development and research of robot motions, especially walking motions, has advantages like saving real hardware, being able to replay specific situations or logging various data. Unfortunately research in this area using a simulation depends on transferability of the results to reality, which is not given for common robotic simulators. This paper presents extensions to a basic rigid body physics simulation leading to more realism. Parametrization matching a particular real robot is done using Evolutionary Strategies. Using stable walking and kicking motions as reference for the ES the newly developed MoToFlex simulator is able to reflect typical walking issues which can be observed in reality using different walking motions.

1 Motivation

Humanoid robots are an important research area due to the higher acceptance by human beings and due to the their advantage of being able to operate in environments designed for them [7]. To grant the usability of biped robots in service robotics or the health care sector a stable walk is vital. Many researchers work on this field, not only on dynamic walking based on the Zero Moment Point (ZMP) [19], but on problems such as pushing or carrying objects [15], dealing with uneven surfaces [6] or the need for a fast modification of foot placement [14]. Currently existing biped robots of human size, e.g. the HRP-4C developed by the AIST Institute in Japan [8] or the Honda Asimo [5], are too expensive to be used widely. More favorable robots like the Nao by Aldebaran Robotics [1] reveal significant inaccuracies in joint movements resulting in an unstable walk even of theoretically correct walking motions. Authors dealing with these issues are Kim et. al [9], Meriçli et. al [12] and Gouaillier et. al [2]. Studying the occurring errors during the walk by means of a real robot has some disadvantages like wearing out gears, noisy and delayed sensor data, not replicable reactions and time-consuming experiments. Therefore working with a simulation has advantages, but because of the big gap between simulations and reality adapting a walk developed using a simulation to a real robot is a difficult task.

---

[1] http://www.aldebaran-robotics.com/

T. Röfer et al. (Eds.): RoboCup 2011, LNCS 7416, pp. 126–136, 2012.

Thus the goal of this work is to downsize the difference between a simulation and reality. A simulation, called MoToFlex simulator [2] , that reflects some typical problems of a real walking Nao robot is proposed here.

The next chapter describes the fundamentals of humanoid robot simulations and related work. In section 3 the basic simulation is extended by a motor model, flexible gears with tolerance and flexible bodies. An appropriate parametrization of the simulation is crucial to reach the goal of realistic dynamical and kinematical behaviour. This is described in section 4. Finally section 5 displays that the objective has been fulfilled.

2 Related Work

The base of a simulation for humanoid robots is a rigid body physics engine, like the Open Dynamics Engine (ODE) [18]. Physics simulations implementing such an engine, like SimRobot [10], Webots [13] or SimSpark[16], allow to simulate the kinematics and dynamics of rigid bodies, represented by their center of mass and inertia tensor. This includes joints, collision detection and dealing with friction in case of contact.

To interact with the simulated model torques and forces can be applied. For example, it is possible to set the torque T_j which is applied by a joint to the connected bodies. Another common way to modify the joint angles is to directly set the desired joint speed ω_j along with a maximum torque T_{max} which can be applied to reach this speed. The torque T_j and the desired joint speed ω_j can be set simultaneously. This is useful for simulating friction when a torque T_j is applied by the joints. To do so, the desired speed of the joint is set to 0 ($\omega_j = 0$). The chosen maximum torque T_{max} is the friction torque added by the ODE to T_j.

A common way to control the joints of a robot is to set the desired joint angles. To set ω_j the joint angle positions can be differentiated. The resulting angle speeds are executed as given, resulting in a walk close to the expected one. From this it follows that a simulation based on a rigid body physics engine has to be extended by elements that imitate the real pipeline from the desired joint angles to the reached positions of the bodies. Hein et al. extended the ODE model of a robot by elements simulating the gear tolerance [4]. This extension leads to an unstable simulation and is therefore not useful. Lima et al. [11] implemented a simulation of the motors including PID controller. This is also one important element in the MoToFlex simulator but more elements will be needed to simulate flexibility and tolerance. The following section illustrates the extensions of the ODE based MoToFlex simulator and starts with an explanation how this motor simulation is adapted.

3 Development / Elements of Simulation

The previous chapters point out that the development of a walking algorithm starts using a simulation, but its usefulness is limited since common walking

---

[2] http://www.irf.tu-dortmund.de/nao-devils/download/2011/MoToFlex.zip

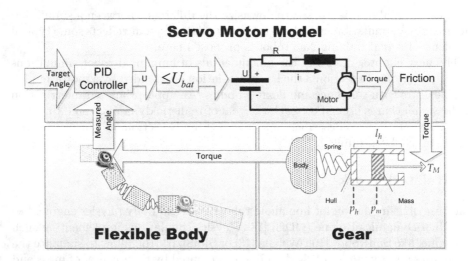

Fig. 1. Pipeline of the simulation

issues are not reflected by ordinary robotic simulations. In this chapter new extensions, arranged as a pipeline, are presented that are able to make simulations more realistic. They are implemented in a newly developed ODE based simulation, the MoToFlex simulator. Figure 1 shows the pipeline. Input of the system is the desired angle position. The pipeline consists of three phases. The first two stages, the servo motor and gear simulation, are independent from the ODE. The third stage is realized as an ODE model.

3.1 Motors

Algorithm 1 explains the servo motor simulation which bases on the work of Lima et al., see [11] for details. A PID controller uses the desired joint angles q_T and the actual joint angles q_A to calculate the voltage U, which is limited by the voltage U_{bat} of the battery. Due to the inductivity and resistance of the motor it has to be modeled as an RL circuit, see figure 1. Using the values given by the manufacturer for the equivalent series resistance R and inductor L the voltage U_r at the resistor R is close to U after only 1 ms. For the sake of simplicity U_r is therefore calculated iteratively without significant errors.

Hereafter the torque can be determined by multiplying U_r with $\frac{S \cdot K_t}{R}$ where S is the stiffness factor of the Nao joints [3], R the value of the resistor and K_t the torque constant of the motor. The latter factors are given by the manufacturer of the motor. An unknown quantity is the back electromotive force (EMF) constant K_e, which connects the motor speed with the back EMF voltage U_e induced by the motor: $U_e = K_e \cdot w_m$. To reduce the amount of variables with unknown

---

[3] By setting the stiffness parameter $S \in [0, 1]$ it is possible to reduce the torque applied by the motor.

Algorithm 1. Algorithm to simulate joint motors

1: $U = PID\left(q_T, q_A\right)$
2: **if** $U > U_{bat}$ **then**
3:     $U = U_{bat}$
4: **end if**
5: $U_R\left(t + \Delta t\right) = U_R\left(t\right) + \left(U - U_R\left(t\right)\right) \cdot \left(1 - e^{-\frac{t}{L/R}}\right)$
6: $T_M = \underbrace{\frac{U_R}{R} \cdot S \cdot K_t}_{\text{Torque due to power}} \underbrace{- \left(B_v + K_e\right) \cdot \omega_M}_{\text{Torque due to friction and back EMF}}$

---

values the back EMF is treated as a speed dependent friction since both reduce
the speed depending only on the joint/motor speed.

Besides the back EMF two types of friction have to be taken into account,
the already mentioned speed dependent friction B_v and the deceleration by the
EMF on the one hand and the friction not depending on the motor speed on the
other. The former are combined to one friction parameter and multiplied with
the motor speed. The later is handled by the ODE, as described in section 2.

3.2   Gears

Tolerances are a wanted property of gears to compensate production inaccuracy
and expansion due to warming. Besides the tolerances there can also be a flexibility caused by the used materials. Hein et. al attempted to reproduce tolerance
within their ODE based simulation which thereby got unstable. Therefore this
is handled outside the ODE here. A model of gears is developed consisting of
two elements: a spring to simulate the flexibility and a hull containing a mass to
simulate the tolerance. This model is realized as an algorithm which calculates
the outcome of flexibility and tolerance before sending the torque to the ODE.
This has the advantage, that there is no need to change the ODE model of the
robot to simulate it, resulting in more stable physics simulation.

Algorithm 2 depicts the proposed method. See table 1 for a description of the
used variables. A linear model is used which is equivalent to a rotational model
but more illustrative. The conversion between torque/angle to force/distance is
done with a radius of 1. First we apply the gear transmission ratio. To achieve a
stable simulation the algorithm for simulating the tolerance and flexibility has to
work with a high frequency. To speed up the overall simulation the ODE works
with a lower frequency. The increase of the gear simulation frequency is done by
the loop in line 2 of algorithm 2.

The torque T_M exerted by the motor accelerates a small mass, which can be
imagined as the mass of the gears next to the motor. In case of contact with the
hull the torque exerted by the spring also accelerates the mass. For a realistic
and stable simulation a speed dependent friction is also required. In position
$p_h = 0$ the spring is not stretched by definition and no torque is exerted by the
spring. Since the hull has no mass it has no contact to the hull if and only if
the mass is located between 0 and 1. Otherwise the hull and the mass are in

Algorithm 2. Algorithm for simulating gears with flexibility and tolerance

1: $T'_M = T_M \cdot \sigma$
2: **for** $i = 0$ **to** $\frac{\Delta t_s}{\Delta t_g}$ **do**
3:      $p_h = 0$
4:      **if** $p_m > l_h$ **then** {The mass is in contact with the hull.}
5:          $p_h = p_m - l_h$
6:      **end if**
7:      **if** $p_m < 0$ **then** {The mass is in contact with the hull on the other side.}
8:          $p_h = p_m$
9:      **end if**
10:      $T_s = D_g \cdot p_h$ {Calculate the torque exerted by the spring.}
11:      $p_m = p_m + \dot{p}_m \cdot \Delta t_g$ {Integrate the speed of the mass to get the position.}
12:      $\dot{p}_m = \dot{p}_m + \frac{(T'_M - T_s - B_g \cdot \dot{p}_m)}{m_g} \cdot \Delta t_g$ {Integrate the acceleration (exerted torque divided by the mass) to get the speed.}
13:      $T_j = T_j + T_s \cdot \frac{\Delta t_g}{\Delta t_s}$
14: **end for**

Table 1. Variables and constants

p_h	Position of upper border of the hull.
p_m	Position of mass.
$\dot{p}_m$	Speed of mass.
D_g	Spring constant.
B_g	Coefficient of viscous friction.
Δt_s	Length of a time step of the ODE physic simulation.
Δt_g	Length of a time step with $\Delta t_s \geq \Delta t_g$.
l_h	Length of the hull.
T'_M	Torque excerted by the motor after gear ratio is applied.
T_j	Torque output to be applied by the joint on the bodies.
σ	Gear ratio.
T_s	Torque exerted by the spring on the body.
T_f	Dry friction torque handled by the ODE.
m_g	Weight of the mass.

contact and have the same speed. If, for example, the mass starts between 0 and 1 and moves towards the springs it gets in contact with the hull at position 0. It then compresses the spring and even in case of an abrupt change in the moving direction of the mass the hull remains in contact with the mass. Once the position of the hull is determined the torque T_s exerted by the spring can be calculated and is then added up to the output torque of the gears T_j (see section 2).

3.3   Flexible Bodies

This torque T_j can not only accelerate the bodies connected by the joint, it can also cause a deformation of the bodies, depending on their rigidity. Some robots

can measure the angle errors of the gears since they have sensors measuring
the angle between the two bodies, not only the angle of the motor. The Nao
is an example for such a robot. Therefore errors due to flexible bodies have to
be handled separately. As figure 1 indicates, a flexible body is realized here as
smaller bodies connected by springs. This is known as the mass-spring model to
realize soft body dynamics. In fact these springs are ODE double joints where a
torque T_B is set:

$$T_B = -D_B \cdot \alpha \qquad (1)$$

where D_B is the spring constant and α the actual angle of the axis. To dampen
the spring a Coulomb friction is incorporated which is realized by the ODE itself
by using the method mentioned above.

3.4 Nao Setup

The most considerable goal of the simulation, besides a realistic simulation, is
computational stability and efficiency. Therefore not all bodies are flexible, but
only those whose flexibility has the greatest effect on walking motions, i.e. upper
and lower legs (see figure 2). Positions of the centers of mass, sizes and masses of
the boxes are chosen according to the official Nao documentation (Version 1.0).
Missing dimensions were measured by hand.

The Nao has two joints with 3 axes, the hip joints, consisting of two perpendicular axes and a third axis crossing the intersection of the other, see figure 2.
The ODE lacks joints with more than two axes. Implementations by connecting a single axis joint to the double axis hip using a body of size and mass 0
would result in unstable simulations. Therefore the HipYawPitch axis is not implemented so far which results in the inability to walk curves but ensures higher
computational stability.

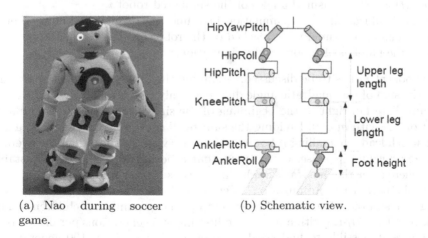

(a) Nao during soccer game.   (b) Schematic view.

Fig. 2. Nao by Aldebaran Robotics

As the Nao documentation clarifies two gear types are used in the legs. Hence two different sets of gear parameters are used. Most of the parameters are unknown, only gear ratios are given by the manufacturer. Therefore an optimization method is needed to find parameters such that the simulated robot walks like a real one.

4   Parametrization

Besides the unknown ODE parameters like friction the extensions presented in the previous chapter also contain some variables with unknown values. To find reasonable values a black box optimization method is applied. Evolutionary Algorithms, in particular the Evolution Strategies proposed by Schwefel et al. [17] evinced good results in terms of optimization in the robotic domain. Hebbel et al. utilizes Evolution Strategies to find simulation parameters that minimizes the difference between the way walked by a four legged robot in the simulation and the way walked in reality [3]. Here they are used to minimize the difference between the measured joint angles in reality and the measured joint angles in the simulation to achieve an overall realistic kinematic and dynamic behaviour of the simulated Nao. To get more generalized simulation parameters there are several robots in an evaluation of one parameter set, namely four walking robots using different walking parameters and one kicking robot. The walking motions are generated using the walking control developed by Czarnetzki et al. [1]. This leads to the following fitness function, which has to be minimized:

$$F\left(q_r, q_s\right) = \frac{\sum_{t=1}^{T} \sum_{i=1}^{n} \left(q_{r,i}\left(t\right) - q_{s,i}\left(t\right)\right)^2}{T^m} \qquad (2)$$

where:

- $q_{r,i}\left(t\right)$ are the measured angles of the real robot i
- $q_{s,i}\left(t\right)$ are the measured angles of the simulated robot i
- T is the duration of the simulation in frames (the simulation ends when a maximum frame number is reached or the robot has fallen down)
- n is the number of robots in the simulation.

The exponent m has to be discussed. Assuming the fitness function would comprise the sum of the quadratic angle difference only ($m = 0$). This would lead to an immediate fall right at the beginning of the simulation since this minimizes equation (2) with $m = 0$. Dividing the sum by T ($m = 1$) gives the mean angle error which leads to a falling robot when it starts walking since then the average angle error would increase. It turns out that choosing $m > 1$ leads to a stable walk where after the quadratic angle difference is further minimized.

For the parameter optimization different evolution strategies are compared. Using a cluster of 200 nodes for the evaluation a large number of children can be used. A $(10 + 800)$-ES with a maximum lifetime of 3 generations performed best and makes it possible to find good parameters after only ca. 100 generations. Using a simulation time step length of $0.001s$ and a maximum simulated walk duration of $10s$ this can be done within one day. The found set of parameters is optimized for a particular robot. Even though the experiments in the next section compares the simulation with this robot to show that it leads to a more realistic simulation, other Naos show very similar walking problems.

5   Experiments

In the context of walk development a useful simulation is able to reflect common walk issues. In this chapter the usefulness of the simulation is shown by presenting typical walk issues using real Naos compared to occurring walk issues in the MoToFlex simulation. Here again the walking control by Czarnetzki et al. generates the walking motions without sensor feedback. Two different setups are used.

In the first setup the walking motions are generated at a target speed of $200\frac{mm}{s}$ and a step duration of $1s$. In reality the orientation of the body is measured using the gyroscope and accelerometer of the robot. In the simulation the orientation of the body is directly given by the ODE. The speed is measured by walking a predefined distance. Using this setup we encounter an oscillation during the walk along the x axis[4]. This results not only in an unstable walk but also causes the swing leg to touch the ground while the body is rotated to the back. This results in a higher step length as desired and thus a higher walking speed is measurable. Figure 3 shows the pitch of the body measured on the real robot using the gyroscope and accelerometer. Compared to the simulation the amplitude is larger but frequency and phase are the same. Apart from that it is also noticeable that every second oscillation is further forward and backward respectively which occurs in reality as well as in the simulation. For comparison the orientation of the body in the simulation without any extensions (ODE only) is also shown. Besides the orientation the increased walking speed appears in reality as well as in the MoToFlex simulator. While the real robot walks at $250\frac{mm}{s}$ when a target speed of $200\frac{mm}{s}$ is set, the simulated one reaches $209\frac{mm}{s}$. Turning off the extensions leads to a walking speed of $200\frac{mm}{s}$.

In the second setup the target speed is set to $50\frac{mm}{s}$ at a step duration of $2s$. Disturbances along the y axis appear mainly at this step duration. This is caused by the joints of the standing leg failing under the strain in the single support phase causing the body to rotate towards the swinging leg. The manufacturer of the Nao, Aldebaran Robotics, also describes this problem and explains it with a sudden change of the desired torque during a single support phase [2]. In consequence the swing leg erroneously touches the ground whereby the robot is pushed to the wrong side. This leads to an overlarge lateral oscillation and a falling robot after the first step. Figure 4 depicts the roll angle of the real robot compared to the simulated one. Here again, the orientation differs quantitatively, but the body inclines to the left at the same time, and after the erroneous contact to the right. Moreover the robot falls down at a similar point of time. The problem is resolvable by changing the reference ZMP. It is worth mentioning

---

[4] The x axis is defined here as the frontal axis and the y axis as the lateral axis.

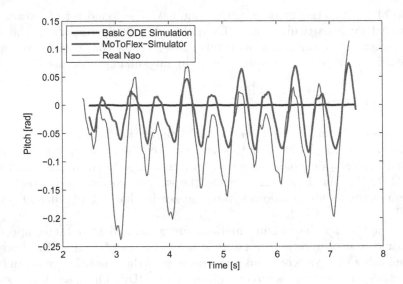

Fig. 3. Pitch of the Nao during a walk at a target speed of $200\frac{mm}{s}$ and step duration of $1s$

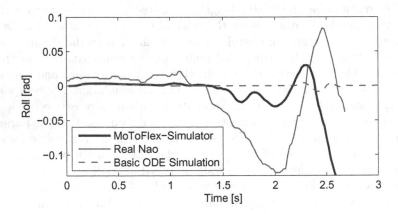

Fig. 4. Roll angle of the Nao during a walk at a target speed of $50\frac{mm}{s}$ and step duration of $2s$

that the tested modified reference ZMPs lead to a stable walk in the simulation if and only if they lead to a stable walk in reality.

6   Conclusion

In this paper, three extensions for a simulation based on the Open Dynamics Engine are proposed: a model of a servo motor including a PID controller, flexible gears with tolerance and flexible bodies used for the legs. Using Evolution

Strategies appropriate parameters for the simulation can be found. Using this parameter set the MoToFlex simulator reflects some common walk issues of a real robot enabling to develop walking motions in a more realistic simulation. Nevertheless the simulation accuracy might be improved to reduce the difference between the simulated and real posture. To do so the parameter optimization could be done by evaluating with more than 5 simulated robots. Beside four walking and one kicking robots other motions like standing up or simple leg movements could be useful.

An open question is the reason for the walking issues. Some of the simulated elements has the same effect than other. E.g. the simulation of a PID controller simulates with its proportional and differential part a spring with a damper, which is very similar to the simulation of the flexible gears. In fact the parameters vary from optimization to optimization. E.g. some have a higher proportional value with lower spring constant. Another interesting fact is, that the simulation of the flexible boxes only (ODE without motor and gear simulation) can lead to an unstable walk using the setup of the second experiment. Therefore it is difficult to use the simulation to find the parts of the real robot that are responsible for the walking issues. Despite this, the simulation can be used to develop algorithms dealing with the inaccuracies. For developing closed looped walking algorithms simulated sensors need to be integrated into the simulation. Then also typical difficulties of real measurements can be simulated, like sensor noise and different measurement delays.

References

1. Czarnetzki, S., Kerner, S., Urbann, O.: Applying Dynamic Walking Control for Biped Robots. In: Baltes, J., Lagoudakis, M.G., Naruse, T., Ghidary, S.S. (eds.) RoboCup 2009. LNCS (LNAI), vol. 5949, pp. 69–80. Springer, Heidelberg (2010)
2. Gouaillier, D., Collette, C., Kilner, C.: Omni-directional closed-loop walk for Nao. In: 2010 10th IEEE-RAS International Conference on Humanoid Robots (Humanoids), pp. 448–454 (December 2010)
3. Hebbel, M., Nistico, W., Fisseler, D.: Learning in a High Dimensional Space: Fast Omnidirectional Quadrupedal Locomotion. In: Lakemeyer, G., Sklar, E., Sorrenti, D.G., Takahashi, T. (eds.) RoboCup 2006: Robot Soccer World Cup X. LNCS (LNAI), vol. 4434, pp. 314–321. Springer, Heidelberg (2007)
4. Hein, D., Hild, M.: Simloid - research on biped robots controller, using physical simulation and machine learning algorithms. In: Concurrency, Specification and Programming, CSP 2006. Informatik-Berichte, vol. 1, pp. 143–151. (2006)
5. Hirai, K., Hirose, M., Haikawa, Y., Takenaka, T.: The development of Honda humanoid robot. In: ICRA, pp. 1321–1326 (1998)
6. Kajita, S., Kanehiro, F., Kaneko, K., Fujiwara, K., Yokoi, K., Hirukawa, H.: Biped walking pattern generator allowing auxiliary ZMP control. In: IROS, pp. 2993–2999. IEEE (2006)
7. Kanda, T., Ishiguro, H., Imai, M., Ono, T.: Development and evaluation of interactive humanoid robots. Proceedings of the IEEE 92(11), 1839–1850 (2004)
8. Kaneko, K., Kanehiro, F., Morisawa, M., Miura, K., Nakaoka, S., Kajita, S.: Cybernetic human HRP-4C. In: 9th IEEE-RAS International Conference on Humanoid Robots, 2009, Humanoids, pp. 7–14 (December 2009)

9. Kim, J.-Y., Park, I.-W., Oh, J.-H.: Experimental realization of dynamic walking of biped humanoid robot KHR-2 using ZMP feedback and inertial measurement. Advanced Robotics 20(6), 707–736 (2006)
10. Laue, T., Spiess, K., Röfer, T.: SimRobot – A General Physical Robot Simulator and Its Application in RoboCup. In: Bredenfeld, A., Jacoff, A., Noda, I., Takahashi, Y. (eds.) RoboCup 2005. LNCS (LNAI), vol. 4020, pp. 173–183. Springer, Heidelberg (2006)
11. Lima, J.L., Gonçalves, J.A., Costa, P.G., Moreira, A.P.: Humanoid Realistic Simulator - The servomotor joint modeling. In: ICINCO-RA, pp. 396–400 (2009)
12. Meriçli, Ç., Veloso, M.M.: Biped walk learning through playback and corrective demonstration. In: AAAI 2010: Twenty-Fourth Conference on Artificial Intelligence (2010)
13. Michel, O.: Webots: Professional mobile robot simulation. Journal of Advanced Robotics Systems 1(1), 39–42 (2004)
14. Morisawa, M., Harada, K., Kajita, S., Nakaoka, S., Fujiwara, K., Kanehiro, F., Kaneko, K., Hirukawa, H.: Experimentation of humanoid walking allowing immediate modification of foot place based on analytical solution. In: Proceedings 2007 IEEE International Conference on Robotics and Automation, pp. 3989–3994 (April 2007),
http://ieeexplore.ieee.org/lpdocs/epic03/wrapper.htm?arnumber=4209709
15. Nishiwaki, K., Kagami, S.: Online walking control system for humanoids with short cycle pattern generation. The International Journal of Robotics Research 28(6), 729–742 (2009), http://ijr.sagepub.com/cgi/doi/10.1177/0278364908097883
16. Obst, O., Rollmann, M.: SPARK – A generic simulator for physical multiagent simulations. Computer Systems Science and Engineering 20(5), 347–356 (2005)
17. Schwefel, H.P.: Evolution and Optimum Seeking. Sixth-Generation Computer Technology. Wiley Interscience, New York (1995)
18. Smith, R.: Open dynamics engine (2002), http://www.ode.org
19. Vukobratovic, M., Borovac, B.: Zero-moment point – Thirty five years of its life. International Journal of Humanoid Robotics 1(1), 157–173 (2004)

Towards Robust Object Categorization
for Mobile Robots with Combination
of Classifiers

Christian A. Mueller, Nico Hochgeschwender, and Paul G. Ploeger

Bonn-Rhein-Sieg University of Applied Sciences, Germany
{christian.mueller@smail.inf.,nico.hochgeschwender@,paul.ploeger@}h-brs.de

Abstract. An efficient object perception is a crucial component of a mobile service robot. In this work we present a solution for visual categorization of objects. We developed a prototypic categorization system which classifies unknown objects based on their visual properties to a corresponding category of predefined domestic object categories. The system uses the Bag of Features approach which does not rely on global geometric object information. A major contribution of our work is the enhancement of the categorization accuracy and robustness through a selected combination of a set of supervised machine learners which are trained with visual information from object instances. Experimental results are provided which benchmark the behavior and verify the performance regarding the accuracy and robustness of the proposed system. The system is integrated on a mobile service robot to enhance its perceptual capabilities, hence computational cost and robot dependent properties are considered as essential design criteria.

Keywords: object categorization, Bag of Features, feature extraction, clustering, machine learning, classifier combination.

1 Introduction

Due to the social phenomenon of a greying society it can be expected that elderly people will be assisted by service robots in their everyday activities at home more and more. In many cases objects play a central role in these activities and thus a service robot must be able to detect and classify known and unknown objects in such domestic environments. Most of the current approaches for object recognition have two characteristics, namely firstly the perception is instance based and secondly the instances of the objects have to be known in advance via some teach-in process. Yet the identification of a particular object instance is not necessarily required or sometimes even not feasible, e.g. serving and delivering drinks require the detection of glasses as such, the recognition of a particular glass is often not needed. It would be very wasteful or even infeasible to teach this large set of individual glass instances in the individual home when just any element from the category glass will do the job. This shift of focus may also be illustrated in the development of the current rule set in the *RoboCup@Home* competition,

T. Röfer et al. (Eds.): RoboCup 2011, LNCS 7416, pp. 137–148, 2012.

which is a well-established international benchmark for service robots in domestic environments [9]. For instance, in the recently established *supermarket* scenario, the service robot should fetch e.g. a cup – as a missing item of a shopping list – from a shelve of the supermarket. Here a object categorizing capability is needed since any cup will resolve the problem.

The presented work develops an object perception system that categorizes unknown domestic object instances like cups, glasses, bottles or cell-phones in their respective category. Our system addresses robustness and reliability in several ways. It can cope with object categorization challenges like perspective variations due to object-rotation-angle and robot-object-distance variations or intra category variations i.e. deformations of object instances of a common category. The system extracts expressive visual properties of example objects and *generalizes* these visual properties according to their respective object category in order to categorize *unknown* objects. In the remaining paper we discuss the related work and our contribution, followed by the requirements and assumptions we made. Further on the components of the system are explained followed by the evaluation. Finally, a discussion and conclusion are given.

2   Related Work and Contribution

The presented work is grounded on 2D image information. Two common approaches are available, geometric- and geometric-free based approaches. *Geometric-based* approaches work on the global geometric appearance of objects. These approaches rely e.g. on shape models or shape descriptors. They show robustness to strong shape deformations. Hence, they provide the capability to make a reliable decision about a corresponding shape class; however a reliable decision about the shape class of an object is not sufficient for object categorization purposes, since objects from *different* object categories with *similar shapes* might fall into the same shape class and thus become indistinguishable. Geometric-based approaches demand a precise image segmentation algorithm, since precisely extracted object boundaries are required as input for an optimal performance. Sophisticated image segmentation algorithms are often computationally expensive ($\gg 1sec$), and therefore unattractive to be applied to a mobile service robot where a short-response-time is required.

Consequently, our work relies on a *geometric-free* approach (see Fig. 1): called *Bag of Features(BoF)* [1,6,8]. This approach has shown its reliability and robustness to object occlusions, illumination changes and especially to geometric deformations of objects which belong to a common category, since the *BoF* approach does not rely on global geometric information; instead it relies on the extraction of local invariant features. Categories which are hardly distinguishable by shape-based approaches, become distinguishable (e.g. cups and glasses) since the extracted features provide information about the structure and texture of objects. The *BoF* approach is based on the assumption that each object category is distinguishable by its individual independent statistical appearance of salient-invariant-local features which are extracted from images. The idea is by comparing the features of a query object to the distribution of those features from previously analyzed objects, to infer the category of the query object. A so-called *visual dictionary* of generalized features is generated based on extracted features from a training set. These generalized features are expressive features which provide a high discriminability regarding the categories. The extracted features of a query object are mapped to these generalized features; the generalized features can be seen as visual words in the visual dictionary whose presence or absence lead to a decision about the actual category.

In the first step of the *BoF*-based object categorization process, an *extraction of invariant features* from images is exploited to transform the visual image information into a compact representation, which provides rich recallable information of the image, i.e. if the image content is transformed by scale, shift or rotation, similar information are extracted. Commonly Scale-Invariant-Feature-Transform (*SIFT*) has been often successfully applied [5]; however our experiments have shown that Speeded-Up-Robust-Features (*SURF*) performs a better feature extraction, due to its feature recallability and computational cost. Next, the *visual dictionary* is created, which analyzes the feature frequencies for a set of images that have passed the feature extraction process. Therein, the features are grouped by similarity, in order to generate clusters of similar features. Based on a cluster, a generalized feature is constructed which represents a *visual word*. Mostly k-means-based algorithms are applied for clustering due to its simplicity and low computational cost [5,8,1]. Other contributions group the features e.g. by randomized cluster trees [7] or mean-shift clustering [6]. After the dictionary is generated, the extracted features of a query image are assigned to the nearest visual words by, e.g. nearest-neighbor-search. The comparison between the visual word frequencies, i.e. distribution of the visual words, of a query image and of labeled example images leads to a decision about the corresponding category of the query image. Often supervised machine learning approaches like Support Vector Machines(*SVM*) are applied [1,8], since they have shown an enhanced robustness to discriminate sets of categories. The learners are trained with the visual word frequencies of examples objects to generate a *prediction model*.

In our work we do not rely on the decision of a single classifier, since a single classifier provides a certain accuracy and also a high risk of a misclassification bias for specific categories. To enhance the accuracy and to reduce the influences of those biases, a set of classifiers is trained and their outcomes are *combined* to make a more *robust* and *reliable* decision about a category. Additionally the performance of each classifier is improved by different *feature-selection algorithms*. Moreover our approach does not completely neglect the object shape information, since it provides a useful indication about a corresponding category. We combine the set of feature-based classifiers with an additional *shape-based classifier* in order to support an appropriate final decision. An appropriate number of clusters (*dictionary size*) is a crucial factor which influences the categorization performance. The discriminability is decreased if a too small or too large dictionary is used; in both cases the efficiency of the dictionary is negatively influenced. Most approaches heuristically examine the dictionary size or they set the dictionary size to a fixed number [8,7]. In contrast, we systematically analyze the dictionary size by the *examination of structure and relation* of the generated visual words to each other, in order to generate a discriminative dictionary; also the *importance* and *relevance* of each visual word is analyzed with respect to the object categories. In many approaches[1,8,5] the entire image is examined as a single entity. Image distortions like cluttered backgrounds, which do not contribute meaningful information, influence the categorization result; moreover multiple object occurrences are generally not considered. We show a basic, but sufficient approach which allows to *detect multiple objects* in an image by an image segmentation algorithm; afterwards the detected objects are classified by the feature- and shape-based classifiers. Most object categorization approaches are stationary and less concerned about the response time. Our system is integrated on a mobile service robot, hence issues like the computational cost or the robustness towards the large variety of perspective variations to objects have to be considered in the development of the system. Further a monocular camera is applied, rather than a cost intensive Time-of-Flight- or Stereo-Vision-Camera.

3   Requirements and Assumptions

The object categorization system acquires 2D images from the camera as input. The camera is mounted on the top of the robot and provides images from a typical service-robot-height of about *100-120 cm* in a resolution of *640×480* pixels. The system is trained for a robot-object distance of about *30-40 cm*. The robot camera has to be focused on a plane surface, as for instance a table. The system is supposed to perceive the objects on the table top and to classify them to a supported object category; up to four different object categories (cup, cell-phone, bottle, glass) are supposed to be classifiable. We verify the system performance and behavior using these four categories in certain constellations; however our system can easily be *extended with additional categories*. Due to the current object detection approach (see section 4.4), the background is assumed to be less cluttered and the objects are mainly present on a uniformly colored surface. The objects are positioned – completely visible and not occluded – in a reasonable distance from the camera and to each other. The system is applied under artificial light conditions. Further on, an *object database* is required in order to supply the system with sufficient object-related-information for an efficient prediction-model generation. It consists of a set of images of each category which is used as training set. Each image contains a single object, thereby different images are taken from different randomly chosen positions and orientations, and on varying uniform backgrounds. Also a test set of images for each category is provided. Both sets are mutually exclusive in terms of particular object instances.

4   Object Categorization with Bag of Features

The system is divided into two phases, namely training phase and evaluation phase. These phases and the involved components with their parameters are depicted in Fig. 1 and discussed in the remaining section.

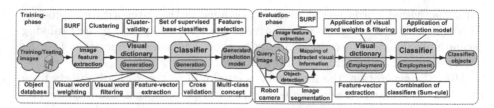

Fig. 1. Illustration of the system design. The components shown, are involved during training- and evaluation phase.

4.1 Image Feature Extraction

Firstly, during training phase features are extracted from images of the provided object database. Note that, only object-related features are extracted since a single object is positioned on a uniform background. During evaluation phase the features are extracted of a query image which is acquired of the robot camera. In this case, object detection has to be applied to find features related to each presented object in the query image (see section 4.4).

The extraction (detection and description) of the features is performed by *SURF*: in the given conditions of our application *SURF* shows a up to *4%* lower classification error and it tends in average to a faster feature extraction than *SIFT*. Each (*SURF-*)feature is described by a vector of *64* elements, moreover up to *120* features per object are extracted since they have proofed to provide sufficient object-related-information for further processing purposes. After the extraction of the features from each object of the object database and the query image, the extracted feature set of each object and query is called *BoF*.

4.2 Visual Dictionary Generation and Employment

During training phase the visual dictionary is generated from the set of *BoFs* of the *training set*. The visual dictionary provides a representation which is able to describe - in a compact and efficient way - the visual information. The dictionary contains a set of visual words, which are generated through grouping the features from the *BoFs* by similarity. The fast k-means clustering algorithm is applied to group and find similarities in the extracted features. In succession of the grouping of the features, the center of each group represents a *visual word*. The goal is to find a discriminative dictionary: the appropriate k(*dictionary size*) is determined by the *Dunn-validity-Index*[3]. Thereby in our experiments the dictionary size is varied from *100*(min) to *1000*(max) visual words with an increment of *10*; the *validity value* and the *classification accuracy* of each dictionary size is examined. An indication of an appropriate dictionary size which leads to an enhanced classification performance, is found by the identifications of local maxima of the global *validity values*. That *local maximum* whose corresponding dictionary size leads to the lowest classification error is selected. In addition, a modified *soft-assignment weighting scheme*[5] based on the examination of the feature frequency related to particular visual words is applied, in order to give particular visual words less or more importance in the categorization process. Also a *filtering scheme* of less informative visual words is applied by first ranking ascendingly the visual words according to their frequency proportions to each object category, and later neglecting the lower ranked words.

After the generation, the dictionary is employed to the *BoF* of objects from the object database (*training and test set*) and of *query objects*. Through the dictionary employment, each *BoF* is represented as a composition of the previously generated visual words in a histogram which is also denoted as *feature-vector*. Thereby the visual word occurrences in each *BoF* are examined by the nearest-neighbor-search: the nearest visual word to each feature is determined.

4.3   Supervised Classifier Generation and Employment

Through the representation of each object as a *feature-vector*, the categorization problem has been converted to a pattern matching problem which we handle as a machine learning problem. Hence, efficient and powerful techniques from the machine learning field are exploited.

As a preprocessing step to enhance the quality of the feature-vectors, *feature-selection algorithms*[4] are applied to identify discriminative features of the vectors during *training phase* and filter out those features during *evaluation phase*. Three filters are applied, based on *Principle Component Analysis* (*PCA*), *Entropy* and *Iterative Adaptive Feature Selection* (*IAFS*); *IAFS* iteratively adds a feature from a ranked set of features[1] and trains accordingly a classifier; if an improvement is achieved, a new feature from the set is selected, else the last added feature is removed, and replaced with a new feature.

We experimented with two popular supervised machine learning techniques, namely *SVM* and *AdaBoost* which we combined with the feature-selection algorithms in order to generate a pool of classifiers. Six (base-)classifiers[2] were defined which are *independently* trained during the training phase by the generated feature-vectors of the training set in order to learn a *prediction model*. The accuracy is verified by a *10-fold cross validation* and the test set. Since the categorization problem is a multi-class problem, we decided to apply the majority-voting-strategy-based *One-vs.-One* multi-class concept, due to its misclassification recovery property. We also experimented with one extra *shape-based* base-classifier to support the final category decision. It is trained with feature-vectors which are based on shape descriptor results[10] of extracted contours[3] of objects from the object database: to each object contour descriptors are applied as statistical moments, Hu-moments, direction, eccentricity, normalized central moments, spatial moments, contour area, contour length, and Fourier descriptor.

During the evaluation phase, a set of the trained base-classifiers is employed to a feature-vector of a query image; the constellation of the set is evaluated regarding its classification accuracy (see section 5). The classification outcomes of the

---

[1] The features are ranked by their discriminability through feature-selection algorithms like *PCA* or *Entropy*.

[2] In Table 1 of section 5 the actual set of base-classifiers is listed.

[3] The contours are extracted during the object detection process (see section 4.4).

base-classifiers are combined with an extended *sum-rule*[2], in order to classify reliably the feature-vector to an appropriate category and to be robust to misclassification of base-classifiers. The *sum-rule* is extended with four *weighting-factors* which are applied to the outcomes: firstly, the outcome of each base-classifier is weighted by the base-classifier's *classification accuracy(self-confidence)*, which is gathered during training phase. Secondly, each outcome is weighted by the *confidence in the actual outcome* of the base-classifier. Thirdly, biases of correct classifications and misclassifications towards particular categories of each base-classifier are considered. Hence, a *penalty to each outcome* is applied according to the biases of the base-classifier. Fourthly, *classification majorities of the categories* from the outcomes are weighted in order to overcome misclassifications by base-classifiers. The sum-rule involves two factors: an implicit *voting* to the most probable outcome and giving a higher importance - by *weighting* - to accurate base-classifiers and less importance to less accurate base-classifiers.

4.4   Multiple Object Detection

During the evaluation phase a query image of the robot camera is acquired. The system has to classify single or multiple occurrences of object instances. Initially the features of the entire image are extracted. However the entire set of features cannot be classified like a single object as during the training phase, because of the probable presence of cluttered background or multiple object occurrences. Hence, the detection of potential objects in the given image and the correspondence of features to the respective object is needed to be determined. An accurate image segmentation algorithm is not the scope of our work and is not necessary, since only regions of interests of potential objects are required to be found. A sufficient segmentation algorithm based on contour extraction is

2D contour based segmentation   3D depth information based segmentation

Fig. 2. Left: object detection result is shown at a distance of ≈30cm to the robot with the extracted object boundaries and detected features. Right: 3D depth-based detection result is shown (point cloud is randomly colored for each detected object).

applied to gather the boundaries of potential objects in the image and to map the corresponding features to the object boundaries as shown in Fig. 2(left). Alternatively, a detection based on 3D depth information is under development. This approach segments objects on different planes in cluttered and occluded environments with a higher reliability compared to the contour-based approach, see

Fig. 2(right). The single plane extraction is based on surface normals extraction and *RANSAC(Random Sample Consensus)* plane fitting. Afterwards, a hierarchy of different heights of overlapping planes is created. Potential objects are extracted via grouping points above each plane by euclidean clustering. Due to the infancy of this approach the following results are based on the contour-based approach.

Afterwards the visual dictionary is employed to the extracted features from each object boundary. So, the content of each object is projected to a feature-vector which is classified to an appropriate category by the employment of the trained classifiers whose outcomes are combined to a final decision.

5   Experimental Evaluation

The following first part of the evaluation is based on the training- and test set of the object database. Our experiments have shown that *190* images as training set and *55* images as test set, deliver sufficient object category related information for an efficient prediction model generation.

The identification of an appropriate dictionary size leads to an enhanced classification accuracy: Fig. 3 shows the visual word dedications with respect to the categories, if an appropriate dictionary size is chosen. The top-ranked *20%*

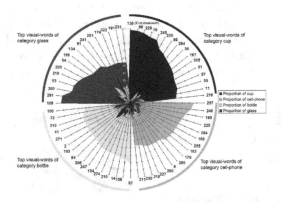

Fig. 3. The top discriminative visual words with their proportions of frequency regarding each category (*16* selected visual words of each category). The visual words are sorted in ascending order of their proportion in the respective dedicated category. E.g. in case of the cup category: the most discriminative visual word is the one with *ID-136* which has a proportion of *97.7%* in the cup category, *0.8%* in the cell-phone category and *1.4%* in the bottle category.

visual words (*16* selected visual words of each category) are shown of a dictionary which contains in total *325* words and supports the four categories (cup, cell-phone, bottle, glass). The selected set of *16* visual words for each category does not intersect with the sets for the other categories; also the top *20%* visual words are strongly dedicated to a specific category. Even lower ranked visual words from the top *20%* show a dedication of ≈*50%* to a certain category, which is still discriminative, since this proportion has the majority; in that case not a single proportion of another category reaches a level of more than *30%* – with one exception: visual word with ID-*251*.

Further on, Table 1(green) shows that dictionary sizes indicated by the local maxima of the Dunn-validity-index values provides in average a *lower classification error* than randomly chosen dictionary sizes: an appropriate dictionary size with the lowest classification error has been found for *2*-categories with *270* words, respectively for *3*-categories *400* words and for *4*-categories *325* words. These dictionary sizes are chosen as the base-classifiers return the lowest

Table 1. The average classification error (%) regarding the test set is shown of each base-classifier which is trained with randomly chosen and Dunn-validity-index indicated dictionary sizes. The classification error in the brackets shows the error if an appropriate dictionary size is chosen: 2-cat.=*270* words, 3-cat.=*400* words, 4-cat.=*325* words.

Base-classifier approach	Number of supported object categories					
	2		3		4	
	Rand.	Dunn.	Rand.	Dunn.	Rand.	Dunn.
SVM	1.91	0.23(0)	5.14	2.04(2.4)	8.01	6.65(5.9)
SVM+Entropy	1.71	0.45(0)	4.84	2.29(1.2)	7.38	6.28(2.7)
SVM+PCA	1.91	0.67(0.9)	3.93	2.78(1.2)	6.73	5.87(4)
AdaBoost	5.34	3.85(3.6)	7.78	7.50(9)	15.1	13.17(12.7)
AdaBoost+PCA	3.62	2.49(2.7)	7.27	7.59(7.8)	10.98	9.30(9)
AdaBoost+PCA+IAFS	4.23	3.17(2.7)	6.66	6.18(6)	10.37	9.85(9.5)

classification error compared to other dictionary sizes of their respective number of supported categories. Note that, the dictionary size *does not increase proportionally* with the increase of supported categories: the discriminability between categories is a decisive factor that affects the dictionary size.

In the following, the *combination* of the feature-based base-classifiers is evaluated. In the experiment, all 63^4 combinations of the six base-classifiers are analyzed (see Fig. 4). The plot in Fig. 4 illustrates that the addition of base-classifiers *generally improves* the classification accuracy, in other words the average classification error of four supported categories has dropped from *8.7%* (single classifier) to *3.1%* (six combined classifiers). Further, in this case of a 4-category trained system, the combination of three base-classifiers, namely *SVM*, *SVM+Entropy* and the *AdaBoost+PCA* leads to the lowest classification error of *2.2%* (see in Fig. 4 right-side), i.e. the application of a *particular sub-set* of the six base-classifiers leads to a *-0.9%* decreased classification error compared to the classification error if *all* six base-classifiers are combined. Also the combination of those three classifiers shows a lower classification error than the *most accurate single base-classifier* (*SVM+Entropy=2.7%* – see Table 1 in brackets). The same behavior is observed for 2- and 3-category trained systems. However in case of the

---

[4] In total *63* combinations of *6* base-classifiers: summed by the number of combinations of one(*6*), two(*15*), three(*20*), four(*15*), five(*6*) and six(*1*) applied classifiers.

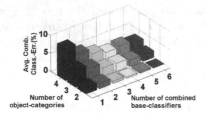

Num. of supported categories	Base-classifier comb. with the lowest classification error	Error(All classifiers combined)
2	SVM, S.+Entropy, S.+PCA, AdaBoost, A.+PCA+IAFS	0(0)
3	S.+Entropy, S.+PCA, AdaBoost	0.6(1.8)
4	SVM, S.+Entropy A.+PCA	2.2(3.1)

Fig. 4. Left: average classification error according to the number of base-classifiers which are combined and the number of supported categories. Right: combinations of base-classifiers which result to the lowest classification error(%) of test set.

2-category trained system the combination of all six classifiers has been chosen, which lead to the same accuracy than the most accurate single classifier(0%); this combination has been chosen, due to the robustness against misclassifications. It is worth to mention, that the single *AdaBoost* base-classifier which has the *lowest accuracy*, still contributes to the combination with the *lowest combined classification error* in case of 3-categories (see in Fig. 4 right-side). These results show that our method of combining sets of particular base-classifiers, effectively reduce the classification error.

The shape-based classifier has shown reasonable results for distinctive categories regarding the object shape, e.g. in system configurations where cups and cell-phones (*5%*) or cups, cell-phones, and bottles (*10%*) are categorized; when additionally glasses are involved (4-categories), the classification error increases (*27.5%*) due to shape similarities between instances of the categories like of cups and glasses. We conclude, that this classifier is helpful to support the combined classification but it is a matter of future work to enhance it.

In the following evaluation, the trained system is treated as a black-box: images of the robot camera are acquired and evaluated. The Table 2 presents

Table 2. The classification accuracy (%) regarding the four categories. The system is trained to support 2-, 3-, and 4-categories. 10 objects of each category are involved in this experiment. (x=category is not applied in respective system configuration).

Actual Category	Number of supported object categories											
	2				3				4			
	Cup	Cell-ph.	Bot.	Gla.	Cup	Cell-ph.	Bot.	Gla.	Cup	Cell-ph.	Bot.	Gla.
Cup	95	5	x	x	92.2	5.6	2.2	x	96.7	1.7	0.8	0.8
Cell-phone	3.3	96.7	x	x	4.4	95.6	0	x	3.4	95.8	0.8	0
Bottle	x	x	x	x	7.8	0	92.2	x	8.3	0	91.7	0
Glass	x	x	x	x	x	x	x	x	2.5	7.5	0	90

the classification accuracy regarding the four categories: certain misclassification biases for particular categories are observed. Further we investigated the categorization behavior depending on the *robot-object-distance* and *object-rotation-angle* – see Fig. 5. The classification accuracy among the supported categories behaves differently under the same experimental setup i.e. *object-robot-distance* and *object-rotation-angle*. Several factors are responsible for this behavior:

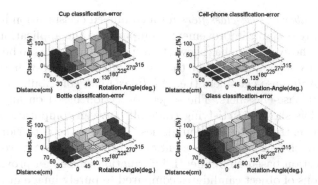

Fig. 5. The classification error results of each category with respect to object-robot-distance and object-rotation-angle. *10* objects of each category are involved in this experiment. The system is trained to support *4* categories.

robot-object-distance implies that the farther the object is positioned the fewer descriptive features are extracted which lead to an increase of the classification error; the *object-material* with respect to the contrast between the object and the background, can increase the classification error because of a weak object detection in farther distance like in the case of glasses (partial transparency). Also the *object-rotation-sensitivity* of object influences the classification accuracy. A sensitivity to the absence of descriptive features due to the object-rotation-angle is observed for cups (e.g. cup-handle is not visible). Bottles and glasses show less rotation sensitivity due to the symmetry of the extracted features from different rotation angles. In case of cell-phones mostly all possibly available features are extracted – regardless the object-rotation-angle – due to its general flat shape and the upper robot-camera-perspective.

6   Discussion and Conclusion

The evaluation has shown that the number of extracted features and especially the *presence* and *absence* of descriptive features due to the distance and viewing angle to the objects from the robot have an important impact on the classification result. Also it was observed that categories have *biases* for being misclassified to particular categories. The construction of an efficient visual dictionary is a crucial factor for the classification performance: the determination of an appropriate number of visual words (dictionary size) plays an important role in order to generate discriminative visual words which show a bias for certain categories. The evaluation of an appropriate dictionary size by exploiting the *Dunn-validity-index* as an indicator for the size plus *visual word weighting* and *filtering* have shown in the experiments reasonable results. The choice of an appropriate machine learning technique and feature-selection algorithm is important: it is observed that a more accurate classification is achieved if additionally a feature-selection algorithm (e.g. *PCA* or *Entropy*) is applied than the application of basic *AdaBoost* or *SVM* learning approaches. Moreover the combination of a *certain number of base-classifiers* for a combined classification has shown in the evaluation reasonable improvements compared to the application of a single classifier. Also the evaluation has shown that a combination has to be determined of *particular base-classifiers*, rather than to combine the top-n base-classifiers; even *less accurate* base-classifiers (i.e. *AdaBoost*) can contribute to an efficient combination of base-classifiers. The object detection based on an basic image segmentation has shown a satisfying trade-off between computational cost and accuracy of the extracted object boundaries. However, the detection has shown its limitation in the evaluation. As mentioned previously we are working on an enhanced detection based on 3D depth information which can provide more reliable indications of object candidates compared to purely image intensity based approaches; we focus on the detection of objects on multiple planes in varying constellations as for instance on shelves in a supermarket. Additionally in the future work, we focus on to increase the number of supported object categories.

Further experiments have shown an average execution time of $\approx 2.6s$ for the categorization of six concurrently present objects ($\approx 802ms$ for a single object). This execution time provides the feasibility of the system to be applied on service tasks, which require *occasional* to *frequent* categorization of objects.

In this paper a prototypic object categorization system has been described which is based on *BoF*. The presented evaluation has shown the behavior and the competitive categorization performance. This system equips a service robot with an ability which supports the application of advanced object-related tasks.

References

1. Dance, C., Willamowski, J., Fan, L., Bray, C., Csurka, G.: Visual categorization with bags of keypoints (2004)
2. Duin, R.P.W.: The combining classifier: To train or not to train? In: International Conference on Pattern Recognition, vol. 2 (2002)
3. Dunn, J.C.: A fuzzy relative of the isodata process and its use in detecting compact well-separated clusters. Cybernetics and Systems 3, 32–57 (1973)
4. Guyon, I., Elisseeff, A.: An introduction to variable and feature selection. Journal of Machine Learning Research 3, 1157–1182 (2003)
5. Jiang, Y.G., Ngo, C.W., Yang, J.: Towards optimal bag-of-features for object categorization and semantic video retrieval (2007)
6. Jurie, F., Triggs, B.: Creating efficient codebooks for visual recognition. In: ICCV 2005: Tenth IEEE Intl. Conf. on Computer Vision, vol. 1, pp. 604–610 (2005)
7. Moosmann, F., Triggs, B., Jurie, F.: Fast discriminative visual codebooks using randomized clustering forests. In: Advances in Neural Information Processing Systems 19, pp. 985–992 (2006)
8. Nowak, E., Jurie, F., Triggs, B.: Sampling Strategies for Bag-of-Features Image Classification. In: Leonardis, A., Bischof, H., Pinz, A. (eds.) ECCV 2006. LNCS, vol. 3954, pp. 490–503. Springer, Heidelberg (2006)
9. Wisspeintner, T., van der Zant, T., Iocchi, L., Schiffer, S.: Robocup@home 2008: Analysis of results. Tech. rep. (2008)
10. Zhang, D., Lu, G.: Review of shape representation and description techniques. Pattern Recognition 37(1), 1–19 (2004)

Learning Visual Obstacle Detection
Using Color Histogram Features

Saskia Metzler, Matthias Nieuwenhuisen, and Sven Behnke

Autonomous Intelligent Systems Group, Institute for Computer Science VI
University of Bonn, Germany

Abstract. Perception of the environment is crucial in terms of successfully playing soccer. Especially the detection of other players improves game play skills,
such as obstacle avoidance and path planning. Such information can help refine
reactive behavioral strategies, and is conducive to team play capabilities. Robot
detection in the RoboCup Standard Platform League is particularly challenging
as the Nao robots are limited in computing resources and their appearance is pre-
dominantly white in color like the field lines.

This paper describes a vision-based multilevel approach which is integrated
into the B-Human Software Framework and evaluated in terms of speed and accuracy. On the basis of color segmented images, a feed-forward neural network is
trained to discriminate between robots and non-robots. The presented algorithm
initially extracts image regions which potentially depict robots and prepares them
for classification. Preparation comprises calculation of color histograms as well
as linear interpolation in order to obtain network inputs of a specific size. After
classification by the neural network, a position hypothesis is generated.

1 Introduction

In RoboCup Standard Platform League (SPL), two teams of three Nao robots compete
in the game of soccer. For them as autonomous systems, acquiring information on the
current state of the environment is essential for playing successfully. In particular the
detection of other robots is important for successfully planning upcoming actions, dribbling the ball along the field, and scoring goals. It is also conducive in terms of reactive
obstacle avoidance and team play skills, such as passing the ball to a team mate. However, given the limited computational resources of the Nao robot and the impossible
task of discriminating robots from field lines by their color, visual robot detection is a
demanding task.

The approach presented here is a multilevel analysis of visual sensor data. This includes the steps of selecting interesting regions within an image, reducing their dimensionality, and finally classifying them to decide if there is a robot. The classification step
is accomplished using an artificial neural network. In this paper, the implementation of
the detection approach is described and evaluated in terms of speed and accuracy.

After discussing related work, the hardware and software prerequisites are described
in Sect. 3. In Sect. 4 the robot detection process is described in detail. Subsequently, the
results of the evaluation of the detection process are presented in Sect. 5. This comprises
simulated as well as real-robot experiments.

T. Röfer et al. (Eds.): RoboCup 2011, LNCS 7416, pp. 149–161, 2012.

2   Related Work

Before 2008, the SPL was staged on four-legged Sony AIBO robots [12]. Among others, Fasola and Veloso [5] as well as Wilking and Röfer [15] established object detection mechanisms for these robots. Nao robot detection approaches significantly depend on the competition rules. Until 2009, the current robots' gray patches were either bright red or blue according to the team color. Whereas now, only the waist bands denote the team color.

Daniş et al. [3] describe a boosting approach to detect Nao robots by means of their colored patches. It is based on Haar-like features as introduced in [13] and is conducted using the Haartraining implementation of the OpenCV [2] library. A different technique for Nao robot detection was proposed by Fabisch, Laue and Röfer [4]. The approach is intended for team marker-wearing robots. A color-classified image is scanned for the team colors. If a spot of interest is found, heuristics are applied in order to determine whether it belongs to a robot. Ruiz-del-Solar et al. [11] detect Nao and other humanoid soccer robots using trees of cascades of boosted multiclass classifiers. They aim at predicting the behavior of robots by determining their pose.

In the context of RoboCup Middle Size League Mayer et al. [9] present a multistage neural network based detection method capable of perceiving robots that have never been seen during training. And as Lange and Riedmiller [6] demonstrate, it is also possible to discriminate opponent robots from team mates as well as from other objects with no prior knowledge on their exact appearance. Their approach makes use of Eigenimages of Middle Size robots and involves training of a Support Vector Machine for recognition.

3   Robot Platform

Humanoid Aldebaran Nao robots are equipped with a x86 AMD Geode LX 800 CPU running at 500 MHz. It has 256 MB of RAM and 2 GB of persistent flash memory [1]. This means, the computational resources are rather limited and low computational complexity is an important demand to the robot detection algorithm. The sensor equipment of the robots includes, among other devices, two head cameras pointing forward at different angles. These are identical in construction and alternately provide images at a common frame rate of 30 fps. The image resolution is 640×480 pixels, however the first step of image processing is a reduction to 320×240 pixels.

The software development of the robot detector is based on the B-Human Software Framework 2009 [10]. This framework consists of several modules executing different tasks. Additionally, a simulator called SimRobot is provided. The robot detection process is integrated as a new module and makes use of already processed image data.

4   Robot Detection Process

The objective of finding robots in an image is to find their actual position on the field. Thus, not the complete robot is relevant but only its foot point. The new module provides the positions of other players on the field by processing visual information and comprises the following stepwise analysis:

- Pre-selection of interesting areas out of a whole image making use of the region analysis of the B-Human framework.
- Calculation of color histograms for the pre-selected areas.
- Down-scaling histograms to a fixed size which reduces their dimensionality.
- Utilization of a neural network to classify the reduced data.
- Consistency checks ensure the final representation only consists of the bottommost detection at a certain x-position assuming that this position refers to the feet of the robot feet whereas the ones above most likely belong to the same robot.
- Transformation of the center of the areas where robots are detected into the actual field position.

Subsequently, the steps of processing are described in further detail and the preparation of training data is stated.

4.1 Finding Potential Robot Locations

During the region analysis within the B-Human system, white regions are classified whether they potentially belong to lines or not. Those regions which do not meet the criteria for lines, such as a certain ratio length and width, a certain direction, and only little white in the neighborhood, are collected as so called non-line spots. The region classification has a low false positive rate, hence it takes most of the actual line fragments but no more. This means, the non-line spots include regions belonging to robots, to lines, in particular crossings of lines, and sometimes also to the net and to the boards. Non-line spots, associated to robots, usually belong to the lower part of the robot body as only objects below the field border are considered for region building. In the case where a robot is standing, the upper body part normally appears above the field border and thus cannot cause non-line spots except the field border is distorted.

As the classification whether a spot is a robot or not is done as often as there are potential robot positions, these non-line spots are merged in advance if they are close to each other. Proximity is defined relative to the expected width of a robot at the same location. Hence less classifications are needed which increases efficiency. The result of two merged locations is represented by a point with the average x-coordinate and the maximum y-coordinate of the original locations. The origin of the image coordinate system is at the upper left corner. The y-direction is not averaged, as the foot points of the robots are most important because they are needed to project the image position to the field. Non-line spots that cannot be merged are reduced to their own maximum y- and average x-coordinate.

This merging reduces the number of potential positions immensely, so that unless a robot is very close, it is usually represented by a single potential spot located between its feet. Importantly, from one frame to another, the potential robot locations deviate slightly. This is caused by deviations in the exact positions of the non-line spots to merge. As a consequence, the detection algorithm is required to be robust against such displacements.

4.2 Histograms and Linear Interpolation for Complexity Reduction

The neural network detection algorithm expects all classifier input to have the same dimension, as is the case for most classifiers. Additionally, the complexity of the

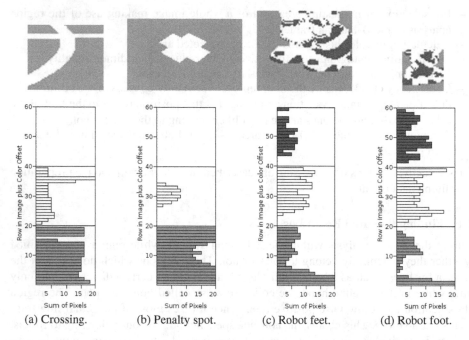

(a) Crossing.   (b) Penalty spot.   (c) Robot feet.   (d) Robot foot.

Fig. 1. Horizontal color histograms of potential robot positions. The original windows are shown at the top. Note that they are not necessarily quadratic due to overlay with the image border.

algorithm heavily depends on the dimensionality of the input. Thus, some effort is made for preparing the input data accordingly.

For each potential robot position, a quadratic window of the expected robot-width at the respective position is extracted out of the color-classified image. More precisely, the window is quadratic unless there is an image border which crops the window.

The first step of dimensionality reduction is to obtain the color histogram of each window. To this end, every window is traversed pixel-wise. While traversing, the sum of pixels of each color is recorded for each row. As this summation is done on an already color-classified image, the number of different colors is usually three: white, green and "none". There exist some more colors which do not occur in the majority of windows of interest, such as orange, blue and yellow, which are disregarded.

The second step is to scale each histogram to a common length of 20, which is a sufficient size. Thereby, histograms of non-quadratic windows are brought to a consistent size. For scaling, linear interpolation is applied to each color of the histogram separately. Hence, the final input vector is of dimension 60.

Figure 1 shows a variety of windows obtained from potential robot positions as well as their respective scaled histograms.

4.3   Classification of Potential Robot Locations

The scaled histograms serve as input to the a neural network which is implemented and trained to decide whether a histogram originates from a robot or not. The utilized network implementation follows a fully connected feed-forward architecture. It has 60 input neurons, as this is the size of the histograms to classify, and 2 output neurons representing the classes "robot" and "non-robot". All non-input neurons use the Fermi function as non-linear activation function. Training is accomplished by backpropagation of error. In order to find a good network configuration, several architectures have been explored empirically.

4.4 Preparation of Training Data

The training data sets are derived from a simulated as well as a real scene. The simulated data is obtained from the camera of one robot out of three robots moving around the field. For the samples, taken from a real scene, only the recording robot is moved around while the two others are standing still at different positions on the field. The color-classified windows of potential robot locations are sorted manually into three subgroups. One group is formed by positive data, meaning the pictures which clearly show the feet of a robot. One group consists of pictures showing for instance line fragments or parts of the board, which is the negative data. The third group contains all pictures for which both answers are valid. For example, if a robot hand is shown in the picture, this is considered neither positive nor negative. Either result is acceptable, because hands usually occur above the feet in about the same x-position and the detection module only considers the windows with maximum y-coordinates for each position. Excluding such ambiguous samples keeps the learning task simple and thus allows for a rather simple network architecture.

Out of the sets of positive and negative data, training patterns are generated. For each picture, the 3×20-dimensional color histogram is calculated. This histogram serves as input pattern, whereas the expected output is defined by means of a 1-of-2 encoding.

The prepared training patterns from the real scene as well as from the simulated scene comprise about 1000 positive examples and 1000 negative examples each which are used for training. The remaining patterns are retained for testing the trained networks: For the samples derived from simulation, there is the same amount of test patterns as for training. The amount of test patterns for the reality-derived samples is 250.

5 Evaluation

5.1 Choice of Network Structure

With respect to the contrary requirements of maximal accuracy and minimal computation time, it is worth choosing a network architecture which is as cheap as possible regarding time consumption while providing a reasonable capability to classify the potential robot locations. On the basis of training data obtained from the simulation, two out of the possible architectures are studied in detail regarding their performance for different variants of the network input.

The types of network input analyzed are horizontal as well as vertical color histograms. Vertical histograms are computed by the amount of pixels of each color per column unlike those described in Sect. 4.2 for the case of horizontal histograms. Also

Table 1. Accuracy and computational cost of different networks with different types of input. All networks are trained and tested using data obtained from simulation. Learning the classification task with horizontal histograms as input yields the highest accuracy. Especially the generalization capability is enhanced compared to all other variants with at least equally high accuracy on the training set. The number of multiplications is derived assuming a fully connected feed-forward network and a bias neuron in each non-output layer.

Architecture	Input type	Accuracy (%) Training set	Test set	Computational cost (# multiplications)
60 - 2	Vertical histograms	90.4	88.6	122
60 - 20 - 2	Vertical histograms	98.9	95.1	1262
60 - 7 - 3 - 2	Vertical histograms	99.1	94.0	459
60 - 8 - 2	Vertical histograms	98.3	93.2	506
60 - 7 - 3 - 2	Horizontal histograms	98.8	96.5	459
60 - 8 - 2	Horizontal histograms	99.0	96.4	506
60 - 7 - 3 - 2	Normalized horizontal histograms	98.4	91.0	459
60 - 8 - 2	Normalized horizontal histograms	98.6	91.7	506
40 - 7 - 3 - 2	Two-colored vertical histograms	82.7	82.5	319
400 - 7 - 3 - 2	Full color-classified image	99.1	84.8	2839

the benefit of normalizing this histogram data by subtracting the mean before presenting it to the network is examined. Furthermore, the use of only two-colored histograms is considered. This is motivated by the fact that detection windows mostly consist of exactly three colors. Thus, in three-colored histograms, one color can be expressed by subtraction of the two others from the maximum histogram height. Omitting one color yields a histogram of size 2×20 referring to the column-wise amount of green and white pixels, and accordingly, the network input is of dimension 40. Moreover the complete color-classified detection windows scaled to a size of 20×20 are taken as 400-dimensional network input in order to determine whether the use of histograms is at all preferable over larger input dimensions.

The two network architectures for which different input kinds are analyzed are built up as follows: The first network has a 60-dimensional input layer followed by one hidden layer with 8 neurons and an output layer of 2 neurons. This architecture is denoted as 60-8-2. The second one has two hidden layers, the first one with 7 neurons and the second with 3 neurons, which is hereafter referred to as 60-7-3-2. In both architectures neighboring layers are fully connected. In order to justify the choice of these two architectures for detailed analysis, also the network architectures 60-2 and 60-20-2 are evaluated in terms of their ability to solve the classification task. Accuracy as well as the computational complexity are compared. The comparison is based on an input of vertical histograms and is summarized in Table 1.

The most accurate networks are obtained by utilizing three-colored horizontal histograms. Utilizing these, the test data can be classified 96.5% accurate by a 60-7-3-2 network as well as a 60-8-2 network.

Table 2. Accuracy for different input data sets. Test data obtained from the simulation can be classified more accurate by a network trained on simulation data than by a network trained with real data and vice versa. If pattern of both, the real and the simulated data are presented during training, the resulting network can perform equally well on both types of data.

Architecture	Training set	Accuracy on test set (%)	
		Real	Simulated
60 - 7 - 3 - 2	Simulated	69.9	96.5
60 - 8 - 2	Simulated	71.8	96.4
60 - 7 - 3 - 2	Real	95.5	85.0
60 - 8 - 2	Real	96.0	83.9
60 - 7 - 3 - 2	Mixed	95.9	95.8
60 - 8 - 2	Mixed	93.4	93.0

5.2 Application on the Real System

For analyzing the performance on real data networks with 8 and with 7-3 hidden neurons and horizontal three-colored histograms as input type are considered. Training is repeated with samples from the real system and with a set of mixed samples from the real as well as the simulated environment. An overview on the results is given in Table 2 where cross tests between simulated, real and mixed training and validation sets are conducted.

The best 60-7-3-2 network obtained after training on a mixed data set can classify unknown real data with an accuracy of 95.9% and performs equally well on simulated data. This shows that detection of robots is transferable between the real and the simulated system and also suggests that due to the color space discretization, the robot detection is fairly independent from the lighting conditions on the field if the network has learned the concept of what a robot is in a sufficiently abstract manner.

5.3 Evaluation of Speed

For measuring the average processing speed of the robot detector a real scene is considered. The setting resembles the one shown in Fig. 2 except that real robots are used.

If the robot detector utilizes a 60-8-2 network, about 1.1 ms of computation time are needed for evaluation of one image. For a 60-7-3-2 network, the processing time is only 1 ms. For comparison, the processing takes on average about 4.7 ms when utilizing a

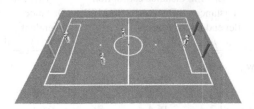

Fig. 2. Reconstruction of the setting for evaluating speed. The experiment has been conducted on the real system. The blue robot records data. All robots are standing still.

400-7-3-2 network. Hence, the robot detector is usable in the real-time vision system which provides 30 frames per second.

5.4 Evaluation of Accuracy

Accuracy of the robot detection module is accessed on two different levels. One is to measure the quality of the classification provided by the neural network. The other is to measure how accurate position of other robots can be estimated with the developed robot detector.

Detection Rates in Comparison to k-Nearest Neighbors. k-Nearest Neighbors (kNN) is a popular classification algorithm since it is straightforward and easy to implement. Like in the context of handwritten digit recognition [8,7], kNN is incorporated as a benchmark in order to rate the performance of the neural network based robot detection.

For the comparison, the kNN algorithm is initialized with the mixed set used to train the 60-7-3-2 network in Sect. 5.2 and $k = 1$. It yields an accuracy of 96.4% on the test data. The trained network can classify the same set of samples with an accuracy of 95.8% which shows that the performance of the network is similar, if not equal the benchmark.

Accuracy of Positions of Detected Robots. The accuracy of the position estimation of the robot detector is determined by comparing the detected positions to independently derived position information. For this purpose, a scene with a defined setting is examined in simulation and on the real field. For the latter, a motion capture system is used to obtain an independent measurement of the robot positions. The scene itself is constituted by two robots on the field. One is standing still on the penalty spot, which is at position $(1.2\,\mathrm{m}, 0\,\mathrm{m})$. The other starts at the opposite goal line, coordinates

Table 3. Position estimation error in each experiment. The overall observed error in distance and angle is subdivided into three components. Distance dependent refers to the offset in slope of the fit of the detections compared to the reference. The distance independent error refers to the y-intercept of the fit, i. e. the permanent offset towards the reference. The RMSD value yields from analysis of the deviation of detections towards the fit.

Setting	Error in distance estimation			Error in angle estimation		
	Distance dependent ($\%/\mathrm{m}$)	Distance independent (%)	RMSD	Distance dependent (deg/m)	Distance independent (deg)	RMSD
Simulation, frontal view	4.9	-2.0	7.1	0.29	-0.87	1.44
Simulation, back view	1.3	11.7	16.5	0.97	-2.14	2.35
Simulation, side view	14.3	-8.8	11.9	0.57	-2.01	1.85
Simulation, lying robot	-1.2	16.4	6.6	-0.00	3.74	2.90
Real scene, frontal view	3.8	2.4	11.5	-0.61	0.82	3.58

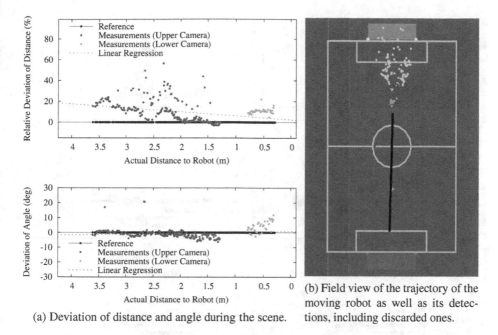

(a) Deviation of distance and angle during the scene.

(b) Field view of the trajectory of the moving robot as well as its detections, including discarded ones.

Fig. 3. Real field, frontal view: Position estimation accuracy of the robot detector in a real scene with frontal view of the robot to detect. The plots in (a) show the deviation of measured distances and angles towards the standing robot throughout the captured scene. The deviation in distance measurements is depicted as a percentage of the actual distance while negative distances refer to measurements shorter than the reference. The deviation of the angle is depicted in degrees relative to the reference. Colors encode which camera a measurement originates from. Detections in the blind spot between the fields of view of the cameras have been discarded. The dotted line in each plot denotes the fit obtained by linear regression on all depicted data points. (b) provides a field view of the perceptions. The color encoding refers to the cameras as in (a). Additionally, pale red spots indicate locations of discarded perceptions. The movement of the recording robot yields the trajectory visualized in black. The robot to be detected is located at the upper penalty spot, in the real scene this is position $(1.24\,\mathrm{m}, 0.07\,\mathrm{m})$.

are $(-3\,\mathrm{m}, 0\,\mathrm{m})$, and moves towards the standing robot. Meanwhile, its estimated distances to the standing robot are recorded. In order to minimize distortion, the head of the recording robot is kept still at zero degrees. This scene is played in the simulator as well as on the real field. In the simulation, to confirm that the detection is view-independent, the simulated scene is replayed with the standing robot oriented to the side as well as the back. Also, the scene is recorded with this robot lying on the penalty spot.

Importantly, although the detection algorithm does not involve filtering, some detections are not considered for the analysis. The robot does not move the head while recording in the experiments. Thus there is a blind spot between the image of the upper and the lower camera. Detections are considered not meaningful if they occur at the lower border of the upper camera image which traces back to the feet of the standing robot being in the blind spot. Likewise, if detections originate from the upper camera

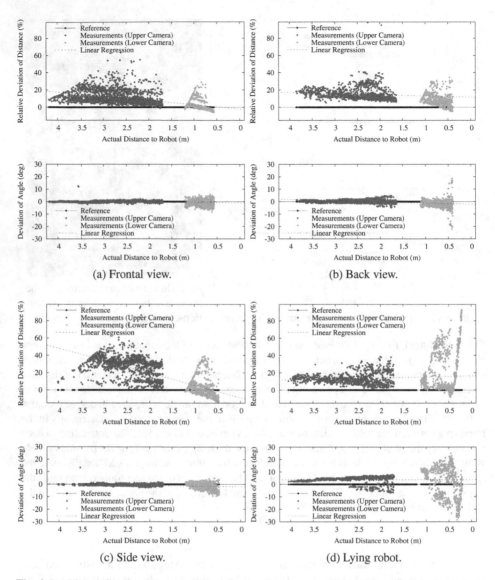

Fig. 4. Position estimation accuracy during the simulated scenes with the standing robot oriented to different directions. For plot details see Fig. 3a. In (d), the linear regression is applied on upper camera measurements only. Perceptions from the lower camera are considered not meaningful, as the lying robot horizontally fills the complete image and hence there is no specific foot point.

while the lower camera provides perceptions, they are discarded. Such detections often refer to the upper body parts of the robot of which the feet are perceived through the lower camera. The majority of discarded perceptions originates from the hands of the robot which do not look too different from the feet.

Overall, the position estimation is found to provide a reasonable amount of accuracy for any perspective. As summarized in Table 3, distance estimations deviate by at most 16.5%. The angle deviates by 2.9° in the worst case. Yet, for the case of side view, the deviation in distance is enlarged due to detections of the body instead of the feet (see Fig. 4c). Such larger deviations for far-away robots are acceptable as they most likely have no impact during play. Detecting a robot at a distance of 4.0 m which is in fact only 3.5 m away under an accurate angle will usually not make any difference to a player's behavior. In case the detected robot is lying, the position estimation seems to provide no meaningful results if the distance is shorter than 1 m (see Fig. 4d). This imprecision however is not necessarily a drawback. As a lying robot actually covers more ground than a standing one, the curve necessary to pass this robot might need to be larger than if it was standing. The amount of deviation in the angle estimation could be used as a hint whether the detected robot is lying on the ground.

In the experiment conducted on the real field, the recording robot is not moving autonomously but slided along the route in order to minimize distortion factors in this experiment. As depicted in Fig. 3, the obtained results correspond to the findings in the simulated experiments. In total, there is only a small error, in particular a distance independent offset of 2.4% and additionally an error of 3.8 %/m dependent on the distance. The root mean square deviation (RMSD) of the data towards the fit is 11.5% as there is a number of detections which deviate by approximately 30% from the reference. These detections mainly occur at a distance of 2 m to 3 m and probably refer to the upper legs or chest of the robot like observed in simulation. The angle estimation as well matches the results from simulation. Though, the RMSD of 3.58° is remarkably larger due some outliers which this measure accounts for.

Notably, the angle estimations which originate from the lower camera only deviate to one direction unlike observed in the experiments before. This might be caused by inaccurate calibration of either the motion capture system or the transformation matrix of the robot. Another reason could be that the position of the standing robot changed slightly between measuring its position and capturing the scene. Likewise, it is possible, that the detection window is not central on the feet for most of the perceptions. But as this issue has not been encountered during simulation and the distance estimation is for the same perceptions as well more than 10% too large, a calibration issue is the more likely explanation.

6   Conclusion and Future Work

The presented neural network based algorithm is suitable for the robot detection task. It provides reasonable accuracy and is sufficiently efficient in terms of computational cost. The major contributions to efficiency are the pre-selection of potential robot positions, the reduction of image regions to color histograms and the use of a network with a small hidden layer.

Still, there is room for improvements. The most obvious one is a filtering algorithm such as a Kalman filter [14]. During evaluation, perceptions from the upper camera have been omitted if there are results from the lower one and also perceptions from the lower border of the upper camera have been considered as invalid. Including these criteria into the algorithm will also be an enhancement. Additionally, as robots are never detected closer than they actually are but sometimes further away, a confidence factor could weight perceptions more the closer they are.

In terms of accuracy, possible improvements can be made to the overall detection rate as well as to the precision of estimated positions of robots. The latter could be enhanced by explicit calculation of the foot point of the detected robot. Currently, the place where a detected robot meets the ground is assumed to be the same as the center of the detection window. This assumption holds as long as the feet are actually detected. But if knees, waist, shoulders or arms yield positive detections, this assumption is no longer valid. As the image segmentation already exists, the exact foot point could be derived by traversing continuous white segments within the detection window downwards until a green region is found.

In this work, the robot detection approach has been considered in an isolated way. The next steps would be to integrate the resulting new perceptions into the behavior control system and to combine them with other perceptions.

A promising combination is to merge the robot detections with data obtained from the ultrasonic devices. At least within the range of up to 1.5 m the ultrasonic distance measure is very accurate and thus can refine the distance estimation. At the same time, the angle estimation which the ultrasonic sensors provide with an uncertainty of $60°$ can be refined by the neural network based detector.

Regarding behavior control, robot perceptions definitely conduce to reactive obstacle avoidance as well as to planning paths on the field. In order to improve team play, perceptions of robots could be combined with localization information. The self localization is usually propagated via WLAN among the players of one team. As yet, it is rather error prone and thus cannot be used to precisely pass the ball between players. If the propagated position information can be verified and further refined by a robot detection in the same place, passing the ball with sufficient precision becomes possible.

Acknowledgement. This work was partially funded by the German Research Foundation (DFG), grant BE 2556/2-3.

References

1. Aldebaran Robotics: Nao Robot Reference Manual, Version 1.10.10, Internal Report (2010)
2. Bradski, G.R.: The OpenCV Library (2000), http://opencv.willowgarage.com/
3. Daniş, F.S., Meriçli, T., Meriçli, Ç., Akın, H.L.: Robot Detection with a Cascade of Boosted Classifiers Based on Haar-Like Features. In: Ruiz-del-Solar, J., Chown, E., Ploeger, P.G. (eds.) RoboCup 2010. LNCS (LNAI), vol. 6556, pp. 409–417. Springer, Heidelberg (2010)
4. Fabisch, A., Laue, T., Röfer, T.: Robot Recognition and Modeling in the RoboCup Standard Platform League. In: Proc. 5th Workshop on Humanoid Soccer Robots at Humanoids (2010)
5. Fasola, J., Veloso, M.M.: Real-time Object Detection using Segmented and Grayscale Images. In: IEEE International Conference on Robotics and Automation, pp. 4088–4093 (2006)
6. Lange, S., Riedmiller, M.: Appearance-Based Robot Discrimination Using Eigenimages. In: Lakemeyer, G., Sklar, E., Sorrenti, D.G., Takahashi, T. (eds.) RoboCup 2006. LNCS (LNAI), vol. 4434, pp. 499–506. Springer, Heidelberg (2007)

7. Lee, Y.: Handwritten Digit Recognition Using K Nearest-Neighbor, Radial-Basis Function, and Backpropagation Neural Networks. Neural Computation 3, 440–449 (1991)
8. Liu, C.-L., Nakashima, K., Sako, H., Fujisawa, H.: Handwritten Digit Recognition: Benchmarking of State-of-the-art Techniques. Pattern Recognition 36(10), 2271–2285 (2003)
9. Mayer, G., Kaufmann, U., Kraetzschmar, G.K., Palm, G.: Neural Robot Detection in RoboCup. In: Wermter, S., Palm, G., Elshaw, M. (eds.) Biomimetic Neural Learning for Intelligent Robots. LNCS (LNAI), vol. 3575, pp. 349–361. Springer, Heidelberg (2005)
10. Röfer, T., Laue, T., Müller, J., Bösche, O., Burchardt, A., Damrose, E., Gillmann, K., Graf, C., de Haas, T.J., Härtl, A., Rieskamp, A., Schreck, A., Sieverdingbeck, I., Worch, J.H.: B-Human Team Report and Code Release 2009 (2009)
11. Ruiz-del-Solar, J., Verschae, R., Arenas, M., Loncomilla, P.: Play ball! fast and accurate multiclass visual detection of robots and its application to behavior recognition. IEEE Robotics Automation Magazine 17(4), 43–53 (2010)
12. Sony Corporation: AIBO (1999), http://support.sony-europe.com/aibo/
13. Viola, P.A., Jones, M.J.: Rapid Object Detection using a Boosted Cascade of Simple Features. In: CVPR, vol. 1, pp. 511–518 (2001)
14. Welch, G., Bishop, G.: An Introduction to the Kalman Filter. Tech. Rep. 95-041, University of North Carolina at Chapel Hill, Chapel Hill, NC, USA (1995)
15. Wilking, D., Röfer, T.: Realtime Object Recognition Using Decision Tree Learning. In: Nardi, D., Riedmiller, M., Sammut, C., Santos-Victor, J. (eds.) RoboCup 2004. LNCS (LNAI), vol. 3276, pp. 556–563. Springer, Heidelberg (2005)

Gradient Vector Griding:
An Approach to Shape-Based Object
Detection in RoboCup Scenarios

Hamid Moballegh, Naja von Schmude, and Raúl Rojas

Institut für Informatik, Arbeitsgruppe Künstliche Intelligenz,
Freie Universität Berlin, Arnimallee 7, 14195 Berlin, Germany
moballegh@gmail.com, {schmude,rojas}@inf.fu-berlin.de
http://www.fumanoids.de

Abstract. This paper describes a new method of extraction and clustering of edges in images. The proposed method results a graph of detected
edges instead of a binary mask of the edge pixels. The developed algorithm contains a sequential pixel-level scan, and a much smaller second
and third pass on the results to determine the connectivities. It is therefore significantly faster than Canny edge detector, performing both edge
detection and grouping tasks. The method is developed for a RoboCup
scenario, however it can also be applied to any other image as long as
the prerequisites are met. The paper explains the idea, discusses the
prerequisites and finally presents the implementation results and issues.

1   Introduction

The vision system is the most important module of a robot to perceive its environment in RoboCup competitions. The task of the vision system is to analyze
an incoming image for detecting important structures or areas (features) and
combine these features to meaningful objects according to the needs of the other
modules of the control software. As vision-based measurements are often used in
closed loop control of the robots, a realtime performance is required. Due to the
fact that on-board CPU power is strongly limited, high-performance and very
specialized algorithms are needed.

Common approaches use image segmentation by color to deal with the object
extraction[11,2]. The reason is that the relevant objects like goals, robots and
the ball can be distinguished by their color. Acceptable detection rates can however only be achieved in presence of a constant and well conditioned lighting.
Moreover, almost all color-based object recognition methods require a manual
or automatic calibration, which makes them hard to use[5,6,7].

Shape-based approaches are rare but are getting more popular in RoboCup
because the RoboCup rules change more and more from the highly specialized
scenario to a more general one[8,9]. Therefore object recognition can't rely only
on the color of the objects in the future, but needs to take the shape of the

T. Röfer et al. (Eds.): RoboCup 2011, LNCS 7416, pp. 162–173, 2012.

objects into account. Many methods have been introduced to detect RoboCup field objects based on their shapes. A majority of these methods focus on the field lines since the uncertainties of other landmarks are prohibitively large[9]. Different techniques have been proposed to extract the field lines. Some examples are the use of kernels[10], Hough transformation[1,10] and edge detection techniques[8].

In this paper we introduce a shape-based object detection scheme, which results a directed graph of detected edges. The graph isn't build on pixel basis but on cell basis; that means the image is overlaid with a grid formed of equal sized cells. A number of graph nodes is calculated for each cell. The nodes are then connected corresponding to the connectivity in the 8-neighborhood of each cell. This is similar to the technique suggested by A. Farag and E. Delp[4]. The algorithm allows adjusting the maximum curvature of the result paths. The idea of griding is inspired by the Histogram of Oriented Gradients (HOG) introduced by Dalal and Triggs[3]. However the final implementation differ greatly from the original work.

The paper is structured as follows: At first the four steps of the method are described. Several techniques are explained to increase the performance of the method as well as to improve the quality of the results. Finally achieved results from the algorithm are presented and the implementation issues are discussed.

2   Object Detection Using Gradient Vector Griding

The main idea of the Gradient Vector Griding (GVG) method is to reduce computation cost by reducing the image dimension to a much smaller grid. An important prerequisite for the method is that the information loss according to this stage remains small. This is provided in RoboCup case. For each cell, one or more features are calculated from the original image, where each feature represents an edge passage through the cell. These features are the so-called *edge representers* (ER). This is unlike HOG where a histogram of orientations is extracted for each cell. The next stage finds connected edges passing through neighboring cells; the connectivity is calculated. In the last step, complete edge traces are extracted out of the cell connectivity graph.

Figure 1 shows the basic implementation of the GVG algorithm. All stages are explained in detail in the next subsections.

2.1   Gradient Vector Calculation

Like every shape based object recognition method, the procedure begins with the calculation of the gradient vector for each pixel in the image. Sobel operators can be used to compute the components of the gradient vector in each pixel. The better alternative is Robert's cross operator described in equations 1 to 4. Note the gradient vector provided by this method is $45°$ rotated.

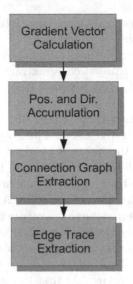

Fig. 1. Stages of Gradient Vector Griding

$$G_x : \begin{bmatrix} 1 & 0 \\ 0 & -1 \end{bmatrix} \tag{1}$$

$$G_y : \begin{bmatrix} 0 & 1 \\ -1 & 0 \end{bmatrix} \tag{2}$$

$$|G| = \sqrt{G_x^2 + G_y^2} \text{ or } |G| = |G_x| + |G_y| \tag{3}$$

$$\theta = \arctan(G_x/G_y) + \frac{\pi}{4} \tag{4}$$

Sobel operators are less sensitive to noise, however the accumulation in the next stage of the algorithm makes the Robert's cross version pretty noise tolerant.

There are two important points to consider when calculating the gradient vector for GVG. First, it is necessary to calculate the direction in a -180° to 180° range, i.e. using the function $\arctan 2(y, x)$ instead of $\arctan(y/x)$. This prerequisite is discussed in detail in the next subsection. The second point is that the method requires the edge direction which is perpendicular to the gradient direction. The following convention is used in the algorithm to convert gradient vector **V** to edge vector **E**, which is a simple 90° rotation.

$$\mathbf{E} = \begin{bmatrix} 0 & -1 \\ 1 & 0 \end{bmatrix} \mathbf{V} \tag{5}$$

If Robert's cross operator is used for differentiating, edge directions should be calculated as follows, which is a rotation of 135°.

$$\mathbf{E} = \begin{bmatrix} -\frac{\sqrt{2}}{2} & -\frac{\sqrt{2}}{2} \\ \frac{\sqrt{2}}{2} & -\frac{\sqrt{2}}{2} \end{bmatrix} \mathbf{V} \tag{6}$$

2.2 Position and Direction Accumulation

A selective accumulation method is used to calculate a set of *ER*s for each cell. As mentioned above, an *ER* should refer to an edge passing through the cell. In the case that the cell contains a single linear edge (shown in figure 2(a)), the *ER* is simply extracted by averaging the position and edge vectors of the pixels inside the cell belonging to the edge. As the complexity of the cell contents increases, simple averaging fails as shown in figure 2(b). It is then required to distinguish between multiple edges using a more general clustering algorithm.

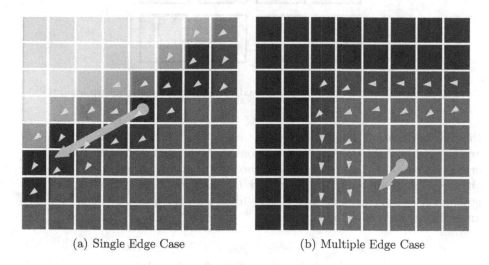

(a) Single Edge Case   (b) Multiple Edge Case

Fig. 2. Average gradient direction and position in a cell. Averaging fails when multiple edges appear inside a cell (b) but works in the single edge case (a).

The problem cannot be solved in general form without a noticeable increase in the amount of calculations. We noticed that a partial solution should be enough for RoboCup use. Figure 3 presents some examples commonly observed in RoboCup scenario. Samples are overlaid with edge direction vectors. An often observed case is a field line or a side pole included with both side edges inside a single cell. The other but less frequent observation is a more or less 90° corner as a result of either an intersection between two lines or a part of a rectangular object. Thanks to 360° representation of the gradient direction, it is possible to separate the edges in a majority of the cases based only on the direction information. Note that the edges of an object such as a field line form two complement directions although they are parallel in the image.

The algorithm described in figure 5(a) is developed based on this idea. As a matter of optimization it is preferred to have a one-pass algorithm so that it can also be implemented without an entire image buffering. The algorithm functions as follows: The image is scanned pixel by pixel. Two *ER*s are used in

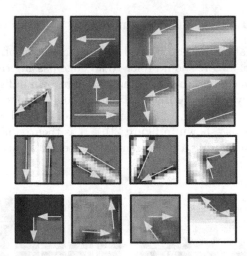

Fig. 3. Common examples of multiple edge distribution in RoboCup scenario

this implementation. Each ER contains a position, an orientation accumulator and a pixel counter. An edge pixel is joined to the first ER when its gradient orientation is closer than a certain distance to the average orientation of the ER. Otherwise it is compared to the second ER, and in the case of no match it is ignored. An empty ER will obviously be filled with the first edge pixel met. The angle distance is defined in equation 7.

$$|\theta_1 - \theta_2| = \begin{cases} |\theta_1 - \theta_2 + 360| & \theta_1 - \theta_2 < -180 \\ |\theta_1 - \theta_2 - 360| & \theta_1 - \theta_2 > 180 \\ |\theta_1 - \theta_2| & otherwise \end{cases} . \tag{7}$$

As a further optimization to the algorithm it is possible to replace the polar representation of the gradient vector with a cartesian one, i.e. using directly the edge vector (e_x, e_y) instead of its orientation θ.

The implementation of angle distance thresholding will then be replaced with thresholding the inner product $\mathbf{V}_1 \cdot \mathbf{V}_1 = |\mathbf{V}_1||\mathbf{V}_2| \cos\theta = e_{x_1} e_{x_2} + e_{y_1} e_{y_2}$ of the vectors $\mathbf{V}_1$ and $\mathbf{V}_2$ as shown in equation 8.

$$\mathbf{V}_1 \cdot \mathbf{V}_2 > |\mathbf{V}_1||\mathbf{V}_2| \cos(thr) \tag{8}$$

$$\Leftrightarrow e_{x_1} e_{x_2} + e_{y_1} e_{y_2} > \sqrt{e_{x_1}^2 + e_{y_1}^2} \sqrt{e_{x_2}^2 + e_{y_2}^2} \cos(thr) \tag{9}$$

This saves the dynamic calculation of the arctan 2 function needed in the polar presentation. According to the CPU documentation, 32 bit multiplication can be performed in one cycle. However the normalization of the vectors $|\mathbf{V}_1|$ and $|\mathbf{V}_2|$ could cause performance problems, because the square root has to be calculated.

This can also be optimized away using the z component of the cross product $\mathbf{V}_1 \times \mathbf{V}_2 = |\mathbf{V}_1||\mathbf{V}_2| \sin\theta \mathbf{n}$, where $\mathbf{n}$ represents the normal vector to $\mathbf{V}_1$ and

$\mathbf{V}_2$. As the vectors are defined in $\mathbb{R}^2$, we use zero as z component of $\mathbf{V}_i$, so the calculation of the cross product results in

$$\mathbf{V}_1 \times \mathbf{V}_2 = \begin{pmatrix} e_{x_1} \\ e_{y_1} \\ 0 \end{pmatrix} \times \begin{pmatrix} e_{x_2} \\ e_{y_2} \\ 0 \end{pmatrix} = \begin{pmatrix} 0 \\ 0 \\ e_{x_1}e_{y_2} - e_{y_1}e_{x_2} \end{pmatrix}. \tag{10}$$

We can now write $|(\mathbf{V}_1 \times \mathbf{V}_2)_z| = |\mathbf{V}_1||\mathbf{V}_2|\sin\theta = |e_{x_1}e_{y_2} - e_{y_1}e_{x_2}|$. As $\tan\theta = sin\theta/\cos\theta$, the sinus and cosine function can be replaced with the cross and inner product described above.

$$\tan\theta = \frac{\frac{|(\mathbf{V}_1 \times \mathbf{V}_2)_z|}{|\mathbf{V}_1||\mathbf{V}_2|}}{\frac{\mathbf{V}_1 \cdot \mathbf{V}_2}{|\mathbf{V}_1||\mathbf{V}_2|}} = \frac{|(\mathbf{V}_1 \times \mathbf{V}_2)_z|}{\mathbf{V}_1 \cdot \mathbf{V}_2} = \frac{|e_{x_1}e_{y_2} - e_{y_1}e_{x_2}|}{e_{x_1}e_{x_2} + e_{y_1}e_{y_2}} \tag{11}$$

To avoid the division, we multiply the tangents with the cosine part and the final angle thresholding of the vectors is then given in equation 12.

$$|(\mathbf{V}_1 \times \mathbf{V}_2)_z| < \mathbf{V}_1 \cdot \mathbf{V}_2 \tan(thr) \tag{12}$$

$$\Leftrightarrow \quad |e_{x_1}e_{y_2} - e_{y_1}e_{x_2}| < e_{x_1}e_{x_2} + e_{y_1}e_{y_2}\tan(thr) \tag{13}$$

Using this technique, vector angle thresholding is done with only four integer multiplications having a great impact on the performance of the algorithm.

The grid structure can be chosen overlapped, i.e. each cell also covers half of every neighboring cell. This increases the smoothness of the results but is computationally more expensive.

2.3 Connection Graph Extraction

A list of *ER*s is produced through a single image scan. A connection graph should be calculated by scanning the grid and comparing *ER*s of neighboring cells. The graph is implemented using an extra field in the cell structure pointing to the following neighbor *ER* called *"outbound"* and a boolean field indicating that the *ER* is added to the trace called *"inbound"*.

The algorithm is shown in figure 5(b). Upon two conditions two neighbor *ER*s are marked as connected. The first condition verifies that both *ER*s are in the same direction. This condition is however not enough as it also holds true for separate parallel edges. Therefore the second condition verifies that the vector connecting the *ER*s is also in the same direction as both ERs. By adjusting the thresholds the maximum accepted curvature of the edge trace can be determined. Both conditions can be optimized using the technique described in the last section. Note that for the second condition the edge direction should be used. Figure 4 demonstrates different examples of neighboring *ER*s. A connection is only accepted in example 4(d).

The internal loop of the algorithm breaks as soon as a connection is found. This guarantees that every node in the connection graph has an out degree of at most one. It is encouraged also to reduce the in degree of the nodes to at most

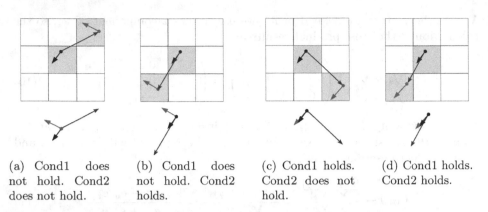

(a) Cond1 does not hold. Cond2 does not hold.

(b) Cond1 does not hold. Cond2 holds.

(c) Cond1 holds. Cond2 does not hold.

(d) Cond1 holds. Cond2 holds.

Fig. 4. Examples of connected and unconnected neighbors

one. This is implemented by refusing a connection if the destination node has already been connected by checking its *inbound* property.

It is theoretically not guaranteed that the connection graph becomes free from loops, however if a loop exists it should be the result of an uninterrupted semi-circular edge in the image which does not usually appear in RoboCup images.

2.4 Edge Trace Extraction

The final stage of the algorithm produces an array of edge traces. Each component of the array is a connected *ER* chain. The algorithm is demonstrated in figure 5(c). It searches the graph for source nodes, which are simply *ER*s with the *inbound* property not set. Upon a match, the trace is followed using *outbound* pointers and the matching *ER*s are pushed in the trace.

3   Implementation and Experimental Results

The algorithm was implemented as the low level part of the vision software for participation in RoboCup 2010. Two *ER*s are calculated per cell. The grid contains 40×30 non-overlapping cells, each covering 256 pixels of the captured image.

To achieve the required frame rate and still have enough CPU power free for other processes running, the following optimizations are applied to the algorithm.

Quarter Resolution Scan. Color digital cameras usually provide images with a so called *"Bayer pattern"*. A VGA image contains therefore 640×480 single channel values. This is 1/3 of the information recorded in common RGB pattern. The remaining values are interpolated in such representations. Hence it is possible to skip every other pixel and every other image line without a distinct loss of information.

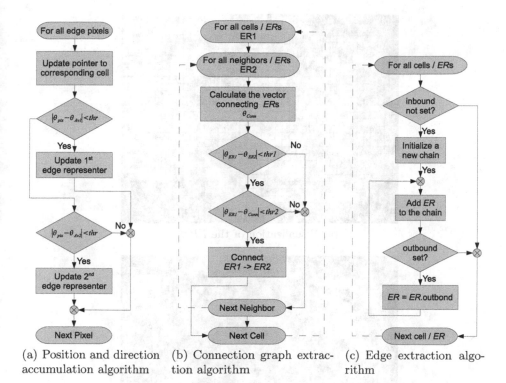

(a) Position and direction accumulation algorithm

(b) Connection graph extraction algorithm

(c) Edge extraction algorithm

Fig. 5. Algorithms

Over Horizon Skip. The only objects which partly appear over the horizon are the landmarks and other robots. To detect and calculate the position of these objects it is always enough to find the lowest point belonging to them. Therefore there is no need to detect objects over the horizon. As the camera is equipped with a wide angle lens, the horizon is projected as a curve in the captured image. This can however be estimated with a horizontal line touching the actual projection in the top-most point. There are different methods to calculate the horizon in the image. Two solutions currently used in FUmanoids are IMU sensors and field range detection. Upon detection of the horizon, scanning the image can vertically begin from this line.

Out of Circle Skip. Fish eye optics used in FUmanoid robots project the image inside a circle. The rest of the imaging surface is covered with black pixels. This effect can be observed in figure 6. It is possible to skip these pixels by calculating the horizontal extents for each image line.

Random Line Skip. Apart from the fish-eye optics deformation, the camera provides a perspective view of the field. Therefore much more information is available from near objects, observed in the lower area of the image, than from far ones appearing in the upper area. Due to this fact it is possible to ignore more and more lines as the scan gets closer to the bottom of the image. This is done by comparing a uniformly distributed random number

(a) Visualization of the *ER*s.

(b) Extracted edges overlaid on the test image.

(c) Typical results of the test case.

Fig. 6. Implementation results of GVG

with a dynamic threshold calculated form the scan lines Y coordinate. The following formula is used to skip a line:

$$r < Y/640 \tag{14}$$

Here r is a random real number between 0 and 1 and Y is the vertical component of the scan line coordinate.

Implementation results are summarized in table 1 on two available processor platforms. The first platform is Gumstix Verdex Pro which is equipped with a 600 MHz PXA270 and the second platform is a Gumstix Overo with a 720 MHz OMAP3 3530 processor. Each row of the table shows the optimization added to the last state. The last row shows the highest optimization level achieved by applying all described techniques.

A standard test procedure is used for this performance measurement. The robot is placed upright on one of the penalty points directed to the goal placed on the other side of the field as shown in figure 6(c). Two parameters are measured, which together show the performance of the algorithm. These are processed frame rate and CPU usage. Since the camera delivers a limited number of frames per second and the CPU is also capable of performing a limited amount of processing, there are two possible scenarios.

1. A frame can be entirely processed before the next frame is available. In this case the processor goes idle and the CPU shows less than 100% of usage. Frame rate remains constant in this mode and CPU usage is used as the indicator of the performance.
2. A new frame gets available while the last frame is still being processed. This leads to a frame buffer overflow, which is in turn handled with dropping frames. CPU is never idle in this case and the usage indicator shows always 100%. Frame rate is then used as the performance indicator.

Using the Canny algorithm from the OpenCV implementation as edge detection, the described test procedure shows the results listed in table 2. Compared to the GVG, the Canny edge extraction is slower by a factor of 10. The Canny algorithm also only results a bitmap of the edges and doesn't perform any edge

Table 1. GVG Implementation Results

	Gumstix Overo		Gumstix Verdex	
	Frame rate (FPS)	CPU usage (%)	Frame rate (FPS)	CPU usage (%)
No optimization	12	100	10	100
Quarter res. scan	18	100	14	100
+Over horizon skip	20	100	17	80
+Out of circle skip	26	80	17	75
+Random line skip	26	65	17	60

Table 2. Canny Implementation Results

	Gumstix Overo		Gumstix Verdex	
	Frame rate (FPS)	CPU usage (%)	Frame rate (FPS)	CPU usage (%)
Canny	2.7	100	1.5	100

grouping, so this has to be done additionally. Therefore we can conclude that the GVG algorithm significantly outperforms the standard Canny procedure in the RoboCup scenario. Figure 6(a) shows the visualization of the ERs for a typical image captured by the robot. Despite the rough grid a relatively high level of detail is detected. In figure 6(b) the result of the edge trace extraction is demonstrated for field lines.

4   Conclusion

In this work we introduced an edge detection and grouping method based on gradient vector directions. The method was described and some key ideas to increase the performance of the method in each stage were presented. Implementation results and issues were given and discussed in the paper. It was shown as well that the new algorithm outperforms the Canny edge detector in the RoboCup scenario.

The developed method was successfully used as the pixel-level processor in the vision system of the RoboCup team FUmanoids in 2010. The results provided by this method facilitated robust object recognition as well as rapid localization of the robot. The team won the second place of the Humanoid Kid-Size competition.

References

1. Bais, A., Sablatnig, R., Novak, G.: Line-based landmark recognition for self-localization of soccer robots. In: Proceedings of the IEEE Symposium on Emerging Technologies, pp. 132–137 (2005)
2. Bandlow, T., Klupsch, M., Hanek, R., Schmitt, T.: Fast Image Segmentation, Object Recognition, and Localization in a RoboCup Scenario. In: Veloso, M.M., Pagello, E., Kitano, H. (eds.) RoboCup 1999. LNCS (LNAI), vol. 1856, pp. 174–185. Springer, Heidelberg (2000)
3. Dalal, N., Triggs, B.: Histograms of oriented gradients for human detection. In: IEEE Computer Society Conference on Computer Vision and Pattern Recognition, vol. 1, pp. 886–893 (2005)
4. Farag, A.A., Delp, E.J.: Edge linking by sequential search. Pattern Recognition 28(5), 611–633 (1995)
5. Gunnarsson, K., Wiesel, F., Rojas, R.: The Color and the Shape: Automatic On-Line Color Calibration for Autonomous Robots. In: Bredenfeld, A., Jacoff, A., Noda, I., Takahashi, Y. (eds.) RoboCup 2005. LNCS (LNAI), vol. 4020, pp. 347–358. Springer, Heidelberg (2006)

6. Heinemann, P., Sehnke, F., Streichert, F., Zell, A.: Towards a Calibration-Free Robot: The ACT Algorithm for Automatic Online Color Training. In: Lakemeyer, G., Sklar, E., Sorrenti, D.G., Takahashi, T. (eds.) RoboCup 2006. LNCS (LNAI), vol. 4434, pp. 363–370. Springer, Heidelberg (2007)
7. Mayer, G., Utz, H., Kraetzschmar, G.: Towards autonomous vision self-calibration for soccer robots. In: Proceedings of the 2002 IEEE/RSJ Intl. Conference on Intelligent Robots and Systems (2002)
8. Röfer, T., Jüngel, M.: Fast and Robust Edge-Based Localization in the Sony Four-Legged Robot League. In: Polani, D., Browning, B., Bonarini, A., Yoshida, K. (eds.) RoboCup 2003. LNCS (LNAI), vol. 3020, pp. 262–273. Springer, Heidelberg (2004)
9. Schulz, H., Liu, W., Stückler, J., Behnke, S.: Utilizing the Structure of Field Lines for Efficient Soccer Robot Localization. In: Ruiz-del Solar, J., Chown, E., Plöger, P. (eds.) RoboCup 2010. LNCS (LNAI), vol. 6556, pp. 397–408. Springer, Heidelberg (2010)
10. Strasdat, H., Bennewitz, M., Behnke, S.: Multi-cue Localization for Soccer Playing Humanoid Robots. In: Lakemeyer, G., Sklar, E., Sorrenti, D.G., Takahashi, T. (eds.) RoboCup 2006. LNCS (LNAI), vol. 4434, pp. 245–257. Springer, Heidelberg (2007)
11. Wasik, Z., Saffiotti, R.: Robust color segmentation for the robocup domain. In: Proc. of the Int. Conf. on Pattern Recognition (ICPR), pp. 651–654 (2002)

AnySURF: Flexible Local Features Computation

Eran Sadeh-Or and Gal A. Kaminka

Computer Science Department,
Bar Ilan University, Israel

Abstract. Many vision-based tasks for autonomous robotics are based on feature matching algorithms, finding point correspondences between two images. Unfortunately, existing algorithms for such tasks require significant computational resources and are designed under the assumption that they will run to completion and only then return a complete result.

Since partial results—a subset of all features in the image—are often sufficient, we propose in this paper a computationally-flexible algorithm, where results monotonically increase in quality, given additional computation time. The proposed algorithm, coined AnySURF (Anytime SURF), is based on the SURF scale- and rotation-invariant interest point detector and descriptor. We achieve flexibility by re-designing several major steps, mainly the feature search process, allowing results with increasing quality to be accumulated.

We contrast different design choices for AnySURF and evaluate the use of AnySURF in a series of experiments. Results are promising, and show the potential for dynamic anytime performance, robust to the available computation time.

1 Introduction

The use of computer vision in autonomous robotics has been studied for decades. Recently, applications such as autonomous vision-based vehicle navigation [1], 3-D localization and mapping [11,4,3] and object recognition [10] have gained popularity due to the combination of increased processing power, new algorithms with real-time performance and the advancements in high quality, low-cost digital cameras. These factors enable autonomous robots to perform complex, real-time, tasks using visual sensors.

Such applications are often based on a local feature matching algorithm, finding point correspondences between two images. There are many different algorithms for feature matching, however in recent years there is a growing research on algorithms that use local invariant features (for a survey see [16,13]). These features are usually invariant to image scale and rotation and also robust to changes in illumination, noise and minor changes in viewpoint. In addition, these features are distinctive and easy to match against a large database of local features.

Unfortunately, existing algorithms for local feature matching [2,11,12] are designed under the assumption that they will run to completion and only then

T. Röfer et al. (Eds.): RoboCup 2011, LNCS 7416, pp. 174–185, 2012.
© Springer-Verlag Berlin Heidelberg 2012

return a complete result. Many of these algorithms therefore require significant computational resources to run in real-time. As we show in the experiments, this prohibits some of the algorithms from being used in current robotic platforms (where computation is limited). For instance, a Nao[1] humanoid robot computing the full set of features in an image of size 640×480 requires 2.4 seconds using a state-of-the-art implementation of the SURF algorithm [2,15].

Note, however, that for many robotics applications, even partial results—a subset of all features in the image—would have been sufficient (for example, to estimate the pose of the robot for obstacle detection). On the other hand, being able to invest computation time in getting higher-quality results is also important, e.g., in object recognition or in building accurate maps. Indeed, robots can benefit from computationally-flexible algorithms, where the computation time is traded for the accuracy requirements of the task. To do this, simply interrupting the algorithm when needed is not enough: We need to guarantee that the results of the algorithm would necessarily monotonically increase in quality, given additional computation time. This class of algorithms is called *Anytime* [17].

In this paper we present AnySURF, an anytime feature-matching algorithm, which can accumulate results iteratively, with monotonically increasing quality and minimal overhead. We achieve flexibility by re-designing several major steps in the SURF algorithm [2], mainly the feature search process and the order of interest point detection. We additionally discuss the design choices underlying AnySURF.

We evaluate the use of AnySURF in a series of experiments. We first demonstrate that non-anytime feature matching indeed suffers from significant computation time on limited platforms (including, in particular, the Nao humanoid robot). Then, we contrast different design choices for AnySURF, and analyze its performance profile under different image types. We also demonstrate the usability of AnySURF in computing approximate homography.

2   Related Work

Image matching using local features (or interest points) has been around for almost three decades – the term "interest point" was first introduced by Moravec in 1979 [14]. A decade ago Lowe [10] introduced Scale Invariant Feature Transform (SIFT), which had a significant impact on the popularity of local features. Since SIFT was published, several new algorithms inspired by SIFT have emerged, including PCA-SIFT [8], GLOH [12] and SURF [2].

SURF [2] is a state of the art algorithm for local invariant feature matching - a scale and rotation invariant interest point detector and descriptor. SURF is composed of three steps similar to SIFT, however it uses faster feature detection / extraction algorithms (approximation of the Hessian matrix and using the distribution of Haar-wavelet responses within the interest point neighborhood, relying on integral images to reduce computation time). SURF is faster to compute than SIFT, while allowing for comparable results.

---
[1] http://www.aldebaran-robotics.com

SIFT, SURF and other such algorithms are not anytime algorithms. Although several authors did accomplish complex real-time visual tasks such as Visual SLAM, using SIFT-like features [3] and correlation with reference templates [4], these implementations were tailored for a specific platform and are not computationally flexible. Therefore, they do not answer out research goals.

3   Methodology

The proposed AnySURF algorithm is based on SURF, which was selected over SIFT and SIFT-like algorithms since it is more suitable for an anytime implementation while also having an excellent quality/run-time ratio. It is more suitable for a flexible implementation because whereas SIFT begins with the computationally expensive operation of constructing several scale space representations (DoG, Difference of Gaussians approximation), SURF is based on a basic approximation of the Hessian matrix where an integral image is computed once for all scales and only the filter size changes when working on each scale. This means that in SURF there is very little overhead for working with specific scales or areas and this is very suitable for a flexible algorithm.

3.1   How Does SURF Works?

SURF [2] is composed of three steps: Detecting interest points, calculating descriptors and matching them (Alg. 1).

The first step, detection of interest points, starts with scale-space extrema detection: search over all scales and image locations is performed, using an approximation of the Hessian matrix to identify potential interest points that are invariant to scale and rotation. Calculation of the Hessian approximation relies on an integral image to reduce computation time. Interest points are first thresholded so that all values below a predetermined threshold are removed, then a non-maximal suppression is performed to find candidate points (each pixel is compared to its 26 neighbours, comprised of the 8 points in the native scale and 9 in each of the scales above and below) and finally they are localized in both scale and space by fitting a 3D quadratic.

The second step, calculation of the keypoint descriptors, is based on a distribution of Haar-wavelet responses within the interest point neighborhood, again relying on integral images for speed. The SURF descriptor describes how the pixel intensities are distributed within a scale dependent neighbourhood around each detected interest point. It is calculated by first assigning a repeatable orientation via Haar wavelet responses weighted with a Gaussian centered at the interest point, then a square oriented window is constructed around the interest point, divided into 4×4 regular sub-regions. For each sub-region 4 Haar wavelets responses are summed up (d_x, d_y, $|d_x|$, $|d_y|$), so a vector of length $4 \times 4 \times 4 = 64$ is produced.

The third step, matching different descriptors, is done via the Euclidean distance of their feature vectors. A fast nearest-neighbor algorithm is used that can

---

Algorithm 1. Generic SURF
(Input: image; Output: list of matched descriptors)

```
0.   Construct integral image
1.1  Over all octaves (Fine-to-Coarse)
1.2     Pre-calculate discriminants
1.3     Over inner octave layers
1.4        Over all pixels
1.5           Find interest point
2.1  Over all interest points
2.2     Calculate descriptor
3.1  Over all descriptors
3.2     Match descriptor
3.3     Add matched descriptor to list
4.   Return list of matched descriptors
```

---

perform this computation rapidly against large databases. SURF uses the sign of the Laplacian (the trace of the Hessian matrix) to distinguish bright features on dark background from the reverse situation. Since SURF's descriptor uses 64 dimensions, time for feature computation and matching is reduced.

3.2 Making SURF Computationally Flexible

In order to make SURF computationally flexible, several important design decisions had to be made. These are: accumulating results iteratively, using a suitable search strategy and calculating the Hessian in-place. An in-depth explanation of these design decision follows. The impact of these decisions is presented in Section 4.

Guaranteeing Monotonically-Improving Descriptor List. The first step in making an anytime version of SURF is trivial: Accumulate results iteratively. SURF divides the work to several large consecutive steps (get all interest points from all scales, compute descriptors for all interest points, match all descriptors against database - Alg. 1, steps 1–3) and so if the final stage is not reached – there might be no useful results. Contrary to this batch approach, we propose an iterative approach, where results are accumulated during the execution of the algorithm and are returned when the algorithm is interrupted.

This can be achieved by computing a full result, including a descriptor, in each iteration. The new descriptor can immediately be used to match against a database. This change is trivial yet vital as it guarantees good anytime functionality: usable results are generated such that the number of results is monotonically increasing.

Generating Descriptors Faster. Now that we can guarantee that the list of matched descriptors will be monotonically-increasing in length, we can explore design choices that can make sure quality descriptors are generated faster. Below we discuss two such design choices.

Search strategy. Detection of interest points (Alg. 1, step 1) is done by scanning the entire image in multiple octaves. This search usually starts with the smallest kernel and continues applying kernels of increasing size to the image. Since our flexible algorithm accumulates results iteratively, we have an opportunity to select an ordering on the search of octaves for interest points (Alg. 1, step 1.1), thereby allowing detection of more promising features earlier. Note that this search strategy need not be hard-coded, but can be changed according to the image or task at hand.

We considered two types of general search strategies: Coarse-to-Fine and Fine-to-Coarse. Coarse-to-Fine means we start with the largest filter size and continue to use smaller filter sizes so that we find larger features first and smaller ones later, while Fine-to-Coarse means the exact opposite. Note that if the algorithm runs to completion the search order does not matter and exactly the same features are found. Additional search strategies are also possible: order of going over inner octave layers (Alg. 1, step 1.3), order of going over pixels (Alg. 1, step 1.4), however we did not consider them here.

Selecting an appropriate search strategy according to the image type (e.g., blurry image) can maximize the number of features detected during the early phase of the search. However, sometimes the number of features is not what we prefer to optimize. For example, some vision tasks work better when the features have a good spatial distribution over the image (e.g., homography calculation [7]) or when coarse features are first matched and only then fine features are searched for in a limited area to complete the match (e.g., object recognition [10]). In such cases it might be preferable to use Coarse-to-Fine search, even if the initial number of features is smaller when compared to the Fine-to-Coarse strategy. In other cases, such as when we have prior information that there are very few Coarse features or when we know we search for small (Fine) features, Fine-to-Coarse strategy can be used.

Calculating the Hessian discriminants in-place. All SURF implementations we inspected (Pan-o-matic, OpenSURF[5], OpenCV) pre-calculate the determinant of Hessian (discriminant) for each octave, over the entire image (Alg. 1, step 1.2). This step has high initial computational cost, however once calculated, results are faster to compute so the total running time is lower. Since we assume the algorithm might not run till completion, it might be preferable to sacrifice some of the running-time in order to get initial results sooner.

Memory consumption by the pre-calculated arrays is another issue to consider. Pre-calculating the determinant of Hessian requires several 2D arrays to be kept in memory. The SURF implementations we inspected (see above) use arrays the size of a full image to simplify coding (smaller arrays can however be used). So we have number of layers×image_width×image_height, which means multiple arrays each one the size of a full image are saved in memory. For large images or platforms with little memory available, this can be quite problematic. Obviously,

---

Algorithm 2. AnySURF
(Input: image; Output: list of matched descriptors)

```
0.   Construct integral image
1.   While not interrupted
2.1     Over all octaves (Coarse−to−Fine)
2.2        Over inner octave layers
2.3           Over all pixels
2.4              Find interest point
2.5              Calculate descriptor
2.6              Match descriptor
2.7              Add matched descriptor to list
3.   Return list of matched descriptors
```

---

when pre-calculation is not used and the determinant of Hessian is calculated in-place, there is no need to save multiple arrays in memory.

3.3 AnySURF – Anytime SURF

The following algorithm (Alg. 2), coined AnySURF (Anytime SURF), is a computationally flexible SURF algorithm. Results are accumulated iteratively, with a descriptor computed in each iteration. Octaves are searched in Coarse-to-Fine order and the determinant of Hessian is calculated in-place. We believe these design choices are appropriate for a generic Anytime SURF algorithm and an analysis of the Anytime performance profile is performed in Section 4.

A possible variant of AnySURF is to use pre-calculation. Compared to a batch approach such as panosurf, this alternative is more suitable to anytime since results are produced earlier yet the total computation time is exactly the same. Compared to the AnySURF without pre-calculation, the total computation time of this variant is lower yet first results are generated much later since pre-calculation has a high initial computational cost (see Figure 1).

4   Results

A flexible algorithm is required only when the non-flexible algorithm is slow and when partial / low accuracy results are useful. In this section we will show that both criteria are met in SURF. In addition, we analyze the design decisions explained in Section 3.2 and present an example of using AnySURF to approximate homography between 2 images.

To demonstrate that SURF is not fast enough for real-time full image feature search on current robotic platforms which have limited computational power, Table 1 shows computation time on multiple platforms for the same image in different sizes (QVGA: 320×240, VGA: 640×480, 3MP: 2048×1536). Evaluation was done using Pan-o-matic open-source SURF implementation [15] with default parameters, which produces very similar results compared to the published SURF binary [6] to which source code is not available.

Table 1. SURF detector-descriptor computation time (ms) on different image sizes and platforms

Platform	QVGA	VGA	3MP
Desktop PC (Intel Q9400 2.66GHz)	27	103	1021
Mini-ITX (Intel T7200 2.0GHz)	74	249	1599
Nao Robot (x86 AMD GEODE 500MHz)	560	2425	26367
Nokia N900 (ARM Cortex-A8 600MHz)	938	3656	442512

From Table 1 it is clear that in order to run real-time full-image feature search with SURF we need to work on a small resolution image coupled with a powerful platform. In addition, in this test the CPU and memory were devoted entirely to the SURF process, while in robotic applications additional non-vision tasks might also require processing time and memory usage (localization, mapping, motion generation, behavior selection, etc.).

Table 2 shows where the processing time is spent across the different major steps. The Intel Q9400 platform was selected for this test, to eliminate as many bottlenecks as possible and allow the optimal behavior of the algorithm show.

Table 2. Analysis of SURF detector-descriptor computation time (ms), on Intel Q9400 2.66 GHz

Image size	integral image	keypoints	descriptors
QVGA	1 (3.7%)	19 (70.4%)	7 (25.9%)
VGA	3 (2.9%)	80 (77.7%)	20 (19.4%)
3MP	32 (3.1%)	910 (89.1%)	79 (7.8%)

As can be seen in Table 2, calculation of the integral image is minor (~3.2%) and detection of keypoints takes most of the time (~79%). In the context of a flexible algorithm, since detecting keypoints takes most of the time, it seems beneficial to calculate the descriptor immediately upon keypoint detection, thus significantly shortening the time till partial results are available. We now turn to analyzing the impact of the various design choices of AnySURF (presented in Section 3.2).

The image database used in all following figures is a standard evaluation set, provided by Mikolajczyk [12]. It contains 48 images across 8 different scenes. All images are of medium resolution (approximately 800×640 pixels) and are either of planar scenes or the camera position is fixed during acquisition, so that in all cases the images are related by homographies (plane projective transformations). The scenes contain different imaging conditions: viewpoint changes, scale changes, image blur, JPEG compression and illumination changes.

Figure 1 shows the averaged rate of acquiring descriptors (%) as a function of run-time (%). Three alternatives are considered: First, a SURF implementation called Pan-o-matic [15] (henceforth will be referred to as panosurf). This implementation first detects all keypoints at all scales and only then calculates descriptors (as in Alg. 1). In addition, the determinant of Hessian is precalculated. Next, we tested two variants of the AnySURF algorithm (Alg. 2), where descriptors are computed immediately upon keypoint detection. In the first variant pre-calculation of the determinant of Hessian is used and in the other it is calculated in-place.

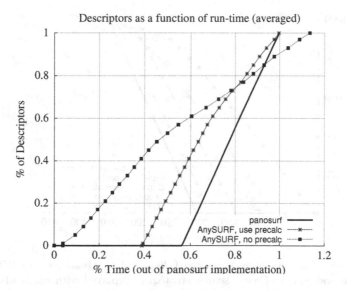

Fig. 1. The descriptors (%) vs. time (%) graph for different algorithms. Data is averaged across all (48) database images. Time (%) is compared to panosurf time (therefore can be ¿ 1.0).

Figure 1 allows us to inspect the impact of calculating descriptors immediately upon keypoint detection and also the effect of pre-calculation. First, let us consider AnySURF with pre-calculation: calculating descriptors immediately is beneficial compared to the original panosurf approach since it does not adversely affect the total computation time while allowing results to start accumulating earlier (after 39% of time passed instead of 57% as in panosurf). Now, let us consider AnySURF without pre-calculation: although the algorithm does take longer to complete (13.5% more on average), we start getting results almost immediately (after 4% of time passed), with a near-linear acquire rate. Note that pre-calculation is only useful when all descriptors are needed or when it can be assumed that the algorithm will run to near completion (AnySURF with pre-calculation supersedes the no pre-calculation version after 80% of the time passed).

Next, let us inspect the impact of the search strategy. As explained in Section 3.2, since AnySURF accumulates results iteratively, we can select an ordering on the search for interest points. The following figure is of a specific images

(1st image of "bricks" sequence), showing the number of descriptors as a function of run-time. All four combinations are shown (with/without pre-calculation, Coarse-to-Fine/Fine-to-Coarse search strategy), compared to the baseline panosurf.

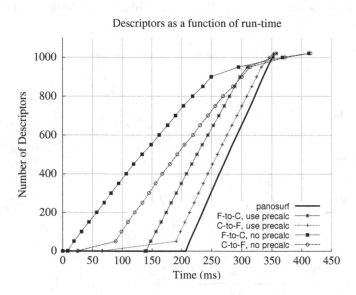

Fig. 2. 1st image of bricks sequence, panosurf displayed as a baseline, AnySURF with Coarse-to-Fine and Fine-to-Coarse search strategies displayed with and without pre-calculation

In Figure 2, the Fine-to-Coarse search strategy produces results much faster than Coarse-to-Fine (in other images the opposite might be true). Note that results start sooner, accumulate faster and the difference in number of descriptors is significant, up to an order of magnitude (e.g., after ~30ms). This means that appropriate selection of the search strategy is vital for an efficient Anytime performance.

As for the use of pre-calculation, it seems that our previous conclusion holds and for an anytime algorithm an in-place calculation is preferred. However, in other images (not shown here due to lack of space) the pre-calculated (Coarse-to-Fine) version supersedes the in-place (Fine-to-Coarse) version. This only stresses that selecting the appropriate search strategy is very important indeed: when selecting the correct search strategy, the no pre-calculation (Coarse-to-Fine) version triumphs again (at least until most of the descriptors are found, as explained earlier).

After witnessing a major difference in the above specific image according to the chosen search strategy, it is interesting how the Fine-to-Coarse vs. Coarse-to-Fine search strategies perform in our full image database, across the different scenes. To test this, Figure 3 shows the effect of search strategy in different scenes on the first 50ms of AnySURF (without pre-calculation).

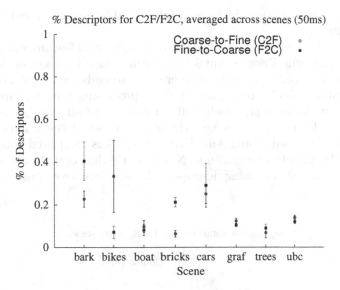

Fig. 3. The descriptors (%) for Coarse-to-Fine and Fine-to-Coarse tested on different scenes (first 50ms of AnySURF). The markers represent the mean for each scene, the error bars are two standard deviation units in length.

Figure 3 demonstrates that it is beneficial to select the appropriate searching strategy according to the type of image at hand if the number of descriptors is to be optimized. The image sequences "bark" and "bricks" are clearly more suited for Fine-to-Coarse search whereas "bikes" is more suited for Coarse-to-Fine search. The reason behind this is that "bikes" is a sequence of blurred images (so there are less fine features and processing time is wasted on searching for them) while "bark" and "bricks" contain images with many fine features and few coarse ones. It is also interesting to note that between 5% to 40% of all descriptors can be acquired within 50ms (however, higher percentages are usually for images with a lower total number of descriptors and a lower total computation time). The average number of descriptors acquired after 50ms on our image database is 119 for Coarse-to-Fine and 138 for Fine-to-Coarse.

A higher number of descriptors is not necessarily better. For example, coarse features are larger, fewer and usually more spread over the image so they might suit some tasks better than fine features. One such task is homography estimation [7]. Figure 4 shows the time passed till the homography between the 1st and 2nd images in each scene could be estimated (to within an order of magnitude from the optimal value). Homography was calculated via the RANSAC approach [7,9].

Figure 4 demonstrates that coarse features enable a quicker homography estimation compared to fine features. The Coarse-to-Fine search strategy took equal or less time to estimate the homography in most scenes (7/8) and more time in just one scene (the bricks scene, which contains very few coarse features and so a lot of time is wasted searching for coarse features). The results for the bark, cars and trees scenes do not differ significantly, while results for others do (two-tailed t-test, p=0.01). Also note that the homography could be estimated within a very short time (˜20 ms).

Finally, we go back to evaluating the use of flexible local feature matching on the Nao robot platform. Prior to AnySURF, estimating a homography between two images on this platform took on average 4 seconds (averaged across all images in database, similar to Figure 4). This processing time was spent not on estimating the homography itself, but on computing all descriptors in the image. However, for estimating a homography a subset of the results suffice, so AnySURF can be used. Using AnySURF, this task is completed within 0.33 seconds, faster by an order of magnitude. Note that this homography can actually assist in computing the remaining descriptors faster, since we can now estimate their location.

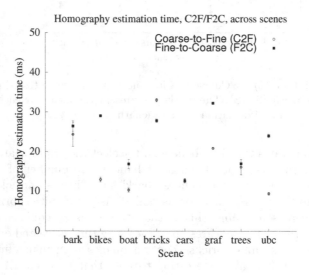

Fig. 4. Homography approximation time for Coarse-to-Fine and Fine-to-Coarse, tested across scenes. In each scene image 1 was processed fully. Image 2 was processed till the homography could be estimated (to within an order of magnitude from optimal value). The markers represent the mean for each sequence, the error bars are two standard deviation units in length.

5   Conclusion

We presented and analyzed AnySURF, a SURF-based anytime local feature matching algorithm, which can trade quality of results for computation time: It guarantees that the number of matched descriptors monotonically increases with computation. For robotics applications that can work with a subset of descriptors, this allows for much faster response times.

We discuss and carefully evaluate design choices in the feature search process, using several computational platforms. We demonstrate that changing the feature search order can significantly impact the rate at which descriptors are generated. Surprisingly, avoiding pre-calculation steps that are intended to optimize the search process, leads to generating results at a faster rate. Future work will focus on the problem of dynamically managing AnySURF within the context of a flexible real-time vision-based autonomous navigation system.

References

1. DARPA grand challenge (2007), http://www.darpa.mil/grandchallenge/index.asp
2. Bay, H., Tuytelaars, T., van Gool, L.: SURF: Speeded Up Robust Features. In: Leonardis, A., Bischof, H., Pinz, A. (eds.) ECCV 2006. LNCS, vol. 3951, pp. 404–417. Springer, Heidelberg (2006)
3. Chekhlov, D., Pupilli, M., Mayol-Cuevas, W., Calway, A.: Real-Time and Robust Monocular SLAM Using Predictive Multi-resolution Descriptors. In: Bebis, G., Boyle, R., Parvin, B., Koracin, D., Remagnino, P., Nefian, A., Meenakshisundaram, G., Pascucci, V., Zara, J., Molineros, J., Theisel, H., Malzbender, T. (eds.) ISVC 2006. LNCS, vol. 4292, pp. 276–285. Springer, Heidelberg (2006)
4. Davison, A.J., Reid, I.D., Molton, N.D., Stasse, O.: MonoSLAM: Real-Time single camera SLAM. IEEE Transactions on Pattern Analysis and Machine Intelligence 29(6), 1052–1067 (2007)
5. Evans, C.: Notes on the OpenSURF library. Tech. rep., University of Bristol (January 2009)
6. Gossow, D., Decker, P., Paulus, D.: An Evaluation of Open Source SURF Implementations. In: Ruiz-del-Solar, J., Chown, E., Ploeger, P.G. (eds.) RoboCup 2010. LNCS, vol. 6556, pp. 169–179. Springer, Heidelberg (2010)
7. Hartley, R., Zisserman, A.: Multiple View Geometry in Computer Vision, 2nd edn. Cambridge University Press (2004)
8. Ke, Y., Sukthankar, R.: PCA-SIFT: a more distinctive representation for local image descriptors. Null 2, 506–513 (2004)
9. Kovesi, P.D.: MATLAB and Octave functions for computer vision and image processing (2000), http://www.csse.uwa.edu.au/~pk/Research/MatlabFns
10. Lowe, D.G.: Object recognition from local Scale-Invariant features. In: Proceedings of the International Conference on Computer Vision, vol. 2, p. 1150. IEEE (1999)
11. Lowe, D.G.: Distinctive image features from Scale-Invariant keypoints. International Journal of Computer Vision 60(2), 91–110 (2004)
12. Mikolajczyk, K., Schmid, C.: A performance evaluation of local descriptors. IEEE Transactions on Pattern Analysis and Machine Intelligence 27(10), 1615–1630 (2005)
13. Mikolajczyk, K., Tuytelaars, T., Schmid, C., Zisserman, A., Matas, J., Schaffalitzky, F., Kadir, T., van Gool, L.: A comparison of affine region detectors. Int. J. Comput. Vision 65(1-2), 43–72 (2005)
14. Moravec, H.P.: Visual Mapping by a Robot Rover (1979)
15. Orlinski, A.: Pan-o-matic - automatic control point creator for hugin
16. Tuytelaars, T., Mikolajczyk, K.: Local invariant feature detectors: a survey. Found. Trends. Comput. Graph. Vis. 3(3), 177–280 (2008)
17. Zilberstein, S.: Using anytime algorithms in intelligent systems. AI Magazine 17(3), 73–83 (1996)

Online Motion Planning for Multi-robot Interaction Using Composable Reachable Sets

Aris Valtazanos and Subramanian Ramamoorthy

School of Informatics, University of Edinburgh
Edinburgh EH8 9AB, United Kingdom
a.valtazanos@sms.ed.ac.uk, s.ramamoorthy@ed.ac.uk

Abstract. This paper presents an algorithm for autonomous online path planning in uncertain, possibly adversarial, and partially observable environments. In contrast to many state-of-the-art motion planning approaches, our focus is on decision making in the presence of adversarial agents who may be acting strategically but whose exact behaviour is difficult to model precisely. Our algorithm first computes a collection of *reachable sets* with respect to a family of possible strategies available to the adversary. Online, the agent uses these sets as *composable behavioural templates*, in conjunction with a particle filter to maintain the current belief on the adversary's strategy. In partially observable environments, this yields significant performance improvements over state-of-the-art planning algorithms. We present empirical results to this effect using a robotic soccer simulator, highlighting the applicability of our implementation against adversaries with varying capabilities. We also demonstrate experiments on the *NAO* humanoid robots, in the context of different collision-avoidance scenarios.

Keywords: Online Motion Planning, Autonomous Decision Making, Composable Behavioural Templates.

1 Introduction

As robots become increasingly more autonomous, they require online decision making skills for adversarial environments, in the presence of other agents with conflicting objectives. Although there are many theoretical frameworks for addressing such decision making problems, nearly all assume significant knowledge on the capabilities and actions of the adversaries. In actual practice, agents must synthesise their beliefs from limited observations of the world and devise plans that are likely to succeed, despite significant imprecision regarding actions and *strategic profiles* corresponding to the adversary. Even though an agent's own goals may be concretely formulated, the behaviours of its adversaries may not be easily characterised to the level required by formal algorithmic tools.

Our proposed approach draws on the notion of a *reachable set*, as used in the literature on hybrid systems in control theory, to encode unsafe sets of state configurations up to infinite time horizons. We extend the basic concept by using a collection of infinite-horizon reachable sets corresponding to coarse profiles.

T. Röfer et al. (Eds.): RoboCup 2011, LNCS 7416, pp. 186–197, 2012.
© Springer-Verlag Berlin Heidelberg 2012

These sets, computed offline, form *composable templates* that can be used online to generate safe trajectories with respect to an adversary's realised behaviour. In partially observable environments, we use a particle filter algorithm for estimating *beliefs* about other agents. These beliefs are used to select the template that most closely matches the perceived coarse capabilities of the adversary, which in turn informs selection of appropriate landmarks for motion planning.

The reachable set formulation has a strong game-theoretic flavour, as it builds on the notion of selecting optimal control inputs with respect to a class of disturbances available to an adversary. However, it also differs from traditional game theory in not assuming any explicit knowledge of payoffs associated with specific actions. When used in conjunction with probabilistic filtering, reachable set composition is less sensitive to errors in the estimation of the adversary's strategy. This leads to the generation of safer trajectories in reactive path planning.

2   Background and Related Work

2.1   Hybrid System Modeling

A key concern in our work is the modeling of an opponent's behaviour, which is dictated by choices over discrete behavioural modes and underlying continuous dynamics. A good framework for thinking about such problems is available within the control theory literature, where systems with joint discrete and continuous dynamics are known as hybrid systems [14].

A major application of hybrid system modeling has been the formal description of aircraft collision avoidance as a pursuit-evasion game between two adversaries [14]. Each aircraft assumes the role of an *evader* seeking to avoid collision with an adversary, who is modeled as a *pursuer* with the exact opposite goal. An evader has a notion of a *target* set of unsafe states, which must be avoided to prevent collision.

A key innovation of this approach is the introduction of *reachable* sets of states, which can be classified in one of two ways. A **forward reachable set** is the set of states that can be reached from some given initial configurations. A **backward reachable set** is the set of states that may give rise to trajectories terminating in a target set of unsafe states. In path planning, backward reachable sets provide an elegant way of determining *the entire set of trajectories* that are likely to lead to the satisfaction of a goal. These sets can be computed directly, without recourse to exhaustive simulation over all possible state transitions.

One popular approach for approximating reachable sets is their estimation as the zero sublevel set of the solution of the Hamilton-Jabobi-Isaacs PDE with respect to the system dynamics [14]. Other works use overapproximations of reachable sets to compute optimal trajectories [7]. A comprehensive toolbox with level set implementations for reachable sets was developed in [11].

2.2   Reactive Path Planning

Several mobile robots require planning routines that can adapt to dynamically changing obstacles and environmental context. *Reactive* path planning refers to the class of algorithms that are used to solve this problem online, based on continuously updated information on a robot's environment.

In *velocity obstacle* algorithms [8], agents estimate the velocities of their surrounding obstacles, which are used to define regions that should be avoided. Reciprocal collision avoidance [4] is a special case of the above, which assumes that all agents follow the same planning procedure. Other state-of-the-art motion planning examples for a variety of multi-agent domains are surveyed in [5].

In comparison to the work in this paper, most of the above examples do not address sensing uncertainty, partial observability, physical capability modeling, and unknown adversarial strategy profiles. The path planning problem is most challenging precisely when all these constraints arise simultaneously, which highlights the need for robust and efficient solutions. This is the focus of our paper.

2.3   Template-Based Planning and Control

Many practical problems in robotics are hard to solve using direct optimisation, due to their high dimensionality or complex dynamics. An approach that is increasingly gaining traction is to devise coarse abstractions that simplify the overall problem. [6] presents a framework of local feedback planning policies, represented as funnels, which can be sequenced through back-chaining to achieve a global goal. Extending this idea, [13] defines these local controllers as linear quadratic regulators. Our approach extends these notions by explicitly considering control in the presence of other, possibly strategic, agents.

A related theme is found in risk-sensitive planning and control [16]. Here, obstacles are associated with the risk they impose on a given trajectory. Potential functions have also been proposed as a means of expressing the desirability of different objects and landmark points for candidate trajectories [12].

3   Algorithm

In this paper, we focus on robotic soccer with humanoid robots, a domain capturing the features we have discussed so far: strategic adversaries with unknown behavioural models, partial observability, uncertainty in sensing and actuation, and dynamically varying goals. We first define the dynamics of our system and the corresponding reachable sets following the conventions of Tomlin et al [14].

3.1   System Dynamics

We consider a system involving two holonomic robots r_A and r_B (Figure 1). As the robots are autonomous and are not provided with external input, all coordinate frames are egocentric. Throughout this section, we consider the case of r_A planning trajectories with respect to the adversarial agent r_B.

The state x of r_B relative to r_A is defined in terms of the vector:

$$x = \begin{bmatrix} x_1, x_2, x_3 \end{bmatrix}^T \tag{1}$$

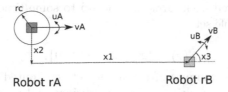

Fig. 1. Coordinate system centred on robot r_A

where x_1, x_2 are the planar coordinates and x_3 is the heading of r_B relative to r_A. Because of the partial observability assumption, we approximate the relative heading x_3 as the difference between the directions of their linear velocities. A particle filter is used to deal with the uncertainty in the estimation of relative positions (Section 3.4). Furthermore, r_c is defined as a fixed-length distance, within which collisions between the two robots occur.

The dynamics of the system can then be defined as

$$\dot{x} = \frac{d}{dt}\begin{bmatrix} x_1 \\ x_2 \\ x_3 \end{bmatrix} = \begin{bmatrix} -v_A + v_B \cos x_3 + u_A x_2 \\ v_B \sin x_3 - u_A x_1 \\ u_B - u_A \end{bmatrix} = f(x, u_A, u_B). \tag{2}$$

where v_A and v_B are the linear velocities of the two robots, and u_A and u_B are their angular velocities. In accordance with most works in the hybrid systems literature (e.g. [14]), linear velocities are treated as *fixed* parameters, whereas angular velocities are *control* inputs selected by the two robots.

3.2   Reachable Set Calculation

In our problem domain, backward reachable sets represent the states the agents must avoid over some time horizon, in order to prevent collisions with each other. We compute these sets following the methodology in [15]. Each robot may select a control input from a set of admissible values:

$$u_A \in \mathcal{U_A} = [u_{Am}in, u_{Am}ax], \ u_B \in \mathcal{U_B} = [u_{Bm}in, u_{Bm}ax] \tag{3}$$

where $\{u_{Am}in, u_{Am}ax, u_{Bm}in, u_{Bm}ax\}$ are predefined upper and lower angular velocity bounds. The backward reachable set is then obtained by solving the Hamilton-Jacobi-Isaacs Partial Differential Equation (HJI PDE):

$$\frac{\partial v(x, t)}{\partial t} + \min[0, H(x, \nabla v(x, t))] = 0, \ v(x, 0) = g(x), \tag{4}$$

with Hamiltonian (replacing p with $\nabla v(x, t)$)

$$H(x, p) = \max_{a \in \mathcal{U_A}} \min_{b \in \mathcal{U_B}} p \cdot f(x, a, b), \tag{5}$$

where $g(x)$ is a scalar function representing the target set:

$$g(x) = \sqrt{x_1{}^2 + x_2{}^2} - r_c. \tag{6}$$

The HJI PDE is solved backwards in time up to some time horizon τ, to give the backward reachable set

$$S(\tau) = \{x \mid v(x, -\tau) \leq 0\}. \tag{7}$$

Figure 2 shows some examples of reachable sets computed for the system dynamics defined above, for various initial conditions and parameters. These sets and their approximations form the basis of our motion planning algorithms, by defining state space regions that should be avoided to prevent collisions. For a more detailed coverage of the convergence properties of this method, see [15].

3.3   Template Estimation

Strategic adversarial domains like robotic soccer require agents to cope with the *unknown* and potentially *non-stationary* behaviour of their opponents. The objective of a soccer-playing agent may dynamically change, for example, from navigating to the ball to making a maneuver to mark an adversary. This is in contrast to the aircraft collision avoidance example described in Section 2.1, where aircrafts are modeled as consistently seeking to avoid collision.

Table 1. HJI PDE running time for various horizon times

Time horizon τ (s)	0.1	0.5	1.0	2.0	3.0	4.0	5.0
Running time (s)	2.94	9.43	15.56	31.30	46.56	60.61	76.01

We model different strategic behaviours as a collection of reachable sets, each corresponding to a hypothesis on the capabilities of the adversaries. Ideally, robots would compute different sets at each time step, depending on their aggregated observations. However, this is not feasible because of the computational expense associated with the solution of the HJI PDE backwards in time (Table 1). Instead, we select a fixed number of hypotheses, for which we compute the reachable sets *offline*. We complement the definitions of Section 3.2, by defining a set of *admissible linear velocities* for the adversarial agent r_B:

$$v_B \in \mathcal{V}_B = [v_{Bmin}, v_{Bmax}] \tag{8}$$

A similar argument applies to the set of admissible bounds $\mathcal{V}_A$ for agent r_A. Both pairs of bounds depend on the *physical* capabilities of the robots (e.g. the maximum velocities permitted by their hardware). We first discretise $\mathcal{V}_A$ and $\mathcal{V}_B$ to obtain a countable set of pairs $\mathcal{VP} = \{v_A \in \mathcal{V}_A, v_B \in \mathcal{V}_B\}$. For each pair $(v_A, v_B) \in \mathcal{VP}$, we compute a backward reachable set as in the previous section, thus obtaining a finite collection of *templates*.

Figure 2 shows an example of estimation of different reachable set templates, using Mitchell's toolbox [11]. The parameters reflect different extrema in the combinations of strategies followed by the two robots. For example, the *red* set depicts the worst case where r_B is moving at full speed and r_A is stationary. The right half of Figure 2 illustrates an irregular set that is not easily approximable.

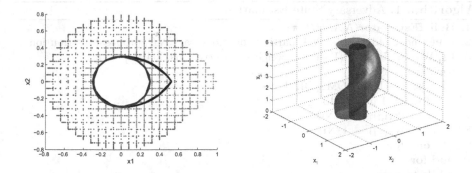

Fig. 2. Left: Two-dimensional projection of reachable set templates. Fixed parameters: $r_c = 0.3m$, $u_A min = u_A max = -u_A min = -u_B min = 1.6 rad/s$. Blue set: $v_A = 0.0m/s$, $v_B = 0.0m/s$. Black set: $v_A = 0.5m/s$, $v_B = 0.0m/s$. Green set: $v_A = 0.5m/s$, $v_B = 0.5m/s$. Red set: $v_A = 0.0m/s$, $v_B = 0.5m/s$. Right: The full three-dimensional version of the red and blue sets.

3.4   State Estimation and Reachable Set Composition

We use a particle filter estimation algorithm, in order to convert *observations* of the adversaries into *beliefs* with associated probabilities (Algorithm 1). The algorithm runs continuously, generating new beliefs based on the most recent observations. At each step, the robot queries its sensors (GETLATESTOBSERVATIONS), and attempts to match them to the best past beliefs (CLUSTEROBSERVATIONS). The clustered observations are then passed through the particle filter to update the likelihood of each belief. We have implemented a variant of the Sampling-Importance-Resampling algorithm introduced in [10]. The particle filter also helps alleviate some of the oscillation problems that arise in reactive path planning.

Beliefs are used to compute the relative state x of the adversary, as defined in Section 3.1. These estimates are then used to select the best reachable set **dynamically**, and plan trajectories that are optimal with respect to the current beliefs. When a robot is *outside* the chosen reachable set, it plans trajectories with respect to its boundaries. Similarly, when it is *inside*, it seeks to find the nearest point on its boundary, so it can exit and reach a safe configuration again.

4   Results

We demonstrate the application of our algorithm in the context of various robotic soccer scenarios. First, we present results on a 3-D strategy simulator, which was developed to comply with the specification of the RoboCup Standard Platform League [3]. Then, we present experiments on real NAO robots [2].

4.1   Simulation

We have developed a MATLAB 3-D strategy simulator for the NAO robots. The soccer field (Figure 3-top left) has dimensions of 6x4m, and the ball is

---

Algorithm 1. Adversary State Estimation

1: **BeliefEstimator**(NA, NP, π)
2: INPUT:no. of adversaries NA, particles per adversary NP, proposal distribution π
3: Ps, Ws, Os, Bs ← ∅ {Particles, weights, observations, beliefs}
4: **for** $i = 1$ to NA **do**
5:       Bs(i) ← RANDOM_POINT
6:       **for** $j = 1$ to NP **do**
7:             Ps(i,j) ← RANDOM_POINT
8:             Ws(i,j) ← 1/NP
9:       **end for**
10: **end for**
11: **while** true **do**
12:       Os ← GetLatestObservations {vision}
13:       Os ← ClusterObservations(Os, Bs) {find nearest past beliefs}
14:       **for** $i = 1$ to NA **do**
15:             **if** Os(i) == NULL **then**
16:                   Os(i) ← Bs(i) {no current observation, use last belief}
17:             **end if**
18:             Bs(i), Ps(i), Ws(i) ← ParticleFilter(Os(i), Ps(i), Ws(i))
19:       **end for**
20: **end while**

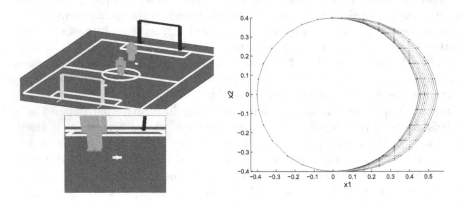

Fig. 3. *Top Left:* Simulated field with two robots. *Bottom Left:* The view of the blue robot. *Right:* Two-dimensional projection of reachable set templates for the NAO humanoid robots. The selected parameters are $r_c = 0.4m$, $u_A min = u_A max = -u_A min = -u_B min = 1.7 rad/s$, $v_A = 0.1 m/s$, $v_B \in \{0, 0.1\}$, with v_B sampled every $0.01 m/s$.

an orange sphere with a radius of 3cm. Each simulated robot is equipped with a 58° field-of-view camera (Figure 3-bottom left) with which it perceives the world. Sonar sensing is not modeled in this setup. Robots detect each other through waistbands, which are coloured light blue and pink for the two teams. Locomotion dynamics are not modeled explicitly, although noise can be added to executed movements. Uncertainty in sensing arises from the *egocentric* nature of belief estimation, which impacts the visibility of adversarial robots.

We first compute reachable set templates based on the hardware specifications
[2] of the NAO (Figure 3-right). The velocity bounds were selected to comply
with the physical limits of the NAO. The outer sets represent conservative hypotheses, where adversary r_B moves with high velocity towards r_A.

One-versus-one Game. We first consider the case of a one-versus-one game of
soccer between two robots (Figure 4(a)). We assign a different role to each robot
- the goal of the *attacker* is to navigate to the ball, while avoiding the *defender*.
In a reachable set context, the defender selects the template which best models
his adversary, and then tries to move towards its boundary.

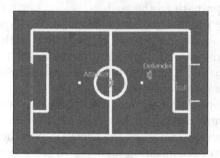
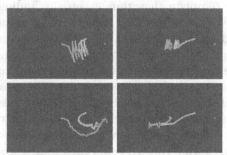

(a) Initial configuration. The attacker (blue robot) is located on the
halfway line, with the defender further away.

(b) 1-v-1 examples. Format: (ATTACKER,
DEFENDER). *Top left:* (RSC,RSC).
Top right: (TSB,TSB). *Bottom left:*
(RSC,TSB). *Bottom right:* (TSB,RSC).

Fig. 4. One-versus-one experiments

The reachable set composition heuristic is compared against a benchmark
inspired from the collision-avoiding velocity and dynamic window ideas presented
in [9] and [4]. This algorithm allows the robot to move with its current velocity
v_A, as long as it is outside the *target* set of unsafe velocities (defined in terms of
the adversary's current *velocity* v_B). When inside, it computes the nearest point
p_b on the boundary of the unsafe set, and uses it to define a set of safe velocities:

$$\mathcal{V}_S = \{v \mid (v - (v_A + p_b)) \cdot \frac{p_b}{|p_b|} \geq 0\} \tag{9}$$

that may lead it out of the unsafe set. The robot then selects the velocity $v_S \in \mathcal{V}_S$
that will minimise the distance to its current goal. This variant was chosen to
reflect the difference between planning with respect to a *target* set and planning
with respect to multiple *reachable* sets. We refer to the two algorithms as **RSC**
(Reachable Set Composition) and **TSB** (Target Set Benchmark) respectively.
In summary, the two algorithms differ in the state set they use when planning
(reachable vs target), as well as the space they operate in (position vs velocity).

Strategies are initially evaluated in a **fully observable** setting. Agents are
provided with exact information on the location of the ball and their adversary.

Table 2. 1v1, partial observability. Success rate and mean time to goal (MTTG).

(a) Reachable sets vs benchmark

(Att,Def)	Success %	MTTG
(Rsc,Rsc)	72%	63.28 steps
(Tsb,Tsb)	44%	64.5 steps
(Rsc,Tsb)	98%	39.5 steps
(Tsb,Rsc)	16%	60.5 steps

(b) Different reachable set algorithms.

(Att,Def)	Success %	MTTG
(Rsc,Srs-Bc)	38%	58.5 steps
(Srs-Bc,Rsc)	42%	72.5 steps
(Rsc,Srs-Wc)	82%	67.8 steps
(Srs-Wc,Rsc)	40%	60.1 steps

Figure 4(b) shows all strategy combinations for the attacker and the defender. In the Rsc-Tsb combination, the attacker benefits from the adaptive nature of the template algorithm, eventually finding a path to the goal.

We now investigate the more challenging **partially observable** case. We also add random Gaussian noise to the commanded movements of the robot, with mean equal to the magnitude of the command and a standard deviation of 1. We ran 50 trials for each strategy combination. Each trial was simulated for 80 discrete simulation steps, or until the attacker reached the goal.

Table 2(a) summarises these results, where Rsc maintains its superiority over Tsb. As with full observability, this difference is stronger when the attacker and the defender use different algorithms. However, even when the heuristics are the same ((Rsc,Rsc), (Tsb,Tsb)), reachable set composition manages to steer the attacker to the goal more reliably. This also means that Rsc defenders are less successful against Rsc attackers, as the reactive nature of the game prevents them from both tracking and blocking their opponents' moves robustly.

As a further comparison, we also compare the Rsc algorithm to two variants of path planning algorithms using a *single* reachable set. The first variant plans with respect to the *best-case* reachable set, where the adversary is hypothesised as always moving with a minimum linear velocity. The second variant takes into account only the *worst-case*, where the adversary moves with the maximum allowed velocity. We term the two variants **SRS-BC** (Single Reachable Set - Best Case) and **SRS-WC** (Single Reachable Set - Worst Case) respectively.

Table 2(b) summarises the computed metrics for this case. In terms of success rate, the different variants are comparable; this is largely due to the small discrepancy between the best- and the worst-case reachable set (Figure 3). The results also appear to favour optimistic defensive strategies (e.g.(Rsc,Srs-Bc)), which allow defenders to move closer to the attacker when marking. Nevertheless, the Rsc algorithm offers some improvement in the time to reach the goal.

Multi-robot Games. In this example, the blue team now consists of two attackers, and the pink team of two defenders. However, no cooperation or information exchange between teammates is allowed. The attacking team succeeds if at least one of its members reaches the ball; the defenders succeed if they prevent both attackers from reaching the ball. Figure 5(a) shows the initial configuration.

The algorithms are once again evaluated on the four different permutations of the Rsc and Tsb algorithms under full observability. Figure 5(b) displays a pattern similar to Figure 4(b). In most cases, the resulting trajectories vary among

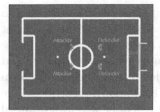

(a) Initial configuration.

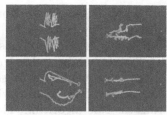

(b) 2-v-2 examples. Format: (ATTACKER, DEFENDER). *Top left:* (RSC,RSC). *Top right:* (TSB,TSB). *Bottom left:* (RSC,TSB). *Bottom right:* (TSB,RSC).

(c) Multiple reachable set defending example - agent plans path with respect to the intersection of the sets.

(Att,Def)	Success %	MTTG
(RSC,RSC)	90%	61.4 steps
(TSB,TSB)	62%	56.0 steps
(RSC,TSB)	94%	46.2 steps
(TSB,RSC)	22%	75.0 steps

(d) Performance statistics.

Fig. 5. Two-versus-two experiments

teammates. This discrepancy is due to the presence of the additional adversary, who influences the selection of appropriate landmarks for path planning.

Table 5(d) summarises the results for 50 runs in the partially observable case. Compared to the one-versus-one experiment, partial observability is more favourable towards attackers, as seen by the higher success rate in most cases. However, the RSC algorithm still behaves more robustly than the TSB variant, despite the additional constraints. The significant improvement in the attackers' success rate is partly explained by the presence of multiple robots in the field, which makes disambiguation and tracking difficult during close interactions.

The robustness of the RSC algorithm under both full and partial observability is a strong indication of its suitability to multi-agent path planning. When planning a path with respect to multiple adversarial agents, it is sufficient to consider the intersection of their reachable sets as a safety criterion. Figure 5(c) illustrates how a defender uses this intersection to plan a path that accounts for both attackers. The plotted sets correspond to the templates computed in Figure 3, which are translated and rotated based on the current output of the particle filter algorithm - this estimate may not always correspond to the exact true state of the adversary. In this noisy setting, RSC may generate safer trajectories than simpler heuristic algorithms that do not explicitly consider the evolution of system dynamics over some time horizon.

4.2 Physical Robot

We conclude by presenting some illustrative applications of our algorithm on the NAO humanoid robots. Similarly to the simulation, robots identify each other using coloured waistbands, with the same observability constraints as in section 4.1. Imperfect sensing arises naturally in this setup, so each robot runs a variant of Algorithm 1 to deal with incomplete and/or noisy vision estimates.

The behavioural models have been adjusted to comply with the hardware and processing limitations of the humanoids. We consider one-versus-one collision avoidance tasks, against a variety of adversaries with unknown strategies. The attacking robot always runs the RSC algorithm. The defender was programmed to execute one of the following behaviours: move towards the attacker, move in a circle around the ball, move vertically with respect to the attacker, also execute the RSC algorithm in order to navigate to a second ball, and move randomly.

Our supporting **video** [1] illustrates the above examples on NAO robots. Screen shots from two examples are given in Figures 6 and 7. In both cases, the robot manages to successfully reach the goal, despite losing track of the ball or its adversary for long periods of time. This demonstrates the robustness of the composable reachable set algorithm under realistic constraints and conditions.

Fig. 6. NAO going to the ball while avoiding an adversary moving directly towards it

Fig. 7. NAO going to the ball while avoiding an adversary moving randomly

5   Conclusions

We have presented an algorithm for autonomous online path planning in realistic adversarial conditions, where the adversary's capabilities and strategy profiles can not be characterised or estimated precisely. Our algorithm extends the concept of reachable set computation for safe trajectory generation, to incorporate bounds and hypotheses on adversarial behaviours. By generating suitable templates offline and composing them dynamically online, the algorithm outperforms simpler modeling heuristics. Thus, our results favour the notion of planning with respect to an infinite-horizon reachable set of states. This superiority is reflected in both the success rate of reaching a given goal and the time taken to reach it.

Our reachable set template computation builds on very simple hypotheses on the adversary's linear velocity bounds. Depending on the parameters of the problem, this may lead to small (Figure 3-right) or large (Figure 2) between best- and worst-case templates. Future work would be to utilise more sophisticated distributions over reachable sets, while also incorporating more elaborate coarse dynamics models that go beyond the scope of rectangular velocity bounds.

Composable reachable sets may also be beneficial in numerous applications beyond multi-robot path planning. One obvious example is Human-Robot Interaction, where the human agent's strategies and behaviours can not be modeled exactly but we still want strategically meaningful behaviour on the part of the robot. This is an area of future work for us.

References

1. Supporting material, `http://www.youtube.com/watch?v=BfJgWz4TwlE`
2. NAO robot documentation, `http://academics.aldebaran-robotics.com/`
3. RoboCup SPL rules, pp. 1–27 (2010), `http://www.tzi.de/spl/`
4. van den Berg, J., Guy, S.J., Lin, M., Manocha, D.: Reciprocal n-body collision avoidance. In: Pradalier, C., Siegwart, R., Hirzinger, G. (eds.) Robotics Research. STAR, vol. 70, pp. 3–19. Springer, Heidelberg (2011)
5. Bruce, J.: Real-Time Motion Planning and Safe Navigation in Dynamic Multi-Robot Environments. Ph.D. thesis, School of Computer Science, Carnegie Mellon University, Pittsburgh, PA (January 2006)
6. Burridge, R.R., Rizzi, A.A., Koditschek, D.E.: Sequential composition of dynamically dexterous robot behaviors. IJRR 18(6), 534–555 (1999)
7. Ding, J., Tomlin, C.J.: Trajectory optimization in convex underapproximations of safe regions. In: CDC, pp. 2510–2515 (2009)
8. Fiorini, P., Shillert, Z.: Motion planning in dynamic environments using velocity obstacles. International Journal of Robotics Research 17, 760–772 (1998)
9. Fox, D., Burgard, W., Thrun, S.: The dynamic window approach to collision avoidance. IEEE RAM 4(1), 23–33 (1997)
10. Gordon, N.J., Salmond, D.J., Smith, A.F.M.: Novel approach to nonlinear/nongaussian bayesian state estimation. Radar and Signal Processing 140(2), 107–113 (1993)
11. Mitchell, I.M., Templeton, J.A.: A Toolbox of Hamilton-Jacobi Solvers for Analysis of Nondeterministic Continuous and Hybrid Systems. In: Morari, M., Thiele, L. (eds.) HSCC 2005. LNCS, vol. 3414, pp. 480–494. Springer, Heidelberg (2005)
12. Rimon, E., Koditschek, D.E.: Exact robot navigation using artificial potential functions. IEEE Trans. Rob. Aut. 8, 501–518 (1992)
13. Tedrake, R.: LQR-trees: Feedback motion planning on sparse randomized trees. In: Proceedings of Robotics: Science and Systems, Seattle, USA (June 2009)
14. Tomlin, C.J., Lygeros, J., Sastry, S.S.: A game theoretic approach to controller design for hybrid systems. Proc. IEEE, 949–970 (2000)
15. Tomlin, C.J., Mitchell, I., Bayen, A.M., Oishi, M.: Computational techniques for the verification of hybrid systems. Proc. IEEE, 986–1001 (2003)
16. Varaiya, P.: Hierarchical control of semi-autonomous teams under uncertainty. Final report of Darpa Contract F33615-01-C-3150 (2004)

Real-Time Trajectory Generation by Offline Footstep Planning for a Humanoid Soccer Robot

Andreas Schmitz, Marcell Missura, and Sven Behnke

University of Bonn,
Computer Science VI, Autonomous Intelligent Systems
Friedrich-Ebert-Allee 144, 53113 Bonn, Germany
{schmitz4,missura,behnke}@cs.uni-bonn.de
http://ais.uni-bonn.de

Abstract. In recent years, humanoid soccer robots improved considerably. Elementary soccer skills, such as bipedal walking, visual perception, and collision avoidance have matured enough to provide for dynamic and exciting soccer games. While the elementary skills still remain hot research topics, it is time to move forward and address higher level skills, such as motion planning and team play. In this work, we present a new method to generate ball approach trajectories by planning footstep sequences offline and training an online policy to meet the real time requirements of embedded systems with low computational power, as typically used for soccer robots. We compare the results with our current reactive behavior that was used in the last RoboCup competitions and show the improvements we achieved.

Keywords: humanoid robots, robot soccer, footstep planning, motion planning, trajectory generation.

1 Introduction

Looking back at nearly a decade of history since the introduction of the Humanoid League to RoboCup in the year 2002, an impressive improvement of humanoid robot soccer can be observed. In the beginning, robots were barely able to walk and only penalty kick competitions were possible. Today, elementary soccer skills, such as bipedal walking, visual perception, and collision avoidance have matured enough to provide for dynamic and exciting soccer games played throughout the KidSize, TeenSize and Standard Platform leagues. Guidance and source code for the implementation of low level skills became freely available through a large number of scientific publications and even open source software released by some of the leading teams. While the elementary skills still remain hot research topics, it is time to move forward and address higher-level skills, such as motion planning and team play.

Typically, the control software of humanoid soccer robots is organized in multi-tiered architectures. On top of the fastest sensorimotor control loop, usually a motion layer is used to generate walking, kicking and get-up motions. These

T. Röfer et al. (Eds.): RoboCup 2011, LNCS 7416, pp. 198–209, 2012.

skills are enforced by the qualification requirements for all participants as they are essential for playing soccer. The motions are controlled by a higher behavior layer that covers more complex actions, such as approaching the ball, avoiding obstacles, dribbling the ball towards the goal and aligning for a kick. In the implementation of all leading teams these behaviors are pure reactive [19,7,2] in the sense that performed actions are direct results of the current sensory input without contemplation on possible future states of the environment. It is remarkable how well robots can already play without planning into the future. We believe it is time to investigate how planning can be incorporated to produce smoother and more intelligent motion trajectories during games to improve the overall performance of soccer robots.

In this work, we demonstrate how ball approach trajectories can be generated by footstep planning with the A* algorithm. As motion planning is a computationally expensive task, we use a set of precalculated footstep sequences to distill an online policy for obstacle-free situations on the soccer field. The recall of the policy is fast, just like a reactive behavior, but it produces trajectories that are implicitly planned into the future.

The remainder of this paper is organized as follows. After reviewing related work, we will outline our implementation of the A*-based footstep planning algorithm. In Section 4, we explain how we used our footstep planning algorithm to precalculate planned trajectories offline for a grid of situations and how we used standard machine learning concepts to learn an online policy that reproduces the planned trajectories. Finally, we integrate the online policy as a ball approach module into our soccer software and compare its performance with our reactive behavior from the last competitions.

2 Related Work

Footstep planning is a fairly new research topic. The most prominent proposals in [3,14,4] and also [8] are based on the A* algorithm. By imposing a strong discretization on the state space and using only a small, discrete set of actions, these online solutions plan a few steps ahead and are able to deal with dynamic environments. Uneven floor plans are also considered, so that the footstep plans can include stepping over obstacles and climbing stairs. An intriguing alternative solution has been recently shown in [13]. Here, a short sequence of future footsteps is considered to be a virtual kinematic chain as an extension of the robot. Their location is determined by inverse kinematics. The configuration space and the action space are not discretized, but the algorithm is computationally expensive. A computationally more promising method that can plan in a few milliseconds, if the environment is not too cluttered, has been suggested in [1]. The idea is to solve the footstep planning problem mostly with a path planning algorithm. Actual footstep locations are only given in key points, where the walking speed of the robot has to be zero, for example when stepping over an obstacle. The majority of the footstep locations are laid out along the planned paths by the motion generator developed for HRP-2 [10,11,12]. The closest related work is [4], where an A*-based footstep planning algorithm was adapted for the humanoid robot ASIMO. As the walking algorithm of ASIMO was not precisely known, the authors were forced to reverse engineer a footstep prediction algorithm from observations with a motion capture system.

Another approach to trajectory planning is the Dynamic Window algorithm [6], but it only plans a small amount of time into the future and therefore it cannot produce optimal trajectories, can get stuck in a local optimum and is likely to produce oscillating behaviors, just like any reactive algorithm.

Since the trajectories described by human walkers appear to have a strong resemblance with those of non-holonomic vehicles [16], global path planning methods that determine a geometrical path that adheres to continuous curvature and minimal turning radius restrictions, such that in principle it can be followed by a vehicle with a steering wheel [15] [20] are also potential candidates to be used for humanoid robot motion planning. But this setting is purely kinematic and ignores important physical aspects of motion planning: velocity and acceleration.

In full-fledged kinodynamic planning [5], the state of a moving object includes not only the Cartesian coordinates and the orientation, but also the translational and angular velocities. Planning is mostly performed directly in control space where velocity and acceleration bounds can inherently be taken into account. The increase in dimensionality makes discrete cell decomposition methods impractical. Thus, randomized approaches [17] are often used that sample possible control inputs and project them into the state space by numerical integration. This way, the open-loop controls to execute a calculated trajectory arise naturally. The avoidance of moving obstacles is also possible, as demonstrated in [9], by extending the state space by the time dimension. The computation times reported by the works cited above do not yet allow the application of kinodynamic planning on embedded systems.

The significant difference between our approach and the approaches discussed above is that we do not attempt online execution of a planning algorithm. Instead, we plan near-optimal solutions for a large number of situations offline and learn a fast-to-evaluate policy. This approach is similar to Plan-Then-Compile architectures [18] in the aspect that the first action of a planned action sequence is "compiled" to a stimulus response of a reactive behavior.

3   Footstep Planing

The foundation of our concept is an implementation of the A* algorithm taylored to the task of planning footsteps for a humanoid robot to reach a goal state by placing its foot on a given target footstep location. In the robot soccer domain this target is in most cases a position behind the ball suitable for kicking. We define the state $s = (s_l, s_v) \in S = L \times V$ in a six dimensional state space S that contains the set of all Cartesian left foot poses $L = \mathbb{R} \times \mathbb{R} \times [-\pi, \pi]$ and velocity vectors $V = [-1, 1]^3$, which satisfy the velocity constraints of the robot. The gait velocity vector describes fractions of the maximum allowed velocities in the sagittal, lateral and rotational directions, respectively. The velocity has a strong influence on the step location of the next state, because the step size of the robot grows with higher velocities. We are able to map the gait velocity vector to a step location in Cartesian space with a motion capture-based transformation that we published last year [21]. To limit the branching factor, we defined only a small discrete set of five actions $A \subset [-1, 1]^3$ as feasible accelerations of the lateral velocity to the left or to the right, the rotational velocity clockwise or counter clockwise, and the sagittal velocity only forward, each with the maximum allowed value as configured for a specific robot. We did not need a sagittal deceleration, because after every action the velocity limits of the robot are enforced and accelerating the lateral or rotational components automatically leads to a decrease of the sagittal velocity. An action is always executed twice, such that after a right-left double step the robot is again in a valid state standing on the left foot. The state transformation $s = t(s, a)$ is given by

$$s_l = m(m(s_l, s_v + a), s_v + 2a) \tag{1}$$
$$s_v = s_v + 2a, \tag{2}$$

where s_l is the Cartesian location of the left foot, s_v is the current velocity of the robot, $a \in A$ is the action to perform, and $m(l, v)$ is the gait control vector to step location transformation as described in [21]. We chose the double step as a unit action to cut the depth of the search tree in half and allow the A* algorithm to find results in shorter time, but as we will show later on, this does not necessarily mean that the robot has to reach the goal with the left foot. Please note that since the state includes the current velocity of the robot and the actions are accelerations, we are performing kinodynamic planning that takes the dynamic properties of the robot into account.

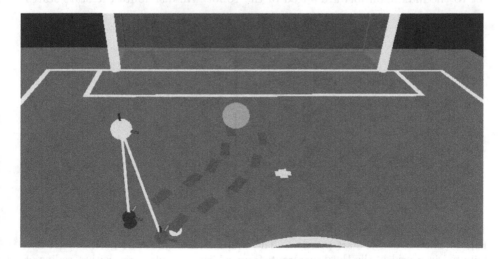

Fig. 1. Visualization of a planned footstep sequence that leads the robot to the ball aiming at the center of the yellow goal

The goal is to reach the set of target states $T \subseteq S$ that are contained by a ball with a radius of 0.1 around the target state s^*

$$T = \{s \in S, \|s - s^*\| \leq 0.1\}. \tag{3}$$

The cost function for every state is simply the number of steps that had to be taken so far. As a heuristic $h(s)$ we used the Euclidean distance on the ground plane between the state s and the target state s^* divided by the largest possible step size d.

$$h(s) = \frac{\|(s_x, s_y) - (s_x^*, s_y^*)\|}{d}. \tag{4}$$

The A* algorithm defined above is able to find nearly optimal dynamic ball approach sequences. An example is illustrated in Figure 1. The problem with this approach is that in many cases its runtime exceeds several minutes of computation and can require gigabytes of memory to keep track of open states. Clearly, it is not suitable to be run on embedded systems with low computational power along with the real time requirements of a soccer game. Thus, we only use it to compute a large set of trajectories offline that we use as training examples to learn a policy.

4   Policy Learning

Using the A* algorithm described in the previous section, we calculate the footstep plans for a set of start states defined as follows. The ball is located in the origin of the coordinate system and the y-axis points towards the target direction (the goal), as illustrated in Figure 2. The target velocity at the ball is set to the maximum forward speed of the robot. We distribute the start states

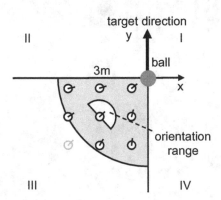

Fig. 2. The domain of start positions. The coordinates of the start positions are distributed in a grid in the third quadrant, from which a circular area with a radius of three meters is selected. Each of these positions has an orientation in the range $\pm\pi$ centered around the direct line to the ball.

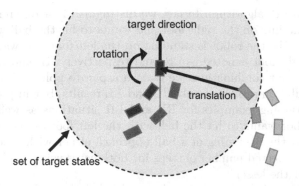

Fig. 3. The last step of precalculated paths does not hit the target exactly. A transformation can correct these inaccuracies. The size of the set of target states is strongly exaggerated in this illustration.

in the third quadrant in a six-dimensional grid layout with ten intersections in each dimension. From these we select only those with a distance no greater than three meters to the ball. The orientations are taken from a range of $\pm\pi$ centered around the direct line to the ball. Altogether we have somewhat less than 10^6 start states due to the selection of the circular area. Any situation can be mirrored between the third and the fourth quadrant, so the fourth quadrant does not need to be calculated explicitly. The first and the second quadrant are excluded from our considerations.

With 24 CPU cores, we managed to calculate approximately 17,000 paths randomly sampled from the defined grid in two days time totaling approximately 400,000 single steps. In post processing we corrected the error A* makes by applying a translation and a rotation to each of the entire footstep sequences such that the final footstep precisely hits the target. This is depicted in Figure 3. Applying a mirroring technique, we also produced example footstep plans that finish with the right foot instead of the left foot and doubled the size of the training set. The idea of designing the policy is that no more than the next single step is needed to be known at any time. After the robot has executed the step, it will evaluate the policy again and obtain a new step. By successive recall of the policy, the same footstep sequence will be reproduced that was planned with the A* algorithm requiring only very little computational power. The policy $\pi : a(s) \in A$ we aim to learn is a mapping from a six dimensional state space to a three dimensional action space. Here the state s is expressed in a reference frame centered on the target. Consequently, we do not need to deal with entire footstep sequences, but every single step can be used as a training example. A convenient property of the method we used to calculate the training data is that since all A* paths lead to the target, the density of the steps increases automatically in the vicinity of the ball. Since the training set contains both, the robot is standing on the right foot and the robot is standing on the left foot states, we can now discard the double step restriction that we introduced to accelerate the A* algorithm. In fact we distinguish four different situations. The robot is standing on the right foot and wants to hit the ball with the right foot denoted as RR, the robot is standing on the left foot and wants to hit the ball with the right foot denoted as LR, and respectively RL and LL situations. For each of the four combinations we train a separate policy. Please note that the A* precalculation produced only RL and LL results, but in post processing we generated training examples for RR and LR situations as well. The choice whether the robot wants to hit the ball with the left foot or the right foot can be made once at the beginning of a ball approach either with a heuristic or by calculating the expected number of steps for both cases using the learned policy and deciding for the lesser.

To perform the actual learning task, we divided the training data into appropriate sets for the RR, LR, RL, and LL policies, each of them containing 200,000 examples. We evaluated three different machine learning approaches: k-nearest neighbor interpolation, piecewise linear approximation and multi-layer perceptrons. First we tried k-nearest neighbor interpolation by retrieving the 200 nearest neighbors for a query point, fitting a six-dimensional hyperplane with a least squares method into the point cloud of the neighbors for each of the three output dimensions. The hyperplanes are evaluated at the location of the query. We found that this method produced smooth step sequences that were able to hit the ball reliably in simulation. However, the policy recall times were not satisfactory due to the overhead of the nearest neighbor retrieval and the hyperplane fitting with each query. Therefore we transformed the data set to a regular grid by evaluating each grid node with the aforementioned k-nearest neighbor technique. In a regular grid the nearest neighbors can be found instantly and no hyperplane fitting is needed, since we can just interpolate between the corners of a grid cell (a six-dimensional grid cell has 64 corners). Using the regular grid we achieved efficient policy recall times without sacrificing precision, but it still has some undesirable memory requirements. We used a grid with 11 intersections in each of the six dimensions resulting in 80 MB of data in double precision. As a third alternative we trained-multi-layer perceptrons. We used a separate network for each of the three output dimensions with one output neuron, 20 sigmoid neurons in the hidden layer and six input neurons. The networks learned the desired functions with a precision comparable to the other two approaches. The evaluation of a network is even faster than the piecewise linear approximation and the memory requirements for storing the network weights are minimal. We present two types of accuracy measures in Figures 4 and 5. The single step accuracy in Figure 4 shows the mean error and standard deviation of the gait velocities produced by the learned policies compared to the corrected output of the A* algorithm. The errors were determined using a test set of approximately 10,000 examples that were not used for training. The single step errors do not accumulate over an entire path, because the policy recall after each step has an error correcting property. Most important is the accuracy at the target point at the end of the entire footstep plan. This is shown in Figure 5. The policy recall times are shown in Figure 6. In light of the low hardware requirements of the neural network we decided to use it for all subsequent experiments.

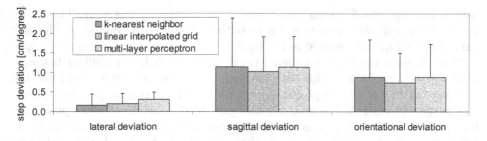

Fig. 4. Evaluation of the single step accuracy of the three machine learning methods. The bar chart represents the mean error and standard deviation of the gait velocities produced by the learned policies compared to the output of the A* algorithm.

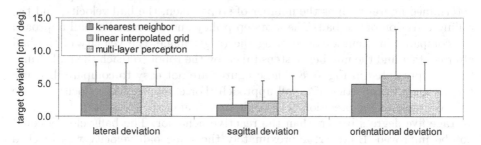

Fig. 5. Evaluation of the path accuracy. Mean errors and standard deviations are measured after the last step of a successive policy recall for an entire footstep plan.

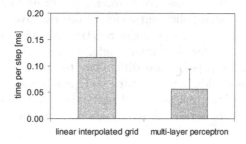

Fig. 6. Policy recall times for a single step when using the regular grid (left) or the multi-layer perceptron (right). The k-nearest neighbor method is omitted from this comparison, because its recall times are much higher. The times were measured on an Intel Core 2 Duo T8300 2.4GHz CPU with Windows 7.

5   Experimental Results

We integrated the trained policies into our soccer software and performed an experiment with a real robot. We compared the performance of the new trained footstep policies with our reactive dribbling behavior that played an important role in our success of winning the TeenSize competitions last year. The same robot "Dynaped" was used to evaluate both ball approach strategies. The experimental setup is illustrated in Figure 7.

Starting from the poses numbered from one through five, dribbling ball approaches were performed with both methods such that the robot would hit the ball located in one of the positions a, b or c with maximum possible speed. The target direction is the center of the goal to the left. Each start pose and ball position combination was repeated seven or eight times, so that at least 100 ball approaches were performed with each, the footstep policy and the reactive behavior. The robot and the ball were equipped with reflective markers and we recorded every approach with a motion capture device. Using the motion capture data we were able to reconstruct the ball approach trajectories and identified the footstep locations of single steps. For a qualitative comparison we determined figures such as the number of steps taken, the ball velocity, and the rolling direction of the ball. The footstep policy and the reactive ball approach are compared in accuracy of meeting the target angle, the velocity of the ball after contact and the number of steps taken by the robot to reach the ball. Numbers are presented in Figure 8. These figures are not easy to compare, because they trade off properties of a ball approach. For example, ball velocity can be gained by sacrificing precision. The angle precision of the footstep policy was in average five degrees better, than the reactive behavior. The ball velocity could not be improved. However, approximately the same ball velocity was reached with a higher precision. The average number of steps taken was also improved by the footstep policy and it is worth taking a closer look at Figure 9, where the number of steps is depicted in detail for each start pose and ball position combination. In easy situations, where the robot basically only has to walk straight forward against the ball, the reactive behavior performs just as well. In hard cases, however, the footstep policy outperforms the reactive behavior by five or six steps. There are also cases, where the reactive behavior just walks straight into the ball without care for the target angle, as shown in Figure 10 part (a). In these cases the footstep policy has difficulties to approach the ball with the same number of steps while giving the target angle a higher priority.

In Figure 10 some of the reconstructed ball approach trajectories are shown. In case (a) the reactive behavior takes only a small amount of steps and moves

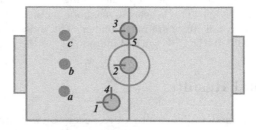

Fig. 7. Arrangement of the robot start poses [1..5] and ball positions [a,b,c]

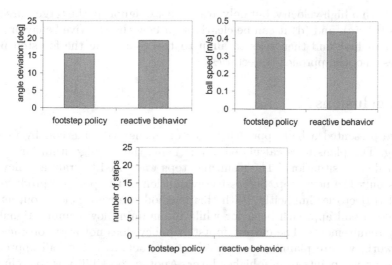

Fig. 8. Comparison of the average angle deviation (a), ball velocity (b), and number of steps (c) of the footstep policy and the reactive behavior

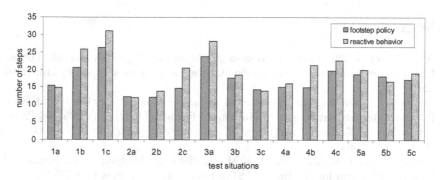

Fig. 9. Comparison of the average number of steps used by the reactive behavior and the footstep policy in the different test situations

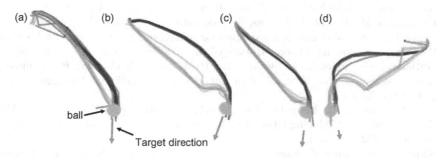

Fig. 10. Reconstructed ball approach trajectories of the reactive behavior (green / light) and the footstep policy (blue / dark)

the ball with a high velocity, but only at the cost of ignoring the target direction. In cases (b), (c), and (d) it can be clearly seen how the reactive behavior heads for the ball first and then tries to align to the goal while the footstep policy produces smooth approach trajectories.

6 Conclusions

We have presented a ball approach trajectory generation method by footstep planning. The plans were calculated offline using the A* algorithm for a set of predefined start situations. The resulting steps were used to train a policy that predicts only the next step. Successive evaluation of the policy reproduces the planned trajectories implicitly. With this method, we were able to outperform our reactive ball approach behavior while maintaining low computational and memory requirements. The current footstep policy does not avoid obstacles. In future work, we are planning to implement obstacle avoiding ball approaches by injecting via points from a higher layer. Another possibility is to include an obstacle in the training data and training the footstep policy to take obstacles into account.

References

1. Ayaz, Y., Owa, T., Tsujita, T., Konno, A., Munawar, K., Uchiyama, M.: Footstep Planning for Humanoid Robots Among Obstacles of Various Types. In: Humanoids (2009)
2. Behnke, S., Stückler, J.: Hierarchical Reactive Control for Humanoid Soccer Robots. International Journal of Humanoid Robots (IJHR) 5(3), 375–396 (2008)
3. Chestnutt, J., Kuffner, J.: A Tiered Planning Strategy for Biped Navigation. In: Humanoids (2004)
4. Chestnutt, J., Lau, M., Cheung, K.M., Kuffner, J., Hodgins, J.K., Kanade, T.: Footstep Planning for the Honda ASIMO Humanoid. In: ICRA (2005)
5. Donald, B., Xavier, P., Canny, J., Reif, J.: Kinodynamic Motion Planning. J. ACM 40, 1048–1066 (1993)
6. Fox, D., Burgard, W., Thrun, S.: The Dynamic Window Approach to Collision Avoidance. IEEE Robotics and Automation Magazine 4(1), 23–33 (1997)
7. Friedmann, M., Kiener, J., Petters, S., Sakamoto, H., Thomas, D., von Stryk, O.: Versatile, high-quality motions and behavior control of humanoid soccer robots. In: Workshop on Humanoid Soccer Robots of the 2006 Humanoids, pp. 9–16 (2006)
8. Gutmann, J.S., Fukuchi, M., Fujita, M.: Real-time path planning for humanoid robot navigation. In: Proc. of 19th Int. Conf. on Artificial Intelligence, pp. 1232–1237 (2005)
9. Hsu, D., Kindel, R., Latombe, J.-C., Rock, S.: Randomized Kinodynamic Motion Planning with Moving Obstacles. The International Journal of Robotics Research 21, 233–255 (2002)
10. Kajita, S., Kanehiro, F., Kaneko, K., Fujiwara, K., Harada, K., Yokoi, K.: Biped walking pattern generation by using preview control of zero-moment point. In: Proc. of the International Conference on Robotics and Automation, pp. 1620–1626 (2003)

11. Kajita, S., Kanehiro, F., Kaneko, K., Yokoi, K., Hirukawa, H.: The 3D linear inverted pendulum mode: a simple modeling for a bipedwalking pattern generation. In: IROS, vol. 1, pp. 239–246 (2001)
12. Kaneko, K., Kanehiro, F., Kajita, S., Hirukawa, H., Kawasaki, T., Hirata, M., Akachi, K., Isozumi, T.: Humanoid Robot HRP-2. In: ICRA, pp. 1083–1090 (2004)
13. Kanoun, O., Yoshida, E., Laumond, J.P.: An Optimization Formulation for Footsteps Planning. In: Humanoids (2009)
14. Kuffner, J., Kagami, S., Nishiwaki, K., Inaba, M., Inoue, H.: Online Footstep Planning for Humanoid Robots. In: ICRA, pp. 932–937 (2003)
15. Lamiraux, F., Laumond, J.-P.: Smooth motion planning for car-like vehicles. IEEE Transactions on Robotics and Automation 17, 498–502 (2001)
16. Laumond, J.-P., Arechavaleta, G., Truong, T.-V.-A., Hicheur, H., Pham, Q.-C., Berthoz, A.: The Words of the Human Locomotion. In: Kaneko, M., Nakamura, Y. (eds.) Robotics Research. STAR, vol. 66, pp. 35–47. Springer, Heidelberg (2010)
17. LaValle, S.M., Kuffner Jr., J.J.: Randomized Kinodynamic Planning. I. J. Robotic Res. 20, 378–400 (2001)
18. Mitchell, T.M.: Becoming Increasingly Reactive. In: Proceedings of the Eighth National Conference on Artificial Intelligence, vol. 2 (1990)
19. Röfer, T., Laue, T., Müller, J., et al.: B-Human Team Report and Code Release (2010), http://www.b-human.de/file_download/33/bhuman10_coderelease.pdf
20. Scheuer, A., Fraichard, T.: Collision-Free and Continuous-Curvature Path Planning for Car-Like Robots. In: Proc. IEEE Int Conf. on Robotics & Automation, pp. 867–873 (1997)
21. Schmitz, A., Missura, M., Behnke, S.: Learning Footstep Prediction from Motion Capture. In: Ruiz-del-Solar, J., Chown, E., Ploeger, P.G. (eds.) RoboCup 2010. LNCS (LNAI), vol. 6556, pp. 97–108. Springer, Heidelberg (2010)

Real Time Biped Walking Gait Pattern Generator for a Real Robot

Feng Xue[1], Xiaoping Chen[1], Jinsu Liu[1], and Daniele Nardi[2]

[1] Department of Computer Science and Technology,
University of Science and Technology of China,
Hefei, 230026, China
[2] Department of Compute and System Science,
Sapienza University of Roma,
Via Ariosto 25, Roma, Italy

Abstract. The design of a real time and dynamic balanced biped walking gait pattern generator is not trivial due to high control space and inherently unstable motion. Moreover, in the Robocup domain, robots that are able to achieve the goal footstep in a short duration have a great advantage when playing soccer. In this paper, we present a new technique to realize a real time biped walking gait pattern generator on a real robot named Nao. A Zero Moment Point (ZMP) trajectory represented by a cubic polynomial is introduced to connect the goal state (the position and the velocity of the CoG) to the previous one in only one step. To apply the generator on the real robot Nao, we calculate the compensation for two HipRoll joints in a theoretical way by modeling them as elastic joints. The nao of version 3.3 is used in the experiments. The walk is intrinsically omnidirectional. When walking with step duration 180ms, the robot can respond to the high level command in 180ms. The maximum forward speed is around 0.33m/s. The maximum backward speed is around 0.2m/s. The maximum sideways speed is around 0.11m/s. The maximum rotational speed is around 90°/s.

Keywords: Biped Walking, ZMP, 3D Inverted Pendulum, Elastic Joints.

1 Introduction

The design of a real time and dynamic balanced biped walking gait pattern generator is not trivial due to high control space and inherently unstable motion. Moreover, in the Robocup domain, robots that are able to respond to higher level commands in a small delay and achieve the goal footstep in a short duration have a great advantage when playing soccer.

Zero Moment Point (ZMP) [15] has been applied widely to ensure dynamic balance in many biped walking gait pattern generators. ZMP is a point about which the sum of the horizontal ground reaction moment due to ground reaction force is zero. It is an effective criterion for measuring instantaneous balance. The state-of-art biped walking gait pattern generators usually plan the Center of

T. Röfer et al. (Eds.): RoboCup 2011, LNCS 7416, pp. 210–221, 2012.

Gravity (CoG) trajectory according to the given ZMP trajectory which satisfies this criterion.

An approach based on control theory has been proposed by Kajita et al. [9] in which the CoG is designed to converge at the end of previewing period. Stefan et al. [3] also used the preview controller and handled the sensor feedback in a different, more direct and intuitive way. Due to the inner tracking error of the preview controller, the magnitude of the preview gain needs around 1.5 seconds to converge. Thus, it is not suitable for a motion that changes its direction or speed in high frequency, such as in RoboCup domain.

Colin et al. [6] proposed a CoG-based gait in which CoG and foot trajectory are represented as functions with lots of parameters, instead of following the dynamic equation of the robot. The notion of assuring a dynamic balanced gait is not discussed. In the following work [5], they used basic 3D-LIPM theory and the ZMP is set at the origin of the foot for each step resulting in discontinuous ZMP trajectory.

Kajita et al. [8] proposed an analytical technique by modeling the robot as a 3D Linear Inverted Pendulum (3D-LIP). Jinsu Liu et al. [10] also uses this technique to achieve an online sampling search to switch between different walking commands. It also proposed a *Simplified Walking* in which a ZMP point namely *ZMP Decision* is used for each step. The previous state and the goal one are connected using two *ZMP Decisions* (two steps). We extend this work and use a ZMP trajectory represented by a cubic polynomial. Thus we do not need two steps, but only one to connect the goal state to the previous one.

In the work [14] and in the walking engine developed by Aldebaran company, the two HipRoll joints are compensated with a sinusoidal offset in an experimental way. In our generator, we consider the two joints as elastic joints which have been widely studied over decades in trajectory tracking tasks and regulation tasks for manipulators [12][17]. To our knowledge, we are the first to propose elastic joints on a humanoid robot. The main contribution of this part is that, we model the robot as a 3D-LIP, so that the computation cost is reduced a lot to attain a real time computation.

The remainder of the paper is organized as follows. Section 2 will present our walking engine focusing on the simultaneously planning of ZMP and CoG. Section 3 gives the theoretical way of compensating the two HipRoll joints. Experiments are shown in section 4. And we conclude in section 5.

2 Walking Engine

A gait pattern is usually a set of trajectories of the desired ZMP, the feet, the arms and the upper body.

Before the gait pattern generator, see Figure 1, a footstep planning algorithm is employed. In the case of a complex environment having obstacles, stairs and slopes, some heuristic or stochastic search algorithms are applied to generate footstep trajectory leading the robot from the start position to the goal position[2][11]. In the case of Robocup domain, it is relatively simple. The input command is a vector, denoted as $[\dot{x}, \dot{y}, \dot{\theta}]$, where $\dot{x}$ is the linear speed along

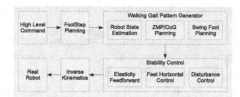

Fig. 1. The Architecture of Our Walking Engine

X axis, $\dot{y}$ is the linear speed along Y axis and $\dot{\theta}$ is the rotation speed around Z axis. Thus if the ith footstep is $\bar{f}s^i = [\bar{x}^i, \bar{y}^i, \bar{\theta}^i]$, then $(i+1)^{th}$ footstep is $\bar{f}s^{i+1} = [\bar{x}^{i+1} + \dot{x}T, \bar{y}^{i+1} + \dot{y}T, \bar{\theta}^{i+1} + \dot{\theta}T]$, where T is the step duration. The walking engine usually handles input of a vector, in which each value specifies the related portion of the maximum speed, ranging between [-1.0, 1.0].

Now we have the footsteps, in another words, a set of support polygons, so the ZMP trajectory for reference is determined. In our previous work[10], a single ZMP namely *ZMP Decision* is placed at the center of each footstep.

We can see that the ZMP based approach is intrinsically omnidirectional. The method using Central Pattern Generator (CPG) suffers from this problem and additional techniques are studied [1].

Following the gait pattern generator is a stability controller which is in charge of rejecting the external disturbance applied from outside, and also the internal one due to the model error.

Then, an inverse kinematics solver is applied to generate the joint angles which are used to command the joint actuators. We solve it analytically in a standard way which is also reported in [6].

Given the cycle duration T as known constraint, we have N control cycles. Taking the Nao robot as an example, it can be controlled every 10ms, so, N is 50 if T equals to 0.5s. In our walking engine, if the robot reaches the last control cycle, it will read the cached user command, call footstep generation algorithm, set ZMP points for reference and call walking gait pattern generator.

2.1   Overview of Biped Walking Gait Pattern Generator

As mentioned above, the state-of-art biped walking gait pattern generator deals with a given ZMP trajectory which is in the support polygon set to guarantee a dynamic balanced gait. Then, the upper body trajectory which satisfies the desired ZMP trajectory is calculated using an approximate dynamics model.

In our walking gait pattern generator, as shown in Figure 2, the i_{th} step begins with a single support phase named $ss1^i$, following a double support phase named ds^i, and ends up with a single support phase named $ss2^i$. In $ss1^i$, the foot reaches the goal footstep of $(i-1)^{th}$ cycle. Compared to our previous work [10], the *Simplified Step* is divided into two swing phases: $ss2^{i-1}$ and $ss1^i$. As mentioned above, for the reason of highly response, we read the cached command

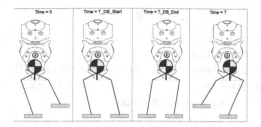

Fig. 2. A Step Is Formed by Two Swing Phases and One Double Support Phase

at the last control cycle of $ss2^i$, so the time duration for the next step is not known and the duration for $ss1^{i+1}$ could be fast or slow. Consequently, we set $ss2^{i-1}$ for lifting and $ss1^i$ for landing.

In our generator, we connected the goal CoG state (the position and the velocity of CoG) to the previous one in one step, so the goal user command can be reached in a small time which is determined by the duration of a step. A ZMP trajectory represented by a cubic polynomial is introduced to do this, which is talked about in Subsection 2.3 after the dynamic model in the next subsection.

2.2 Dynamics

In our point of view, no matter how accurate the dynamic model is, there will still be difference compared to the real robot, not to mention the accuracy of joint actuators which are usually not modeled. And, the model error can be treated as inner disturbance and handled together with the external disturbance. That's what the stability controller does. Extension of this controller is reported in detail in [16].

Fig. 3. Physics Model: The 3D Inverted Pendulum

The 3D Inverted Pendulum (3D-IP) proposed by Kajita et al. [8] is used to describe the approximate movement of a biped walking when the robot is supporting its body on one leg. A 3D-IP is an inverted pendulum which moves in 3D space. Assuming the current ZMP of the pendulum is at (p_x, p_y, p_z), the physics model shown in Figure 3 can be described as follows according to the definition of ZMP:

$$\ddot{x}(z - p_z) = (x - p_x)(g + \ddot{z}) \tag{1}$$

$$\ddot{y}(z - p_z) = (y - p_y)(g + \ddot{z}) \tag{2}$$

where g is the acceleration due to gravity, (x, y, z) and $(\ddot{x}, \ddot{y}, \ddot{z})$ are the position and the acceleration of the CoG.

In order to get a linear equation, a horizontal plane with intersection with Z axis z_p is applied. And the 3D-IP becomes a 3D Linear Inverted Pendulum (3D-LIP) [8]. The following equations hold:

$$\ddot{x} = \frac{g}{z_p}(x - p_x) \tag{3}$$

$$\ddot{y} = \frac{g}{z_p}(y - p_y) \tag{4}$$

In the case that the ZMP of X axis p_x is considered as a constant value p_x^*, the analytical solution of Equation 3 is (solution of Y axis is similar):

$$\begin{bmatrix} x_f(t) \\ \dot{x}_f(t) \end{bmatrix} = A(t) \begin{bmatrix} x_i \\ \dot{x}_i \end{bmatrix} + [I - A(t)] \begin{bmatrix} p_x^* \\ 0 \end{bmatrix} \tag{5}$$

where $[x_i, \dot{x}_i]^T$ is the initial state, $[x_f(t), \dot{x}_f(t)]^T$ is the final state at time t, I is a 2×2 identity matrix and $A(t)$ is a state transition matrix which only depends on the duration t:

$$A(t) = \begin{bmatrix} cosh(qt) & \frac{1}{q}sinh(qt) \\ qsinh(qt) & cosh(qt) \end{bmatrix}, \quad q = \sqrt{\frac{g}{z_p}} \tag{6}$$

We now give the analytical solution when ZMP of X axis P_x is not constant, but represented by a cubic polynomial: $P_x = \sum_{i=0}^{3} a_i t^i$. The analytical solution of Equation 3 is:

$$\begin{bmatrix} x_f(t) \\ \dot{x}_f(t) \end{bmatrix} = A(t) \begin{bmatrix} x_i \\ \dot{x}_i \end{bmatrix} + [I - A(t)] \begin{bmatrix} a_0 + 2a_2/q^2 \\ a_1 + 6a_3/q^2 \end{bmatrix} + \begin{bmatrix} \sum_{i=0}^{3} a_i t^i + 6a_3 t/q^2 - a_0 \\ 3a_3 t^2 + 2a_2 t \end{bmatrix} \tag{7}$$

2.3   Simultaneously Planning of ZMP and CoG

The length of the foot is around 14cm which is large compared to the short leg. Usually, the distance between each footstep in sagittal direction is lower than 8cm due to kinematics constraint caused by the relatively short leg. Thus, the heel of one foot will never exceed the toe of the other one, which is making a large support polygon. This inspires us of not using two *ZMP Decisions* to connect the desired CoG state to the previous one.

In consequence, we need to plan ZMP and CoG trajectory simultaneously. We first introduce the sagittal direction. The initial ZMP denoted as PY_i and the initial state of CoG denoted as $(y_i, \dot{y}_i)$ are decided by the last control cycle of ith

footstep. The goal ZMP denoted as PY_{i+1} is determined by the $(i+1)^{th}$ footstep. In order to make a quick stop where the two foot positions along sagittal direction are the same, we put the goal CoG position denoted as y_f at the center of the $(i+1)^{th}$ footstep. In the work proposed by Harada et al. [7], they use this similar technique to connect the new trajectories to the current ones. However, in our case, the robot could change speed rapidly, so without control of the target CoG velocity may lead the robot to fall down. Thus, we introduce another parameter denoted as α which is the percentage of the user required speed Y_{user}. And let $\dot{y}_f$ is equal to αY_{user}. Then, according to Equation 7, the four coefficients $a_{0..3}$ in the cubic polynomial are determined using the following equations:

$$PY_i^* = \sum_{i=0}^{3} a_i(iT)^i$$

$$PY_{i+1}^* = \sum_{i=0}^{3} a_i(i(T+1))^i$$

$$\begin{bmatrix} y_f \\ \dot{y}_f \end{bmatrix} = A(T) \begin{bmatrix} y_i \\ \dot{y}_i \end{bmatrix} + [I - A(T)] \begin{bmatrix} a_0 + 2a_2/q^2 \\ a_1 + 6a_3/q^2 \end{bmatrix} + \begin{bmatrix} \sum_{i=0}^{3} a_i t^i + 6a_3 t/q^2 - a_0 \\ 3a_3 t^2 + 2a_2 t \end{bmatrix}$$

In the case of lateral direction, the initial state of CoG and the initial and goal ZMP are already known like the case of sagittal direction. The goal CoG velocity is known too, which is equal to zero, see Figure 2. We now explain how the goal CoG position can be determined using *Simplified Walking*. Note that the goal CoG state will become the initial CoG state of the next step. And we know that the initial CoG velocity ($\dot{x}_i$) of the next step is zero and the CoG position (x_f) of the next step after half step is determined by offset of Hip Joint. The two unknown variables are the initial CoG position (x_i) of the next step and the CoG velocity ($\dot{x}_f$) of the next step at the time of half step duration, and they are determined using Equation 5. Thus, the goal CoG position is known. So, like the case of sagittal direction, the coefficients in the cubic polynomial are determined.

By applying the cubic ZMP trajectory, a CoG state (both in X and Y axis) can be achieved in one step as long as the computed ZMP trajectory are in the support polygons. The preview control based walking needs around 1.5s

Fig. 4. Schematic Representation of An Elastic Joint

to converge. Our previous work *Simplified Walking* needs two steps. Thus, this technique has a great advantage for soccer players that change speed or direction rapidly.

3   Elasticity Modeling Control

The compensation to the HipRoll Joint has beed discussed in the previous works. However, a lack of theory or model leads the compensation to an empirical way. Here we model the joint as elastic joint, see Figure 4. Under the assumption proposed in [13] by Spong, we have the following dynamic equations:

$$M(q)\ddot{q} + S(q, \dot{q})\dot{q} + g(q) + K(q - \theta) = 0 \qquad (8)$$

$$I\ddot{\theta} + K(\theta - q) = \tau \qquad (9)$$

Where q is the (n × 1) vector denoting the link positions. θ is the (n × 1) vector denoting the motor positions. Note that q_i is not equal to θ_i because of elasticity. M(q) is the (n × n) robot link inertia matrix. $S(q, \dot{q})\dot{q}$ is the (n × 1) vector of centrifugal and Coriolis torques. K is the (n × n) diagonal matrix of joint stiffness coefficients. g(q) is the (n × 1) vector of gravitational torque. I is an (n × n) constant diagonal matrix including rotor inertia and the gear ratios. τ is the (n × 1) vector of the torque acting on the elastic joints.

Vector q is given by the Inverse Kinematics. Vector θ is the final joint commands that will be sent to the robot. However, the computation power of Nao does not allow for the above heavy computation. Thus, we again take the 3D-LIP as a simplified model.

Let $^{i-1}T_i$ denote the transformation matrix from link frame i to link frame i-1, for the Nao robot, $^1T_2 = Rot_y(q_2)$, $^2T_3 = Rot_x(q_3)$, $^3T_4 = Trans_z(-R)Rot_x(q_4)$, $^4T_5 = Trans_z(-S)Rot_x(q_5)$, $^5T_6 = Rot_y(q_6)$. Note that, the coordinate is rotated - 90 degrees compared to the coordinate used in the document provided by Aldebaran [4]. The transformation matrix from CoG to the first link of left leg is: $^GT_1 = Trans_x(-H)Trans_z(-N)$, where H is the Hip offset along X axis and N is the Hip offset along Z axis. The first joint is not considered, so $^1T_2$ is an identity matrix. Following the standard Denavit-Hartenberg method, we have:

$$^GT_6 = {}^GT_1 \prod_{2}^{6} {}^{i-1}T_i \qquad (10)$$

Thus, ^6T_G is obtained by computing the inverse matrix. The 3rd column of this matrix is the CoG position relative to the foot coordinate, denoted as $P(q_2, q_3, q_4, q_5, q_6)$, where q_i is the i-th link position. So, the Lagrange function is:

$$L = \frac{1}{2}M\dot{P}^T\dot{P} - MgP_z - \frac{1}{2}(q - \theta)^T K(q - \theta) \qquad (11)$$

Following the Lagrange approach, we have:

$$\frac{d}{dt}\frac{\partial (\Gamma - V_{grav})}{\partial \dot{q}_i} - \frac{\partial (\Gamma - V_{grav})}{\partial q_i} + k_i(q_i - \theta_i) = 0 \qquad (12)$$

Compared to Equation 8 and 9, Equation 12 does not contain the term of centrifugal and Coriolis torques. Thus, the computation for elasticity is simplified quite a lot. By using the symbol differential tool developed by us, the final equation for one joint is formed by, approximately, one thousands of terms (in average 10 multiplications per term). That's nearly one third of the computation required by Equation 8 and 9 induced from full dynamic model. However, this is still relatively heavy for Nao.

In the Robocup domain, the robot is walking on a planar ground, and the foot is always horizontal to the ground. Thus, we have:

$$q_2 + q_6 = 0$$
$$q_3 + q_4 + q_5 = 0 \tag{13}$$

By, substituting q_6 with q_2 and q_4 with $-q_3 - q_5$ in Equation 12, we have 167 terms for computing θ_3, 471 terms for computing θ_4, 517 terms for computing θ_5. The equations for computing θ_2 and θ_6 are simpler and can be written down:

$$
\begin{aligned}
k_2\theta_2 = {}& \ddot{L}_z MN \sin(q_2) + 2\dot{L}_z MN\dot{q}_2 \cos(q_2) - L_z MN\dot{q}_2{}^2 \sin(q_2) + \\
& L_z MN\ddot{q}2 \cos(q_2) + H^2 M\dot{q}_2{}^2 \cos(q_2) \sin^3(q_2) + HMN\dot{q}_2{}^2 - \\
& 2HMN\dot{q}_2{}^2 \sin^2(q_2) - HMN\dot{q}_2{}^2 \cos^4(q_2) + HMN\dot{q}_2{}^2 \sin^4(q_2) + \\
& HMg - MN^2\dot{q}_2{}^2 \cos(q_2) \sin^3(q_2) + k_2 q_2
\end{aligned}
$$

$$
\begin{aligned}
k_6\theta_6 = {}& -\ddot{L}_z L_z M \sin(q_2)\cos(q_2) - L_z Mg \sin(q_2) + \ddot{L}_z MN \sin(q_2) + \\
& 2\dot{L}_z MN\dot{q}2 \cos(q_2) + L_z^2 M\dot{q}_2{}^2 \cos(q_2)\sin(q_2) - L_z^2 M\ddot{q}_2 \cos^2(q_2) + \\
& L_z MN\ddot{q}2 \cos(q_2) - L_z MN\dot{q}_2{}^2 \sin(q_2) - 2H^2 M\dot{q}_2{}^2 \sin(q_2)\cos(q_2) + \\
& H^2 M\dot{q}_2{}^2 \sin^3(q_2)\cos(q_2) + 2H^2 M\dot{q}_2{}^2 \cos^3(q_2)\sin(q_2) - \\
& 2\dot{L}_z L_z M\dot{q}_2 \cos^2(q_2) + HMg - k_6 q_2
\end{aligned}
$$
$$\tag{14}$$

where $\dot{L}_z$, $\ddot{L}_z$, $\dot{q}_2$ and $\ddot{q}_2$ are determined by:

$$
\begin{bmatrix} \dot{X}_g \\ \dot{Z}_g \end{bmatrix} = \begin{bmatrix} -\sin(q_2) & -L_z \cos(q_2) \\ -\cos(q_2) & L_z \sin(q_2) \end{bmatrix} \begin{bmatrix} \dot{L}_z \\ \dot{q}_2 \end{bmatrix} \tag{15}
$$

$$
\begin{bmatrix} \ddot{X}_g \\ \ddot{Z}_g \end{bmatrix} = \begin{bmatrix} -\sin(q_2) & -L_z \cos(q_2) \\ -\cos(q_2) & L_z \sin(q_2) \end{bmatrix} \begin{bmatrix} \ddot{L}_z \\ \ddot{q}_2 \end{bmatrix} + \begin{bmatrix} -2\dot{L}_z\dot{q}_2 \cos q_2 & L_z\dot{q}_2{}^2 \sin q_2 \\ 2\dot{L}_z\dot{q}_2 \sin q_2 & L_z\dot{q}_2{}^2 \cos q_2 \end{bmatrix} \tag{16}
$$

Thus, the robot is able to do the compensation online. We now explain how to determine the stiffness coefficients. When the robot is standing with one of its leg, we have $\dot{X}_g = \ddot{X}_g = 0$, Equation 14 are:

$$\theta_2 = HMg/k_2 + q_2 \tag{17}$$

$$\theta_6 = -L_z Mg \sin(q_2)/k_6 + HMg/k_6 - q_2 \tag{18}$$

The deflection of HipRoll ($\theta_2 - q_2$) is constant, which is equal to HMg/k_2. So we conducted an experiment in which the robot is standing with one of its leg, and k_2 is adjusted online according to the readings from inertial sensor. The rest joint stiffness coefficients are simply set to the same value.

4   Experiments

All the experiments are conducted on a real robot named Nao with the version of 3.3.

4.1   Simultaneously Planning CoG and ZMP Trajectory

The width of the foot is around 8cm and the length of the foot is around 12cm (actually longer). ZMP trajectory should be kept in the support polygon decided by the foot dimension.

We first let the robot walk left with step duration equaling 480ms (100ms for double support phase), step length equaling 4cm. The left part of Figure 5 shows the results, where the black line is the ZMP trajectory and the green line is the CoG trajectory. The red and blue lines show the allowed support polygon. The shadow indicates the double support phase. We can see that with a double support of 17%, the ZMP trajectory is kept in the support polygon, thus making a dynamic balanced gait.

Next, we let the robot walk forward with step duration equaling 400ms (100ms for double support phase), maximum step length equaling 8cm and the portion of the user required speed equaling 50%. At first, the robot is walking with step length 0.8cm, that is 10% of the maximum speed 0.2m/s. Then it switches to full speed 0.2m/s at 10.5s. The right part of Figure 5 shows the result. The CoG state is achieved in one step without breaking the dynamic balance law. Note that at the time of 10.5s, the ZMP trajectory first moves ahead of the CoG to decelerate and then moves back in a short time to accelerate. This effect is due to the cubic polynomial.

4.2   Elasticity Control

The compensation to the elastic joints can obey Equation 14. However, in our dynamic balance control module, the AnkleRoll joint will be modified, indicating

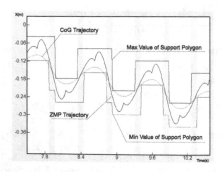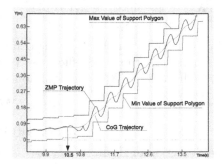

Fig. 5. Simultaneously Planning of ZMP/CoG Trajectory. The Left One Is along Lateral Axis and the Right One Is along Sagittal Axis.

it does not need a feed forward compensation. Besides, the compensations to the HipPitch, KneePitch and AnklePitch are relatively small and computation for these three joints are relatively heavy, so they are not compensated too.

However, the compensations to the two HipRoll joints can not just obey Equation 14, because in the double support phase it is a closed-loop chain, not an open-loop chain as in the single support phase. The key part of this compensation is to guarantee that the landing foot will not hit the ground hard, or else the robot will be in a unbalanced state. So, for the stance leg, the compensation only obeys Equation 14 at the second single support phase of a step. The rest of it is compensated using interpolation to keep the curve continuous, as shown in Figure 6. The step duration is 400ms (100ms for double support phase), and the stiffness coefficient for each HipRoll joint is 41.

Figure 7 shows the trunk angle around sagittal direction. Before the time of 2335s, elasticity is not enabled and the total amplitude is around 0.15 rad. After elasticity control is enabled at the time 2335s, the total amplitude is deduced to 1 rad. When walking in high frequency, if the elasticity is not enabled, the robot will not be able to walk. Our elasticity control algorithm by modeling the joints as elastic ones is effective.

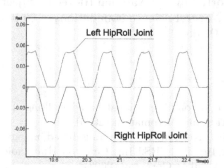
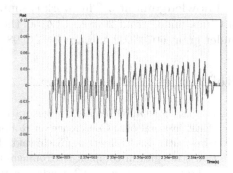

Fig. 6. Trajectories of HipRoll Joints1   **Fig. 7.** Trunk Angle Around Sagittal Axis

4.3   Fast Walking on Nao

Together with the technique in terms of dynamic balance control proposed in [16], we achieved a fast walk with step duration 180ms (20ms for double support phase). The parameter that specifies the percentage of the user required speed is 0.8. The maximum forward speed is around 0.33m/s. The maximum backward speed is around 0.2m/s. The maximum sideways speed is around 0.11m/s. The maximum rotational speed is around 90°/s. The video is at http://ai.ustc.edu.cn/en/demo/index.php.

Team HTWK achieved a forward speed of 0.32m/s based on machine learning approach and its sideways motion is not good. Team Dortmund achieved a very fast walk with forward speed of 0.44m/s. However it is not stable. Team

B-Human achieved a walk with maximum forward speed 0.28m/s, maximum backward speed 0.17m/s, maximum sideways speed 0.07m/s and maximum rotational speed 90°/s.

5   Conclusion

we presented a new technique to achieve a real time biped walking gait pattern generator. By simultaneously planning the ZMP trajectory represented by a cubic polynomial and the CoG trajectory, only one step is needed to connect the goal state (the position and velocity of the CoG) to the previous one.

To apply the generator on Nao used in Standard Platform League (SPL), we calculate the compensation for two HipRoll joints in a theoretical way by modeling the leg joints as elastic ones. Note that, the elasticity control is not just for walk, but also for other motions such as kick when the robot is standing on one foot.

Demonstrations are performed on Nao of version 3.3, showing that the proposed method is effective on a soccer player robot.

Acknowledgements. This work is supported by the National Hi-Tech Project of China under grant 2008AA01Z150, the Natural Science Foundations of China under grant 60745002 and the USTC 985 Project.

References

1. Behnke, S.: Online trajectory generation for omnidirectional biped walking. In: IEEE International Conference on Robotics and Automation (2006)
2. Chestnutt, J., Kuffner, J., Nishiwaki, K., Kagami, S.: Planning biped navigation strategies in complex environments. In: IEEE/RSJ Int. Conf. on Humanoid Robotics, Humanoids 2003, Karlsruhe & Munich, Germany (2003)
3. Czarnetzki, S., Kerner, S., Urbann, O.: Observer-based dynamic walking control for biped robots. Robotics and Autonomous Systems 57(8) (2009)
4. Gouaillier, D., Hugel, V., Blazevic, P., Kilner, C., Monceaux, J., Lafourcade, P., Marnier, B., Serre, J., Maisonnier, B.: Mechatronic design of NAO humanoid. In: IEEE International Conference on Robotics and Automation (2009)
5. Graf, C., Röfer, T.: A closed-loop 3D-LIPM gait for the RoboCup Standard Platform League Humanoid. In: Zhou, C., Pagello, E., Behnke, S., Menegatti, E., Röfer, T., Stone, P. (eds.) Proceedings of the Fourth Workshop on Humanoid Soccer Robots in Conjunction with the 2010 IEEE-RAS International Conference on Humanoid Robots (2010)
6. Graf, C., Härtl, E., Röfer, T., Laue, T.: A robust closed-loop gait for the Standard Platform League Humanoid. In: Proceedings of the 4th Workshop on Humanoid Soccer Robots. A Workshop of the 2009 IEEE-RAS Intl. Conf. on Humanoid Robots (2009)
7. Harada, K., Kajita, S., Kaneko, K., Hirukawa, H.: An analytical method on real-time gait planning for a humanoid robot. Journal of Humanoid Robotics 3(1) (2006)

8. Kajita, S., Kanehiro, F., Kanako, K., Yokoi, K., Hirukawa, H.: The 3D Linear Inverted Pendulum Model: A simple modeling for a biped walking pattern generation. In: IEEE/RSJ Int. Conf. on Intelligent Robots and System, IROS 2001, Hawaii, USA (2001)

9. Kajita, S., Kanehiro, F., Kaneko, K., Fujiwara, K., Harada, K., Yokoi, K., Hirukawa, H.: Biped walking pattern generation by using preview control of Zero-Moment Point. In: IEEE Int. Conf. on Robotics and Automation, ICRA 2003 (2003)

10. Liu, J., Manuela, V.: Online ZMP sampling search for biped walking planning. In: 2008 IEEE/RSJ International Conference on Intelligent Robots and Systems, IROS (2008)

11. Liu, J., Xue, F., Chen, X.: A universal biped walking generator for rough terrain with pattern feasibility checking. International Journal of Humanoid Robotics (to be appeared in 2011)

12. Luca, A.D., Siciliano, B., Zollo, L.: PD control with on-line gravity compensation for robots with elastic joints: theory and experiments. Automatica 41(10) (2005)

13. Spong, M.W.: Modeling and control of elastic joint robots. ASME Journal of Dynamic Systems, Measurement, and Control 109 (1987)

14. Strom, J., Slavov, G., Chown, E.: Omnidirectional Walking Using ZMP and Preview Control for the NAO Humanoid Robot. In: Baltes, J., Lagoudakis, M.G., Naruse, T., Ghidary, S.S. (eds.) RoboCup 2009. LNCS (LNAI), vol. 5949, pp. 378–389. Springer, Heidelberg (2010)

15. Vukobratovic, M., Borovac, B.: Zero-Moment Point - thirty five years of its life. International Journal of Humanoid Robotics 1(1), 157–173 (2004)

16. Xue, F., Chen, X., Liu, J.: Integrated balance control on uneven terrain based on inertial sensors. Submitted to IEEE/RSJ Int. Conf. on Intelligent Robots and System, IROS 2011 (2011)

17. Zollo, L., Siciliano, B., Luca, A.D., Guglielmelli, E., Dari, P.: Compliance control for an anthropomorphic robot with elastic joints: theory and experiment. ASME Transactions: Journal of Dynamic Systems, Measurements, and Control 127(3) (2005)

Efficient Multi-hypotheses Unscented Kalman Filtering for Robust Localization

Gregor Jochmann, Sören Kerner, Stefan Tasse, and Oliver Urbann

Robotics Research Institute
Section Information Technology
TU Dortmund University
44221 Dortmund, Germany

Abstract. This paper describes an approach to Gaussian mixture filtering which combines the accuracy of the Kalman filter and the robustness of particle filters without sacrificing computational efficiency. Critical approximations of common Gaussian mixture algorithms are analyzed and similarities are pointed out to particle filtering with an extremely low number of particles. Known techniques from both fields are applied in a new combination resulting in a multi-hypotheses Kalman filter which is superior to common Kalman filters in its ability of fast relocalization in kidnapped robot scenarios and its representation of multi-modal belief distributions, and which outperforms particle filters in localization accuracy and computational efficiency.

1 Introduction

Localization is a central aspect for autonomous robots playing soccer as well as in most other application fields. The higher level decision making relies on an accurate knowledge of the robot's position, e.g. positioning a defending player to block its own goal or to support a team mate, or kicking the ball into the right direction even when goal posts are temporarily occluded.

Most localization algorithms which have been applied in RoboCup contexts are Bayesian algorithms such as particle or Kalman filters [11]. In general, particle filters are favored when belief distributions are expected to be multi-modal and sensor information is uncertain and ambiguous, while otherwise Kalman filters are expected to produce more accurate and smooth results. Gutmann and Fox express this common consensus in [4]: "Markov localization is more robust than Kalman filtering while the latter can be more accurate than the former". While hybrid solutions, special variants and problem-specific adaptations always have the potential to outperform pure implementations of the general methods, this seems to be the general trend and particle filters are often the method of choice for handling multi-modal belief distributions [11]. However, multiple model Kalman filters have been well established in other fields of research [1] and are well suited for such tasks.

The Kalman filter as an estimator for linear Gaussian systems has been adapted with variants such as the Extended or Unscented Kalman filter. Both

T. Röfer et al. (Eds.): RoboCup 2011, LNCS 7416, pp. 222–233, 2012.

maintain the convenient Gaussian representation throughout the filter update steps by linearizing the process and measurement functions. However, such Gaussian approximations perform unsatisfactory or even diverge completely when the non-linearity in the system becomes too severe. This is obvious especially for multi-modal systems.

Weighted sums of Gaussians offer an approximation for those non-linear non-Gaussian systems [1]. The resulting filters are referred to as Gaussian sum or Gaussian mixture filters. The high number of Gaussians necessary to appropriately approximate any given belief distribution leads to an increase in computational complexity so that pruning of the belief representation becomes an important issue in practical real-time implementations on mobile platforms. Due to this, multiple-model Kalman filter implementations loose some of their generality, computational efficiency and theoretical elegance.

The same is true for particle filter implementations which aim at high performance and applicability on limited platforms. Those usually operate with an extremely low number of particles to be efficient enough to operate for example on the Aibo or the Nao. This paper's main contribution is to point out and apply techniques originally introduced in particle filtering contexts to multiple-model Kalman filtering.

This has been implemented for a RoboCup Standard Platform League scenario, i.e. for humanoid robots with highly uncertain odometry in a dynamic soccer scenario where most landmarks are ambiguous field features, occlusion frequently occurs and false positives are likely to be generated from observations of the audience. Situations with various degrees of similarity to robot kidnapping happen due to frequent struggles and shoving among the robots and interventions by the referees. This work is related to [10] in terms of the application scenario and the general idea of using a multiple-model Kalman filter to address the correspondence problem for ambiguous landmarks and false positive observations, but the proposed solution differs in the overall approach as well as various implementation details.

2   Gaussian Mixture Filtering

An overview of Gaussian Mixture filtering and its applicability to state estimation in domains of multi-modal probability distributions is given in this chapter. The general Bayes filter convention under the Markov assumption [11] is used in the following discussion in order not to commit to any specific form of non-linearity approximation to the Kalman filter.

The initial probability distribution for the n-dimensional state x_0 is expressed by the prior belief $bel(x_0)$. Each discrete time step the estimation is updated using the following equations.

$$\overline{bel}(x_t) = \int p(x_t|u_t, x_{t-1})bel(x_{t-1})dx_{t-1} \tag{1}$$

$$bel(x_t) = \eta p(z_t|x_t)\overline{bel}(x_t) \tag{2}$$

Equation 1 describes the prediction step or *process update*, in which the past belief is updated using the known control input u_t and the process model, which is expressed by the conditional probability $p(x_t|u_t, x_{t-1})$. This predicted posterior belief $\overline{bel}(x_t)$ is corrected in equation 2 using the measurement z_t and the sensor model $p(z_t|x_t)$. This is commonly referred to as *measurement* or *sensor update*.

The standard Kalman algorithm provides an optimal estimation in case of linear Gaussian systems, which however is rarely given for real problems. Non-linearity in the process and sensor model is handled by linearization in order to apply the familiar Kalman filter equations, which then do not yield the optimal solution but only an approximation. This can be done either using Jacobians in the Extended Kalman filter or the unscented transform in the Unscented Kalman filter.

If the model shows only certain non-linear characteristics around the current estimation, this linearization is enough to allow an appropriate estimation. Representing the belief state with Gaussian mixtures as in equation 3 can also improve the estimation in those cases, but the real benefit becomes obvious in cases where the models show multi-modal characteristics.

$$bel(x_t) = \sum_{i=1}^{N} \alpha_i \frac{1}{(2\pi)^{n/2}|P_i|^{1/2}} e^{\left(-1/2(x_t-\mu_i)^T P_i^{-1}(x_t-\mu_i)\right)} \qquad (3)$$

Here μ_i and P_i are the means and covariances of the individual Gaussians and α_i are the weights which sum up to 1 over all N models. Note that N might change during operation of the filter, and the explicit dependence on time in the indexes of those parameters is dropped for simplicity.

Gaussian mixtures introduce several changes into the filtering process. Note that in most applications only a subset of these aspects is actually implemented [10,9,3,12,5], i.e. the one or two aspects most crucial to the estimation process, and a strict limit to the number of separate Gaussians is enforced to maintain acceptable processing time.

The initial belief is obviously easier represented by Gaussian mixtures than by a single Gaussian. This can be done either by a regular distribution parameterized to fit an a-priori belief as proposed in [1], or by generating the initial belief from the first sensor information as done in [3] and [5]. Re-localization from kidnapped robot scenarios is closely related, since allowing for the possibility of robot kidnapping means assigning a small probability to the case that the robot is repositioned without any knowledge, which is similar to the initial global position finding. Gaussian mixtures might also be applied for a better representation of highly uncertain odometry as for walking robots, but due to the drawbacks of the increase of terms in the mixture this is rarely implemented.

2.1   Sensor Update and Correspondence Problem

The most focused on filtering aspect for applying Gaussian mixtures is the sensor update. While Gaussian mixtures can be beneficial in modeling a spread-out

belief for uncertain measurements, the major reason for using them is the update with ambiguous landmarks. In single-Gaussian Kalman filters this case is
handled by choosing the correspondence with maximum likelihood and proceeding by linearizing a sensor model for a unique landmark update. This results in
a Kalman filter operating "more as a maximum likelihood estimator than as a
minimum variance estimator and the mean follows (hopefully) one of the peaks
of the density function" [1].

Gaussian mixtures allow to construct a sensor model with one Gaussian term
for each possible correspondence. In [9] for example, Gaussian mixture sensor
models are used to avoid an exclusive correspondence choice for the likelihood
calculation in Monte Carlo localization. In the Kalman approach the sensor
update using this model results in applying all possible correspondences to all
hypotheses maintained by the current belief prediction $\overline{bel}(x_t)$ as done in [10].
The terms in $bel(x_t)$ therefore increase by the factor M_s which is the number of
terms in the sensor model.

The weights α_i are recursively updated according to [1] by multiplication with
the probability of the measured m-dimensional innovation $\eta = (z_t - \hat{z}_i)$ by

$$\alpha_i = \nu \alpha_i' \left(\frac{1}{\sqrt{(2\pi)^m |P_\eta|}} e^{-\frac{1}{2}\eta^{-1} P_\eta^{-1} \eta} \right) \tag{4}$$

where $\hat{z}_i$ is the expected observation for the fixed correspondence according to the
i'th model, P_η is the sum of the measurement and the prediction covariance, and
ν is a normalization factor. To improve the lack of robustness to outliers, Quinlan
and Middleton add a term ϵ_0 expressing a static probability of the observation
being an outlier, i.e. a false positive in the measurement process such as echoes
in sonar data or incorrectly classified objects in a vision system [10]. A further
discussion of the implications of this will be given in section 3.

2.2 Pruning the Belief Representation

The multiplicative increase per time step in the number of terms in the Gaussian
mixture introduced by the sensor update and potentially also the process update
is countered by pruning the resulting belief representation, for which several
different methods have been proposed. In general, a suitable Gaussian mixture
representation with a specified maximum number of terms needs to be found
which approximates a given probability distribution. This is similar to finding
the initial parametrization, only in this case the target distribution is already
given as a Gaussian mixture, only with a higher number of terms.

In most practical applications, sophisticated re-parametrization strategies
as described in [12] and [6] achieve excellent results, but are often too time-
consuming. Instead of applying iterative optimization or regression procedures,
simple heuristics are used to reduce the number of terms in the Gaussian mixture. This is done by combining multiple terms into one and by neglecting terms
with small enough weighting factors whenever possible [1,10].

3 Problems Resulting from Efficiency-Related Trade-offs and Model Limitations

Many of the problems classical Kalman filters face in realistic application scenarios can be addressed by the extension to use sums of Gaussians. Ensuring efficient computation however introduces compromises which together with some of the common assumptions described in the previous section potentially nullify certain aspects which make the Gaussian mixture representation desirable in the first place. Two potential problems related to pruning and to the handling of false positives shall be analyzed in the following as a motivation for the alternative approach described in section 4 which differs in contrast to common implementations like [1,10].

3.1 Influence of Pruning on Quality of Estimation

The exponential growth of Gaussian terms necessitates aggressive pruning as described in section 2. This pruning however does not only lead to a slight loss of accuracy, but may potentially undo some of the central benefits of using Gaussian mixtures.

One significant problem shall be illustrated using the following simplified example. Consider a robot which is well localized with a current belief state consisting of a single Gaussian, and whose position is altered at a time t_0, e.g. by a collision with another agent changing the robot's orientation significantly without it perceiving it. In the following time steps the robot repeatedly makes ambiguous observations which correspond to one of two possible landmarks in its map, as visualized in figure 1(a). In this scenario, let each blue link correspond to the correspondence choice c_1 and each red link to choice c_2. c_1 is the real observed landmark, but c_2 is the choice which fits the (wrong) initial belief state $bel(x_{t_0})$ better. Figure 1(b) visualizes the model splitting resulting from each multi-modal sensor update without any pruning.

In this situation, updates based on the correspondence choice c_2 result in small innovations, while those based on choice c_1 initially produce large innovations, which only decrease when frequent updates using c_1 shift the corresponding mean closer to the true position. In this scenario, continuous multi-modal updates obviously do not resolve the localization ambiguity, but lead to a belief state $bel(x_{t_0+n})$ which includes a term $\alpha_i N_i$ which is the result of a series of "right" choices and which, after the corresponding mean is close to the true position, also has a high weight. In a situation like this, few additional complementary observations of different landmarks might resolve the ambiguity and leave the true position as the most probable estimate in the belief. Thus the Gaussian mixture filtering allows for more than just mere maximum likelihood estimation if extensive pruning does not interfere with this characteristic.

Assuming equal a-priori probabilities for both data associations and assuming simple uni-modal process updates, the weight factors α_i of the different terms in the belief of the following time steps $t_0 + \delta$ is changed exclusively by the sensor update in equation 4. Since initially the correspondence choices c_1 do not fit the

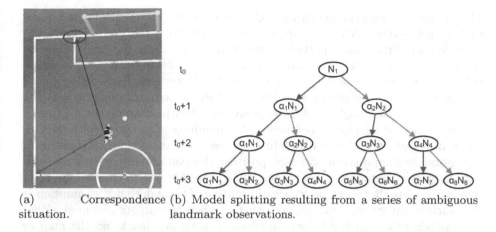

(a)   Correspondence (b) Model splitting resulting from a series of ambiguous
situation.   landmark observations.

Fig. 1. Example situation with ambiguous observation correspondence

belief as the alternative choices do, the weight α_8 in figure 1(b) is much smaller
than α_1 at time step $t_0 + 3$. This makes paths in a tree of correspondence choices,
which do not instantly lead to maximum likelihood estimates, ideal candidates
for pruning techniques as described in [10]. So this kind of aggressive pruning
does not only decrease the estimation quality, but actually removes one of the
most significant advantages of Gaussian mixture filtering, i.e. the possibility to
maintain different hypotheses of which some may temporarily be unlikely, but
still allow the observation of the influence of new measurements on regions of
the state space away from the maximum likelihood estimate. Note that this is
also essential for re-localization in kidnapped-robot scenarios.

3.2   Integrations of Explicit False Positive Handling

False positives are incorrect measurements that violate the assumption of Gaus-
sion distributed errors insofar as they are not only inaccurate measurements
from a known feature, but originate from some other unmodeled source in the
environment unrelated to any known feature and its position. In scenarios such
as the RoboCup Standard Platform League those false positive observations are
quite common since no barrier exists between the field and its surroundings.
This frequently causes false perceptions from the robot's image processing when
the clothings of people standing close to or directly on the edge of the field show
the same outline and color of expected features such as goal posts.

The common compensation is to enlarge the tail-end of an otherwise Gaussian distribution as done in [10] where a probability is assumed of ϵ_0 that an
observation is an outlier, and equation 4 is adapted as follows:

$$\alpha_i = \nu \alpha_i' \left((1 - \epsilon_0) \frac{1}{\sqrt{(2\pi)^m |P_\eta|}} e^{-\frac{1}{2}\eta^{-1} P_\eta^{-1} \eta} + \epsilon_0 \right). \qquad (5)$$

This change prevents the weighting of Gaussians to drop too much by an update with a single outlier. While this approach seems intuitive and is often applied to the weighting functions in particle filters, it does not affect the actual Kalman update with the outlier observation itself. Instead the implicit assumption is that outliers appear randomly and therefore do not systematically influence the system estimation. As can be seen from the SPL scenario example, this is not necessarily true and might lead to seriously biased estimation errors.

The means for an alternative method for handling such false positives is already provided by the sensor update step described in section 2.1. Instead of incorporating the possibility for false positive observations into each sensor update for each correspondence, a more natural alternative is to handle false positives as just another correspondence alternative. The underlying assumption is that each landmark observation, not just the inherently ambiguous ones, either corresponds to one of the known locations of such landmarks on the map or to another source not included in the map. In SLAM contexts these additional observation origins might be mapped and used for further localization purposes, resulting in a multiple-hypothesis SLAM approach similar to [2]. In localization tasks however such observations corresponding to unmapped landmarks can simply be discarded, i.e. no update is performed at all to the term generated by this correspondence choice.

This approach provides the possibility of a position tracking unbiased by false positives, thus more robust especially in situations where false positive observations do not occur randomly.

4   An Alternative Approach to Multiple-hypotheses Kalman Filtering

As argued in section 3.1, operating Gaussian mixture filters with a strictly low limit on the number of terms and the consequential aggressive pruning deprives such filters of much of their multiple-hypothesis tracking potential. This situation shows parallels to certain particle filter implementations. Both algorithms have originally been designed under the assumption of enough hypotheses to appropriately cover the state space. For particle filters with extremely low numbers of particles several techniques have been proposed to compensate for the low state space coverage [7,8]. The established policy in this case consists of two measures. The first one is to limit the influence of single inconclusive measurements (for details see for example [8]). The second is to accept the impossibility to track all important hypotheses in the exponentially growing number of paths, while at the same time providing the means to recover from the neglect to model those which rise in importance again in the future. Particle filters add new particles which are uncorrelated to the current belief during the resampling step, either randomly distributed over the state space, or more efficiently by drawing few particles directly from the sensor model, called *sensor resetting* [7].

In the context of Gaussian mixture Kalman filters this mainly means a modification of the sensor update step. First of all, only the maximum likelihood

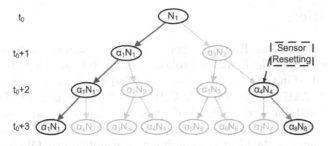

Fig. 2. Tracking only maximum-likelihood correspondence choices for ambiguous landmarks, relying on sensor resetting to generate terms close to those corresponding to the neglected paths leading to conclusive estimates

update is applied, explicitly taking into account also false positive measurement possibilities, in which case the current mean stays unchanged. Consequently it is obvious that the weight update needs to be adjusted. Following the temporal smoothing idea for particle filters [8] the weight update for hypotheses can be adjusted so as not to be influenced that much by different degrees of outlier measurements. Instead it is possible to weight different models based exclusively on how many observations can be conclusively explained by them. But more importantly an additional new sensor update in parallel to the old one will perform the same function as the sensor resetting part of the particle filter's resampling step, i.e. introducing new Gaussian terms based not on the previous state estimate, but on the recent sensor measurements' sensor model only. Note that this is not equivalent to model splitting since the new terms are not correlated to the old ones.

The resulting filter frequently injects new low-weighted models into regions with high probability based on the last observations, which might be expected to rise in weight in case theses hypotheses also fit future observations. This however relieves the necessity to track multiple paths of correspondence choices for updates of existing models as described in section 2.1 and illustrated in figure 1(b). In [10] many multi-modal sensor updates basically result in uni-modal ones due to the aggressive pruning. While this already corresponds to implicit maximum-likelihood updates, it is now possible to explicitly do exactly this. As a consequence both the process update and the old sensor update can be applied with the uni-modal maximum-likelihood choice, explicitly neglecting alternative paths in the decision tree of correspondence choices, but relying on the sensor resetting functionality to pick up those paths which lead to conclusive estimates as illustrated in figure 2 relating to the example in section 3.1.

This application of the sensor resetting concept also solves the problem that common Kalman implementations have concerning the kidnapped robot problem. Both a sudden change of robot orientation with consequential wrong data association as described in the example in 3.1 and a real teleportation event will be handled accordingly.

5   Evaluation

The multiple-hypotheses Kalman filter approach described in this paper has been implemented for the humanoid robot Nao which is produced by Aldebaran Robotics and used in the RoboCup *Standard Platform League* (SPL). This robot contains a x86 AMD Geode 500 MHz CPU and 256 MB SDRAM. The SPL environment consists of a soccer field of specified dimensions and colored goals, which can be used for localization together with the field markings, i.e. lines, corners, and center circle. Most measurements are ambiguous: Observing a single goal post leaves at least two choices, while a field line crossing can be associated with 6 true positions on the field in case of a T-crossing or with 8 positions in case of an L-crossing. Even more correspondences are possible when allowing incompletely/uncertainly classified crossings, e.g. in case of occlusion or for the observation of two perpendicular lines whose crossing is outside of the image. One important feature of the SPL environment is that no barrier exists between the soccer field and its surroundings, which frequently includes colorfully clothed audience. So a localization algorithm needs to be robust not only to noisy measurements due to staggering robots but also to frequent false positives. This is the setup in which the proposed multiple-hypotheses Kalman filter is evaluated.

The software running on the robot includes the manufacturer's middleware NaoQi as well as the robot control software consisting of a motion thread running at 100 Hz and a cognition thread running at 15 Hz, both with all regular SPL soccer modules activated. To evaluate the localization quality a camera system is mounted above the field, detects markers attached to the robot, and provides ground truth data to an additional module in the robot's software.

The presented approach is evaluated in experiments on real robots in a typical SPL scenario and compared against a particle filter solution utilizing sensor resetting, temporal smoothing, and lazy resampling, and which has been in use in RoboCup competitions up to the development of the presented multiple-hypotheses Kalman filter. For all experiments both localization algorithms run in parallel on the robot, thus working with the exact same input from image processing and ensuring the comparability of the results.

A first experiment evaluates the re-localization ability after several teleportation events. In figure 3 each red mark on the time axis indicates that the robot has been picked up to be placed at a random new position on the field. The illustrated performance represents typical behavior for both filters, which expectedly show similar behavior since both get their relocalization ability from the same principles. Between seconds 90 and 110 the multiple-hypotheses Kalman filter generates a hypothesis close to the correct position, but does not rate it high enough in those 20 seconds to output it as the likeliest location. Similarly the particle filter jumps between different clusters after second 170. This shows that the proposed filter is able perform at the same level as a particle filter implementation tuned especially to handle such situations. Note that this is classically regarded as a particle filters specialty and similar performance has not been reported for related Kalman variants.

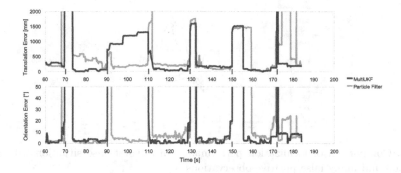

Fig. 3. Ability for relocalization from kidnapped-robot situations

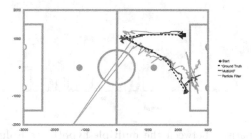

Fig. 4. Comparison of localization performance in an SPL scenario

Additional experiments have been set up to quantitatively evaluate the localization quality. In all experiments the robot is placed on the field without prior knowledge of its position and the movement is started after a fixed time. Figure 4 shows a ground truth paths on the field and the localization result of the multiple-hypotheses unscented Kalman filter and the particle filter. The Kalman filter's characteristic of smooth and accurate position tracking is clearly visible and the filter's output in form of its strongest hypothesis is superior to the particle filter's most probable cluster.

Figure 5 shows one run where many persons walked around the field thus occluding field features and provoking false positives at the same time, e.g. blue jeans sometimes are falsely recognized as blue goal posts. While both localization approaches show diminished results, the multiply-hypotheses UKF clearly produces estimations closer to the real robot path.

A comparison of the different algorithms' runtime is given in figure 6. It has to be noted that the presented measurement is not unbiased, since larger runtimes tend to be influenced more by random threading issues and thus the larger module update time of the particle filter might include more motion thread update cycles. Nevertheless, the multiple-hypotheses Kalman filter is clearly much more efficient. While this does not allow to compare this implementation of a Gaussian mixture Kalman filter to the one proposed by Quinlan and Middleton [10], it can be argued to be more efficient since a similar comparison to a state of

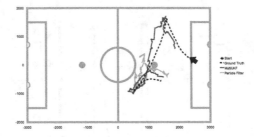

Fig. 5. Comparison of localization performance in a worst case situation with much occlusion and many false positive observations

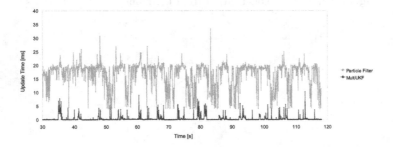

Fig. 6. Runtime comparison between the multiple-hypotheses Kalman filter and the particle filter

the art particle filter in [10] showed only slightly better runtime and averaged around a third of the image processing time, which would still be in the range of multiple milliseconds. The average runtime of the approach presented here is 0.4 ms, which obviously eliminates the localization problem as a computational bottleneck. Most of the time is spend on image processing, the remaining time difference goes to infrastructure and other tasks such as ball tracking or behavior decisions. The periodic tendencies in the measurements in figure 6 are a result of the robot's head motion which also searches for the ball in front of the robot's feet where only little localization information can be extracted.

6   Conclusion

This paper presents an approach for Gaussian mixture filtering which utilizes techniques from particle filtering to incorporate valuable aspects of both filter strategies. The resulting multiple-hypotheses unscented Kalman filter is superior to common Kalman filters in its ability of fast relocalization in kidnapped robot scenarios and its representation of multi-modal belief distributions, and it outperforms particle filters in localization accuracy and computational efficiency. A direct comparison to the approach proposed in [10] or similar classical Gaussian mixture filters has yet to be done.

Future work will focus on a collaborative localization and tracking strategy for a team of robots. First results have been presented in [2], but those were particle filter based and have not been efficient enough to run at high frame rates on the Nao. Building a similar system with a multiple-hypotheses Kalman filter as a basis promises real-time performance, but involves additional difficulties concerning the stochastic soundness of the overall filtering scheme.

References

1. Alspach, D., Sorenson, H.: Nonlinear Bayesian Estimation using Gaussian Sum Approximations. IEEE Transactions on Automatic Control 17(4), 439–448 (1972)
2. Czarnetzki, S., Rohde, C.: Handling Heterogeneous Information Sources for Multi-Robot Sensor Fusion. In: Proceedings of the 2010 IEEE International Conference on Multisensor Fusion and Integration for Intelligent Systems, MFI 2010, Salt Lake City, Utah, pp. 133–138 (September 2010)
3. Duckett, T., Nehmzow, U.: Mobile Robot Self-Localisation using Occupancy Histograms and a Mixture of Gaussian Location Hypotheses. Robotics and Autonomous Systems 34(2-3), 117–129 (2001)
4. Gutmann, J.S., Fox, D.: An Experimental Comparison of Localization Methods continued. In: 2002 IEEE/RSJ International Conference on Intelligent Robots and Systems, vol. 1, pp. 454–459 (2002)
5. Kaune, R.: Gaussian Mixture (GM) Passive Localization using Time Difference of Arrival (TDOA). In: Fischer, S., Maehle, E., Reischuk, R. (eds.) GI Jahrestagung. LNI, vol. 154, pp. 2375–2381. GI (2009)
6. Krauthausen, P., Hanebeck, U.D.: Regularized Non-Parametric Multivariate Density and Conditional Density Estimation. In: Proceedings of the 2010 IEEE International Conference on Multisensor Fusion and Integration for Intelligent Systems, MFI 2010, Salt Lake City, Utah, pp. 180–186 (September 2010)
7. Lenser, S., Veloso, M.: Sensor Resetting Localization for Poorly Modelled Mobile Robots. In: Proceedings of IEEE International Conference on Robotics and Automation, ICRA 2000, vol. 2, pp. 1225–1232 (2000)
8. Nisticò, W., Hebbel, M.: Temporal Smoothing Particle Filter for Vision Based Autonomous Mobile Robot Localization. In: Proceedings of the 5th International Conference on Informatics in Control, Automation and Robotics, ICINCO, vol. RA-1, pp. 93–100. INSTICC Press (2008)
9. Pfaff, P., Plagemann, C., Burgard, W.: Gaussian Mixture Models for Probabilistic Localization. In: IEEE International Conference on Robotics and Automation, ICRA 2008, pp. 467–472 (May 2008)
10. Quinlan, M.J., Middleton, R.H.: Multiple Model Kalman Filters: A Localization Technique for RoboCup Soccer. In: Baltes, J., Lagoudakis, M.G., Naruse, T., Ghidary, S.S. (eds.) RoboCup 2009. LNCS (LNAI), vol. 5949, pp. 276–287. Springer, Heidelberg (2010)
11. Thrun, S., Burgard, W., Fox, D.: Probabilistic Robotics (Intelligent Robotics and Autonomous Agents). The MIT Press (2005)
12. Upcroft, B., Kumar, S., Ridley, M., Ong, L.L., Durrant-whyte, H.: Fast Reparameterisation of Gaussian Mixture Models for Robotics Applications. In: Barnes, N., Austin, D. (eds.) Proceedings of the Australasian Conference on Robotics and Automation, ACRA. Australian Robotics and Automation Association (2004)

A Portable Ground-Truth System
Based on a Laser Sensor

Román Marchant, Pablo Guerrero, and Javier Ruiz-del-Solar

Department of Electrical Engineering,
Universidad de Chile, Avenida Tupper 2007, Casilla 412-3, Santiago, Chile
{romarcha,pguerrer,jruizd}@ing.uchile.cl
http://www.die.uchile.cl/

Abstract. State estimation is of crucial importance to mobile robotics
since it determines in a great measure its ability to model the world from
noisy observations. In order to quantitatively evaluate state-estimation
methods, the availability of ground-truth data is essential since it provides a target that the result of the state-estimation methods should approximate. Most of the reported ground-truth systems require a complex
assembly which limit their applicability and make their set-up long and
complicated. Furthermore, they often require a long calibration procedure. Additionally, they do not present measures of their accuracy. This
paper proposes a portable laser-based ground-truth system. The proposed system can be easily ported from one environment to other and
requires almost no calibration. Quantitative results are presented with
the purpose of encouraging future comparisons among different ground-
truth systems. The presented method has shown to be accurate enough
to evaluate state-estimation methods and works in real time.

Keywords: Ground Truth, Laser.

1 Introduction

In the last decades, a vast work on state estimation and its applications to mobile robotics has been carried out. The existent solutions are able to estimate
non-observable variables from limited and noisy observations. The applications
of state estimation to mobile robotics include robot's self-localization and object
tracking, which are essential when performing complex tasks such as navigation
and interacting with mobile and static objects. In order to quantitatively evaluate the performance of a state-estimation method, a common practice is to
compare the estimates obtained in an experiment by the evaluated method with
a set of data, known as *ground-truth data* (GTD), assumed to be very accurate.
We will call *ground-truth system* (GTS) to any system able to get the GTD in
real time, i.e., a system that is able to give at every instant a very accurate
estimate of the state. Applications of a GTS to the robotics field are on the following areas: (i) Method evaluation: The obtained GTD is used as a target data
set to compare with. (ii) Environment monitor: The obtained GTD represents
the current state of the environment, used for monitoring unattended robots.

T. Röfer et al. (Eds.): RoboCup 2011, LNCS 7416, pp. 234–245, 2012.

(iii) External Control: The obtained GTD is used by an external robot controller for calculating the state error and the corresponding command. (iv) Training: Having a GTS would make easier some automatic training procedures such as gait training, automatic vision calibration, etc.

In most common situations, an environment with a global static origin, O, contains various types of objects, classified into two main classes: (i) Static Objects: Objects with fixed pose relative to O. These objects are usually used as landmarks. (ii) Mobile Objects: Objects with dynamic pose relative to O. In most robotics applications, the goal of a GTS is to achieve a precise estimation of the kinematic state of all the objects in the environment.

Most reported GTSs use information provided by a static camera located above the scene of interest, although some of them employ laser sensors or radars. Usually, camera-based GTSs require large complex structures, containing pillars and a ceiling mesh that holds the sensors. This fact makes the assembly procedure long and complicated and imposes a minimum to the ceiling height.

In this work, a GTS based on a laser sensor is presented. The main feature of the proposed system is its portability. The proposed GTS is said to be portable because its installation is quick and easy and it requires almost no extra space. A tripod, a laser sensor and a laptop are all the required hardware. This makes the system also have a low cost. Additionally, the system requires no extra calibration when changed to a different environment with a priori known map and can recover very fast from movements of the sensor. This feature also improves the portability of the system.

We propose that the different GTS alternatives should be compared from different viewpoints including, of course, their accuracy. To allow such a comparison, the presented GTS is characterized in terms of its accuracy and portability.

The proposed GTS can be applied in any situation, where an object of interest moves along a fixed plane and a single laser scan give enough information to classify them and determine their position (and orientation if needed). In the cases where the orientation is required and the symmetries of the object does not allow the determination of their orientation, a body extension may be added to the robots in order to eliminate the symmetries. The proposed GTS have been tested in a Robocup Standard Platform League (SPL) field and would probably work in a Humanoid League (HL) environment.

This document is structured as follows. Section 2 points out related work. Section 3 makes a brief review of the employed regression methods, while section 4 describes the proposed system in detail. Section 5 presents the experimental results in pose estimation. Finally, section 6 draws some conclusions and suggests some future work.

2   Related Work

The global perception of an interesting scene has been addressed by many authors (e.g. [12,13,7,5,4,3,2]), and it is currently relevant for several applications. The estimation of the state of the environment may be used for monitoring autonomous robots ([1,11]), evaluation of localization methods ([9,10]) and ground-truth database generation ([7]).

Various sensing strategies have been applied, even though most of the existing work on the area uses vision-based systems, placing one or more cameras held by a complex structure on top of the scene ([12,13,5,4,3,2,7]). To the best of our knowledge, a laser-based strategy for a GTS has not been addressed in the literature, except for [9] where no description of the global perception methodology is presented.

At the moment, there are GTSs applied to some Robocup leagues. For instance, the standardized method used at Robocup Small Size League (SSL) is [13]. This work implements an image processing algorithm based on color segmentation that allows pattern recognition with the purpose of estimating each robot's id and pose. In addition, it considers camera-pose calibration, defining the relationship between the field geometry and the image plane. Although no occlusions are possible, this work does not include tracking of detected objects and does not present quantitative results on estimation error, preventing any quantitative comparison with any other methodology. Furthermore, the system requires that each robot has a flat surface in its top that does not rotate with respect to the robot which makes it not suitable for applications with headed robots, which is the case of the Robocup SPL and HL, for instance.

The GTS described on [7] uses several reflectors located in the body of the robot and fifteen infrared cameras to detect them. Due to the redundancy of the sensors, the system accuracy is in the order of millimeters for each reflected point, and occlusions are unlikely to affect. Another positive feature is that it can independently track the movement of the head, allowing it to calculate the orientation of the robot's cameras. An important fact is that this work uses a tracking system and matches space state information with local information on the interesting robot, generating a GTD database with global and local information. However, that GTS requires a very complex hardware structure, hardly reproducible. Furthermore, [7] does not contain any detail on the system algorithms to estimate the robot's pose, neither presents quantitative results on pose estimation error.

The main drawback of the existing GTSs is that they make use of large structures, minimizing portability and disabling its utilization on multiple scenarios. Furthermore, they do not present error statistics thus preventing numerical comparisons among different GTSs. The proposed system differs from other reported GTSs because of its portability. Additionally, an estimation of the measurement errors in position and angle of the presented GTS for fixed robot positions on a Robocup SPL field is presented. As mentioned before, we believe that comparing different methodologies is fundamental, and we expect to encourage other authors to report the error statistics of their system.

3   Employed Regression Techniques

Since several parts of the presented methodology employ some regression techniques, we will briefly review them. Linear regression and *Gaussian Process* (GP) regression are the utilized techniques and thus will be reviewed.

The regression problem may be formulated in the following terms. There is a training set of N_t samples $\{\mathbf{x}_i\}$ in an N_x-dimensional input space $\mathcal{X}$ and a correspondent set of targets $\{y_i\}$.[1] We want to learn a function, f, from the training data and be able to evaluate it at any test input $\mathbf{x}_* \in \mathcal{X}$. For convenience we will define the matrix $X = [\mathbf{x}_0, \ldots, \mathbf{x}_{N_t-1}]^T$ and the vector $\mathbf{y} = [y_0, \ldots, y_{N_t-1}]^T$. The here-described regression techniques assume that the set of targets $\{y_i\}$ have zero mean. If that is not the case, the mean of $\{y_i\}$ must be subtracted to each y_i and then added to any test output.

a. Linear Regression

The linear regression consists in predicting any test output as a linear function of the test input:

$$f(x_*) = Ax_*$$
(1)

The here-employed linear regression mechanism is known as *ordinary least squares* (OLS) which is the most used choice because of its great simplicity. OLS assumes also the training input samples $\{\mathbf{x}_i\}$ to have zero-mean. Again, if that is not the case, the mean of $\{\mathbf{x}_i\}$ must be subtracted from each $\{\mathbf{x}_i\}$ and from any test input. OLS estimates the linear transformation matrix, A, as:

$$A = \left(X^T X\right)^{-1} X^T \mathbf{y}$$
(2)

b. Gaussian Process Regression

Gaussian Processes are a non-parametric tool for non-linear regression and classification. For space considerations, we will make a extremely summarized review of GPs for regression from a practical viewpoint. An excellent review that includes both theoretical and practical aspects and references to deeper theoretical insights can be found in [8].

A key component of GPs are the so-called *covariance functions*. Covariance functions encode the information of the kind of functions that can be learned by a GP. They also give restrictions to the possible measures of proximity that are necessary for the regression mechanism to operate. Usually covariance functions have parameters, called *hyperparameters* and the GPs theory bring a method to calculate them from data. We use a *squared-exponential covariance function* for the regression:

$$k\left(\mathbf{x}, \mathbf{x}'\right) = \sigma_f^2 \exp\left(-\frac{1}{2}\left(\mathbf{x} - \mathbf{x}'\right) W\left(\mathbf{x} - \mathbf{x}'\right)^T\right)$$
(3)

Where W is a diagonal matrix with scaling factors in its diagonal and σ_f^2 is the so-called *signal variance*. Then, for the squared exponential covariance function, the hyperparameters vector is $\mathbf{h} = \left(W_{(0,0)}, \ldots, W_{(N_x, N_x)}, \sigma_f^2, \sigma_n^2\right)$,

---

[1] Note that, for simplicity, every y_i is a scalar but there is a very simple procedure for generalizing the reviewed methods to multidimensional outputs that consists of learning an independent one-dimensional function for each dimension of the output

where σ_f^2 is the so-called *signal variance*. The optimal hyperparameters, $\mathbf{h}^*$, are learned by maximizing the *log marginal likelihood*:

$$\mathbf{h}^* = \arg\max_{\mathbf{h}} \left(\log p\left(\mathbf{y}\,|X\right)\right) \tag{4}$$

$$\log p\left(\mathbf{y}\,|X\right) = -\frac{1}{2}\mathbf{y}^T\left(K_X + \sigma_n^2 I\right)^{-1}\mathbf{y} - \frac{1}{2}\log\left|K_X + \sigma_n^2 I\right| - \frac{N_t}{2}\log 2\pi \tag{5}$$

Where $K_X = K\left(X, X\right)$ and the matrix function K can be defined in a component-wise fashion:

$$K(X', X'')_{(i,j)} = k\left(\mathbf{x'}_i, \mathbf{x''}_j\right) \tag{6}$$

With $\mathbf{x'}_i$ the ith column of X' and $\mathbf{x''}_j$ the jth column of X''. Finally, the output for any testing point is predicted using the GP predictive equation:

$$f\left(\mathbf{x}_*\right) = K_*^T\left(K_X + \sigma_n^2 I\right)^{-1}\mathbf{y}, K_* = K\left(X, \mathbf{x}_*\right) \tag{7}$$

4   Methodology

The presented system consists of a precise laser sensor that provides data input, and a software module that processes information and delivers GTD output, represented as a 2D object map. The system's main goal is to detect objects in the environment and to estimate their pose relative to the environments global origin.

A laser sensor, with field of view with size 2α, is manually placed as close as possible to an initial theoretical pose $\mathbf{s} = (s_x, s_y, s_\theta)$ with respect to O. To prevent reference-system transformation errors due to the initial error and/or changes in the 3D pose (6dof) of the laser, the system uses at every instant the perceived relative landmark poses to estimate a linear transformation that moves the raw data to a reference system with known pose in the environment. The linear transformation parameters are initialized with their theoretical values when the sensor is exactly on $\mathbf{s}$ in order to deal with possible environment symmetries. This procedure enables the system to work in a new environment without needing any calibration to determine the laser's pose.

Every time step, the laser sensor performs a scan consisting of an array of measurements, $\mathbf{m} = (\mathbf{m}_1, \ldots, \mathbf{m}_N)$. The N measurements of each step are equally spaced in angle by $(2\alpha)/N$. For the ith measurement, $\mathbf{m}_i = (r_i, \theta_i, I_i)$, the laser acquires distance (r_i), angle (θ_i) and intensity (I_i) values.

4.1   System Structure

Each laser scan is processed by the GTS using the following stages (See Fig. 1):

a. Segmentation Process
  In this stage, raw measurements are grouped into object candidates. A criterion based on the Euclidean distance between two consecutive points is used

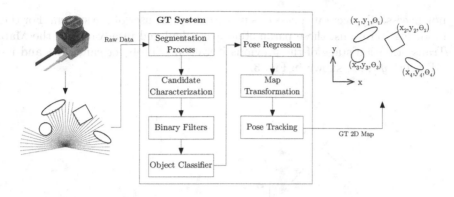

Fig. 1. System Structure Block Diagram

to determine whether these points belong to the same object or not. The output of the process consists of an array $\{S_j\}$ of object candidates, each of them containing a group $\{\mathbf{m}_{j,i}\}$ of M_j measurements. Fig. 2 shows an example of the result of the described process.

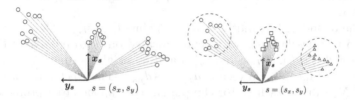

Fig. 2. Measured Data Segmentation

For each object candidate, an *object reference system* (ORS) is established. The origin, $\mathbf{o}'_j = (x'_j, y'_j, \theta'_j)$, of the ORS of the jth object candidate with respect to $\mathbf{s}$ is located in $(\bar{r}_j \cos \bar{\theta}_j, \bar{r}_j \sin \bar{\theta}_j, \bar{\theta}_j)$. With,

$$\bar{r}_j = \frac{\max r_{j,i} - \min r_{j,i}}{2}, \bar{\theta}_j = \frac{\max \theta_{j,i} - \min \theta_{j,i}}{2} \tag{8}$$

Additionally, the ORS-relative position of each measurement is calculated:

$$\begin{pmatrix} x'_{j,i} \\ y'_{j,i} \end{pmatrix} = Rot\left(-\bar{\theta}\right) \begin{pmatrix} r_{j,i}\cos\theta_{j,i} - \bar{r}\cos\bar{\theta} \\ r_{j,i}\sin\theta_{j,i} - \bar{r}\sin\bar{\theta} \end{pmatrix}, Rot(\theta) = \begin{pmatrix} \cos\theta & -\sin\theta \\ \sin\theta & \cos\theta \end{pmatrix} \tag{9}$$

Finally, the origin $\hat{\mathbf{o}}_j = (\hat{x}_j, \hat{y}_j, \hat{\theta}_j)$, of the ORS with respect to O is calculated using the linear transformation:

$$(\hat{x}_j, \hat{y}_j)^T = \mu_o + A\left(\left(x'_j, y'_j\right) - \mu_s\right)^T, \hat{\theta}_j = atan2\left(A_{1,0}, A_{0,0}\right) \tag{10}$$

In the first iteration, the parameters of the linear transformation are set to: $A = Rot(s'_\theta)$, $\mu'_s = (0,0)^T$ and $\mu_o = (s'_x, s'_y)$, with $\mathbf{s}' = (s'_x, s'_y, s'_\theta)$ an initial not-necessarily-accurate guess of **s** provided by the user of the system. For the following iterations, these parameters are calculated in advance in the Map Transformation stage of the previous iteration. An object candidate and its ORS, (x', y'), are shown in Fig. 3.

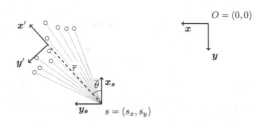

Fig. 3. (x', y') Candidate's Object Reference System

b. Candidate Characterization

This stage generates characteristics for each object candidate. For every object candidate, a characteristic vector, $\mathbf{v}_i$, is built using data from three sources:

(i) The bounding box width, w, of the points $(x'_{j,i}, y'_{j,i})$. The bounding box is a rectangular area, with center located on $(0,0)$ relative to the ORS. Its width is $(\max y'_{j,i} - \min y'_{j,i})$ and its height is $(\max x'_{j,i} - \min x'_{j,i})$.

(ii) The distance profile vector, $d_{j,p} = \{d_{j,k}\}$. This vector has a fixed size of N_d values. Each vector component corresponds to a profile cell, and it is calculated with the following equation:

$$d_{j,k} = \frac{\sum_i^{N_c} x'_{j,i} e^{dist(k,i)}}{\sum_i^{N_c} e^{dist(k,i)}} \qquad \forall k \in \{0, N_d - 1\} \tag{11}$$

(iii) The intensity profile vector, $I_p = \{I_k\}$. It consists of N_I cells, which values are obtained using the following equation:

$$I_{j,k} = \frac{\sum_i^{N_c} I_{j,i} e^{dist(j,i)}}{\sum_i^{N_c} e^{dist(j,i)}} \qquad \forall k \in \{0, N_I - 1\} \tag{12}$$

The expression $dist(j, i)$ is a one dimensional distance over the y' axis between profile cell j and measurement i.

The array of characteristic vectors $\{\mathbf{v}_j\}$ is the output of the stage. The characteristic vector (with $N_d = N_I = 10$) of an example object candidate is graphically shown on Fig. 4.

c. Binary Filters

Every object that does not fulfill a minimum heuristic condition to be considered as an object candidate is filtered. A threshold on the bounding box width is applied on every candidate. As a result, the characteristic vectors of the filtered objects are deleted from $\{\mathbf{v}_j\}$.

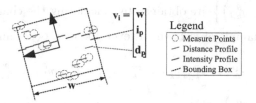

Fig. 4. Characteristic Vector Generation: Points are direct measurements, red lines are distance profile values d_p and blue lines are intensity profile values i_p

d. Object Classifier

This stage classifies each object candidate into an object class. As a result, an object class, c_j, is associated to each candidate characteristic vector $\mathbf{v}_j$. For every characteristic vector in $\{\mathbf{v}_j\}$ and for every object class, c_k, the probability, $p_{j,k}$, of the jth object candidate belonging to c_k is estimated. This estimation is performed by means of the GP regression prediction (see equation 7):

$$p_{j,k} = f_k^{class}\left(\mathbf{v}_j\right) \tag{13}$$

Then the class, $c_j = \arg\max_k (p_{j,k})$, with the maximum probability is assigned to the jth candidate. If p_{j,k^*} is below a threshold, the candidate is filtered, i.e., it is deleted from both $\{\mathbf{v}_j\}$ and $\{c_j\}$.

In a prior off-line process, for every object class, k, the training samples $\left\{\left(\mathbf{x}_{k,l}^{class}, y_{k,l}^{class}\right)\right\}$ are obtained by calculating the characteristic vectors $\left\{\mathbf{v}_{k,l}^{class}\right\}$ of several known objects and making:

$$\mathbf{x}_{k,l}^{class} = \mathbf{v}_{k,l}^{class}, y_{k,l}^{class} = \begin{cases} 1 \text{ if the } l\text{th object belongs to class } k \\ 0 \text{ else} \end{cases} \tag{14}$$

Then, the GP, GP_k^{class}, associated to f_k^{class}, is trained, i.e., its hyperparameters, $\mathbf{h}_k^{class}$, are learned using equations 4 and 5.

e. Pose Regression

The ORS-relative pose, $\tilde{\mathbf{p}}_j$, of each object with characteristic vector $\mathbf{v}_j$ and object class c_j is estimated using an independent GP regression for each of the three pose components:

$$\tilde{\mathbf{p}}_j = \left(\tilde{x}_j, \tilde{y}_j, \tilde{\theta}_j\right)^T = \left(f_{c_j}^x\left(\mathbf{v}_j\right), f_{c_j}^y\left(\mathbf{v}_j\right), f_{c_j}^\theta\left(\mathbf{v}_j\right)\right)^T \tag{15}$$

Then, the obtained pose estimate is moved from the ORS to O:

$$\mathbf{p}_j = \begin{pmatrix} x_j \\ y_j \\ \theta_j \end{pmatrix} = \left(\begin{pmatrix} \hat{x}_j \\ \hat{y}_j \end{pmatrix} + \begin{pmatrix} \cos\hat{\theta}_j & -\sin\hat{\theta}_j \\ \sin\hat{\theta}_j & \cos\hat{\theta}_j \end{pmatrix}\begin{pmatrix} \tilde{x}_j \\ \tilde{y}_j \end{pmatrix}\right) \\ \hat{\theta}_j + \tilde{\theta}_j \tag{16}$$

Just like in the previous stage, some training procedure must be performed in a prior off-line process. For every object class, k, several training samples

$\left\{ \left(\mathbf{x}^{\mathbf{P}}_{k,l}, y^x_{k,l}, y^y_{k,l}, y^\theta_{k,l} \right) \right\}$ are obtained by calculating the characteristic vectors $\left\{ \mathbf{v}^{\mathbf{P}}_{k,l} \right\}$ of an example object put in several known poses and making:

$$\mathbf{x}^{\mathbf{P}}_{k,l} = \mathbf{v}^{\mathbf{P}}_{k,l}, y^x_{k,l} = \hat{p}^x_{k,l}, y^y_{k,l} = \hat{p}^y_{k,l}, y^\theta_{k,l} = \hat{p}^\theta_{k,l} \qquad (17)$$

With $\left(\hat{p}^x_{k,l}, \hat{p}^y_{k,l}, \hat{p}^\theta_{k,l} \right)$ the known pose of the object when $\mathbf{x}^{\mathbf{P}}_{k,l}$ was acquired. Then, the GPs, GP^x_k, GP^y_k and GP^θ_k, associated to respectively f^x_k, f^y_k and f^θ_k are trained, i.e., its hyperparameters, $\mathbf{h}^x_k$, $\mathbf{h}^y_k$, and $\mathbf{h}^\theta_k$, are learned using equations 4 and 5.

f. Map Transformation

The detected 2D positions $\{\mathbf{l}^s_i\}$ of the N_l static landmarks and their respective 2D positions $\{\mathbf{l}^o_i\}$ in the a-priori-known map of the environment are used to recalculate the linear regression transformation for the next instant. A linear regression can cope with the effects produced by changes in the 3D pose (6dof) of the laser sensor. The means of the position sets are calculated:

$$\mu_s = \frac{\sum_i \mathbf{l}^s_i}{N_l}, \mu_o = \frac{\sum_i \mathbf{l}^o_i}{N_l} \qquad (18)$$

Then $X = \left[\mathbf{l}^s_0 - \mu_s, \ldots, \mathbf{l}^s_{N_l-1} - \mu_s \right]$ and $\mathbf{y} = \left[\mathbf{l}^o_0 - \mu_o, \ldots, \mathbf{l}^o_{N_l-1} - \mu_o \right]^T$ are calculated to finally calculate A using equation 2.

g. Pose Tracking

The pose of each mobile object is tracked using a Bayesian filter that use the pose observations of each object for the corrective stage. The predictive stage consists only of an increase of the covariance matrix. When there are several objects of the same class, the minimum Mahalanobis distance is used as a criterion for matching the observations with the tracked objects. The Bayesian filter currently implemented for pose tracking is an *extended Kalman filter* (EKF). The output of this stage is the GTD provided by the presented system.

The GTS is designed for tracking multiple objects, although occlusions between moving objects may lead to estimation errors. Extra laser sensors must be placed for dealing with occlusions in a multiple robot environment. The experimental section evaluates GTS estimation error for one mobile object.

5   Results

The system was implemented on a Robocup SPL environment, using goal posts as landmarks and Nao Robots as mobile objects. The used Laser sensor was a Hokuyo URG-04LX, this sensor has $4[m]$ distance range and $240°$ of FOV.

The object classifier was trained using characteristic vector examples of goal posts and robot, with their corresponding targets. The tripod height was set to allow the laser to aim at the body extension, two centimeters below the shoulders of the robot. At least three landmarks are always detected by the system allowing the map transformation stage to be calculated on every method iteration. The GP pose regression training data base was obtained using the experimental setup on Fig. 5.

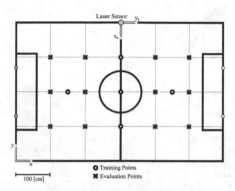

Fig. 5. Experiment setup

For obtaining a quantitative evaluation of the proposed GTS, a static robot was placed on several arbitrarily defined positions, which are specified on Fig. 5 as circles on the field. For each position, all the orientations spaced by $45°$ where tested. The pose estimation error was calculated for several samples on each testing pose.

The mean square error in the estimation of the position and angle are respectively plotted in Fig 6(a) and 6(b), showing the values of the mean square error on each evaluated position. The results show a low error on the position and orientation estimation near training positions. Nevertheless, the evaluated positions $(100, 100)$ and $(500, 100)$ show the biggest error because they have the longest distance to the laser sensor located on $(300, 400)$. The error increases at a larger distance because less sampling points are available to calculate the distance and the intensity profiles, worsening the performance of the regressor.

Although no quantitative results are presented for a dynamic case, the system ability to work in real time and in a dynamic situation was tested by using the provided GTD as a feedback to an external pose controller that allows the robot to follow an arbitrarily defined trajectory on the environment. Fig. 7 shows schematically the experimental setup for this test. In all the performed tests, the robot was able to follow any arbitrary trajectory that was required. A video of the robot following a desired trajectory can be found at http://www.robocup.cl/PapersMedia/GTS.mov

6   Conclusions

A method for generating a laser-based ground-truth map has been established. The results show that a laser sensor in conjunction with the proposed methodology are able to detect an object and then correctly classify it and characterize

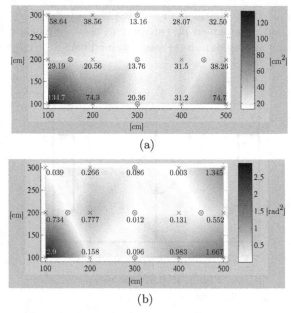

(a)

(b)

Fig. 6. (a) Position Mean Squared Error (b) Angle Mean Squared Error

Fig. 7. Trajectory Experiment on Field

its pose in a known environment. Like in camera-based GTSs, the size of the environment is an issue. It has been shown that a laser-based GTS is able to generate controlling signals to handle a blind robot movement allowing it to execute trajectories, eliminating accumulative odometry errors. Standard deviation of the GTS is much smaller than usual levels of error in odometry and perceptions of an autonomous mobile robot.

As a future work, we expect to test the scalability of the system to larger environments by adding extra laser sensors. Additionally, it could be interesting to analyze a possible multi-sensor (camera and laser) data fusion for applications where the accuracy is very important. In order to eliminate geometric plane movement constrain, the use of a 3D camera instead of a laser sensor could be studied. In order to be able to track a robot soccer match with several robots in the field the system should be enhanced by adding extra sensors and probably finding a way of identifying the robots, such as reflective materials. Finally, testing a particle filter in the Bayesian filter stage may reduce error due to impulsive noise on pose estimation.

References

1. Arenas, M., Ruiz-del-Solar, J., Norambuena, S., Cubillos, S.: A Robot Referee for Robot Soccer. In: Iocchi, L., Matsubara, H., Weitzenfeld, A., Zhou, C. (eds.) RoboCup 2008. LNCS (LNAI), vol. 5399, pp. 426–438. Springer, Heidelberg (2009)
2. Baltes, J., Anderson, J.: Interpolation Methods for Global Vision Systems. In: Nardi, D., Riedmiller, M., Sammut, C., Santos-Victor, J. (eds.) RoboCup 2004. LNCS (LNAI), vol. 3276, pp. 434–442. Springer, Heidelberg (2005)
3. Egorova, A., Simon, M., Wiesel, F., Gloye, A., Rojas, R.: Plug and Play: Fast Automatic Geometry and Color Calibration for Cameras Tracking Robots. In: Nardi, D., Riedmiller, M., Sammut, C., Santos-Victor, J. (eds.) RoboCup 2004. LNCS (LNAI), vol. 3276, pp. 394–401. Springer, Heidelberg (2005)
4. Hayashi, Y., Tohyama, S., Fujiyoshi, H.: Mosaic-Based Global Vision System for Small Size Robot League. In: Bredenfeld, A., Jacoff, A., Noda, I., Takahashi, Y. (eds.) RoboCup 2005. LNCS (LNAI), vol. 4020, pp. 593–601. Springer, Heidelberg (2006)
5. Masutani, Y., Tanaka, Y., Shigeta, T., Miyazaki, F.: Pseudo-local Vision System Using Ceiling Camera for Small Multi-robot Platforms. In: Polani, D., Browning, B., Bonarini, A., Yoshida, K. (eds.) RoboCup 2003. LNCS (LNAI), vol. 3020, pp. 510–517. Springer, Heidelberg (2004)
6. Naruse, T., Masutani, Y., Mitsunaga, N., Nagasaka, Y., Fujii, T., Watanabe, M., Nakagawa, Y., Naito, O.: SSL-Humanoid RoboCup Soccer Using Humanoid Robots under the Global Vision. In: Ruiz-del-Solar, J., Chown, E., Ploeger, P.G. (eds.) RoboCup 2010. LNCS, vol. 6556, pp. 60–71. Springer, Heidelberg (2010)
7. Niemüller, T., Ferrein, A., Eckel, G., Pirro, D., Podbregar, P., Kellner, T., Rath, C., Steinbauer, G.: Providing Ground-Truth Data for the Nao Robot Platform. In: Ruiz-del-Solar, J., Chown, E., Ploeger, P.G. (eds.) RoboCup 2010. LNCS (LNAI), vol. 6556, pp. 133–144. Springer, Heidelberg (2010)
8. Rasmussen, C.E., Williams, C.K.I.: Gaussian Processes for Machine Learning (Adaptive Computation and Machine Learning). The MIT Press (2005)
9. Röfer, T., Jüngel, M.: Fast and robust edge-based localization in the Sony four-legged robot league (2004)
10. Sheh, R., West, G.: Visual Tracking and Localization of a Small Domestic Robot. In: Nardi, D., Riedmiller, M., Sammut, C., Santos-Victor, J. (eds.) RoboCup 2004. LNCS (LNAI), vol. 3276, pp. 410–417. Springer, Heidelberg (2005)
11. Ruiz-del-Solar, J., Loncomilla, P., Vallejos, P.A.: An Automated Refereeing and Analysis Tool for the Four-Legged League. In: Lakemeyer, G., Sklar, E., Sorrenti, D.G., Takahashi, T. (eds.) RoboCup 2006. LNCS (LNAI), vol. 4434, pp. 206–218. Springer, Heidelberg (2007)
12. Umemura, S., Murakami, K., Naruse, T.: Orientation Extraction and Identification of the Opponent Robots in RoboCup Small-Size League. In: Lakemeyer, G., Sklar, E., Sorrenti, D.G., Takahashi, T. (eds.) RoboCup 2006. LNCS (LNAI), vol. 4434, pp. 395–401. Springer, Heidelberg (2007)
13. Zickler, S., Laue, T., Birbach, O., Wongphati, M., Veloso, M.: SSL-Vision: The Shared Vision System for the RoboCup Small Size League. In: Baltes, J., Lagoudakis, M.G., Naruse, T., Ghidary, S.S. (eds.) RoboCup 2009. LNCS (LNAI), vol. 5949, pp. 425–436. Springer, Heidelberg (2010)

Proposal for Everywhere Evacuation Simulation System

Masaru Okaya and Tomoichi Takahashi

Meijo University, Aichi, Japan
m0930007@ccalumni.meijo-u.ac.jp, ttaka@ccmfs.meijo-u.ac.jp
http://sakura.meijo-u.ac.jp/ttakaHP/Rescue_index.html

Abstract. In the aftermath of the September 11 attacks, evacuation simulation has the potential for decreasing the amount of damage resulting from disasters, and, in particular, for saving human lives. Agent based simulation provides a platform for computing individual and collective behaviors that occur in crowds. Such simulations have led to proposals for enhanced prompt public evacuation.

For the public, it is desirable to simulate the behavior of evacuation in any building. We propose an everywhere evacuation simulation system. This system provides a software environment that permits evacuation simulation for any location for which plans are provided on the Web. Three-dimensional (3D) models of buildings and geographic information system (GIS) data for areas that are created for everyday purposes such as sightseeing are used as the environments for simulations. The characteristics of humans are set by users, and their evacuation behaviors are simulated with the relationships among them. The results of simulations can be viewed on the Web by allocating heterogeneous agents inside 2D/3D maps of buildings.

1 Introduction

A number of great disasters have occurred since RoboCup Rescue Simulation (RCRS) started at 2000. Figure 1 shows September 11 attack and disasters

Fig. 1. Many disasters have happened after RoboCup Rescue was started 9.11 WTC (2001), Tohoku earthquake and tsunami (2011), New Zealand earthquake (2011), Australia floods (2011)

T. Röfer et al. (Eds.): RoboCup 2011, LNCS 7416, pp. 246–257, 2012.
© Springer-Verlag Berlin Heidelberg 2012

occured in 2011. There were other disasters such as Sichuan earthquake at 2008, Northern Sumatra earthquake at 2004, etc. Disaster management systems are assumed to be used before, during, and after disasters [5]. Evacuation simulations are becoming a tool for analyzing the level of safety for human life provided in buildings. A fluid-flow model and other macro-level simulations have been modeled on the basis of precedent cases or experiments; however, they do not compute the interpersonal interactions that occur in a crowd or the behavior of people returning to the site. Agent-based simulation (ABS) provides a platform that involves interactions and communications among agents. ABS makes it possible to simulate the following situations; some people may follow emergency announcements, others do not follow the announcements and continue their ordinary work. Moreover, the use of simulations has led to preparations for prompt evacuation in public places.

With regard to RoboCup, ten years have passed since the RoboCup Rescue Simulation (RCRS) league started in 2001. In competitions, rescue agents' performances are compared in various situations in different cities, and many issues have been discussed, presented, and resolved [15]. In 2010, the RCRS platform was revised into a Java based system [3]. The new RCRS(v.1) has adopted an open-space model to represent a map as a set of areas. The space presentation makes it possible to simulate human behaviors inside buildings and egress from buildings that could not be simulated by the previous version that used a network model.

We propose an everywhere evacuation simulation system (EESS). The system provides a software environment that enable evacuation simulation for any location for which plans are available on the Web. Three-dimensional (3D) models of buildings and geographic information system (GIS) data for areas are used as the environments for simulations. The characteristics of humans are set by users and their evacuation behaviors are simulated with the relationships among them. The results of simulations can be viewed on the Web by allocating heterogeneous agents in 2D/3D maps of the interior of buildings. The maps are created for everyday purposes such as sightseeing, information maps, or directions. Section 2 describes key components of our EESS. Section 3 explains our proposals for ESS and Section 4 presents the simulation results. A summary of our proposal and discussions are presented in Section 5.

2   Related Works on Evacuation Simulation

2.1   Agent-Based Evacuation System

Kuligowski et al.[8] reviewed 28 egress models and categorized them into three basic categories: movement models, partial-behavioral models, and behavioral models, and they stated that there is a need for a conceptual model of human behavior in time of disaster. Pelechano et al. suggested that simulations based on grids are limited in term of simulating crowd evacuation and that there is a need for behavior simulations; Pan's framework [12] deals with human and social behavior .

A detailed report on occupant behavior in the World Trade Center (WTC) disaster has been published and a related study has been carried out by Galea et al. [7]. They noted some features that are not included in existing simulations. Table 1 lists the issues cited in Galea and provides a comparison among the systems of Pelechano et al., Pan and RCRSs. Issue groups in the table are as follows:

a) **Travel speed model:** It is well known that congestion occurs as people move [6]. For example, people congest at exits when they evacuate though a narrow space, rescue teams that go to victims collide against people who are evacuating from buildings, and heavy congestion occurs at staircase landings where people from upper and lower floors come together.

b) **Information seeking task:** People who are unfamiliar with buildings are at a loss as to which way they can exit. They look for iconic warning signs, they exchange information with people nearby, or they follow persons who seem to be evacuating. The perception abilities or behavior patterns change according to their psychological states caused by anxiety.

c) **Group formation:** Guidance from well-trained leaders can make evacuation flow more smoothly [10]. Schools drill their students to follow the instructions of teachers and they evacuate together. In the time of disaster, people evacuate under various motivations, and they form a group or break away from the group.

d) **Experience and training:** In the WTC disaster, announcements affected the evacuation behavior of the occupants. Proper announcements save lives; incorrect announcements increase the amount of damage resulting from disasters.

e) **Choosing and locating exit routes:** There are various kinds of obstacles in disaster situations that threaten safe and smooth evacuation. Choosing evacuation places and selecting routes affect the evacuation behaviors.

2.2   Evacuation Scenarios and Space Presentation Model

In disasters, various types of causalities and accidents occur. The following situations become scenarios of evacuation simulation when disasters occur at a school campus.

Outside buildings: Students evacuate from buildings to open-space areas or go to refuges. Rescue teams rush to the buildings to rescue injured students. The injured are transported to hospitals.

Inside buildings: Students go to exits of rooms, look for emergency exits, and take stairs to the ground floor. The rescue teams go to rooms, and check whether occupants remain there.

Outside and inside buildings: Emergency notices are announced to lead people to safe places.

Table 1. Research issues presented in Galea's work[7] and comparison to related works

	issue	Pelechano	Pan	RCRS(v.0)	RCRS(v.1)
a	congestion at exit	✓	✓		✓
	pass the injured (same direction)	✓	✓		
	Meet rescues (counter flow)				✓
	join at staircase landing				
b	sensing model of people			✓	✓
	information share among people	✓	✓	✓	✓
	communication among people			✓	✓
	psychological model of people		✓		
c	group evacuation	✓	✓		
	group formation & break				
	human relation				
d	rescue agents			✓	✓
	rescue headquarters			✓	✓
	announce on evacuation			✓	✓
e	exit routes barred by (debris, smoke, heat, water)	debris	debris	debris (outside)	debris

Two space-presentation models, namely, a network model and an open-space model, have been employed to simulate the movements of agents. Table 2 lists the features of the models. In a network model, intersections and buildings of a map are presented as nodes; edges connecting nodes are shown as roads. The movements of agents flow through the edges and the flows are calculated on the basis of the network model [16] [15]. An open-space model presents all objects on a map, such as intersections, buildings, roads and squares, as areas. Although the open-space model needs more computational resources than the network model, it calculates the position of agents and simulates congested situations in two-way traffic and squares [11].

2.3   Applications of Evacuation Simulation

After disasters, disaster-related measures have been taken in various countries and regions [13][2]. These measures aim to decrease the amount of damage caused by disasters. Some projects and systems have been proposed to ensure prompt planning for disaster mitigation, risk management, and support of IT infrastructures [5][9].

Technologies and concepts cited as Web 2.0 have provided users with the means to interact and collaborate with each other on the Internet. Google Earth is one of such applications; it can let us go anywhere on the earth and see terrain or 3D buildings. Not only can we enjoy them on the Web we can also update data of own concerns to the Web. The following are widely used in our everyday life:

Table 2. Space presentation and features of traffic simulations

real map	network	open space
jam in one-way traffic	✓	✓
jam in two-way traffic		✓
congestion in squares		✓
group behaviors		✓

3D models of buildings: We develop 3D models of buildings, and public facilities are created for landscape simulations or sightseeing guide signs; these are available on the Web.

GIS data of areas: OpenStreetMap (OSM) is a collaborative project to create a free editable map of the world by people's efforts to gather location data with GPS devices [1].

3   Proposal for EESS

3.1   Checking Safety of Buildings Everywhere

Figure 2 shows a snapshot of the Junior League at RoboCup 2010. Team members participate in games, and their mentors watch them from distant places. In the time of emergency, the following situations can be imagined: some rush to exits, or mentors go to their members to evacuate together rather than go to the exits. Some behaviors make the difference between life and death[14].

Before scheduling major events, safety managers of the events are required to assess the levels of safety for human life provided in the buildings and plan effective layout of escape route signs in case of emergency. In emergencies, human behaviors differ from those in ordinary times. There are various kinds of people in the buildings, such as young and old and men and women. It is desirable that the characteristics of occupants are taken into consideration in simulating the evacuation behavior and planning escape routes.

3.2   BDI Models at Evacuation and Motion Model in a Crowd

Evacuation behaviors are based on various intentions, which differ among people. Some people evacuate by themselves and others in groups. The behaviors depend

Fig. 2. Snapshot of an event where many people gather

$$S_t = \{s \mid s \in sensor\ input\}$$
$$B_t = belief(S_t, B_{t-1})$$
$$D_t = option(B_t, I_{t-1})$$
$$I_t = filter(D_t, I_{t-1})$$

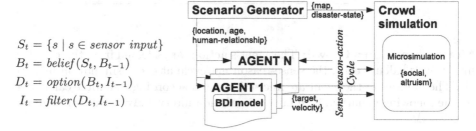

Fig. 3. Architecture of BDI based crowd evacuation system

on physical, mental and social states, as well as on the information that they have. Figure 3 shows the architecture of our system [deleted for blind review]. The properties of agents (in the left part) are sent to the crowd simulator at the start time as well as their targets at every sense-reason-action cycle. The targets present the agent's intentions that are selected from their belief, desire, and intention (BDI) models. The crowd simulator calculates the movements of agents according to equation (1). The results of micro simulation are returned to the agents as their own and other agents' positions that are within their visible area.

BDI Models at Evacuation: In our model, human relations affect the stages of the sense-reason-act cycle. S_t is a set of input that agents receive as visual and auditory information according to their environmental conditions at time step t. B_t, D_t, and I_t are the sets of beliefs, desires, and intentions of the BDI model, respectively. We implement three types of civilian agents with different BDI models (Table 3).

adult agents move autonomously and have no human relations with others. This type of agent can look for exits even when they have no knowledge of escape routes.

parent agents are adult agents who have one child each. They are anxious about
their children and have methods related to *anxiety*.

child agents have no data on escape routes and no ability to understand guidance from others. They can distinguish and follow their parents.

Table 3. Agent types and their behavior in BDI model

type	Belief	Desire	Intention
adult	personal risk	risk avoidance	evacuate to refuge
			hear guidance
parent	personal risk	risk avoidance	evacuate to refuge
			hear guidance
	family context	anxiety removal	seek child
			evacuate with child
child	personal risk	risk avoidance	follow parent

Motion Model in a Crowd: The agent's actions are selected from their I_t. The
intention is calculated in the sense-reason-act cycle at every simulation interval
Δt. The motions of the agent are micro simulated according to a force calculated
by the following equation. The micro simulation interval $\Delta \tau$ is smaller than the
interval Δt.

$$m_i \frac{d\mathbf{v}_i}{dt} = \mathbf{f}_{social} + \mathbf{f}_{altruism} \tag{1}$$

$$\mathbf{f}_{social} = m_i \frac{v_i^0(t)\mathbf{e}_i^0(t) - \mathbf{v}_i(t)}{\tau_i} + \sum_{j(\neq i)} \mathbf{f}_{ij} + \sum_{W} \mathbf{f}_{iW}$$

$$\mathbf{f}_{altruism} = \sum_{j \in G} \mathbf{f}_{ij}$$

$\mathbf{f}_{social}$ is a social force in Helbing's particle model that can simulate jamming by
uncoordinated motion in a crowd[4][6]. The first term is a force that moves the
agents to their target. $\mathbf{f}_{ij}$ and $\mathbf{f}_{iW}$ are repulsive forces that avoid collision with
other agents and walls, respectively. $\mathbf{e}_i^0$ is a unit vector to the targets , $\mathbf{v}_i(t)$ is a
walking vector at t, m_i is the weight of agents i, and v_i^0 is the speed of walking.
m_i and v_i^0 are set according to the age and sex of agent$_i$. In our model,

1. $\mathbf{e}_i^0$ is derived from the agents' intentions I_t. The targets are places or humans.
   When child agents follow their parents, the targets are their parent whose
   positions change during the simulation interval Δt.
2. $\mathbf{f}_{altruism}$ is an attractive force that keeps the agents in a group. It works
   when a parent waits until his or her child catches up. Group G is a unit
   in which members physically recognize each other. Therefore, it becomes
   zero when parents lose sight of their children. In this case, parents intend to
   look for their children. This change of intentions I_{t+1} causes the setting of a
   different $\mathbf{e}_i^0$.

4 Prototype System Based RCRS and Simulation Results

RoboCup Rescue Simulation v.1 (RCRS) is used as a platform [3]. RCRS comprehensively simulates agents' behavior in a simulated disaster world and our architecture can be run with small modifications.

4.1 ABS Example: Evacuation from Event Hall

Figure 4 shows a snapshot of an event site at a hall in which many families participate. Children enjoy the events and their parents watch them from distant places. The scenario is one of which agents evacuate from a hall that is $70m \times 50m$ and has one $4m$ wide exit. The parameters of the scenario are the number of agents and whether they are family members.

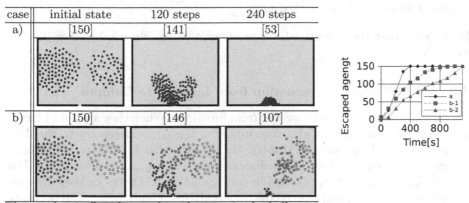

The number in [] is the number of agents in the hall.
case a) all agents are adults without human relations.
case b) 100 of 150 agents are parent-child relations. b-1)without $f_{altruism}$. b-2)with $f_{altruism}$. (bright: child, gray: parent, dark: adult)

Fig. 4. Snapshots of evacuation and change in number of agents who evacuate the room

(a) The 150 agents are all adults. They are divided in two groups; the left group is composed of 100 adults, and the right group, 50 adults.
(b) The 150 agents comprise 50 adults and 50 parent-child pairs (50 parents and 50 children). The left group is composed of 50 adults and 50 parent agents; the right group is composed of 50 children.

The following can be seen from the figure:

- All agents move toward the exit and congest there.
- Where adult agents move toward the exit, parent agents move to their child. When they move toward their children, some parents collide with other agents who move toward the exit.

At 120 steps, there are approximately 140 agents in the hall in both cases. However, it is clear that their behaviors differ. At 240 steps, the number of agents in case (b) is twice that in case (a).

Figures of the right column show the number of agents that evacuate from the hall. The rows correspond to case (a) and (b). The figures show that congestion caused by the behavior of the parent agents takes more steps to evacuate from the hall. Three lines in the right column shows case (a) and two following settings;

b-1) without $f_{altruism}$: When the parents lose their child, they look for their children at the level of the BDI cycle. The ◇ marked line shows a case in which the v_i^0s of parent agents and child agent are the same. The □ marked line is that in which the v_i^0s of child agents are 0.8 of those of the parent agents.

b-2) with $f_{altruism}$: The parent - child pair moves together keeping with their distance constant (△ marked line).

It is shown that the parents who care about their children take more time to evacuate.

4.2   EESS Example: Evacuation from Library on Campus

People exit from rooms and evacuate from buildings. When they are out of buildings, they go to refuges. A global coordinate, Σ_{Global}, is necessary to simulate such evacuation behaviors and display. On the other hand, 3D data of buildings or agents are created in each coordinates system, $\Sigma_{Buildingk}$. (Figure 5). The data in $\Sigma_{Buildingk}$s are represented in corresponding KMZ files or they are converted ones in Σ_{Global} and merged into one KMZ file. Figure 6 shows an overview of EESS which consists of three phases.

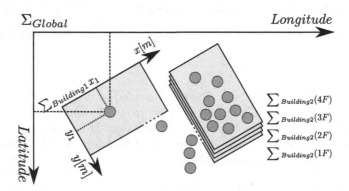

Fig. 5. Global and local coordinate systems in evacuation simulation

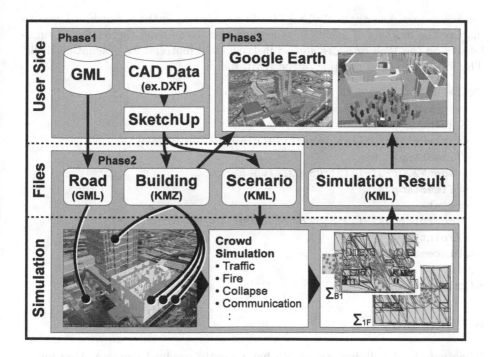

Fig. 6. Three phases of EESS. Users prepare 3D models of buildings and set properties of agents (Phase 1), ABS evacuation simulation (Phase 2), and display on the Web (Phase 3).

Phase1: Modeling 3D Buildings and Web World. 3D models of buildings or GIS data of towns are created and used for sightseeing or guidance every day. They are represented in several formats such as DXF and are converted into a format that can be input to a Web browser. The EESS system uses two types of files, namely, KMZ and GML files. KMZ files have the geographical data and location data of objects in their own Σ_{Global}, and GML files contain map data of the area.

Phase2: Evacuation Simulation on Converted Pseudo-3D World. Evacuation from buildings is simulated using the 3D data created in Phase 1 and an ABS described in Section 4.1. Inside building, the movements of agents are simulated and their motions are calculated in an open space in Σ_{Local} of the building. Figure 6 shows the simulation results of an evacuation behavior in a two-story library. Two figures correspond to two floors, i.e., B1F and 1F, which are connected with a stairway that is also represented in an open-space models.

Phase3: Displaying Simulation on the Web. The results of evacuation simulation are embedded in <Placemark> tag in KML files. The tag consists of three sub-tags, i.e., `<styleUrl>`, `<TimeStamp>`, and `<Point>`, which are used to display the motions of agents in Google Earth. Displaying the results of simulations on the Web provides a good instructional material for evacuation drills. The following shows a listing of the KML file.

```
<kml>
  ...
  <Placemark>
   <styleUrl>#civilian-parent</styleUrl>
   <TimeStamp>
     <when>2007-01-14T21:05:02Z</when>
   </TimeStamp>
   <Point>
     <coordinates>-122.536226,37.86047,0</coordinates>
   </Point>
  </Placemark>
  ...
</kml>
```

5 Summary

The analysis of building evacuation has recently received increase attention as people are keen to assess the safety of occupants. Agent-based simulation provides a platform for computing individual and collective behaviors that occur in crowds. The report on occupant behavior in the WTC disaster noted some features that are not included in existing simulations.

This paper proposes an agent-based evacuation system that simulates evacuation behaviors in any location where 3D models have been created for everyday usage such as sightseeing and guidance. The evacuation behaviors are simulated with a BDI model in which the mental disposition of each agent is represented. The BDI-based evacuation system solves the group formation issues cited in the WTC disaster report. The results are embedded into a KML file and can be viewed on the Web.

Our prototype system shows that the simulation results can be displayed on the Web and can be used for public relations to illustrate how proper behavior will save human lives and decrease the harm from disasters.

References

1. http://wiki.openstreetmap.org/wiki/Main_Page (January 30, 2011)
2. Abe, K., Misumi, F., Okabe, K.: Behavior Science at Natural Disasters. Fukumura Shuppan (1988) (in Japanese)
3. Cameron Skinner, S.R.: The RoboCup Rescue simulation platform. In: Proc. of 9th Int. Conf. on Autonomous Agents and Multiagent Systems, AAMAS 2010 (2010)
4. Kaup, T.I.L.D.J., Finkeistein, N.M.: Modifications of the Helbing-Molnar-Farkas-Vicsek social force model for pedestrian evolution. Simulation 81(5), 339–352 (2005)

5. de Walle, B.V., Turoff, M. (eds.): Emergency response information systems: Emerging trends and technologies. Communications of the ACM 50(3), 28–65 (2007)
6. Farkas, I., Helbing, D., Vicsek, T.: Simulating dynamical features of escape panic. Nature 407, 487–490 (2000)
7. Galea, E.R., Hulse, L., Day, R., Siddiqui, A., Sharp, G., Boyce, K., Summerfield, L., Canter, D., Marselle, M., Greenall, P.V.: The UK WTC9/11 Evacuation Study: An Overview of the Methodologies Employed and Some Preliminary Analysis. In: Klingsch, W.W.F., Rogsch, C., Schadschneider, A., Schreckenberg, M. (eds.) Pedestrian and Evacuation Dynamics 2008, pp. 3–24. Springer, Heidelberg (2010), http://www.springer.com/mathematics/ applications/book/978-3-642-04503-5
8. Kuligowski, E.D., Gwynne, S.M.V.: The Need for Behavioral Theory in Evacuation Modeling. In: Klingsch, W.W.F., et al. (eds.) Pedestrian and Evacuation Dynamics 2008, pp. 721–732. Springer, Heidelberg (2008), http://www.springer.com/mathematics/ applications/book/978-3-642-04503-5
9. Mehrotra, S., Znatri, T., Thompson, C.W. (eds.): Crisis management. IEEE Internet Computing 12(1), 14–54 (2008)
10. Pelechano, N., Badler, N.I.: Modeling crowd and trained leader behavior during building evacuation. IEEE Computer Graphics and Applications 26(6), 80–86 (2006)
11. Okaya, M., Yotsukura, S., Sato, K., Takahashi, T.: Agent Evacuation Simulation Using a Hybrid Network and Free Space Models. In: Yang, J.-J., Yokoo, M., Ito, T., Jin, Z., Scerri, P. (eds.) PRIMA 2009. LNCS, vol. 5925, pp. 563–570. Springer, Heidelberg (2009)
12. Pan, X.: Computational Modeling of Human and Social Behaviors for Emergency Egress Analysis. PhD thesis, Stanford (2006)
13. Perry, R.W., Mushkatel, A. (eds.): Disaster Management: Warning Response and Cummunity Relocation. Quorum Books (1984)
14. Ripley, A.: The Unthinkable: Who Survives When Disaster Strikes - and Why. Three Rivers Press (2008)
15. Takahashi, T.: RoboCup Rescue: Challenges and Lessons Learned, ch. 14, pp. 423–450. CRC Press (2009)
16. Yamashita, T., Soeda, S., Noda, I.: Evacuation Planning Assist System with Network Model-Based Pedestrian Simulator. In: Yang, J.-J., Yokoo, M., Ito, T., Jin, Z., Scerri, P. (eds.) PRIMA 2009. LNCS, vol. 5925, pp. 649–656. Springer, Heidelberg (2009)

Line Point Registration:
A Technique for Enhancing Robot Localization in a Soccer Environment

Thomas Whelan, Sonja Stüdli, John McDonald, and Richard H. Middleton

Department of Computer Science, NUI Maynooth, Maynooth, Co. Kildare, Ireland
Hamilton Institute, NUI Maynooth, Maynooth, Co. Kildare, Ireland
thomas.j.whelan@nuim.ie, stuedlis@ee.ethz.ch,
johnmcd@cs.nuim.ie, richard.middleton@nuim.ie

Abstract. The Standard Platform League (SPL) provides an environment that is essentially static; with the exception of other robots and the audience, the area in which a robot is expected to localise itself is quite favourable. However, a large number of the predefined landmarks in the given world model can be perceived as ambiguous in many scenarios, with the prime example being field line markings. In this paper a technique is presented that implicitly disambiguates these detected field line objects in order to use them for localization purposes.

1 Introduction

The environment in which a robot in the SPL must localise in is very well defined; pitch dimensions, field markings and landmark objects are fully described previous to any competition [11]. As a result we are reliably presented with a constrained environment. While other robots and audience members may introduce some dynamics into the environment, in general it is static and rich in predefined information.

1.1 Basics of Localization

In order to localise within this fully defined world model, we need only identify a small number of the known fixed features before we can employ a number of traditional methods for maintaining robot localization. A typical localization system used in humanoid robotic soccer would make use of either a Kalman filter[8] or a Particle filter[12,4]. Either of these filters use odometry and visual information to provide an accurate estimate of a robot's global position. Speaking in even simpler terms, only two known field landmarks are needed for triangulation of one's position. The goal posts are a perfect example of two such landmarks[1]. However, it is not always possible to maintain two goal posts in a robot's field of vision. Either of the filters listed previously are designed to account for this and aided by reasonable odometry measures can give a moderately accurate estimate of a robot's position in such scenarios. However, given

T. Röfer et al. (Eds.): RoboCup 2011, LNCS 7416, pp. 258–269, 2012.

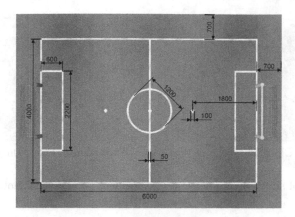

Fig. 1. Given Standard Platform League pitch description in rule book [11]

the relatively fast pace of a robotic soccer game, it is desirable to be accurately localised at all times.

The remaining known landmarks that exist on the field are the field line markings. As can be seen in Figure 1 these include a large number of corners (line intersections) and a centre circle. Considering the number of these landmarks one would expect to be able to triangulate on a position in the majority of view points. However, this is where our two main issues arise.

1.2   Computational Load

The first issue, albeit slightly more accessible than the second, is associated with computational performance. A typical method for the detection of lines in an image is to perform an operation such as the Hough Transform and extract lines from Hough Space[3]. Unfortunately this operation is quite expensive and cannot be realistically used in realtime on the current SPL hardware. A more basic approach, given the constrained environment, is to scan a given image for green-white-green transitions in order to detect what may be points on a field line. Then, clustering can be performed on these detected line points that should yield line segments within the image. This approach does yield good results, as shown in Figure 2, but still requires a significant amount of computation. For example, this technique makes up 50% of the processing time of the vision component of our SPL system.

1.3   Ambiguity

The second issue associated with field line markings is the ambiguity of corners and curves. While many heuristics can be applied to attempt to uniquely identify a detected corner, there are scenarios where it is impossible unless information about a robot's currently estimated position is used. This can lead to a nasty feedback loop - if the estimate is incorrect, a field line landmark may be

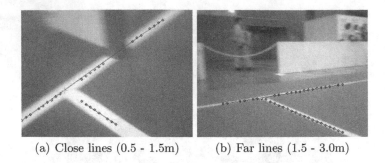

(a) Close lines (0.5 - 1.5m)   (b) Far lines (1.5 - 3.0m)

Fig. 2. Results of simple line point and line segment detection on close and far away field lines

incorrectly identified which would further the corruption of the current position estimation. A curve, the centre circle to be specific, is also ambiguous. From as many as 4 different view points any of these field line landmarks may appear the same. This presents us with a significant ambiguity problem.

The work presented in this paper provides a solution which removes the need for the initial stage of line detection and explicit disambiguation. However, it does make use of the information provided by processing an image containing field lines. In effect, by combining Cox's algorithm with a Kalman Filter based system, the aim is to rapidly and continually use the algorithm to keep close to the best fit. In other words, disambiguation on our proposed system is a consequence of continually maintaining an estimate of the range of possible robot locations. From this range of possible locations, Cox's algorithm is used to enhance the accuracy of estimation, and thereby narrow the range of possibilities.

2   Background

The process of matching detected points to a predefined map of line segments was first detailed by Cox in 1991 [2]. A robotic system was described that used a laser scanner to detect points on solid objects in the surrounding. These detected points were then subsequently matched in a local search fashion to an *a priori* map of the environment, producing an estimated pose. Cox treated the problem as least-squares linear regression with an analytical solution, successfully demonstrating that this technique was both accurate and practical.

Lauer et al. described a similar algorithm in 2006 for robots in the RoboCup Midsize league using points detected on field lines instead of laser points [7]. They introduced an alternative error function for minimisation that was more robust to outliers when compared to the squared error function. Noisy distance estimates were cited as the motivation behind this. They also introduced gradient descent as an alternative optimisation method in place of the least-squares linear regression used by Cox, due to the form of their new error function. This evoked a requirement to calculate the gradient for translation and orientation at each position before descending towards a minimum. *RPROP* was then used to solve the minimisation task [10]. Also in line with Cox's original suggestion, they used a simplified application of the Kalman filter in the form of a stochastic weighted averaging approach for smoothing of the estimated pose.

In 2010, Rath implemented an adaptation of the algorithm for use on a Nao robot in the SPL [9]. This adaptation included most of the methods used by Lauer et al. but due to the hardware constraints of the Nao some small changes were introduced. The most significant modification was the pre-calculation of gradients with respect to the translation for a 5×5cm grid. The motivation behind this was to reduce computational load, the result being that only the gradient for the orientation would need to be calculated at run time.

In 2009, Inam [4] presented a related method where a Particle filter could be used in conjunction with an image database. The matching technique described by Inam, unlike the previous examples, functions in image space rather than world space and uses correlation in place of an optimisation. It therefore uses substantially more computational power than required for the algorithm presented here.

2.1 Cox's Algorithm

Given a set of detected line points in an image, the basic process of Cox's algorithm involves 3 main steps, as outlined by both Cox and Rath; This description is taken mainly from [9].

2.1.1 Line Point Transformation from Image to World Coordinates
Transformation from image coordinates to world coordinates is achieved using typical back projection associated with the extrinsic and intrinsic camera parameters. In this regard the camera location based on the geometry of the robot and the current joint sensor readings.

2.1.2 Selecting the Closest Line for Each Point
This is carried out for each transformed line point by calculating the shortest Euclidean distance from each line segment in the world model to the point, and selecting the one with the shortest distance.

2.1.3 Finding a Correction for the Current Pose
In this final step, a correction to the current robot pose, described by $l_t = (x, y, \theta)^\top$ where x and y describe the estimate of the robot's global position and θ describes the estimated orientation of the robot, is calculated. We wish to calculate $b = (\Delta x, \Delta y, \Delta \theta)^\top$ such that a new estimate, $l'_t = l_t + b$, gives a pose which better matches observed line points to field line markings.

The aim of Cox's Algorithm is to minimise the squared distances associated with each point to its closest line segment that we calculated in 2.1.2. To achieve this, the problem is linearised into a least-squares linear regression problem and each line segment is treated as an infinite line with orthogonal unit vector $u_i = (u_{ix}, u_{iy})^\top$ and offset r_i such that $u_i \cdot z_i = r_i$ holds for all arbitrary points z_i on the line.

Let the ith transformed line point be $z_i = (z_{ix}, z_{iy})^\top$ and the current position of the robot be $c = (l_{tx}, l_{ty})^\top$. The transformation of each line point z_i can be described as:

$$t(b)(z_i) = \begin{pmatrix} \cos \Delta\theta & -\sin \Delta\theta \\ \sin \Delta\theta & \cos \Delta\theta \end{pmatrix} (z_i - c) + c + \begin{pmatrix} \Delta x \\ \Delta y \end{pmatrix} \tag{1}$$

Cox suggests that the correction angle $\Delta\theta$ should be sufficiently small such that we can approximate the transformation to:

$$t(b)(z_i) \approx \begin{pmatrix} 1 & -\Delta\theta \\ \Delta\theta & 1 \end{pmatrix} (z_i - c) + c + \begin{pmatrix} \Delta x \\ \Delta y \end{pmatrix} \tag{2}$$

Next, the squared distance of each line point z_i can be found as:

$$d_i^2 = (t(b)(z_i)^\top u_i - r_i)^2 \approx ((x_{i1}, x_{i2}, x_{i3}))b - y_i)^2 \tag{3}$$

Where:

$$x_{i1} = u_{ix} \tag{4}$$

$$x_{i2} = u_{iy} \tag{5}$$

$$x_{i3} = u_i^\top \begin{pmatrix} 0 & -1 \\ 1 & 0 \end{pmatrix} (z_i - c) \tag{6}$$

$$y_i = r_i - z_{ix} u_{ix} - z_{iy} u_{iy} \tag{7}$$

Now we can calculate the sum of squared distances for all points z_i:

$$E(b) = \sum_i ((x_{i1}, x_{i2}, x_{i3}))b - y_i)^2 = (Xb - Y)^\top (Xb - Y) \tag{8}$$

Where:

$$X = \begin{pmatrix} x_{11} & x_{12} & x_{13} \\ \vdots & \vdots & \vdots \\ x_{n1} & x_{n2} & x_{n3} \end{pmatrix} \qquad Y = \begin{pmatrix} y_1 \\ \vdots \\ y_n \end{pmatrix} \tag{9}$$

The correction $\hat{b}$ that minimises $E(b)$ can then be solved by:

$$\hat{b} = (X^\top X)^{-1} X^\top Y \tag{10}$$

And finally a new pose is given by:

$$l_t' = l_t + b \tag{11}$$

3   Extensions to the Algorithm

Lauer et al. described a method for matching detected field line points to a global map of curves that overcame the two main shortcomings of Cox's original method. (i) The restriction to straight lines, and (ii) the lack of robustness against distance measure outliers. In 2010, Rath described an adaptation to the method introduced by Lauer et al. that functioned on a smaller, less powerful system - Aldebaran's Nao. However, this adaptation was not without a cost; it required the pre-calculation of the gradients for translations. This resulted in what was effectively a drastically down-sampled representation of all possible translation gradients, from an almost continuous (limited only by computation precision) 6 × 4m area, to a discrete 120 × 80 size area, with each component of the area measuring 5 × 5cm. The hardware used by Lauer et al. contained a 1GHz CPU, which was more than adequate for evaluating gradients for all three components of the optimisation online. However implementing this full online gradient computation on Nao hardware is impractical, which justifies Rath's pre-calculated gradient look up table.

Here we present a set of adaptations to Cox's original algorithm that solve both the non-straight line and distance measurement error limitations without the need for gradient calculation or significant offline computation or quantization, while still maintaining reasonable performance for online execution onboard an Aldebaran Nao. Given the relatively small size of the SPL pitch, it is desirable to maintain as high a precision as possible when estimating a robot's pose.

3.1   Voronoi Diagram

In his paper [2] Cox suggests the use of a Voronoi diagram to simplify the determination of the closest line to a given point, but did not implement one himself. Lauer et al. do not mention any preprocessing done to speed up the determination of closest lines and in the approach taken by Rath, the pre-calculated gradient look up table essentially removes the need for a Voronoi diagram.

On-the-fly calculation of the closest field marking feature taking a naive brute force approach is $O(m)$, where m is the number of features. Pre-calculation of the closest line at each point in the form of a Voronoi diagram provides a look-up of $O(1)$, a much more desirable complexity. According to the description of the field markings described in the SPL rule book [11], we pre-calculate a Voronoi diagram of precision 1 × 1cm for the entire pitch, as shown in Figure 3.

Then, the closest line to the transformed line points as described in Section 2.1.1 can be found in a look-up table.

3.2   Inclusion of All Field Markings

The mathematics described by Cox only supports straight line segments as part of the optimisation process. While straight line segments do make up 88.5% of the white field marking area on the SPL pitch there is a lot of useful information in the penalty spots and the centre circle, as shown in Figure 1.

Fig. 3. Voronoi diagram for field markings

3.2.1   Inclusion of Points

In order to include penalty spots in Cox's original algorithm we must devise a method to include single points. Points can be included by modifying the entries made for Equations 4 through 7 in Section 2.1.3 in matrices X and Y:

$$\begin{pmatrix} x_{i,1} & x_{i,2} & x_{i,3} \\ x_{i+1,1} & x_{i+1,2} & x_{i+1,3} \end{pmatrix} = \begin{pmatrix} 1 & 0 \\ 0 & 1 \end{pmatrix} \begin{pmatrix} 0 & -1 \\ 1 & 0 \end{pmatrix} (c - z_i) \tag{12}$$

$$\begin{pmatrix} y_{i1} \\ y_{i2} \end{pmatrix} = \begin{pmatrix} s_{ix} - z_{ix} \\ s_{iy} - z_{iy} \end{pmatrix} \tag{13}$$

Where $s_i = (s_{ix}, s_{iy})^\top$ describes the coordinates of the point, or penalty spot, in world coordinates.

3.2.2   Inclusion of Circles

In order to include circular curves or the centre circle in particular, similar modifications are required for Equations 4 through 7 in Section 2.1.3. Namely, to include line points on a circle with radius R:

$$r_i = R \qquad u_i = \frac{1}{\sqrt{z_{ix}^2 + z_{iy}^2}} \begin{pmatrix} z_{ix} \\ z_{iy} \end{pmatrix} \tag{14}$$

3.3   Accounting for Distance Outliers

Lauer et al. and Rath used a normalising distance error function for minimising, said to be more robust to distance outliers than the squared error function of Cox's approach. In order to introduce some tolerance for such errors we added a weighting to each detected line point based on its relative distance. The weight for line point i is defined as:

$$W_i = \frac{1}{d^2 + \eta} \tag{15}$$

Where W is a diagonal matrix, d is the relative distance to the point and η is some small offset value.

Another concern for Cox's approach is the issue of singularity occurring during the pseudo inversion, particularly when no line intersections are visible. This is accounted for by adding some small value ζ in the form of a diagonal matrix before carrying out the inversion. The final modified version of Equation 10 is then:

$$\hat{b} = (X^\top W X + \zeta I)^{-1} X^\top W Y \tag{16}$$

4   Kalman Filter Integration

To stabilise the pose estimate given by the modified version of Cox's algorithm the output is smoothed with a specialised Kalman Filter. The unscented transform [13,5] is used, so that the uncertainty of the robot's position estimate is taken into account for the initial pose estimate given by the modified algorithm. Then from the output of the modified algorithm we compute an estimate of the position of the robot and an associated covariance matrix. Afterwards a normal Kalman filtering approach can be applied.

The unscented transform uses a set of points called sigma-points χ_i and associated weights w_i to represent a probability distribution; in our case the pose estimation. The sigma-points and their weights have to fulfill the following characteristics:

1. The weights must sum to unity.
   $\Rightarrow \sum_{i=0}^{n} w_i = 1$

2. The weighted sum must be the mean value of the distribution.
   $\Rightarrow \sum_{i=0}^{n} w_i \chi_i = \mathbf{x}_{k|k-1}$

3. The weighted square error sum must be equal to the covariance of the distribution.
   $\Rightarrow \sum_{i=0}^{n} w_i (\chi_i - \mathbf{x}_{k|k-1})(\chi_i - \mathbf{x}_{k|k-1})^T = P_{k|k-1}$.

There are various selections of sigma-points which fulfill these characteristics. For this implementation a symmetric sigma-point set similar to that of Julier and Uhlmann [5] is chosen with:

$$\chi_0 = \mathbf{x}_{k|k-1}$$
$$w_0 = \frac{1}{n_x + 1}$$

$$\chi_i = \mathbf{x}_{k|k-1} + \frac{1}{\gamma}\sqrt{n_x + 1}(p_{k|k-1})_{(i)}$$
$$w_i = \gamma\frac{1}{2(1+\gamma)(n_x + 1)}$$

$$\chi_{i+n_x} = \mathbf{x}_{k|k-1} - \frac{1}{\gamma}\sqrt{n_x + 1}(p_{k|k-1})_{(i)}$$
$$w_{i+n_x} = \gamma\frac{1}{2(1+\gamma)(n_x + 1)}$$

$$\chi_{i+2n_x} = \mathbf{x}_{k|k-1} + \gamma\sqrt{n_x + 1}(p_{k|k-1})_{(i)}$$
$$w_{i+2n_x} = \frac{1}{2(1+\gamma)(n_x + 1)}$$

$$\chi_{i+3n_x} = \mathbf{x}_{k|k-1} - \gamma\sqrt{n_x + 1}(p_{k|k-1})_{(i)}$$
$$w_{i+3n_x} = \frac{1}{2(1+\gamma)(n_x + 1)}.$$

The modified version of Cox's algorithm is then applied at each sigma-point, delivering a representation of the distribution of the measurement data. This means that the algorithm is executed thirteen times; once for each sigma-point. In addition to an estimated position of the robot Υ_i, the modified algorithm also computes a goodness of fit between the detected line points and their closest lines ξ_i. These measures are used to modify the weights of the sigma-points, according to:

$$\hat{w}_i = \frac{w_i}{\xi_i^2}. \tag{17}$$

As the weights must sum up to one, they are normalized after modification.

Then these new sigma-points are used to calculate the measurement:

$$\mathbf{y}_k = \sum_{i=0}^{n} \hat{w}_i\Upsilon_i \tag{18}$$

and its square root of the covariance matrix, where HT denotes the Householder Triangularization:

$$R_k = \mathrm{HT}\left(\left[\sqrt{\hat{w}_0}(\Upsilon_0 - \mathbf{y}_k), \ldots, \sqrt{\hat{w}_n}(\Upsilon_n - \mathbf{y}_k), r_0\right]\right). \tag{19}$$

These two values are then used for a measurement update of a Kalman filtering approach. In our case a Covariance Intersection as described by Julier and Uhlmann [6] is used.

5   Results

5.1   Localization System Tests

To fully test the system, the performance of various localization algorithms are compared in moderately realistic game scenarios. The localization algorithms compared are as follows:

- **UKF:** The Unscented Kalman Filter algorithm used by the RoboEireann team in the 2010 Competition.
- **Particle Filter:** A basic particle filter, with 100 particles, implemented for testing purposes.
- **CI (line detection):** A Covariance Intersection alternate to the UKF based on standard vision measurement data, that is, using post measurements, corner points, T points, centre circle etc.
- **CI (Cox's algorithm):** A Covariance Intersection algorithm, using goal post measurements and integrated with Cox's algorithm as described in Section 4.

For each of the localization systems, tests were conducted in four different scenarios. In each of these, there is a start location and orientation and a destination location and orientation. Once the robot reaches its final location, the error in this location was manually recorded. (The coordinate frame is a standard Cartesian system with origin at the centre of the pitch, x-axis directly towards the centre of the yellow goal). The scenarios are described in the Table 1.

Table 1. Scenarios used for localization tests

Scenario	Starting Position			Destination		
	$x(m)$	$y(m)$	$\theta(°)$	$x(m)$	$y(m)$	$\theta(°)$
1	0.0	-2.0	135	-1.2	0.0	180
2	-2.4	-1.1	0	-0.7	0.7	0
3	0.0	2.0	-135	-2.8	0.0	0
4	0.0	0.0	0.0	-1.0	1.0	-135

The first scenario denotes a situation where an attacker must move from the sideline, to a point facing the goal at the penalty spot. The second scenario describes a situation where a robot in the back corner of the pitch must move up to the opposite side of the field, just before the half way line. The third scenario describes a goal keeper moving to position and the fourth is where a robot on the center circle, must turn around, and move to an attacking position slightly to one side of the pitch.

To minimize variability due to changing lighting conditions, the vision system camera calibration was adjusted between each set of trials. The results presented in Figure 4 give data from 5 repetitions of a test in each scenario.

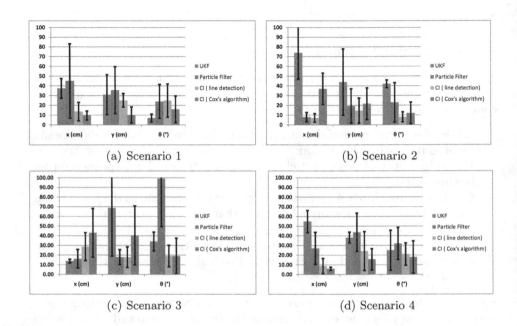

(a) Scenario 1   (b) Scenario 2

(c) Scenario 3   (d) Scenario 4

Fig. 4. Localization error results for tests in four different scenarios. For each scenario 5 tests are conducted, and statistics on the errors are computed. The error bars in the figures indicate standard error distance, and the vertical scale is either cm or degrees.

The implementation chosen for the inclusion of the modified version of Cox's algorithm in the Kalman filter took an iterative approach. Given the size of the pitch and the prevalence of local minima it was decided that each execution of the algorithm would be capped in translation and rotation, reducing the potential for overstepping a local minimum valley in the error minimisation space. This approach turned out to be quite computationally intense, taking between 8 and 30ms per frame, due to the necessity for multiple iterations at each of the 13 sigma-points.

6   Conclusion

In this paper we have considered a number of extensions to a line point registration based algorithm, Cox's algorithm, for utilizing field markings in localization. A number of modifications to the basic algorithm have been examined. These include: (i) use of a Voronoi diagram to reduce computational load in determining the nearest field mark; (ii) extension of the basic algorithm to include all types of field markings (lines, circles and points); (iii) distance based outlier detection; (iv) weighted least square cost minimization; and, (v) integration with unscented Kalman Filter based localization.

The performance of the algorithm achieved so far is similar to or better than any of the other algorithms tested. It also shows promise for further developments, such as modification to learn robot pose from the data. There is also potential in the future to use the algorithm in a non-iterative fashion and more cleverly chose integration points with the Kalman filter, solving any costly computation issues. It would also be interesting to see if the algorithm can be integrated with particle filter approaches, though generally, the algorithm itself is too expensive to be run once for each particle.

References

1. Cañas, J.M., Puig, D., Perdices, E., González, T.: Visual goal detection for the RoboCup Standard Platform League. In: X Workshop on Physical Agents, WAF 2009, Cáceres, Spain, pp. 121–128 (2009)
2. Cox, I.: Blanche - An experiment in guidance and navigation of an autonomous robot vehicle. IEEE Transactions on Robotics and Automation 7(2), 193–204 (1991)
3. Duda, R.O., Hart, P.E.: Use of the Hough transformation to detect lines and curves in pictures. Communications of the ACM 15, 11–15 (1972)
4. Inam, W.: Particle filter based self-localization using visual landmarks and image database. In: Proceedings of the 8th IEEE International Conference on Computational Intelligence in Robotics and Automation, CIRA 2009, pp. 246–251. IEEE Press, Piscataway (2009)
5. Julier, S., Uhlmann, J.: Unscented filtering and nonlinear estimation. Proceedings of the IEEE 92(3), 401–422 (2004)
6. Julier, S., Uhlmann, J.: Using covariance intersection for SLAM. Robotics and Autonomous Systems 55(1), 3–20 (2007)
7. Lauer, M., Lange, S., Riedmiller, M.: Calculating the Perfect Match: An Efficient and Accurate Approach for Robot Self-localization. In: Bredenfeld, A., Jacoff, A., Noda, I., Takahashi, Y. (eds.) RoboCup 2005. LNCS (LNAI), vol. 4020, pp. 142–153. Springer, Heidelberg (2006)
8. Middleton, R.H., Freeston, M., McNeill, L.: An application of the extended Kalman filter to robot soccer localisation and world modelling. In: Proceedings of the IFAC Symposium on Mechatronic Systems (2004)
9. Rath, C.: Self-localization of a biped robot in the RoboCup domain. Master's Thesis. Institute for Software Technology, Graz University of Technology (2010)
10. Riedmiller, M., Braun, H.: A direct adaptive method for faster backpropagation learning: the RPROP algorithm. In: IEEE International Conference on Neural Networks, vol. 1, pp. 586–591 (1993)
11. RoboCup Technical Committee: RoboCup Standard Platform League (Nao) Rule Book, http://www.tzi.de/spl/pub/Website/Downloads/Rules2010.pdf
12. Röfer, T., Laue, T., Thomas, D.: Particle-Filter-Based Self-localization Using Landmarks and Directed Lines. In: Bredenfeld, A., Jacoff, A., Noda, I., Takahashi, Y. (eds.) RoboCup 2005. LNCS (LNAI), vol. 4020, pp. 608–615. Springer, Heidelberg (2006)
13. Van Der Merwe, R., Wan, E.: The square-root unscented Kalman filter for state and parameter-estimation. In: IEEE International Conference on Acoustics Speech and Signal Processing, vol. 6, pp. 3461–3464. Citeseer (2001)

Learning to Discriminate Text from Synthetic Data

José Antonio Álvarez Ruiz

Bonn-Rhine-Sieg University of Applied Sciences
Computer Science Department
Grantham-Allee 20
53757 Sankt Augustin, Germany
jose.alvarez@smail.inf.h-brs.de

Abstract. *Service robots* could use textual information to perform important tasks, like *product identification*. However, *natural scene text* such as found in household environments can be very arbitrary in terms of size, color, font, layout, symbol repertoire, language, etc. This *large variability* makes robust *text information extraction* extremely difficult. Our work on textual information extraction for gray-scale still images uses *adaptive binarization, connected component* classification with a *support vector machine* and filtering based on the proximity of the connected components to their neighbours. The contribution of our approach is the use of a partially *synthetic dataset* for training. This decreases the burden of *ground truth labelling* at the connected component level. Our experiments show that classification generalization on real instances can be attained when training a classifier with synthetic data. We present our results on the ICDAR dataset.

Keywords: Natural scene text, object identification, adaptive binarization, support vector machine, synthetic dataset.

1   Introduction

Text can be found in many forms in household environments, e.g. books, newspapers, product wrappings, etc. Hence, service robots would profit substantially from the ability to read text; one application we can think of is *product identification*. In the *RoboCup@Home*[1] competition this is often done with classifiers trained on *appearance based features*. However, this approach suffers from an inherent *lack of generalization* in the sense that, given that the product wrappings change over time, a well performing classifier can easily be rendered useless. Moreover, the same product can have completely unrelated appearance across different vendors. On the contrary, the text found in the products has *regularities* useful for identification, which are likely to be *consistent* under different brands and in accordance with the *product class*. For example, a ketchup bottle will often exhibit text as "tomato", "ketchup", "sauce" whereas rat poison bottles will

---

[1] http://www.robocup.org/robocup-home/

T. Röfer et al. (Eds.): RoboCup 2011, LNCS 7416, pp. 270–281, 2012.
© Springer-Verlag Berlin Heidelberg 2012

unlike have "delicious" written on them. Beyond our intentions, advances in *text information extraction (TIE)* have other important applications: aiding visually impaired and blind humans, image and video databases indexing, surveillance and tracking systems, etc.

Unfortunately, TIE from *real scene images* is very challenging and to the best of our knowledge, high hit rates as found in other computer vision applications are rarely seen, except under very constrained conditions or when prior information on the targeted text is available. Specifically in the case of product wrappings and book covers, there is a huge variety of fonts, sizes, colors, layouts, languages, symbol repertoires, materials, etc. This issue was analysed on a comparative study, in which principal component analysis (PCA) was applied to text and faces images; text had a much larger number of non-zero eigenvalues and consequently, more features would be needed to preserve a certain level of variance [4]. Nevertheless, we think that TIE is already applicable to the product identification problem, not as an ultimate solution, but as a way of getting more cues for the sake of robustness. TIE can be divided in several sub-problems, among which we can identify[2]:

Detection. To discriminate between images containing text an images without text.

Localization. To identify and delimit areas containing text, in other words, to define regions of interest (ROI).

Extraction. To separate the text from its background, producing a binary image.

Rectification. To correct distortions induced by perspective and simplify the text layout.

Recognition. To transform text in an image into a string of symbols.

Our work addresses detection, localization and extraction; however localization and detection are tackled as a single problem, a common practice. In these steps, *machine learning* is a common practice in one form or another.

Localization methods can be divided into three categories: *region-*, *connected components (CC)-* and *hybrid*-methods. Region based methods assume that text areas[3] in the image have distinctive features respect to non-text areas. They are specially well suited for low resolution text, in which text components are merged with each other. To obtain training data for *supervised learning* it is enough to have the text areas bounding rectangles. For CC methods, those distinctive features are extracted from connected regions of arbitrary shape, defined under some concept of homogeneity. These methods perform well on "big" characters in which individual components do not touch each other. Obtaining training sets involves manually labeling CCs, an extenuating process. Hybrid methods, employ techniques proper from region and CC methods.

Extraction is the segmentation of the text from the background; an important issue because it increases the accuracy of *optical character recognition (OCR)*

---

[2] In practice, some of these steps depend on each other.

[3] Usually rectangular neighbourhoods.

systems [8]. Extraction can be grouped in two broad categories: clustering and binarization. Clustering often involves learning labels for the data without supervision. In contrast, binarization assumes two classes of objects, background and foreground; the classes are assigned using thresholds whether global or local. Regardless of the method, if extraction is done before localization, a large amount of non-text elements will also be usually present, which need to be classified and discarded somehow.

Now, we briefly summarize some important contributions made by other authors. [4] introduced an hybrid localization method based on Adaboost and low entropy features. For region validation the method also acts at the CC level, extracting the CCs using a stroke width invariant version of Niblack binarization. [17] developed a CC based approach (Niblack binarization and mean-shift were explored) and text CCs discrimination using a cascade of classifiers. They used a very rich set of features, including different moments invariants; the target application was container tracking. An hybrid based text localization system based on ensemble learning, probability maps, CC extraction and two stages of classification with a conditional random field is illustrated in[10]. [2] elaborates on the use of synthetic data for training a *multi-layer perceptron* to map features from the gradient histogram into an estimate of the text's pose. Niblack binarization was found to be the *best binarization* technique for *general text* in [14], thus it has been often applied whether "as is" or with some modifications [10,9,4,17,16]. A high performance text extractor based on k-means clustering, denoising using log-Gabor filters and other post processing steps can be found in [8]; one interesting result is that, for robust clustering, the selection of a distance metric is more important than the color-space. A commonly used TIE dataset has been released by [7].

Our method is based on adaptive binarization, CC classification using a support vector machine (SVM) [5] and filtering based on the proximity of the CCs to their neighbours. The contribution of our work is the use of a *partially synthetic dataset*, which greatly alleviates of the burden of datasets preparation. This document is organized as follows: A description of specific problems addressed in this document can be found in Sect. 2. Our approach is briefly described in Sect. 3 and details on image segmentation are given in Sect. 3.1. Training data generation and features used for classification are covered in Sect. 3.2 and Sect. 3.3. Our experiments are illustrated in Sect. 4. Finally, our conclusions and future work can be found in Sects. 5 and 6 respectively.

2   Problem Statement

Currently, TIE remains a general open problem with many associated subproblems. We are interested in providing a robot with the ability to read text from 2D images. Although several approaches have been devised for TIE, integral solutions are still missing. In general, *machine learning* has demonstrated to be useful in this domain. Unfortunately, to apply such techniques, large amounts of training data are necessary. This becomes an important concern specially for

CC based methods because labeling CCs into *text* and *non-text* requires a lot of effort and is error prone. Therefore, we would like to explore the use of synthetically generated text. Besides of the high variability found in natural scene text, common annoyances to other computer vision tasks are also present, e.g. illumination inhomogeneity, acquisition parameters, viewpoint, noise, perspective distortions, background clutter, occlusion, etc. Those variations would be very hard to emulate altogether.

3   Learning to Discriminate Text from Synthetic Data

Our method consists of the following basic steps (see Fig. 1, Fig. 2 and Fig. 3):

- Adaptive binarization. We select pixels found in *homogeneous regions of high contrast (HRHC)*. We convert the input image into two binary images called planes. The "on" pixels in the first and second planes correspond to pixels in HRHCs *darker* and *lighter* than the background respectively.
- Connected components extraction. We group individual pixels into regions. Two "on" pixels belong to the same CC if and only if there is a path of "on" pixels between them.
- Connected component classification. We use an SVM to filter non-text CCs. The classifier was trained using a partially synthetic dataset to ease the acquisition of training data. After the SVM classification, a neighborhood graph of CCs is created using the linkage rule given in Eq. 1 [10]. Those CCs not connected with any other CC are discarded.

$$dist(CC_i, CC_j) < 1.5 \times \min(\max(width(CC_i), height(CC_i),$$
$$\max(width(CC_j), height(CC_j)) \ . \tag{1}$$

3.1   Niblack Adaptive Binarization

For intensity based segmentation we will use the Niblack binarization algorithm and a *stroke width independent* version of it [4]. Some advantages of this method are: its relative *simplicity*, its low *sensitivity to parameter* values and its adequacy for *parallelization*. The algorithm separates an input gray scale image $\mathcal{I}$ into HRHCs and produces an output image $\mathcal{O}$ in which those regions have distinctive intensity values (see Eq. 2).

$$\mathcal{O}(x, y) = \begin{cases} 0 & \text{if } \mathcal{I}(x, y) < T_{r-}(x, y) \\ 255 & \text{if } \mathcal{I}(x, y) > T_{r+}(x, y) \\ 100 & \text{otherwise} \end{cases} \ . \tag{2}$$

The *thresholds* $T_{r\pm}$ are estimated adaptively for each pixel using Eq. 3.

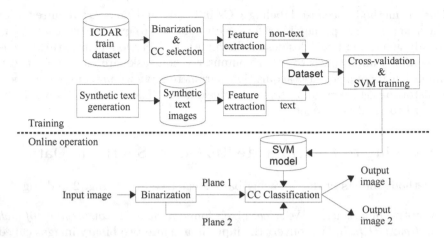

Fig. 1. Method block diagram. We aim at decreasing the effort put on dataset preparation for CC based methods. Training data for the non-text class is extracted from binarized versions of the images in the ICDAR train dataset. Examples of the text class are generated synthetically. Both example sets are joined into a training dataset. During online operation, the image is processed with the Niblack algorithm and each of the produced planes are processed by extracting CCs and features out them that are passed to the classifier. All CCs labeled as text are rendered into the output images.

Fig. 2. Performance of our system on items commonly found in household items. Input images on top and result images on bottom.

$$T_{r\pm}(x,y) = \mu_r(x,y) \pm k\sigma_r(x,y) \ . \tag{3}$$

where $\sigma_r(x,y)$ and $\mu_r(x,y)$ are the standard deviation and mean value of the intensity in a window of size $r \times r$ centered at (x,y) and k weights $\sigma_r(x,y)$. In order to have symmetric windows, r must be odd. Our implementation was

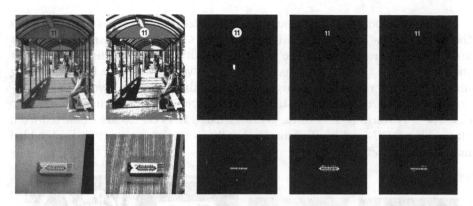

Fig. 3. Example of the performance of our system. From left to right: original, binarized, plane 1 and 2 after SVM classification and plane 1 and plane 2 integrated after CC proximity filtering.

boosted using *integral images* [13] as described in [11]. By using two integral images it is possible to calculate the sum $\mathbb{S}(x, y, r)$ and sum of squared $\mathbb{S}^2(x, y, r)$ intensity values within a squared region of size $r \times r$ centered at (x, y) in constant time, for any r; with those values, the calculation of $\mu_r(x, y)$ and $\sigma_r(x, y)$ is trivial. To efficiently apply the algorithm close to the borders of the image, the integral images are created from an *extended* version of the original image in which the content of the bordering pixels was *replicated* $R/2$ pixels at each border, where R is a necessary upper-bound for r to limit the size of the extend image. In the case of text, r accounts for the *stroke width*, a fixed parameter in the original Niblack binarization algorithm. In [4] the algorithm was further developed by introducing Eq. 4. From now on, we will refer to this variant as *Chen-Niblack* binarization.

$$r(x, y) = \arg\min_r(\sigma_r(x, y) > T_\sigma) \ . \tag{4}$$

3.2   Training Data Acquisition and Generation

To generate training data for the text class, we used a python[4] script that renders *random strings* with random font and size in two kinds of images; a *binary image* which stands for ideally segmented text and a *color image* which is drawn using different colors for background and foreground, the colors are selected randomly from a set of predefined colors. We constrained ourselves to use the standard font set distributed with our OS[5] and removed overly artistic, cursive and non Latin based fonts, besides of adding *rotation* as unique distortion. Our rationale behind this is that, emulating all possible annoyances found on natural scene text would be a quite challenging pursue by itself.

---

[4] http://www.python.org
[5] Mac OS X Snow Leopard.

CCs for the non-text class come from binarizing and separating into planes the images in the ICDAR training dataset. All CCs with overlap with the text areas bounding rectangles in the standard ground truth were dismissed, unless the CCs contain the bounding rectangles they overlap with (see Fig. 4). During binarization, we noted that a large number of tiny CCs were produced. To decrease the amount of training data, those CCs, with height less than 10 were also discarded. *Features vectors* of the features listed in Sect. 3.3 are extracted from all remaining CCs and labelled automatically as text or non-text and assembled into a *training set*.

(a)   (b)   (c)   (d)

Fig. 4. Examples of the images from which training features are extracted. 4(a) and 4(b) are synthetically generated text images for the same random string; 4(a) is an ideally binarized version and 4(b) a color version. 4(c) and 4(d) non-text class examples.

3.3   Features Used

The features to use when planning a classification schema are a key issue. Good features capture regularities within a class and provide separability between classes. Since we are working on synthetically data generated in a simple way, we decided to use shape based features (except for *EdgeContrast*). If you think of the synthetic text instances as perfectly segmented text CCs obtained from a natural image, these features should resemble the features extracted from a manually prepared dataset. We use the following features to train our classifiers:

1. $CCCount = |\{c \in CCs : c \neq CC_i \wedge contains(bb(CC_i), bb(c))\}|$, where CC_i is the CC being evaluated, CCs are all CCs in the same plane, and bb extracts the bounding box of a CC and $contains(b_1, b_2)$ is true if b_1 contains b_2.
2. $HoleCount$. The number of holes in the CC . Characters have low hole count, e.g. "H" contains no holes and "O" contains just one hole [16].
3. $Roughness_{open} = |fill(CC) - (fill(CC) \circ S)|/|CC|$. Using the open morphological operator [16]. Where $fill$, fills the holes in CC and S is a box structuring.
4. $Roughness_{close}= |fill(CC) - (fill(CC) \bullet S)|/|CC|$. Using the close morphological operator [17].
5. $AspectRatio = max(width(CC)/height(CC), height(CC)/width(CC)).$[16].
6. $Compactness = area(CC)/|contour(CC)|^2$ [16].
7. $EdgeContrast = |contour(CC) \cap gradient(CC)|/|contour(CC)|$. We do not use the canny edge detector as in [16], instead we just use a 3×3 Sobel operator. This feature is the only feature extracted out of the synthetic color text images.

8. Hu Moments. Seven linear combinations of central moments [6]. They are invariant to several distortions but are not orthogonal.
9. Zernike and Pseudo-Zernike invariants. These features [12] are derived out of the Pseudo-Zernike and Zernike moments and are calculated after fitting the image to a circle. These moments are *orthogonal* within the *unit circle* and their *moduli* (called invariants) are rotation invariant and are what we use as *features*. Translation and scale invariance are achieved with the help of *geometric moments* $m_{n,m} = \int \int x^n y^m f(x,y) dx dy$. Translation invariance is accomplished by translating the CC's image, so that its center of mass (COM) $(\bar{x}, \bar{y})$ is at the origin, where $\bar{x} = \frac{m_{10}}{m_{00}}$ and $\bar{y} = \frac{m_{01}}{m_{00}}$. For scale invariance, we simply divide the moments by the CC's image mass m_{00} [1]. In general, both kinds of moments come from the same basic equation, which states that the moment of order n and repetition m is defined as [6]:

$$C_{n,m} = \alpha \int_0^{2\pi} \int_0^1 R_{n,m}(r) e^{-im\theta} f(r,\theta) dr d\theta \ . \tag{5}$$

Where α is a factor, $R_{n,m}(r)$ is a radial polynomial, i is the imaginary unit , $f(r,\theta)$ is the intensity of the image at polar coordinates (r,θ), $r = \sqrt{(x-\bar{x})^2 + (y-\bar{y})^2}/R$, $0 \le r \le 1$, $\theta = \arctan(\frac{y}{x})$, x and y denote the position of a pixel in the image and R is the distance between the image COM and the farthest "on" pixel from it. The difference between Zernike and pseudo-Zernike moments lies in how $R_{n,m}(r)$ and α are defined. For Zernike moments:

$$R_{n,m}(r) = \sum_{s=0}^{\frac{n-|m|}{2}} \frac{(-1)^s (n-s)!}{s!(\frac{n+|m|}{2})!(\frac{n-|m|}{2})!} r^{n-2s} \ . \tag{6}$$

Subject to the constraints $|m| \le n$, $n - |m|$ and $n - |m|$ is odd, we can calculate up to $\frac{(n+1)(n+2)}{2}$ moments for a given maximum order n, $\alpha = \frac{n+1}{\pi}$. In the case of Pseudo-Zernike moments:

$$R_{n,m}(r) = \sum_{s=0}^{\frac{n-|m|}{2}} \frac{(2p+1-s)!}{s!(n+|m|+1-s)!(n-|m|-s)!} r^{n-s} \ . \tag{7}$$

Subject to the constraints $|m| \le n$ and $n > 0$, we can calculate up to $(n+1)^2$ moments for a given maximum order n, $\alpha = \frac{2(n+1)}{\pi}$. Moments for which $m < 0$ are not used since $C_{n,m} = C_{n,-m}^*$, thus $|C_{n,m}| = |C_{n,-m}|$, where * denotes the complex conjugate.

4   Experimental Evaluation

We present some experiments to asses the performance of our method. We begin by training four SVMs using different combinations of features and 14 000 training examples, 50 % for the text and 50 % for the non-text class. The training

---

[6] The integrals are replaced by sums in the case of digital images. Factorial calculations are cached using a recursively fill-up-on-demand look up table.

data generation was explained in Sect. 3.2. The four SVMs use the first seven features described in Sect. 3.3. One of the SVMs also uses Hu moments, whereas the others use Zernike and Pseudo-Zernike invariants.

H7 Hu moments.
Z10 Zernike invariants up to the 10th order.
P10 Pseudo-Zernike invariants up to the 10th order.
Z10-P10 Zernike and Pseudo-Zernike invariants, both up to the 10th order.

The results are measured in terms of precision p, recall r and their harmonic mean $f = \frac{1}{\alpha/p+(1-\alpha)/r}$, for $\alpha = 0.5$. We measure p and r using the ICDAR [7] and Wolf definitions [15] [7]. Since our approach lacks word grouping, using word based ground truth, as currently available, would not be fair. For this reason we labelled the *ICDAR test* dataset trying to adhere as good as possible to the bounding boxes of the text CCs and used the resulting ground truth [8] for evaluation. Table 1 illustrates the performance of the different classifiers. The images were binarized by the original Niblack algorithm with $k = 0.5$ and $r = 223$. In general, the classifiers using Zernike and Pseudo-Zernike invariants performed better than H7.

Table 1. Classification performance. Note that the classifiers trained using Zernike and Pseudo-Zernike invariants produce very similar results and outperform H7. The results are very dependent on the quality of the image binarization.

Classifier	ICDAR			Wolf		
	p	r	f	p	r	f
H7	0.57	0.55	0.56	0.53	0.51	0.52
Z10	0.68	0.56	0.62	0.63	0.52	0.57
P10	0.76	0.52	0.61	0.70	0.48	0.57
Z10-P10	0.77	0.51	0.61	0.71	0.48	0.57

Our second experiment is concerned with the classification performance using Chen-Niblack binarization, for $R = 223$. Fig. 5(a) shows the performance change for $H7$ given different values of T_σ using the same values for the other parameters as we did with Niblack binarization. As we can see, larger values of T_σ improve the classification, although after $T_\sigma = 70$ this tendency is less pronounced. With the tested parameters, using Chen-Niblack binarization produced no classification improvement. On a final experiment we set $T_\sigma = 100$ and use the P10 classifier with the Niblack and Chen-Niblack algorithm without significant changes in performance (see Fig. 5(b) and Fig. 5(c)) .

---

[7] For evaluation, we use the software found at
http://liris.cnrs.fr/christian.wolf/software/deteval/index.html.
The default values have been used for the Wolf metric.

[8] Available at http://home.inf.h-brs.de/~jalvar2s/

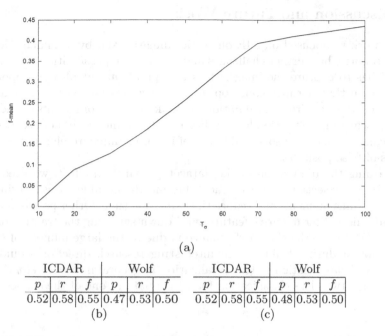

(a)

ICDAR			Wolf			ICDAR			Wolf		
p	r	f	p	r	f	p	r	f	p	r	f
0.52	0.58	0.55	0.47	0.53	0.50	0.52	0.58	0.55	0.48	0.53	0.50

(b)   (c)

Fig. 5. 5(a) Effect of $T\sigma$ and Chen-Niblack binarization on the classification performance of the $H7$ classifier, f-mean is the mean value of the ICDAR and Wolf f metrics. 5(b) Results using P10 and Chen-Niblack binarization. 5(c) Results using P10 and the standard Niblack algorithm. These results were obtained without the CC proximity filtering.

5   Conclusions

We developed a CC classification schema to discriminate between text and non-text examples, the training was done only with synthetically text. This reduced the amount of work spent on preparing the datasets. Our results show that the SVM classifiers were able to generalize from the features extracted from the synthetic text instances to real scene images. We tried different feature combinations although the differences in performance were minimal for the classifiers Z10, P10, and Z10-P10. This may point out that the first seven features introduced in Sect. 3.3 were too dominant. Besides, the experiments run in less than 10 minutes using $H7$ but for the classifiers using Zernike and Pseudo-Zernike invariants, the execution times rises to up to 6 hours. In this sense, the improvement in performance might not be worth the execution time increase. With respect to binarization, we did not notice a significant improvement using the Chen-Niblack algorithm and observed that the original Niblack algorithm can account for a variety of stroke widths given large values of r.

6   Discussion and Future Work

In this work we focused on TIE on static images taken by a human. However, TIE in robotics has further challenges and offers more possibilities. For example, *a robot* has to capture the images; this is a problem, but also an opportunity in the sense that a robot could optimize the *acquisition parameters* and the *view point*; this requires consideration of the kinematic constraints of the robot, planing, scheduling, etc. Besides, a robot usually counts with more sensors than just a camera. Those sensors could be useful, for example, to eliminate common sources of false positives.

Regarding the diminution of the dataset preparation effort, we need an in-depth feature selection and a cascade classifier to be able to use Zernike and Pseudo-Zernike moments and reduce the execution time. Other possibility would be to add more discriminative features and down-sampling the training datasets of the SVMs, since the slow performance is due to the large number of support vectors found during training. An interesting research direction would be to pose the CC classification or text localization in general into a *novelty detection* problem, i.e. two-class classification for which data of one of the classes is hard to get.

References

1. Bin, Y., Jia-Xiong, P.: Improvement and Invariance Analysis of Zernike Moments using as a Region- based Shape Descriptor. Journal of Pattern Recognition and Image Analysis 12(4), 419–428 (2002),
   http://sibgrapi.sid.inpe.br/col/sid.inpe.br/banon/2002/
   10.21.17.56/doc/2.pdf
2. Bulacu, M., Ezaki, N., Schomaker, L.: Text Detection and Pose Estimation for a Reading Robot. ai.rug.nl (2003),
   http://www.ai.rug.nl/ bulacu/bulacu-ezaki-schomaker-
   reading-robot-2008.pdf
3. Chang, C.C., Lin, C.J.: LIBSVM: a library for support vector machines (2001), software, http://www.csie.ntu.edu.tw/~cjlin/libsvm
4. Chen, X., Yuille, A.: Detecting and reading text in natural scenes. In: IEEE Computer Society Conference on Computer Vision and Pattern Recognition, vol. 2. IEEE Computer Society (1999, 2004),
   http://www.comp.nus.edu.sg/~cs4243/projects2008/text_natural_scene.pdf
5. Lai, J.Y., Sowmya, A., Trinder, J.: Support Vector Machine Experiments for Road Recognition in High Resolution Images. In: Perner, P., Imiya, A. (eds.) MLDM 2005. LNCS (LNAI), vol. 3587, pp. 426–436. Springer, Heidelberg (2005)
6. Hu, M.: Visual Pattern Recognition by Moment invariants. IRE Transactions on Information Theory 8(2), 179–187 (1962),
   http://ieeexplore.ieee.org/xpls/abs_all.jsp?arnumber=1057692
7. Lucas, S., Panaretos, A., Sosa, L., Tang, A., Wong, S., Young, R., Ashida, K., Nagai, H., Okamoto, M., Yamamoto, H., et al.: ICDAR 2003 robust reading competitions: entries, results, and future directions. International Journal on Document Analysis and Recognition 7(2), 105–122 (2005),
   http://www.springerlink.com/index/R43V711MV0X01U27.pdf

8. Mancas-Thillou, C., Gosselin, B.: Color text extraction with selective metric-based clustering. Computer Vision and Image Understanding 107(1-2), 97–107 (2007), http://linkinghub.elsevier.com/retrieve/pii/S107731420600213X

9. Pan, Y.F., Hou, X., Liu, C.L.: A Robust System to Detect and Localize Texts in Natural Scene Images. In: 2008 The Eighth IAPR International Workshop on Document Analysis Systems, pp. 35–42 (September 2008), http://ieeexplore.ieee.org/lpdocs/epic03/wrapper.htm?arnumber=4669943

10. Pan, Y.F., Hou, X., Liu, C.L.: Text Localization in Natural Scene Images Based on Conditional Random Field. In: 10th International Conference on Document Analysis and Recognition, pp. 6–10 (July 2009), http://ieeexplore.ieee.org/lpdocs/epic03/wrapper.htm?arnumber=5277814

11. Shafait, F., Keysers, D., Breuel, T.M.: Efficient implementation of local adaptive thresholding techniques using integral images. In: Document Recognition and Retrieval XV, San Jose, CA (January 2008)

12. Teh, C.H., Chin, R.: On image analysis by the methods of moments. IEEE Transactions on Pattern Analysis and Machine Intelligence 10(4), 496–513 (1988), http://ieeexplore.ieee.org/lpdocs/epic03/wrapper.htm?arnumber=3913

13. Viola, P., Jones, M.: Rapid object detection using a boosted cascade of simple features. In: Proceedings of the 2001 IEEE Computer Society Conference on Computer Vision and Pattern Recognition, CVPR 2001, pp. 511–518 (2001), http://ieeexplore.ieee.org/lpdocs/epic03/wrapper.htm?arnumber=990517

14. Wolf, C., Jolion, J.M.: Extraction and recognition of artificial text in multimedia documents. Formal Pattern Analysis & Applications 6(4) (February 2004), http://springerlink.metapress.com/openurl.asp?genre=article&id=doi:10.1007/s10044-003-0197-7

15. Wolf, C., Jolion, J.M.: Object count/area graphs for the evaluation of object detection and segmentation algorithms. International Journal on Document Analysis and Recognition 8(4), 280–296 (2006)

16. Zhu, K.h., Qi, F.h., Jiang, R.j., Xu, L.: Automatic character detection and segmentation in natural scene images. Journal of Zhejiang University Science A 8(1), 63–71 (January 2007), http://www.springerlink.com/index/10.1631/jzus.2007.A0063

17. Zini, L., Destrero, A., Odone, F.: A Classification Architecture Based on Connected Components for Text Detection in Unconstrained Environments. Advanced Video and Signal Based Surveillance, 176–181 (2009)

RoboViz: Programmable Visualization for Simulated Soccer

Justin Stoecker and Ubbo Visser

Department of Computer Science
University of Miami, Coral Gables FL
{justin,visser}@cs.miami.edu

Abstract. This work describes RoboViz, a new software program designed to assess and develop agent behaviors in a multi-agent system, the RoboCup Soccer Simulation 3D sub-league. RoboViz is an interactive monitor that renders both agent and world state information in a three-dimensional scene. In addition, RoboViz provides programmable remote drawing functionality to agents or other clients that can communicate over a network. The tool facilitates real-time visualization of agents running concurrently on the SimSpark simulator to provide higher-level analysis of agent behaviors not currently possible with existing tools. Provided appropriate hardware, the monitor and debugging tool can produce high-quality stereo vision images. RoboViz is proposed as a replacement for the current SimSpark 3D league monitor to benefit developers as well as elevate public interest in the 3D simulation league, and it has been used officially at the 2011 German Open in Magdeburg, Germany. RoboViz was released in February 2011 as an open-source project under the Apache 2.0 license.

1 Introduction

The environment of RoboCup Soccer is one of the most difficult for artificial intelligence researchers and presents several problems: an uncertain environment, multiple competitive agents, full physics, and the need for high-level cooperative behaviors. One of the greatest challenges in developing autonomous robotic agents is debugging and analyzing behaviors and algorithms. As such, there is a significant need for tools that assist researchers in understanding and developing their agents.

Presently, there is a lack of generally accessible or effective software tools for analyzing and supporting development of agents for the simulation 3D sub-league. Many researchers in the 3D league are capable of writing specialized programs for their needs, and it is common to see each team develop their own tools. Spending significant time on such specialized tools distracts from the overarching goals of the simulation league; the capabilities and flexibility of team-specific tools are reduced as a consequence. Sifting through the immense amounts of data generated from these simulations remains a challenge. Researchers in the simulation league face a shared set issues while developing agents' reasoning, skills,

T. Röfer et al. (Eds.): RoboCup 2011, LNCS 7416, pp. 282–293, 2012.

and behavior; many of these behaviors possess spatial and temporal properties that are amenable to visual presentation.

Compared to the physical leagues, the simulation league also struggles with a less impressive presentation [10]. The current monitor for rendering the simulation has not seen significant improvements over the past years. Some of the commonly mentioned issues with the SimSpark monitor [1] include poor performance, dated graphics, and an awkward user interface. The simulation league in RoboCup is a useful platform to pioneer and experiment without physically risking robots; this league should be exciting and at the forefront of research before it is deployed in the physical leagues. The existing monitor used to view the simulations does not reflect the state of the art research being conducted in this field, and for the simulation league to attract increased public interest the monitor must be updated accordingly.

This paper proposes that there is an unmet potential for high-quality analysis and development of agents and their algorithms in the 3D simulation league through real-time visualization. This is supported by a software solution that fills the role of a simulation monitor with visual data overlaid and integrated with the 3D scene. Our solution is designed to be accessible to resolve the shared debugging and analysis issues researchers encounter. This program addresses both the visualization needs of developers while providing a significant upgrade over the existing monitor to benefit the community as a whole.

2   Related Work

Several teams competing in the simulation leagues of RoboCup develop their own tools to optimize and debug their agent behaviors. For the most part, these tools are described in team description papers. The 2D simulation team *Mainz Rolling Brains* developed a debug and visualization tool called *FUNSSEL* [3]. *FUNSSEL* acts as a layer between the server and agents to intercept and process communication. The primary features of this software include filtering data, agent training, and graphic overlays for the 2D field in a secondary monitor. Other examples of tools for the 2D league can be found, for example, in the Portuguese team *FC Portugal* [9] and many other team description papers not mentioned here for brevity. For the 3D simulation league, the team *Virtual Werder 3D* utilized an evaluation tool [5] to analyze agent performance. The program also supported basic drawings in a simplified 2D monitor; however, all analysis and visualization was done on server and agent generated logfiles.

There have been few attempts at providing useful analysis and debugging tools for the general community. The *logfile player and analyzer* [8] provided improvements to the log playing capabilities of the 3D monitor; for example, it allows agents to record behaviors as simple drawings displayed in the 3D scene when the log is replayed. Additional features of this program included some basic filtering of logfile data and an improved graphical user interface.

Simulators for multi-agent systems often include some manner of visualization. Usually, these visualizations refer to modeling the robots and environment.

However, a few simulators do provide additional functionality. In Webots [6], for example, user code can initiate the drawing of primitives to further model the robot components. Breve [4] is another simulation program that supports the modeling and simulation of large multi-agent environments. Breve also exposes simple drawing routines to add shapes to the scene.

Despite the importance of processing, visualizing, and understanding data output from 3D league simulations, there is a dearth of solutions to address these needs. This is evidenced by the scarcity of high-quality tools available to the 3D simulation league community. Tools mentioned in team description papers are, unfortunately, not well documented, obsolete, or unsuitable for general use as they may be tightly bound to a team's agent architecture. Simulators such as Webots and Breve, while providing more advanced visualization capabilities, are primarily focused on the simulations themselves. The drawing routines exposed by these simulators are secondary and not integrated with the interface in a meaningful way. The existing SimSpark monitor [1] has a number of limitations as well. In particular, we felt the following issues should be resolved or improved upon:

- *Usability*: the Simspark monitor has a rudimentary interface and the user experience is less polished. For example, the monitor may only be active while the server is online, must be manually restarted with the server, and the window cannot be resized for a higher resolution.
- *Interactivity*: the Simspark network protocol exposes functionality for modifying the game state and moving the players or ball; however, the monitor does not yet make use of these features.
- *Portability*: the monitor is deeply integrated with the Simspark framework making it more difficult to configure, compile, and use.
- *Graphics Quality and Performance*: we also felt the Simspark monitor exhibited suboptimal resource usage and performance. While less pressing as other issues, the graphics effects also have significant room for improvement.

The current 3D monitor fulfills the basic requirement that it is real-time, and it is also marginally interactive by allowing users to control a virtual camera; however, the extent of its visualization capabilities is limited to presenting the physical world state, or ground truth. The SimSpark monitor is functional merely as a passive viewer, and as a visualization tool it is incapable of conveying information other than actual positions of players and the ball.

3   Requirements

Before designing RoboViz, we looked at the needs of developers and the problems with current solutions. The most crucial issue we looked into was determining how researchers view and process data from the simulations. Teams in the 3D simulation league typically output values specific to their implementation to the terminal or a file. This form of data can be used for correcting the underlying algorithms when the output doesn't match expectations, and the code required is minimal. The obvious problem with text is that it does not provide any higher-level understanding of behaviors; by itself, textual data does not reveal patterns or other interesting situations that may lead to unexpected results. A deeper analysis is only achievable by processing this data, which requires further tools of some kind.

Given the nature of the soccer simulation, a greater amount of data is geometric and spatial and better suited to a visual representation. A large portion of research efforts may also be spent on algorithms that are inherently spatial or temporal: path planning, localization, and obstacle avoidance are a few examples. Visualization is a natural choice for understanding and debugging these types of algorithms. The aforementioned *logfile player and analyzer* attempted to provide a visual overlay or drawing component for analyzing logfiles; however, the implementation was crude and limited to logged games. The following observations may be made concerning the requirements for effectively visualizing the general scenario of soccer-playing robots:

- Logging data and analyzing it post-simulation is useful, but it is better to possess real-time information. An online system allows researchers to influence and analyze behaviors interactively, which is essential in testing various cases.
- Agents are not omniscient, and their belief state is just as relevant as their actual state. The visualization must be capable of presenting both forms of information simultaneously and effectively.
- It is ideal to abstract the visualization in such a way that is suitable for multiple agent architectures. However, there should be no compromise in functionality to achieve this. The visualization should be easily programmable and flexible.
- Agent behaviors are reliant upon the state of the environment as much as the decision-making architecture. Viewing a single agent's status, while useful, may not be sufficient to understand why a particular action was performed. It is necessary to provide multiple perspectives of the simulation.

The above observations indicate that a suitable visualization solution runs in real-time with the Simspark simulation, is interactive, and flexible in its presentation of data. We also wanted a tool that has few dependencies, can be setup with minimal effort, and can be used on many operating systems.

4 Approach

While the Simspark monitor does not fulfill the requirements for a visualization tool, it is effective in illustrating the true world model of the simulation. Initially there was no intention of reproducing this functionality, and it was expected that a visualization tool could independently complement the monitor. The earliest prototype for RoboViz was a program detached entirely from the Simspark framework; it required a tight coupling with an agent architecture that supplied all the data needed for visualization. This was done with the belief that a debug tool should not have any dependency on the Simspark framework, which was likely to change over time. Since the agent code would need to be updated to conform to an updated Simspark simulation, this would centralize code modification to the agents.

There are serious drawbacks to the previously mentioned design that appeared during early prototyping. Unless the agent architecture has access to the world model of the simulation, there is no way to visualize both believed and actual world models simultaneously. Viewing the separate models side-by-side, using both the monitor and visualization tool concurrently, is ineffective in situations where there are discrepancies between what an agent believes and the truth. Furthermore, such a design requires much more effort on the part of a team hoping to utilize the visualization features with this interface. These are the reasons that necessitated a revised design.

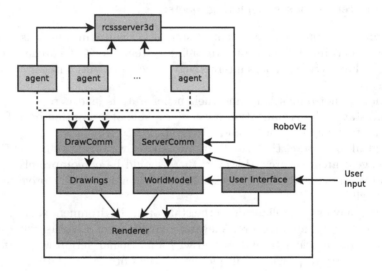

Fig. 1. Architecture for RoboViz, SimSpark server (rcssserver3d), agent, and user interaction

It was concluded that RoboViz would still need to communicate with agents to access their internal states, but it must also use the simulation server's scene graph to render the actual world model. With such a design, RoboViz essentially performs the roles of both the 3D monitor and a visualization tool (Fig. 1). This approach provided an opportunity to address many of the SimSpark monitor's deficiencies.

4.1   Visualization

To facilitate support for multiple agent architectures, it is better to assign the task of generating useful debug information to the agents themselves. The more traditional approach for visualization software is for agents to expose their state, and have the debug tool poll this information and render it appropriately. However, this approach tends to split development into two tasks: developing agent behaviors and then reconfiguring the debug tool to visualize it. This generates unnecessary work, and can easily introduce additional complexity in an environment where multiple individuals are working on different modules of the agent.

A preferred approach is to delegate the responsibility of producing useful visualizations to the agent. With this approach, the visualization tool acts as a passive receiver and simply renders what the agents would like displayed. Agents have the option of adding several simple shapes, such as line segments, spheres, and points, to the rendering. Each shape can be configured to have a different color, size, thickness, and is also tagged with a name so it can be identified. This makes for a very small and simple interface for agents to use, and more complicated visualizations can be achieved using these basic shapes. This system enables multiple programmers to visualize the algorithms they are working on without needing to reprogram the visualization tool or spend more time writing their own debug program for an individual problem.

One issue with allowing agents to be in control of what is rendered is that users should ultimately decide what they are seeing. Each agent may be sending localization information, strategy decisions, believed opponent locations, and so forth. Having all of this information presented at the same time can be overwhelming, and it can be useful to isolate drawings. This issue can be addressed by allowing the user to selectively filter content agents are providing. Filtering is done using regular expressions on drawing names or toggling individual drawings' visibility.

5   Implementation

RoboViz is a real-time 3D application that connects with the Simspark server using its network protocol and has a simple drawing interface to interface with agents . It relies entirely upon communication over a network (Fig. 1) to process information from live simulations. RoboViz possesses all of the existing functionality of the SimSpark monitor while providing the visualization described in section 4.1. Some of the other features that the program implements:

- *Graphics*: Advanced visual effects including variance shadow mapping [2] and bloom post-processing have been added to improve overall presentation quality. Stereoscopic 3D rendering is also supported on systems with the necessary hardware.
- *Controls*: Enhanced interactivity allows users to move the ball or agents with the mouse, and the user interface includes more intuitive controls.
- *Camera*: In addition to the standard user-controlled camera, there is an automated camera for tracking the ball as it moves around the field. Users may also view the scene from the perspective individual agents.
- *Logging*: RoboViz now takes over the role of generating logfiles, freeing valuable resources for the machine running simulation server. These logs are also

Fig. 2. The current user interface includes the same game state information as the 3D monitor; however, it also has an optional 2D aerial view, a panel for filtering drawings, an agent-perspective camera

    recorded at a greater framerate than the server currently generates logs, resulting in much higher quality logfiles.

– *Interface*: A 2D top-down view is provided as an ancillary means of keeping track of player locations, and a panel is provided to filter visualizations using check-boxes or a regular expression.

As mentioned in section 4, another goal is to achieve a program that is lightweight, configurable, and cross-platform. RoboViz is implemented entirely as a Java application to meet these needs. No external libraries or packages are used except for Java Bindings for OpenGL (JOGL) [7], which provides access to OpenGL rendering. This makes RoboViz especially easy to deploy on any platform, and the source code can be compiled on any Linux, Windows, or OS X system that has Java and recent graphics drivers.

5.1   Drawing Protocol

One of the primary functions of RoboViz is allowing other processes or clients to perform drawing of simple shapes inside of RoboViz. This functionality is referred to as *remote drawing*, and serves as the process by which visualization is achieved. To accomplish *remote drawing*, RoboViz has a simple network protocol for allowing clients to control custom drawing. The primary intention for this is to allow robots running on the SimSpark server to send information useful in debugging their internal state or behavior.

5.2   Commands

Clients interact with RoboViz by issuing commands. Agents submit packets in the format of commands that RoboViz can recognize. For drawing purposes, an example of a command is *draw line*, or *draw sphere*. Each command is formatted in a specific way that is suitable for it to be sent using UDP packets over a network (Fig. 3). We chose UDP primarily because we expect for RoboViz to be run on the same host or local area network (where packet loss is highly uncommon) as the agents and simulation. UDP also simplifies the connections between RoboViz and agents, which may crash unexpectedly.

Each packet may contain one or more commands, though the maximum size of a packet containing commands is currently 512 bytes; a drawing command is typically around 50 bytes, depending on the length of its name. Each command has a specific format (Fig. 3) so that its parameters can be identified unambiguously, even if it is grouped in the same packet with other commands. Commands should not be split between two or more packets. The most important command our visualization and debugging tool recognizes is the *draw shape* command. This command appends a primitive shape to a list of drawings in RoboViz and can be used by individual agents to visualize a particular algorithm in action: locations of opponents, believed world state information, and so forth.

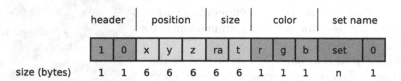

Fig. 3. Example of a *draw shape* command. The header section tells RoboViz the type of command it should parse; the first byte indicates a *draw command*, and the second indicates the shape is a line. For a line, the parameters include position (x,y,z), size (radius, thickness), color (r,g,b), and set name. All floats are stored as 6 bytes of text (ex. "3.1456"). Strings, such as the set name, are terminated with a 0 byte.

5.3   Shapes and Sets

To make remote drawing simple and flexible, all drawing is accomplished by working with a small set of shapes. More complex shapes can be constructed from these primitives, so we avoid introducing an overly verbose drawing interface by permitting only basic shapes. A shape's description has properties such as position, color, and scaling. Clients send commands to RoboViz that request it to add such shapes to its rendering; these commands are called *draw shape* commands.

Each shape is also part of a group of shapes called a *shape set*; these sets can contain one or more shapes. Every *shape set* is identified by a unique name, which is a string. When RoboViz receives *draw shape* commands from the client, it parses the individual shapes and adds it to the appropriate set. During rendering in RoboViz, each set is visited and its shapes are drawn.

Shapes are grouped into sets mainly so they can be filtered inside RoboViz. For example, each robot may have its own set of shapes so that the user of RoboViz can view only that agent's shapes and hide everything else. A robot may also have many sets for each type of behavior. How the sets are organized is entirely up to the client. However, it is recommended to use a hierarchical naming pattern. For instance, a shape set name such as "TeamName.1.Behaviors.PathPlanning" might be used to indicate all shapes belonging to the first agent on a team called "TeamName" that pertain to path planning behavior. With this naming pattern, all of this particular agent's sets can be referenced as "TeamName.1".

A unique class of shapes are *annotations*, which are text billboards displayed inside the 3D scene. There are two types of *annotations*: positional annotations are strings rendered at a specific location in the scene; agent annotations are attached to a specific agent in the simulation, and always rendered over the agent's head (Fig. 4).

5.4   Rendering Control

While *draw shape* commands request that RoboViz add shapes, they are not enough to sufficiently control the remote drawing process. Commands that influence the rendering process of shapes are called *draw option* commands. These commands are used for achieving animation and resolving concurrency issues.

Problems arise if a shape set is rendered at the same time RoboViz is receiving new *draw shape* commands for that set. To resolve this problem, each shape set has two buffers: a front and back buffer. When a client sends a *draw shape* command, RoboViz parses the shape and add it to the back buffer. These shapes will not be visible until the client sends a *swap buffer* command, at which point the back and front buffers swap roles. In other words, the front buffer always contains a set's shapes which may be rendered in RoboViz. The *swap buffer* command is synchronized inside of RoboViz, and will block while shapes are being rendered.

Fig. 4. An agent annotation that displays a numeric value overhead of a goal keeper. Also shown is an upright vector and localization information as a line and circle, respectively.

It is important to note that as soon as RoboViz executes a buffer swap, the shape set's new back buffer is cleared of all shapes. If multiple agents try to swap the buffers of the same shape set, flickering will occur in RoboViz. For this reason it is expected that agents will submit drawings relevant to their own state or behavior within their own shape sets, but this is not strictly required. A team may, for instance, designate a captain that sends drawings relevant to the team as a whole. We did not wish to impose a set of rules on how drawings should be assigned to RoboViz, so there is nothing to prevent agents from sending drawings identified by names other than their own.

5.5 Static and Animated Shapes

The type of shapes a client can add to RoboViz may be categorized into two types: static and animated. Static shapes are those that persist after being added and do not need to be refreshed or updated. For example, a grid on top of the field or 3D coordinate axes do not require the shapes to change over time. Animated shapes are those that need to have their properties updated; a vector representing a robot's forward direction is an example.

Inside *draw shape* commands, there is absolutely no distinction between static and animated shapes. These types merely refer to how a client program treats the *draw commands* for a shape set. Static shapes have the advantage of only needing to be transmitted a single time and only require a single swap buffer command. Animated shapes must have their values sent repeatedly; each time values are updated the shape set must also have its buffers swapped to display the changes.

6   Results and Future Work

While RoboViz is still being developed, it has been successfully received by the simulation 3D community. The tool has generated a great deal of interest from several teams, and has been suggested as a replacement for the current 3D monitor in the 2011 RoboCup in Istanbul, Turkey; RoboViz was used as the official monitor during the 2011 German Open in Magdeburg. We have not only seen the tool adopted by other teams, but also developers. For instance, the TinMan [11] framework has integrated support for the visualizations. Finally, RoboViz has been used at the University of Miami as a demonstration tool to promote awareness of the RoboCup events.

Applications: We have used the tool to iron out many bugs and optimize several of our agent behaviors. In particular, the tool has been of great use in visualizing our agents' localization routines, path planning, decision-making, and behavior modeling. We expect the usefulness of this tool to expand as we continue to refine existing features and develop our agents. This section provides only a few samples to illustrate how the RoboViz tool can be used to visualize various algorithms and routines.

Localization concerns an agent's knowledge of where it is positioned in a the environment. An agent will have a difficult time with formations, kicking the ball, or avoiding obstacles if it thinks it is in a location it is not. This is one of the most basic and obvious things to visualize, and we represent an agent's believed position as a yellow circle at its base that is drawn just atop the field (see Fig. 5). We also include each agent's perceived upright vector and a sphere where

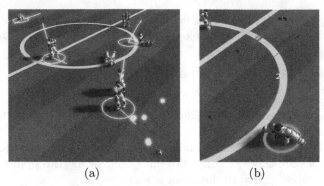

(a)   (b)

Fig. 5. Localization: In **(a)**, each agent's upright vector and position are represented by lines and circles, respectively. In **(b)**, individual particles from Monte Carlo localization are seen, scattered around, after an agent is moved unexpectedly. The speed with which these particles converge at the agent's base indicates how quickly it has determined its position.

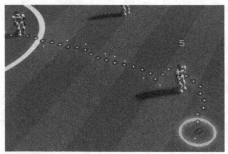 

(a) **Path planning:** A single agent's path plan is laid out as a series of blue circles; the white points indicate other considered routes. In this example, the agent anticipates a collision with an opponent player and schedules the path accordingly.

(b) **Decision making:** Potential tasks, such as field positioning, are illustrated by green circles on the field. Agents are matched with tasks, and their allocated assignments are visualized as connecting black line segments.

Fig. 6.

the agent thinks the ball is. This information can be quickly used to gauge the effectiveness and accuracy of the localization routines and explain unexpected behavior.

Another type of behavior that is well-suited to this type of visualization is path planning and obstacle avoidance. This can be directly visualized by the vertices on an agent's path (see Fig. 6(a)). Some other behaviors, such as allocating tasks to agents, is not as immediately obvious to visualize. We show a basic example of player positioning, where choices and assignments are drawn as connected nodes of a graph (Fig. 6(b)). However, a more sophisticated visualization might include weighting tasks by importance using color or size.

Future Work: There are a few areas where RoboViz is being improved. While a greater emphasis is placed on real-time visualization, analysis capabilities (such as game statistics) may be provided to provide users with more information. Furthermore, drawings are not logged and cannot be used while RoboViz is playing a logfile. Another area that we intend to improve is the user interface and viewing functionality. Currently, RoboViz retains a camera system that is similar to the SimSpark monitor to avoid a learning curve for users. A semi-automated camera mode is available that tracks the ball around the field; ultimately, a more intelligent and fully automated camera would be useful for presenting matches. Finally, the drawing panel that allows users to filter shapes needs to be integrated directly into the main window.

References

1. Boedecker, J., Dorer, K., Rollmann, M., Xu, Y., Xue, F.: SimSpark User's Manual (June 2008)
2. Donnelly, W., Lauritzen, A.: Variance shadow maps. In: Proceedings of the 2006 Symposium on Interactive 3D Graphics and Games (I3D), pp. 161–165. ACM Press (2006)
3. Arnold, A., Flentge, F., Schneider, C., Schwandtner, G., Uthmann, T., Wache, M.: Team Description Mainz Rolling Brains 2001. In: Birk, A., Coradeschi, S., Tadokoro, S. (eds.) RoboCup 2001. LNCS (LNAI), vol. 2377, pp. 531–534. Springer, Heidelberg (2002)
4. Klein, J., Spector, L.: 3D Multi-Agent Simulations in the breve Simulation Environment. In: Komosinski, M., Adamatzky, A. (eds.) Artificial Life Models in Software, pp. 79–106. Springer, London (2009)
5. Lattner, A.D., Rachuy, C., Stahlbock, A., Warden, T., Visser, U.: Virtual Werder 3D Team Documentation 2006. Tech. Rep. 36, TZI, Universitaet Bremen (August 2006)
6. Michel, O.: Webots: Professional Mobile Robot Simulation. Journal of Advanced Robotics Systems 1(1), 39–42 (2004)
7. Java Bindings for OpenGL (JOGL), http://www.jogamp.org
8. Planthaber, S., Visser, U.: Logfile Player and Analyzer for RoboCup 3D Simulation. In: Lakemeyer, G., Sklar, E., Sorrenti, D.G., Takahashi, T. (eds.) RoboCup 2006. LNCS (LNAI), vol. 4434, pp. 426–433. Springer, Heidelberg (2007)
9. Reis, L.P., Lau, N.: FC Portugal Team Description: RoboCup 2000 Simulation League Champion. In: Stone, P., Balch, T., Kraetzschmar, G.K. (eds.) RoboCup 2000. LNCS (LNAI), vol. 2019, pp. 29–40. Springer, Heidelberg (2001)
10. Shahri, A.H., Monfared, A.A., Elahi, M.: A Deeper Look at 3D Soccer Simulations. In: Visser, U., Ribeiro, F., Ohashi, T., Dellaert, F. (eds.) RoboCup 2007. LNCS (LNAI), vol. 5001, pp. 294–301. Springer, Heidelberg (2008)
11. TinMan: c-Sharp framework for 3D simulation league, http://code.google.com/p/tin-man/

A Generic Framework for Multi-robot Formation Control

Tiago P. Nascimento[1,*], André Gustavo S. Conceição[2],
Hugo P. Alves[1], Fernando A. Fontes[1], and António Paulo Moreira[1]

[1] FEUP - INESC Porto, University of Porto,
4200-465 Porto, Portugal
tiagopn@ieee.org,
{hugo.alves,faf,amoreira}@fe.up.pt
[2] Federal University of Bahia - Brazil,
40210-630 Salvador, Bahia, Brazil
andre.gustavo@ufba.br

Abstract. This paper describes a novel approach in formation control
for mobile robots. A Nonlinear Model Predictive Controller (NMPC) is
used to control the formation of a heterogeneous mobile robots group.
The desired formation is formed by an holonomic robot and a nonholonomic robot. The same nonlinear controller is used in both robots with
the same cost function. The details of the controller structure are presented in order to track a fixed target departing from different positions
in the field avoiding collisions with each other. A soccer robot competition field is used to present the simulations to evaluate the performance
of the controller.

Keywords: Formation Control, Nonlinear Model Predictive Controller,
Mobile Robots.

1 Introduction

An adaptive framework based in predictive control for creation and maintaining
of a mobile robot team formation was conceived as main objective of this work.
A formation is usually defined as the special arrangement of a set of agents of
the same type, where the relative positions of its elements are steady even if the
formation is moving. The used formation differs from the usual rigid formations
where the relative position of a team element must be precisely maintained.
Here, the ideal formations are the ones that maximize the team perception of
the environment or of an element that can be a leader robot or of a moving
target.

The three major approaches used in multi-robot formation are: Virtual Structure, Behavior-Based and Leader-Following. This last one being one of the most

---

* The authors thank the FCT (Fundação para Ciência e Tecnologia) from Portugal for
  supporting the project PTDC/EEA-CRO/100692/2008 - "Perception-Driven Coordinated Multi-Robot Motion Control".

T. Röfer et al. (Eds.): RoboCup 2011, LNCS 7416, pp. 294–305, 2012.
© Springer-Verlag Berlin Heidelberg 2012

studied in multi-robot formation [3], [7] and [4]. Nevertheless, it's important to mention that different techniques forming the Decentralized approach have always been sustained using artificial potential fields [1], constrained force [14] or path planner strategies [9]. Nevertheless, a good review of the three major approaches in formation control can be seen in [2]. In this paper, the leader-following approach will be used. The leader here will be the target (in this case the ball). Nevertheless, here the target is fixed as the objective is the convergence of the formation.

One of the most used controllers in the leader-follower approach is the Model Predictive Controller. It has been the target of study in multi-robot motion control in almost a decade [5]. In 2008, the first use of a MPC applied in the leader-following approach using holonomic robots was done by [8]. In these approaches only the formation maintenance is discussed. In these works the circle trajectory and the eight trajectory were used. The robots should follow the paths in a pre-set time while changing their formation (column or triangle). It's known that in a highly dynamic environment, if the trajectory is pre-defined, the linear MPC, even though applied to a non-linear system, can control the system maintaining the set-point. This was the exact result given by the authors.

In the following year, [6] applied a Linear Model Predictive Controller in the leader-following approach. They applied it in nonholonomic robots using the separation principle to make a NMPC control the trajectory while a MPC would be used to formation control. Both works only had simulation results and no obstacles were considered.

Finally, departing from the idea of using the separation principle done by [6], this paper presents a generic Nonlinear Model Predictive Controller (NMPC) framework to converge the formation using a holonomic and a nonholonomic robot around a tracking target.

2 Formation Control

The controller used in this work to formation control was a Non-linear Model Predictive Controller (NMPC). The general structure of this controller can be classified in three types: distributed, centralized, or hybrid. These categories are based on the way the control signals of each robot are calculated.

Here, the distributed architecture was chosen as can be seen in Fig. 1. In this case, each one of the robots calculates the total control inputs U_n solving its own optimization problem. This takes away the dependency from a central processing unit, guaranteeing the functioning of the formation even in cases of communication failure. Therefore, each robot must have information about the state X_n (position and speed) of each mate of its team. Also, in case of the communication failure or supervisor failure, the robot uses its predicted open-loop strategy to determine these informations, having, therefore, a tolerance degree to failure. Nevertheless, it has the disadvantage of putting a cost in computing the simulation of the entire formation progression, which is done by each one of the robots. However, this was not a problem, for the robots only calculate their own

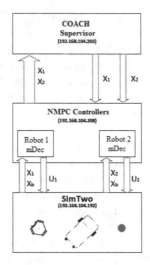

Fig. 1. Distributed Architecture of NMPC Controller

control inputs. As each robot solves its optimization problem in a decentralized architecture, the formation becomes difficult to stabilize.

In this multi-robot formation control case of study, a simulation software called *SimTwo* was used to simulate the formation [10]. In this simulation, the *SimTwo* has the job of another software called HAL (Hardware Abstraction Layer), which is an application that receives the sensor signals and communicates with the actuators, and then with the mDec (software of control of the real robots) by UDP protocol. In the real robots, the HAL sends to the robot's mDec the state of the other robots and the state of the ball. Then, each mDec sends to the *SimTwo* the control references of its robot. Each mDec also communicates with another central computer (the supervisor) that contains the Coach software, sending its own state and the state of the ball while observing it. Finally, the Coach sends to each mDec individually the state of the other robots in formation, in a way that each robot has the information of position and velocity of its mates. It can be noticed that this arrangement is similar the one used in real experiments, where the only difference is the replacement of the *SimTwo* for the HAL in each robot.

The capacity of the NMPC controller to create and maintain a formation comes from the fact that cost functions used by the controllers of each robot in the team formation are coupled. This coupling is done while the information about the position and speed of the other robots are used in the cost function of each robot to penalize the geometry or desired objective deviation. This turns the entire group formation stable where the actions of each robot affect the other mates. Fig. 2 exposes the structure of the used controller. This controller can be divided in three parts:

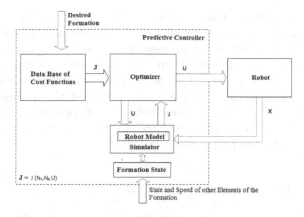

Fig. 2. Structure of the Formation Controller Projected

- **State of the Formation** - The controller contains structures to keep the formations state (position and speed of each other robot in the formation or of any target that should be followed), updating them in each control loop. These informations can be received by a supervisor or by other robots from the team, or even by the robot itself using its own resources;
- **Optimizer** - This part uses a numeric minimization method to optimize the cost function and obtain the signals of optimal control. Here it is used a method called Resilient Propagation (RPROP), which guaranties quick convergence;
- **Simulator** - This part does the simulation not only of the robot state evolution but also the state evolution of the other elements in the formation (other robots or targets). This element uses a dynamic simplified model to emulate the robot evolution. The speeds of the other robots or targets are assumed during the entire horizon of prediction as being constant and equals to the actual speed.

Fig. 3. The used holonomic robot (left) and nonholonomic robot (right)

The Resilient Propagation algorithm (RPROP) appeared in the learning algorithms category used in neural networks [11], being adapted to this application. This is an adaptive method where the step value is not proportional to the gradient function value to be minimized in a desired point (as it happens in the Steepest Descent algorithm), but it keeps adapting with the function behavior. Therefore, it becomes immune to the uncertainties of the derivative function value, depending only on the temporal behavior of its signal. This algorithm was tested initially with the values suggested by [11] and it revels to be capable to converge where the Steepest Descent failed.

3   Problem Formulation

The developed framework was applied to a formation with one holonomic mobile robot from the FEUP's 5DPO team and one nonholonomic mobile robot also from FEUP (Fig. 3). These robots can fulfill one main objective: the optimization of the target relative state perception (ball relative state perception) using the nonholonomic robot as observer while the holonomic robot places itself in an ideal position to receive the ball (receiver). The robots should converge to this formation departing from different positions and avoiding the collisions between themselves or with the target.

The mathematical definition of the system can be understood as having two robots and a ball (target). One of the robots is an omnidirectional robot composed by three holonomic wheels and the second robot is a differential robot with two normal wheels aligned in the center of the robot. Taking as base for this formation definition the elements presented in Fig. 4. The ball position and speed vectors in global coordinates are respectively:

$$X_{ball}(k) = \left[\, x_{ball}(k)\ y_{ball}(k)\,\right]^T,\tag{1}$$

$$v_{ball}(k) = \left[\, vx_{ball}(k)\ vy_{ball}(k)\,\right]^T.\tag{2}$$

It is considered also that the unit vector of the ball's velocity

$$\hat{v}_{ball}(k) = [\hat{vx}_{ball}(k), \hat{vy}_{ball}(k)]^T.\tag{3}$$

is such that:

$$\hat{v}_{ball}(k) = \frac{v_{ball}(k)}{\sqrt{vx_{ball}^2(k) + vy_{ball}^2(k)}}.\tag{4}$$

For each robot n, its state is represented by:

$$X_n(k) = \left[\, x_n(k)\ y_n(k)\ \theta_n(k)\,\right]^T,\tag{5}$$

$$V_n(k) = \left[\, vx_n(k)\ vy_n(k)\ w_n(k)\,\right]^T.\tag{6}$$

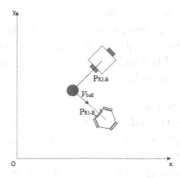

Fig. 4. The desired formation

Note that for the nonholonomic robot, $vy_n(k) = 0$. The position of the ball with respect to robot n is given by $P_{Rn-B}(k)=[x_{Rn-B}(k), y_{Rn-B}(k)]$, where:

$$P_{Rn-B}(k) = \left[(x_{ball}(k) - x_n(k)) \, (y_{ball}(k) - y_n(k)) \right]. \tag{7}$$

Then, it shall be defined the unit vector $\hat{P}_{Rn-B}(k)=[\hat{x}_{Rn-B}(k), \hat{y}_{Rn-B}(k)]$, which indicates the direction of the ball with respect to the robot, and its angle θ_{Rb-B}:

$$\hat{P}_{Rn-B}(k) = \frac{P_{Rn-B}(k)}{\sqrt{x_{Rn-B}^2(k) + y_{Rn-B}^2(k)}}, \tag{8}$$

$$\theta_{Rb-B}(k) = atan2(y_{Rn-B}(k), x_{Rn-B}(k)). \tag{9}$$

Finally, there is also the definition of the positions of each robot n with respect to its mates y, given by $P_{Rn-Ry}(k)=[x_{Rn-Ry}(k), y_{Rn-Ry}(k)]$, where:

$$P_{Rn-Ry}(k) = \left[(x_n(k) - x_y(k)) \, (y_n(k) - y_y(k)) \right]. \tag{10}$$

3.1   The Observer Robot

The estimation of the quality of the ball state is a function of its moving direction with respect to the robot and the distance in between. This estimation is done by using an omnidirectional vision system. Therefore, it's clear that in this case, the robot's direction is irrelevant. If the ball is in movement, the robot should be as indicated in Fig. 4. As in this case the ball is not moving, the robot's orientation becomes also irrelevant.

Nevertheless, big distances between the robot and the ball results in failure of the ball's detection. Consequently, this leads to the failure to estimate its velocity. When the distance is too small, it can occur that the robot cannot see the entire ball and, therefore, become incapable to detect correctly its position increasing the risk of undesired collisions. Take as an example the case in which the distance between the ball and the robot decreases with time in a straight line. In this case the robot only sees the ball increasing in size, making it difficult to estimate its velocity. In the ideal case the ball should move perpendicular to its position with respect to the robot. Therefore, the desired formation for the observer robots to be around the ball in a way to better estimate the ball velocity possesses the following characteristics:

- The observer robot puts itself by the side of the ball, maintaining a parallel velocity with respect to the ball, v_{ball}, with the same modulus. In this case, as the velocity of the target is zero, the position around the target is irrelevant;
- The robot position vector with respect to the ball, P_{Rx_B}, must be perpendicular to the ball's velocity vector, v_{ball}. Again, in this case this product is null for the target is fixed;
- The robot must maintain a distance $|P_{Rx_B}|$ from the ball;
- The robot must not collide between them or with the target.

Therefore, taking into account all the elements previously described, the weights given to each one of them, and a penalization term to the variation of control effort, the cost function that represents all this, embedded in the observer robot is as follows:

$$J(N_1, N_2, N_c) = \sum_{i=N_1}^{N_2} \lambda_1 (d_{setpoint} - |P_{Rn-B}(i)|)^2 +$$

$$\sum_{i=N_1}^{N_2} \lambda_2 (\hat{P}_{Rn-B}(i) \cdot \hat{v}_{ball}(i))^2 +$$

$$\sum_{i=N_1}^{N_2} \lambda_3 ((\frac{1}{-d_{min} + |P_{Rn-Rm_1}(i)|})^2 + \tag{11}$$

$$(\frac{1}{-d_{min} + |P_{Rn-Rm_2}(i)|})^2) +$$

$$\sum_{i=1}^{N_c} \lambda_4 (\Delta U(i))^2,$$

Where N_1, N_2 is the prediction horizon limits, in discrete time, so that $N_1 > 0$ e $N_2 \leq N_p$, where N_p is the desired prediction horizon. Also, N_c is the control horizon, λ_1, λ_2, λ_3, λ_4 are weights for each component of the cost function and $\Delta U(k) = [v_r(k) - v_r(k-1)] + [vn_r(k) - vn_r(k-1)] + [w_r(k) - w_r(k-1)]$ is the variation of the control signals, with $U(i)$ being the reference velocities vector with respect of the center of mass of the robot.

3.2   The Receiver Robot

The ideal position of the receiver robot with respect to the ball to have a good reception of it corresponds to the one in which the robot velocity vector is collinear with the ball velocity vector, with the same modulus. Also, the robot orientation should be such that the front of the robot is turn towards the ball. Therefore, the robot can then slowly decelerated and the distance between it and the ball can be decreased in a way to receive the ball in ideal conditions.

Summarizing it, the formation here should possess the following characteristics:

- The robot's velocity has to be equal in modulus and direction to the ball's velocity v_{ball};
- The robot's position vector with respect to the ball, P_{Rn-B}, must be collinear to the ball's velocity vector, v_{ball};
- The robot's orientation θ_n must be at all times equal to the vector P_{Rn-B}'s angle, defined by θ_{Rn-B}, in a way that the kicker of the robot is always turn towards the ball;
- The robot must be at a distance $|P_{R-B}|$ from the ball.

Finally, joining all the elements previously described, the weights given to each one of them, and a penalization term to the variation of control effort, the cost function that represents all this, embedded in the receiver robot is as follows:

$$
\begin{aligned}
J(N_1, N_2, N_c) = & \sum_{i=N_1}^{N_2} \lambda_1 (d_{setpoint} - |P_{Rn-B}(i)|)^2 + \\
& \sum_{i=N_1}^{N_2} \lambda_2 (-1)^2 + \\
& \sum_{i=N_1}^{N_2} \lambda_3 ((\frac{1}{-d_{min} + |P_{Rn-Rm_1}(i)|})^2 + \\
& (\frac{1}{-d_{min} + |P_{Rn-Rm_2}(i)|})^2) + \\
& \sum_{i=N_1}^{N_2} \lambda_4 (diffAngle(\theta_n, \theta_{Rn-B}))^2 + \\
& \sum_{i=1}^{N_c} \lambda_5 (\Delta U(i))^2,
\end{aligned}
\tag{12}
$$

Where N_1, N_2 is the prediction horizon limits, in discreet time, so that $N_1 > 0$ e $N_2 \leq N_p$, where N_p is the desired prediction horizon. Also, N_c is the control horizon, λ_1, λ_2, λ_3, λ_4, λ_5 are weights for each component of the cost function and $\Delta U(k) = [v_r(k) - v_r(k-1)] + [vn_r(k) - vn_r(k-1)] + [w_r(k) - w_r(k-1)]$ is the variation of the control signals, with $U(i)$ being the reference velocities vector with respect of the center of mass of the robot.

4   Results

Once the formation algorithm was implemented, some simulations were made to validate the proposed controller and to test its performance under different conditions.

There are many variables that influences the quality of the result. Among them are the weights (the λ_i) of each cost function and the optimizer parameters. The cost function values for both observer and receiver robots can be seen in table 1. In the minimization of the cost function values, only the relationship between the weights given to each element that meters to the final result. Therefore, the final values were a result of an iterative process where the weights are slightly different from one robot to another due to the existing physical differences. This process did not need to be very precise, due to the fact that there were a very large range of weights that could give similar results. Nevertheless, the NMPC controller parameters were $N_p = 10$, $N_c = 2$ and the used reference trajectory to find them was an gate signal extracted in a previous work done by [12].

Table 1. Weights for the Observers and Receiver

Weight	Observer Value	Receiver Value
λ_1	30	5
λ_2	10	10
λ_3	100	100
λ_4	10	5
λ_5	-	10

The initial parameters used on the RPROP optimization algorithm were the ones suggested by [11] ($\eta^+ = 1.5$, $\eta^- = 0.5$, $\Delta_0 = 0.1$) where the algorithm description can also be found. The fist tests resulted in a very satisfactory performance by the controller. After 20 interactions, some changes made on these values were tested ($\eta^+ = 1.2$, $\eta^- = 0.8$, $\Delta_0 = 0.05$) and produced visible improvements.

Therefore, the following simulation results made with the formation control framework evaluate the proposed controller. The simulations for formation convergence are shown to evaluate the formation controller in the following subsection.

4.1   Formation Convergence Results

The following results show the trajectories followed for each one of the robots when, starting from different positions, converge to a preset formation. The target in these cases is stationary during the simulation making the internal product of any vector with the ball velocity vector equals to zero. It is important to notice that the robot number 2 is the observer robot (the nonholonomic AGV) which means that it can be placed in any position exactly 2m far from the ball, while the robot number 1 is the receiver robot (the holonomic 5DPO) which means that it must be placed in front of th ball. The desired distance between both robot to the ball was defined to be 2 m.

Simulation 1. In this simulation the robots start at positions perfectly opposites and far from the ball (Robot 1 in position (-7,0) and Robot 2 in position (7,0)). For having less risk of collision or probability of the robots to interfere with each other, this became the simplest case. The results can be seen in Fig. 5. From this simulation on it can be noticed some interaction between the robots. The robot 1 goes to the front of the ball while the robot 2 places itself anywhere at the desired distance. As the robot 2 arrives first and places itself in front of the target, the robot 1 stops near the robot 2 avoiding the collision and placing itself as close as possible to the desired position (front of the ball) at the desired distance. The convergence is made in 15 seconds due to the attempt of robot number 1 to position itself in front of the target.

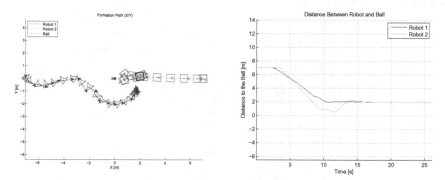

Fig. 5. Convergence into formation, simulation 1

Simulation 2. Here all robots start from the same side of the ball (Robot 1 in position (-7,4) and Robot 2 in position (-7,-4)), thought separated by a distance of 8 m. The results can be seen in Fig. 6. The robots converge perfectly to their positions in formation, making simple trajectories towards the target. As can be seen in the plot of the distance with respect to the time, it can be estimate that the robots have converged to the desired formation in approximately ten seconds.

Simulation 3. The third simulation shows a more complex situation, where the robots start from the same alignment with respect to the ball (Robot 1 in position (-7,0) and Robot 2 in position (-5,0)). The results can be seen in Fig. 7. It is important to notice that the robots cross it other paths but do not collide, for the instant of time is different when passing though that specific

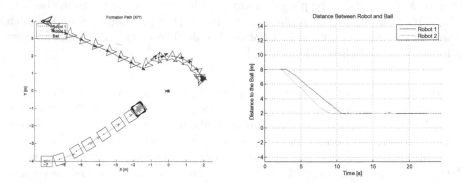

Fig. 6. Convergence into formation, simulation 2

position. The robot 2, even being closest to the ball, had to do much more turns to position itself in the desired position and orientation due to its nonholonomic constrains. While the robot number 1 goes smoothly to its position. This process takes about 22 seconds.

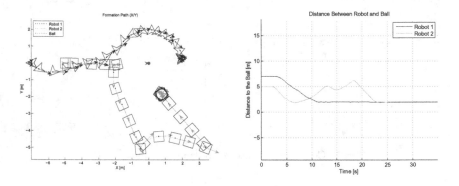

Fig. 7. Convergence into formation, simulation 3

5   Conclusions

In this paper a novel approach of a Non-linear Model Predictive Controller was presented used for multi-robot formation control. The developed framework showed to be very flexible and easily adaptable being used in holonomic and nonholonomic robots. The projected controller is capable of making a team of different robots to converge to a desired position around the target, even if the robots are very far apart. This framework could be also applied for both types of robots due to the fact that the effort of control does not take into account the tensions on the motors or the actuator input signals, but the velocities of center of mass. The results showed that the robots converged to the desired distance from the target (in this case 2m).

References

1. Chen, F., Chen, Z., Liu, Z., Xiang, L., Yuan, Z.: Decentralized formation control of mobile agents: A unified framework. Physica A: Statistical Mechanics and its Applications 387, 4917–4926 (2008)
2. Chen, Y., Wang, Z.: Formation Control: A Review and A New Consideration. In: 2005 IEEE/RSJ International Conference on Intelligent Robots and Systems, pp. 3181–3186 (August 2005)
3. Daigle, M.J., Koutsoukos, X.D., Biswas, G.: Distributed Diagnosis in Formations of Mobile Robots. IEEE Transactions on Robotics 23, 353–369 (2007)
4. Ding, Y., He, Y.: Flexible Leadership in Obstacle Environment. In: International Conference on Intelligent Control and Information Processing, pp. 788–791. IEEE Press, New York (2010)
5. Dunbar, W.B., Murray, R.M.: Model predictive control of coordinated multi-vehicle formations. In: Proceedings of the 41st IEEE Conference on Decision and Control, pp. 4631–4636. IEEE Press, New York (2002)
6. Fontes, F.A.C.C., Fontes, D.B.M.M., Caldeira, A.C.D.: Model Predictive Control of Vehicle Formations. In: Hirsch, M.J., Commander, C.W., Pardalos, P.M., Murphey, R. (eds.) Optimization & Cooperative Control Strategies. LNCIS, vol. 381, pp. 371–384. Springer, Heidelberg (2009)
7. Gu, D.: A Differential Game Approach to Formation Control. IEEE Transactions on Control Systems Technology 16, 1988–1993 (2008)
8. Kanjanawanishkul, K., Zell, A.: A model-predictive approach to formation control of omnidirectional mobile robots. In: 2008 IEEE/RSJ International Conference on Intelligent Robots and Systems, vol. 2, pp. 2771–2776. IEEE Press, New York (2008)
9. Kloder, S., Hutchinson, S.: Path planning for permutation-invariant multi-robot formations. IEEE Transactions on Robotics 22, 650–665 (2006)
10. Costa, P.: Paco Wiki - SimTwo,
http://paginas.fe.up.pt/~paco/wiki/index.php?n=Main.SimTwo
11. Riedmiller, M., Braun, H.: A direct adaptive method for faster backpropagation learning: the rprop algorithm. In: IEEE International Conference on Neural Networks, pp. 586–591. IEEE Press, New York (1993)
12. Ferreira, J.R.A.: Controlo Coordenado de Equipas de Robots Móveis. Master Thesis. University of Porto, Porto, Portugal (2010)
13. Xie, F., Fierro, R.: On Motion Coordination of Multiple Vehicles with Nonholonomic Constraints. In: 2007 American Control Conference, pp. 1888–1893. IEEE Press, New York (2007)
14. Zou, Y., Pagilla, P.R., Ratliff, R.T.: Distributed Formation Control of Multiple Aircraft Using Constraint Forces. In: 2008 American Control Conference, Seattle, pp. 644–649 (2008)

Real-Time Plane Segmentation
Using RGB-D Cameras*

Dirk Holz[1], Stefan Holzer[2], Radu Bogdan Rusu[3], and Sven Behnke[1]

[1] Autonomous Intelligent Systems Group, University of Bonn, Germany
holz@ais.uni-bonn.de, behnke@cs.uni-bonn.de
[2] Department of Computer Science,
Technical University of Munich (TUM), Germany
holzers@in.tum.de
[3] Willow Garage, Inc., Menlo Park, CA, USA
rusu@willowgarage.com

Abstract. Real-time 3D perception of the surrounding environment is a
crucial precondition for the reliable and safe application of mobile service
robots in domestic environments. Using a RGB-D camera, we present a
system for acquiring and processing 3D (semantic) information at frame
rates of up to 30Hz that allows a mobile robot to reliably detect obstacles
and segment graspable objects and supporting surfaces as well as the
overall scene geometry. Using integral images, we compute local surface
normals. The points are then clustered, segmented, and classified in both
normal space and spherical coordinates. The system is tested in different
setups in a real household environment.

The results show that the system is capable of reliably detecting obstacles at high frame rates, even in case of obstacles that move fast or do
not considerably stick out of the ground. The segmentation of all planes
in the 3D data even allows for correcting characteristic measurement
errors and for reconstructing the original scene geometry in far ranges.

1 Introduction

Perceiving the geometry of environmental structures surrounding the robot is a
crucial prerequisite for the autonomous operation of service robots in human living environments. These environments tend to be cluttered and highly dynamic.
The first property requires for three-dimensional information about the environmental structures and objects contained therein, whereas the latter necessitates
real-time processing of the acquired spatial information.

Recent RGB-D cameras, such as the Microsoft Kinect camera, acquire both
visual information (RGB) like regular camera systems, as well as depth information (D) at high frame rates. In terms of measurement accuracy in low ranges
(up to a few meters), the acquired depth information does not rank behind the
accuracy achieved with 3D laser range scanners. With respect to the frame rate,

---

* This research has been partially funded by the FP7 ICT-2007.2.1 project ECHORD
(grant agreement 231143) experiment ActReMa.

T. Röfer et al. (Eds.): RoboCup 2011, LNCS 7416, pp. 306–317, 2012.

RGB-D cameras clearly outperform commonly used 3D laser range scanners, e.g., 30Hz vs. 1Hz. However, for applying and making use of these cameras in typical mobile manipulation problems like detecting objects and collision avoidance, the acquired RGB-D camera data needs to be processed in real-time (possibly with limited computing power). We present a system of algorithms for processing 3D point clouds in real-time which explicitly makes use of the organized structure of RGB-D camera data. It allows for

1. reliably detecting obstacles,
2. detecting graspable objects as well as the planes supporting them, and
3. segmenting and classifying all planes in the acquired 3D data.

An important characteristic of the proposed system is that the all of the above outcomes can be obtained at frame rates of up to 30Hz. In addition to the organized data structure, we exploit a specific characteristic of man-made environments, namely being primarily composed of connected planes like walls, ground floors, ceilings, tables etc. In fact, the work presented in this paper is solely based on a fast segmentation of all planes in 3D point clouds.

The remainder of this paper is organized as follows: After giving an overview on related work in Section 2, we discuss methods for computing local surface normals and present a fast variant using integral images in Section 3. Clustering these normals and segmenting all planes in acquired point clouds is described in Section 4. Detecting graspable objects and obstacles based on this information is discussed in Section 5. Section 6 summarizes experimental results.

2   Related Work

Autonomous robot operation in complex real-world environments requires for powerful perception capabilities. 3D semantic perception has seen considerable progress recently. Here, we focus on two aspects – extracting semantic information from 3D data and using 3D data for obstacle detection and collision avoidance.

Safety-critical tasks like collision avoidance necessitate a fast perception of obstacles and processing acquired sensor data in real-time. A common way of using 3D data for collision avoidance in the navigation domain is to extract relevant information from the 3D input and to project it down to *virtual* 2D laser range scans for which navigation and collision avoidance are well studied topics. Wulf *et al.* use these virtual scans for both collision avoidance and localization in ceiling structures [12]. Holz *et al.* distinguish two types of virtual scans, virtual structure and obstacle maps. The first type models environmental structures such as walls in a virtual 2D laser range scan, the latter information about closest obstacles [3]. Yuan *et al.* follow this approach in [13] to fuse sensor information from a Time-of-Flight (ToF) camera with that of a 2D laser range scanner to compensate for the smaller field of view of ToF cameras. Droeschel *et al.* only use a ToF camera for obstacle detection, but mount this camera on a pan-tilt-unit and use an active gaze control to keep relevant regions in sight [1]. Measured points sticking out of the ground are considered as obstacles. This information is then fused with other 2D laser range scanners on the robot, again, by projecting them into a virtual 2D laser scan. Problems arise in regions where the ToF data is highly affected by noise and in case of motion blur. In both cases, floor points may be considered as obstacles. Furthermore, obstacles whose size lies below the measurement accuracy and the used thresholds to compensate for noisy data cannot be detected. In contrast, we consider as obstacles both points sticking out of segmented planes as well as points with different local surface orientations. This allows for detecting even small objects reliably as obstacles.

Nüchter *et al.* extract environmental structures such as walls, ceilings and drivable surfaces from 3D laser range scans and use trained classifiers to detect objects like, for instance, humans and other robots [8]. They examine range differences in consecutive laser scan points to get an approximate estimate if a point is lying on a horizontal or on a vertical surface. Triebel *et al.* use Conditional Random Fields to segment and discover repetitive objects in 3D laser range scans [11]. Rusu *et al.* extract hybrid representations of objects consisting of detected shapes, as well as surface reconstructions where no shapes have been detected [9]. Endres *et al.* extract objects and point clusters from the background structure in range scans (assuming objects being spatially disconnected) [2]. Using latent Dirichlet allocation, they can derive classes of objects from these clusters that are similar in shape. Lai and Fox use data from Google's 3D Warehouse to classify clusters of points in 3D laser range scans [5]. They detect walls, trees and cars in street scenes. Pathak *et al.* decompose acquired 3D data into plane segments and use this information for registering point clouds. Steder *et al.* compute range images from point clouds, extract borders and key-points and use 3D feature descriptors to find and match repetitive structures [10]. The above approaches show good results when processing accurate 3D laser range data, but tend to have high runtime requirements. Here we focus on less complex but fast methods to obtain an initial segmentation of the environment in real-time that can be used in a variety of applications.

3   Fast Computation of Local Surface Normals

Local geometric features such as surface normal or curvature at a point form a fundamental basis for extracting semantic information from 3D sensor data. A common way for determining the normal to a point p_i on a surface is to approximate the problem by fitting a plane to the point's local neighborhood $\mathcal{P}_i$. This neighborhood is formed either by the k nearest neighbors of p_i or by all points within a radius r from p_i. Given the $\mathcal{P}_i$, the local surface normal n_i can be estimated by analyzing the eigenvectors of the covariance matrix $C_i \in \mathbb{R}^{3 \times 3}$ of $\mathcal{P}_i$. The eigenvector $v_{i,0}$ corresponding to the smallest eigenvalue $\lambda_{i,0}$ can be used as an estimate of n_i. The ratio between $\lambda_{i,0}$ and the sum of eigenvalues provides an estimate of the local curvature.

Both k and r highly influence how well the estimated normal represents the local surface at p_i. Chosen too large, environmental structures are considerably smoothened so that local extrema such as corners completely vanish. If the neighborhood is too small, the estimated normals are highly affected by the depth measurement noise. A common way to compensate these effects that is also used in [9] is to compute the distances of all points in $\mathcal{P}_i$ to the local plane through $\boldsymbol{p}_i$. These distances are then used in a second run to weight the points in $\mathcal{P}_i$ in the covariance computation. By this means, corners and edges are less smoothened and the estimated normals better approximate the local surface structure. However, with or without this second run, estimating the point's local neighborhood is computationally expensive, even when using approximate search in kD-trees which is $O(n \log n)$ for n randomly distributed data points (plus the construction of the tree). Another possibility to compensate for the aforementioned effects is to compute the normals in different neighborhood ranges or different scales of the input data and to select the most likely surface normal for each point.

Less accurate, but considerably faster is to consider pixel neighborhoods instead of spatial neighborhoods [4]. That is, the organized structure of the point cloud as acquired by Time-of-Flight or RGB-D cameras is used instead of searching through the 3D space spanned by the points in the cloud. Compared to a fixed radius r or a fixed number of neighbors k, using a fixed pixel neighborhood has the advantage of having smaller neighborhoods in close range (causing more accurate normals) and larger neighborhoods smoothing the data in far ranges that is more affected by noise and other error sources (see, e.g., [6] for an overview on Time-of-Flight camera error sources).

By using a fixed pixel neighborhood and, in addition, neglecting pre-computed neighbors outside of some maximum range r as in [4], one can avoid the computationally expensive neighbor search, but still needs to compute and analyze the local covariance matrix. Here, we use an approach that directly computes the normal vector over the neighboring pixels in x and y image space.

The basic principle of our approach is to compute two vectors which are tangential to the local surface at the point $\boldsymbol{p}_i$. From these two tangential vectors we can easily compute the normal using the cross product. The simplest approach for computing the normals is to compute them between the left and right neighboring pixel and between the upper and lower neighboring pixel, as illustrated in Figure 1.a. However, since we expect noisy data and regions in which no depth information is available (a special characteristic of the used cameras), the resulting normals would also be highly affected. For this reason, we apply a smoothing on the tangential vectors by computing the average vectors within a certain neighborhood. To perform this smoothing efficiently we use integral images. We first create two maps of tangential vectors, one for the x- and one for the y-direction (again in image space). The vectors for these maps are computed between corresponding 3D points in the point cloud. That is, each element of theses vectors is a 3D vector. For each of the channels (Cartesian x, y, and z) of each of the maps we then compute an integral image, which leads to a total number of six integral images. Using these integral images, we can compute the average tangential vectors with only $2 \times 4 \times 3$ memory accesses, independent of the size of the smoothing area. The overall runtime complexity is linear in the number of points for which normals are computed (cf. Figure 6.a).

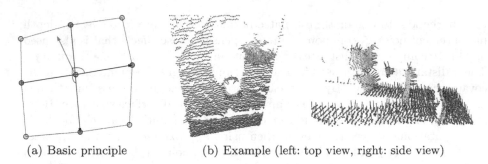

(a) Basic principle   (b) Example (left: top view, right: side view)

Fig. 1. Principle of fast normal computation using integral images (a). Two vectors tangential to the surface at the desired position are computed using the red points. The local surface normal is computed by applying the cross product to them. A typical result of an acquired point cloud with surface normals is shown in (b).

The computation of normals is conducted in the local image coordinate frame ($\hat{Z}$-axis pointing forwards in measurement direction). For further processing, we transform both Cartesian coordinates of the points as well as the local surface normals into the base coordinate frame of the robot (right-handed coordinate frame with the $\hat{X}$-axis pointing in measurement and driving direction, the $\hat{Z}$-axis point upwards representing the height of points). In case this transformation is not known, we only apply the corresponding reflection matrix and a translation by 2cm along the $\hat{Y}$-axis that accounts for the difference in position between the regular camera and the infrared camera that senses the emitted pattern for depth reconstruction. It should be noted that this transformation (or the knowledge of the camera's position and orientation in space) is not necessary for the fast plane segmentation in Section 4, but only for task-specific applications like the extraction of horizontal surfaces.

In addition to the fast normal estimation, we compute spherical coordinates (r, ϕ, θ) of the local surface normals that ease the classification of measured points and the processing steps presented in the following. We define ϕ as the angle between the local surface normal (projected onto the $\hat{X}\hat{Y}$-plane) and the $\hat{X}$-axis, r the distance to the origin in normal space, and θ the angle between the normal and the $\hat{X}\hat{Y}$-plane. (r, ϕ) is of special interest in obstacle detection, as it represents direction and distance of an obstacle to the robot (in the $\hat{X}\hat{Y}$-plane). For plane segmentation, r is abused in our implementation to hold the plane's distance from the origin (in Cartesian space).

4   Fast Plane Segmentation

Man-made environments tend to be primarily composed of planes. Detected and segmented planes already adequately model the surface of most environmental structures. We segment local surface normals in two steps: 1) we cluster (and merge) the points in normal space $(n^x, n^y, n^z)^T$ to obtain clusters of plane candidates and 2) cluster (and merge) planes of similar local surface normal orientation in distance space (distance between plane and origin).

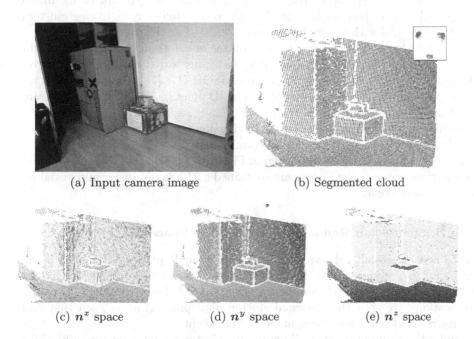

(a) Input camera image   (b) Segmented cloud

(c) n^x space   (d) n^y space   (e) n^z space

Fig. 2. Typical result of the first segmentation step: points with similar surface normal orientation in the input data (a) are merged into clusters (b, shown in both Cartesian and normal space). The components of the normals (n^x, n^y, and n^z) are visualized in (c-e) using a color coding from -1 (red) to 1 (blue). This simple clustering already allows for a fast segmentation of planes with similar surface normal orientation, e.g., extracting all horizontal planes.

4.1   Initial Segmentation in Normal Space

For the initial clustering step in which we want to find clusters of points with similar local surface normal orientations, we construct a voxel grid either in normal space or using the spherical coordinates. Using the spherical coordinates allows for clustering in the two-dimensional (ϕ, θ)-space, but requires a larger neighborhood in the subsequent processing step in which we merge clusters. Both results and processing time do not differ.

For clustering in normal space, we compute a three-dimensional voxel grid and map local surface normals to the corresponding grid cell w.r.t. the cell's size. Points for which the surface normals fall into the same cell, form the initial cluster and potential set of planes with the same normal orientation. Either all non-empty cells or only those with a minimum number of points are considered as initial clusters.

In order to compensate for the involved discretization effects, we examine the cell's neighbors in the three-dimensional grid structure. If the average surface normal orientation in two neighboring grid cells falls below the cluster size (and the desired accuracy), the corresponding clusters are merged. For being able to merge multiple clusters, we keep track of the conducted merges. In case cluster a should be merged with cluster b that was already merged with cluster c, we check if we can merge a and c, or if $a + b$ is a better merge than $a + c$. Although this procedure is less adaptive (and complex) as sophisticated clustering algorithms like k-*Means*, *mean-shift*-clustering or, e.g., ISODATA [7], this simple approach allows for reliably detecting larger planes in 3D point clouds at high frame rates. In all modes and resolutions of the camera, plane segmentation is only a matter of milliseconds. In contrast to region growing algorithms, we find a singe cluster for planes that are not geometrically connected, e.g., parts of the same wall.

An example segmentation is shown in Figure 2. Planes with similar (or equal) local surface normal orientations are contained in the same cluster and visualized with the same color.

4.2 Segmentation Refinement in Distance Space

Up to now the found clusters do not represent single planes but sets of planes with similar or equal surface normal orientation. For some applications like extracting all horizontal surfaces, this information can directly be used. For other applications, we split these normal clusters into plane clusters such that each cluster resembles a single plane in the environment.

Under the assumption that all points in a cluster are lying on the same plane, we use the corresponding averaged and normalized surface normal to compute the distance from the origin to the plane through the point under consideration. Naturally these distances differ for points on different parallel planes and we can split clusters in distance space. For compensating the fact that measurements farther away from the sensor are stronger affected by the different error and noise sources, we compute a logarithmic histogram. Again, points whose distances fall into the same bin form initial clusters. These clusters are then refined by examining the neighboring bins just like in the refinement of the normal segmentation. An example of the resulting plane clusters is shown in Figure 3.

5 Applications

The planes segmented at frame rates of up to 30Hz are useful for a variety of applications. Here, we use the plane segments as well as the computed surface normals and spherical coordinates for detecting obstacles and graspable objects in table top scenes and on the ground floor. In addition, we can compensate for camera-specific noise and error sources by projecting all points onto the planes they belong to.

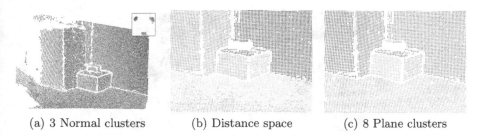

(a) 3 Normal clusters   (b) Distance space   (c) 8 Plane clusters

Fig. 3. Typical result of the second segmentation step: Clusters with similar surface normal orientations (a) are clustered in distance space (b). Clusters with similar normal orientations but varying distances (of the respective planes to the origin) are split. For compensating discretization effects, neighboring clusters are again merged to form the segmented planes (c). Color coding in (a+c) is random per cluster, and in distance space from 0.4m (red) to 1.3m (blue) in (b).

5.1 Obstacle and Object Detection

For detecting obstacles and graspable objects, we first extract all horizontal plane segments, i.e., those with $n^x \approx 1$ (and $\theta \approx +\frac{pi}{2}$ respectively). That is, we exploit the fact that both the robot as well as objects in its environments are standing on (or *supported* by horizontal surfaces). Hence, we need a rough estimate of the camera orientation in order to determine which planes are horizontal. Furthermore, depending on the robot's task, e.g., navigation or manipulation of objects on a table, we limit the height in which we search for horizontal planes.

For navigation purposes, only the ground floor plane ($n^z \approx 1$, $z \approx 0$) is considered safe. All other points and planes including other horizontal surfaces such as tables are considered as obstacles. For object detection, we limit the search space by the height range in which the robot can manipulate. Since our robots are equipped with a trunk that can be lifted and twisted [reference removed], we use a range of $0m - 1.2m$.

We follow a similar approach as in [9] and [4] for detecting objects. The already found plane model (consisting of the averaged normals and plane-origin distances in a plane cluster) is optimized using a RANSAC approach that also sorts out residual outliers. We then project all cluster points onto the plane and compute the convex hull. These steps are repeated for all horizontal planes that have been found in the given height range. For all points from non-horizontal plane clusters, we then check if they lie above a supporting plane (within a range of e.g. 30cm) and within the corresponding convex hull (again with a tolerance of a few centimeters). Points meeting both requirements are then clustered to obtain object candidates. For each of the candidates we compute the centroid and the oriented bounding box in order to distinguish graspable from non-graspable objects. Here we simply assume that the minimum side length of graspable objects needs to lie between 1 and 10cm. Furthermore, we neglect clusters where the number of contained points falls below a threshold (e.g. 50 points). Figure 4 shows a typical result of detecting graspable objects and obstacles in a table top setting.

(a) Example table scene

(b) Detected obstacles   (c) Detected objects

Fig. 4. Typical result of detecting obstacles and objects in a table top setting (a). Even obstacles (b, red) that, in 3D, do not stick out of the supporting surface (b, green) like the red lighter are perceived. Detected objects (c) are randomly colored. For being able to grasp an object, the respective cluster is not considered as an obstacle (in this example the Pringles box).

5.2   Correcting Local Surfaces at Detected Planes

RGB-D cameras suffer from different noise and error sources, especially discretization effects in depth measurements and the fact that the cameras are calibrated for a certain range. Both effects cause considerable measurement errors in far ranges (e.g. >3.5m). Especially for modeling the geometry of environmental structures where planes are of special interest, these measurements hinder from finding accurate surface models. However, with the extracted plane clusters, we can project the contained points onto the corresponding plane to get, at least, an approximate estimate of the true surface geometry. Figure 5 shows a typical result of this naïve measurement correction which considerably increases the quality of acquired geometric information. In fact, the angle between the two walls in Figure 5.b only deviates by 2° from ground truth. However, it is a matter of future work to take this information into account in a precise and adaptive sensor model.

6   Experiments

Both the detection of graspable objects and obstacles as well as the naïve correction using plane segments highly depend on the quality of the estimated surface

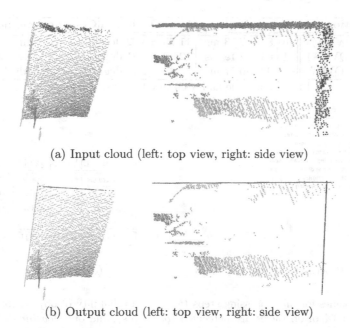

(a) Input cloud (left: top view, right: side view)

(b) Output cloud (left: top view, right: side view)

Fig. 5. Typical result of correcting local surface geometry. Shown are the segmented input cloud (a) and the corrected cloud (b). The views are rotated to be aligned with the plane tangents of a wall and the ceiling. The distance to the wall (magenta) is approximately 4m.

normals. In order to evaluate the accuracy of the estimated normals, we conducted a first sequence of experiments comparing the estimated surface normal at each point with the one computed over the real neighbors using the two-run Principal Component Analysis as described above. For the neighbor search a radius r has been chosen that linearly depends on the measured range to the point under consideration, i.e., a smaller radius for the more accurate close range measurements and a larger radius for measurements being farther apart from the sensor. This radius function has been manually adapted for each of the point clouds used in the experiments in order to guarantee correct (and ground truth-like) normals.

The presented results have been measured over 40 points clouds taken in 4 different scenes in a real house-hold environment: 1) a table top scene with only one object, 2) a cluttered table top scene with >50 objects, 3) a room with a cluttered table top and distant walls, 4) a longer corridor with several cabinets where measurement of up to 6m have been taken. In average, a deviation of roughly 10° has been measured (see Figure 6(a)). This is primarily caused by the fact that the current implementation does not specifically handle edges and corners as is done with the second PCA run on the weighted covariance. Furthermore, using nearest neighbor search better compensates for missing measurements in regions where no depth information is available. However, especially in close range (e.g. up to 2m), the estimated normals are quite accurate and do not deviate from

Resolution	Processing time	Avg. Deviation	Considerable deviations
640 × 480 (VGA)	61.84 ± 7.73 ms	11.75 ± 3.02 deg	roughly 2%
320 × 240 (QVGA)	15.08 ± 2.19 ms	12.39 ± 3.81 deg	roughly 1.2%
160 × 120 (QQVGA)	4.32 ± 0.57 ms	9.32 ± 2.44 deg	roughly 1%

(a) Processing time and accuracy for normal estimation

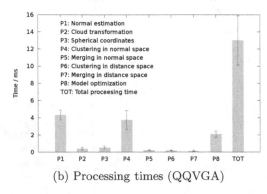

Resolution	VGA	QVGA	QQVGA
obstacles	100%	100%	100%
!obstacles	0.1%	0.15%	0.19%
objects	92%	93%	95%
!objects	0%	0%	0%
Rate	≈7Hz	≈27Hz	**≈30Hz**

(b) Processing times (QQVGA)   (c) Detection rates

Fig. 6. Run-times for all processing steps (a+b) and reliability of object/obstacle detection (c). 100% of obstacles are perceived (*obstacles*), and only !*obstacles* % of the measurements have been incorrectly classified as obstacles. Roughly 93% (*objects*) of the objects have been correctly detected, and only !*objects* % points have been segmented as belonging to a non-existent object. "Rate" is the frequency with which the results are provided to other components in the robot control architecture.

the *true* local surface normals. Only one to two percent of the estimated normals considerably deviated from the true normals (deviations larger than 25°).

In all experiments, processing times have been measured on a Core i7 machine and over several minutes, i.e., several thousand point clouds. No parallel computation has been carried out and all algorithms were run sequentially within a single thread on a single core. That is, processing times should not considerably deviate on (newer) notebook computers. Figure 6.b summarizes all results. Obstacles were detected always (100%) and 93% of the objects have been correctly segmented. The object segmentation gets inaccurate if objects are 1) very small (dimensions falling below the aforementioned thresholds), or 2) being more than 3.5m away from the sensor (where depth measurements are highly inaccurate).

7   Conclusion

We have presented an approach to real-time 3D point cloud processing that segments planes in the space of local surface normals. The detected planes have been used for detecting graspable objects and obstacles as well as for correcting the measured 3D information. With frame rates of up to 30Hz we can reliably detect obstacles in the robot's vicinity as well as objects for manipulation tasks. However, it remains a matter of future work to find reliable and fast approaches

to more complex tasks like autonomous registration of point clouds or recognizing detected objects that make use of the acquired surface information.
Data sets and videos are available at: http://purl.org/holz/segmentation

References

1. Droeschel, D., Holz, D., Stückler, J., Behnke, S.: Using Time-of-Flight Cameras with Active Gaze Control for 3D Collision Avoidance. In: Proc. of the IEEE International Conference on Robotics and Automation, ICRA, pp. 4035–4040 (2010)
2. Endres, F., Plagemann, C., Stachniss, C., Burgard, W.: Unsupervised discovery of object classes from range data using latent dirichlet allocation. In: Proc. of Robotics: Science and Systems (2009)
3. Holz, D., Lörken, C., Surmann, H.: Continuous 3D Sensing for Navigation and SLAM in Cluttered and Dynamic Environments. In: Proc. of the International Conference on Information Fusion (FUSION), pp. 1469–1475 (2008)
4. Holz, D., Schnabel, R., Droeschel, D., Stückler, J., Behnke, S.: Towards Semantic Scene Analysis with Time-of-Flight Cameras. In: Ruiz-del-Solar, J., Chown, E., Ploeger, P.G. (eds.) RoboCup 2010. LNCS (LNAI), vol. 6556, pp. 121–132. Springer, Heidelberg (2010)
5. Lai, K., Fox, D.: 3D laser scan classification using web data and domain adaptation. In: Proc. of Robotics: Science and Systems (2009)
6. May, S., Droeschel, D., Holz, D., Fuchs, S., Malis, E., Nüchter, A., Hertzberg, J.: Three-dimensional mapping with time-of-flight cameras. Journal of Field Robotics, Special Issue on Three-Dimensional Mapping, Part 2 26(11-12), 934–965 (2009)
7. Memarsadeghi, N., Mount, D.M., Netanyahu, N.S., Moigne, J.L.: A fast implementation of the isodata clustering algorithm. International Journal of Computational Geometry and Applications 17, 71–103 (2007)
8. Nüchter, A., Hertzberg, J.: Towards semantic maps for mobile robots. Robotics and Autonomous Systems 56(11), 915–926 (2008)
9. Rusu, R.B., Blodow, N., Marton, Z.C., Beetz, M.: Close-range Scene Segmentation and Reconstruction of 3D Point Cloud Maps for Mobile Manipulation in Human Environments. In: Proc. of the IEEE/RSJ International Conference on Intelligent Robots and Systems, IROS, pp. 1–6 (2009)
10. Steder, B., Rusu, R.B., Konolige, K., Burgard, W.: NARF: 3D range image features for object recognition. In: Workshop on Defining and Solving Realistic Perception Problems in Personal Robotics at the IEEE/RSJ Int. Conf. on Intelligent Robots and Systems, IROS (2010)
11. Triebel, R., Shin, J., Siegwart, R.: Segmentation and unsupervised part-based discovery of repetitive objects. In: Proc. of Robotics: Science and Systems (2010)
12. Wulf, O., Arras, K.O., Christensen, H.I., Wagner, B.: 2D Mapping of Cluttered Indoor Environments by Means of 3D Perception. In: Proc. of the IEEE Intl. Conf. on Robotics and Automation, ICRA, pp. 4204–4209 (2004)
13. Yuan, F., Swadzba, A., Philippsen, R., Engin, O., Hanheide, M., Wachsmuth, S.: Laser-based navigation enhanced with 3D time-of-flight data. In: Proc. of the IEEE Intl. Conf. on Robotics and Automation (ICRA), pp. 2231–2237 (2008)

Catadioptric System Optimisation for Omnidirectional RoboCup MSL Robots

Gil Lopes, Fernando Ribeiro, and Nino Pereira

Industrial Electronics Department, Univ. of Minho,
Campus de Azurém, 4800-058 Guimarães, Portugal
{gil,fernando}@dei.uminho.pt, martins@sarobotica.pt

Abstract. Omnidirectional RoboCup MSL robots often use catadioptric vision systems in order to enable 360° of field view. It comprises an upright camera facing a convex mirror, commonly spherical, parabolic or hyperbolic, that reflects the entire space around the robot. This technique is being used for more than a decade and in a similar way by most teams. Teams upgrade their cameras in order to obtain more and better information of the captured area in pixel quantity and quality, but a large image area outside the convex mirror is black and unusable. The same happens on the image centre where the robot shows itself. Some efficiency though, can be improved in this technique by the methods presented in this paper such as developing a new convex mirror and by repositioning the camera viewpoint. Using 3D modelling CAD/CAM software for the simulation and CNC lathe mirror construction, some results are presented and discussed.

Keywords: Omnidirectional robots, RoboCup MSL, catadioptric system, 3D modelling.

1 Introduction

RoboCup Middle Size league (MSL) robots are able to play soccer games autonomously. It is the major league of the whole RoboCup event. The robot's artificial vision system recognises the surrounding environment, the game field, the ball, the opponents and other obstacles by means of a catadioptric vision system in most participating teams. Fig. 1 shows the images captured by robot vision systems of three different MSL teams. Catadioptric or omnidirectional vision is known in MSL since late nineties when it was introduced for the MSL robots and literature about this technique is widely available. Although this system is producing good results, its efficiency can be increased in order to extract more and better information from the captured image. It is important to extract as much as possible the necessary information from a captured image in order to process it flawlessly.

This paper describes one approach divided in two methods to increase the efficiency of the catadioptric vision system. The paper is organised as follows. Section 2 briefly describes how a catadioptric system for MSL robots works, section 3 describes how this work customises a convex mirror using 3D modelling and simulation, section 4 describes our MSL robot head construction and section 5 draws some conclusions.

T. Röfer et al. (Eds.): RoboCup 2011, LNCS 7416, pp. 318–328, 2012.

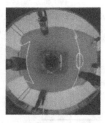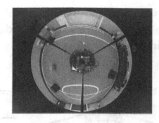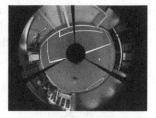

Fig. 1. Three examples of images captured by a catadioptric vision system from different MSL teams. Left – MRT team [1, 2], Centre – Cambada [3], Right – Brainstormers Tribots [4].

2   Catadioptric Vision System for MSL Robots

A catadioptric vision system applied to MSL robots is basically an upright camera facing a convex mirror as it is shown in Fig. 2-a. This makes the robot head. The higher the mirror position, the farther the robot can see. The head's position in relation to the robot can be seen in Fig. 2-b. Fig. 2-c shows a 3D computer model of the robot obtained from CAD software where the whole robot was drawn before being built. Using the same CAD technology, further in this paper it will be shown how a convex mirror was developed and simulated.

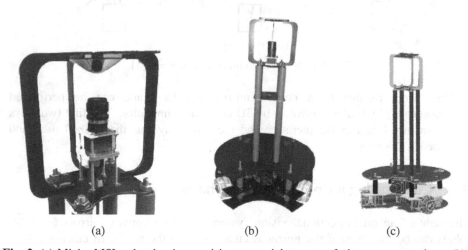

(a)   (b)   (c)

Fig. 2. (a) Minho MSL robot head comprising an upright camera facing a convex mirror. (b) Picture of Minho MSL robot and respective head position. (c) 3D computer model of the robot from CAD software.

2.1   Types of Convex Mirrors

Convex mirrors can be of different types such as conical, spherical, cylindrical, paraboloidal, ellipsoidal and hyperboloidal. Commonly applied to vision systems are spherical [5, 6] and hyperboloidal [7, 8] convex mirrors, and some work was also found on multi-part mirror construction made with a conical, spherical and planar parts [9]. A comparison of a spherical and hyperboloidal mirror types can also be found in [10]. Ishiguro has also reported the use of convex mirrors as it has compared different mirror shapes and their differences [11].

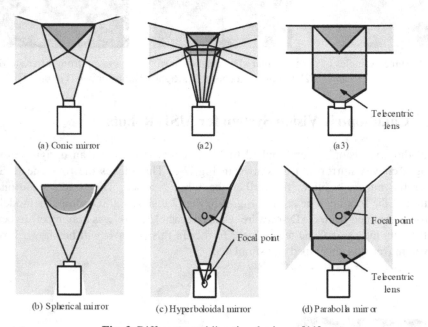

Fig. 3. Different omnidirectional mirrors [11]

The characterisation of a convex mirror and its shape can be performed mathematically; by trial and error or by 3D computer simulation. The first two ways are commonly found in the literature and the third way was the method used and described in this paper.

3   Customising a Convex Mirror Shape

The sight of an omnidirectional vision system based on convex mirrors for MSL robots can be improved if the mirror is customised to the needs. In other words, a proper shape of the convex mirror can improve the robot's sight and therefore, improving its game playing. An empirical method based on 3D modelling is here proposed where the user simulates potential mirror shapes in order to obtain the desired image. After the shape is achieved the next step consists of exporting the data to a milling machine in order to build it. This is based on translating the data to a Computer Numerical Control (CNC) code as it is a direct approach from simulation to the real scenario. This work was carried out using a commercially available 3D modelling package (SolidWorks 2009 © [12]) and a CNC code (G-code) generation package (SolidCAM 2008 © [13]).

3.1   Computer 3D Simulation

A 3D model was developed based on the real size of the game field, ball and robots. The render option of SolidWorks allows the visualisation of the model in shading mode becoming a realistic view of the scene. Objects are rendered with their textures and when the parameters are set to obtain a mirror texture, the rendered image creates a texture in the object as a result of a reflection of the object's surroundings. This allows a more realistic simulation of a mirror object independently of its shape or form.

The texture for the game field was developed to create a checkerboard type in one half of the field with a square side measuring 0.5 m. In this way the distortion produced by the convex mirror can be assessed, measured and analysed directly. The other half is green and with the white field lines visible to simulate the real game field. Fig. 4-a shows an image of the simulated game field with some robots and game balls. SolidWorks also allows the simulation of a camera where the user can select the lens parameters. The rendered image can be generated from the camera perspective. The lens parameters were configured to match the lens used in our MSL robots and the camera was positioned at the same height and facing upwards to an object that simulated the convex mirror. The camera/lens and the object are the head of the robot, as shown in Fig. 4-b.

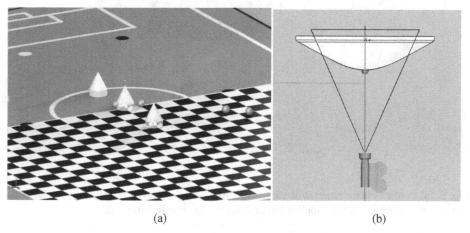

(a)   (b)

Fig. 4. (a) Computer simulated game field with robots and game balls and half of the game field with a checkerboard texture. (b) Simulation of the robot head with a camera/lens and the convex mirror object.

With this setup created, the work was then centred on the development of the mirror profile. This profile is made up of separate parts. A flat circular plate simulates the top of the convex mirror where the fixture is positioned. The mirror itself is based on a line that starts from the mirror centre and moves up to the sides. This line defines the mirror curvature and can be defined with continuous segments of a line each one performing different curvatures. A method called "revolve" from SolidWorks is then applied to the line to perform a 360° turn around a point and in this case, the point is the mirror centre. By selecting a metallic texture with the shininess parameter to its maximum and applying it to the object, a mirror is created. All the reflection surface calculation is then left to the software.   (a)   (b)

Fig. 5-a shows the shaped line made of three segments before the application of SolidWorks 'revolve' method and (a)   (b)

Fig. 5-b shows the final mirror aspect after applying SolidWorks 'revolve' method. The simulation is then applied by rendering or shading the final image from the point of view of the positioned camera. Fig. 6 shows rendered images of different convex mirror shapes and it is clear the differences of the surrounding object positions in relation to the image centre where the robot and camera are. Changing the line shape that defines the mirror profile allows the creation of any type of convex mirror and also allows the user to experience the resulting visualisation, in a few seconds. This technique also allows the creation of multi-part type mirrors as it was used to produce our actual developed mirror as explained next.

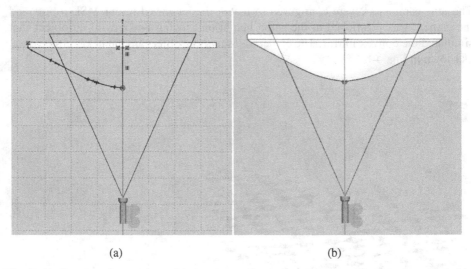

(a)   (b)

Fig. 5. (a) Convex mirror object with the shaped line before SolidWorks 'revolve' method. (b) Convex mirror object after the application of SolidWorks 'revolve' method.

3.2   Simulated Mirrors

Our main goal was to create a multi-part convex mirror that could reflect the robot surroundings in different ways at defined radius. By examining the images obtained from spherical and hyperbolic convex mirrors it was observed that a large area is occupied by the robot itself thus reducing the total amount of valuable and useful information on the image. It was also observed that by convoluting the whole robot image some distortion is created to the game balls near the image centre becoming a drop shape ball. This severe distortion affects the algorithms that detect the game ball by its circular shape. Fig. 7 shows an example of an image strongly convoluted in the centre (left) and with a small convolution (right).

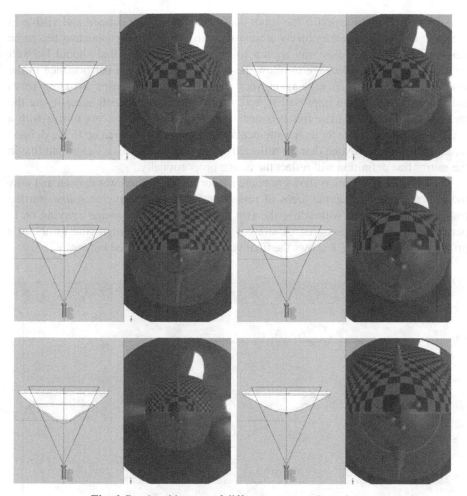

Fig. 6. Rendered images of different convex mirror shapes

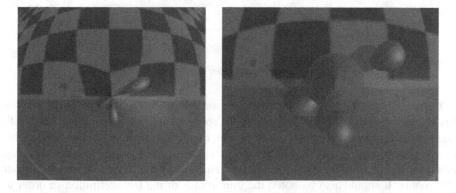

Fig. 7. Distortion of soccer balls produced by a strong image convolution (left) and a small image convolution (right)

Another important aspect is the robot control and how well defined and visible in the captured image are the objects around it, such as the other robots and the game ball. A three meter radius was chosen to be a good distance that should be well defined in the captured image to allow the robot to have a good performance in the control algorithms when moving at very high speeds. This helps avoiding robot collisions and allows an improved representation of the soccer ball, simplifying the tackle and dribble. To attain this it is necessary to provide the convex mirror with a curvature that can create an image augmentation or zoom in the area up to the defined three meter radius. Beyond that the reflected image can be linear, i.e., a straight line in the mirror line definition will reflect the image proportionally.

The previous description shows how the mirror line definition was chosen and how the line segments define the areas of reflection. In brief, a curve or *spline* starting from the mirror centre will reduce the size of the robot on the image carrying on to perform the augmentation area. A straight line is used in the edge side to obtain a proportional image. Fig. 8 shows the final version of the developed mirror.

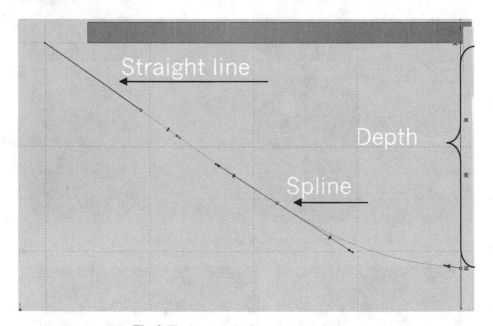

Fig. 8. Final approach of our developed mirror

One important parameter is the convexity or mirror depth (see Depth on Fig. 8). This parameter will define the distance of the robot sight but it must be balanced with other parameters in order to provide a proper image and to avoid major distortions. Some trial-and-error is essential to get the most out of each parameter until the final definition. Table 1 summarizes the final parameters of the developed mirror in our MSL robots. It should also be noted the importance of the lens definition in order to achieve a proper match afterwards when the mirror is machined.

Table 1. Parameters of the final developed convex multi-part mirror

Parameter	Value
Spline – part 1 – X	0.00
Spline – part 1 – Y	-21.60746101
Spline – part 1 – angle	-9.29656199°
Spline – part 2 – X	-17.76355387
Spline – part 2 – Y	-15.42003468
Spline – part 2 – angle	-32.69814149°
Spline – part 3 – X	-30.87114791
Spline – part 3 – Y	-6.47671516
Spline – part 3 – angle	-35.08394856°
Straight line length	11.25296534
Straight line angle	144.86140003°

3.3   Machining the Mirror

When a good result was obtained in the simulation, it was necessary to export the data in a format that could be used by a CNC lathe to produce the convex mirror. SolidCAM software package integrates with SolidWorks when it detects its existence

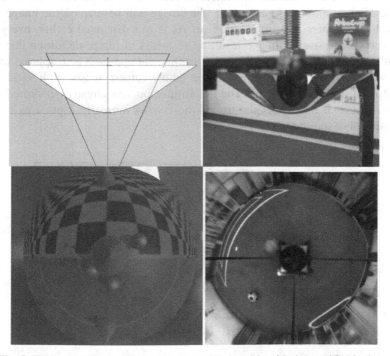

Fig. 9. Final result of the simulated mirror side-by-side with the machined mirror

during the installation. This software establishes a bridge between the 3D models developed by the CAD software and the machining world. It can generate the necessary code to perform the machining and the language chosen was the G-code (language used by the CNC lathe available at the mechanical workshop). Some metals were tested such as stainless steel and aluminium and after polishing to a fine grain size some differences were noticed between the two mirrors. The images from the stainless steel mirror were slightly darker than the ones reflected by the aluminium mirror. The mechanical engineering department suggested the use of brass with a coating of chromium, but that was our decision since the weight of a brass mirror would be equivalent to the stainless steel and they are much heavier than aluminium. In that sense, aluminium was the preferred material to build up the robot mirrors. The final result can be seen in Fig. 9 where the simulated mirror is shown side-by-side with the machined mirror.

4   Robot Head

The robot head is the part where the camera/lens and the convex mirror are fixed. They define how the image will look like. The distance between the lens and the mirror dictates whether the simulated scenario will match the real scenario. This is where our second approach to improve the efficiency takes place. Commonly, teams define the distance from camera and mirror in order to obtain the whole mirror image. Since the camera sensor is square, a circular image will be seen on the image centre and four black corners will fill up the rest. As it was discussed earlier, every pixel counts when processing an image resulting in a loss of image definition due to the resultant lower resolution. To counteract with this, our approach is to approximate the camera with the mirror in order to fill the whole camera sensor with valid image reflected from the convex mirror. The 3D simulation was already performed taking this into account. It maximises the use of the convex mirror to capture as much as possible the robot surroundings.

Fig. 10. Robot head with the thin pillars and corresponding captured image

Another important aspect of the robot head is how the convex mirror is fixed and attached to the robot allowing visibility and sturdiness. Different methods are used by the teams from acrylic pipe to a fixture of three or four pillars that support a flat top. The first option produces image glare and brightness inconsistencies produced by the round shape acrylic. The second option though produces loss of valid image since the pillars can be seen on the image. To reduce this occlusion, the head was drawn with the pillars being as thin as possible and put in a way that would be far from the mirror reducing the occlusion. Three pillars were used and displaced as a cross to avoid any obstruction or occlusion in the 180° angle in front of the robot. The developed robot head is shown in Fig. 10.

5   Conclusions

Computer 3D modelling and simulation of convex mirrors for applications such as omnidirectional robot vision of MSL robots was successfully attained as it was described in this paper. Two methods were proposed to increase the efficiency of omnidirectional vision systems where all the image information is useful with only a few pixels discarded. Mirror simulation and machining enables an easy and faster production of different curvature shapes for customised purposes increasing specific areas of visibility to the vision processing system. Aluminium has also shown to be a better material for the production of convex mirrors due to the final brighter image reflection and to be a lighter material when compared with stainless steel or brass metals. The robot head was developed to reduce the occlusions created with the fixtures that hold the convex mirror to the robot body.

Acknowledgements. The authors wish to thank the ALGORITMI Research Centre for the opportunity to develop this research, and also wish to thank the department of Mechanical Engineering in the person of Prof. Caetano Monteiro for the use of the CNC lathe and help on the mechanical development.

References

1. Bonarini, A., Aliverti, P., Lucioni, M.: An omnidirectional sensor for fast tracking for mobile robots. IEEE Transactions on Instrumentation and Measurement 49(3), 509–512 (2000)
2. Lima, P., Bonarini, A., Machado, C., Marchese, F.M., Marques, C., Ribeiro, F., Sorrenti, D.G.: Omnidirectional catadioptric vision for soccer robots. International Journal of Robotics and Autonomous Systems 36(2-3), 87–102 (2001)
3. Azevedo, J.L., et al.: CAMBADA 2008: Team Description Paper. In: Robocup MSL TDP, vol. 9, Universidade de Aveiro (2008)
4. Lauer, M., Lange, S., Riedmiller, M.: Calculating the Perfect Match: An Efficient and Accurate Approach for Robot Self-localization. In: Bredenfeld, A., Jacoff, A., Noda, I., Takahashi, Y. (eds.) RoboCup 2005. LNCS (LNAI), vol. 4020, pp. 142–153. Springer, Heidelberg (2006)

5. Winters, N., Santos-victor, J.: Mobile robot navigation using omni-directional vision. In: Proc. 3rd Irish Machine Vision and Image Processing Conference (IMVIP 1999), pp. 151–166 (1999)
6. Siemiatkowska, B., Chojecki, R.: Mobile Robot Navigation Based on Omnidirectional Sensor. In: Proc. 1st European Conference on Mobile Robots ECMR 2003, EURON Conference, Radziejowice, p. 6 (2003)
7. Neves, A.J.R., et al.: An efficient omnidirectional vision system for soccer robots: From calibration to object detection. Mechatronics
8. Yi, S., Ahuja, N.: A Novel Omnidirectional Stereo Vision System with a Single Camera in Scene Reconstruction Pose Estimation and Tracking. In: Stolkin, R. (ed.) pp. 454–466. I-Tech Education and Publishing (2007)
9. Marchese, F.M., Sorrenti, D.G.: Omni-Directional Vision with a Multi-Part Mirror. In: Stone, P., Balch, T., Kraetzschmar, G.K. (eds.) RoboCup 2000. LNCS (LNAI), vol. 2019, pp. 179–188. Springer, Heidelberg (2001)
10. Gaspar, J.A.d.C.P.: Omnidirectional Vision for Mobile Robot Navigation, p. 150. IST - Universidade Tecnica de Lisboa (2002)
11. Ishiguro, H.: Development of Low-Cost Compact Omnidirectional Vision Sensors and their applications. In: Panoramic Vision, ch. 3, pp. 433–439. Springer (1998)
12. SolidWorks. Dassault Systèmes - SolidWorks Corp. 2010 (cited 2010), http://www.solidworks.com/
13. SolidCAM. The leaders in integrated CAM. 2010 (cited 2010), http://www.solidcam.com/

Smooth Path Planning around Elliptical Obstacles Using Potential Flow for Non-holonomic Robots

Trenthan Owen, Rebecca Hillier, and Darwin Lau

Department of Mechanical Engineering, The University of Melbourne, Australia
{t.owen,r.hillier2,dtlau}@student.unimelb.edu.au

Abstract. In this paper, an efficient path planning method for non-holonomic robots to avoid elliptical obstacles for RoboCup soccer matches is presented. A hydrodynamic flow field is formulated to model obstacles and target locations. Previous research considers the flow about elliptical and plate obstacles as a superposition of multiple flow fields about circular obstacles. The proposed approach utilises the Joukowsky transform to form a path about an elliptical obstacle with a single flow field. It is shown that the resulting motion satisfies C_∞ continuity at all times, desired for mobile robots. The application of different obstacle shapes in the context of RoboCup soccer matches is also considered and simulated.

Keywords: non-holonomic mobile robots, elliptical obstacles, efficient path planning.

1 Introduction

In recent decades, developments in robotics have significantly enhanced the abilities of mobile robots; becoming more efficient, agile and faster. As a result, path planning has been heavily researched such that robots can autonomously determine a path to a destination whilst avoiding obstacles.

Grid based methods such as the vector field histogram [1], and other techniques using concepts such as electrical resistors [12] and thermodynamic heat flow [10], have been used due to their ease in modelling arbitrarily shaped obstacles. However, there are some inherent drawbacks with these techniques, for example, they are computationally expensive and require knowledge of the entire environment of interest. Another common path planning strategy is artificial potential fields (APF). APF can be implemented using many different potential functions, including hydrodynamic potentials [8][2] and attractive/repulsive forces [9]. The advantages include its computational efficiency and simplicity of implementation. One drawback is that the robot can become trapped within local minima.

The APF method was first introduced by Khatib [4], proposing that the goal and obstacles can be modelled by attractive and repulsive forces, respectively, resulting in a potential field that creates a collision-free path to the target. To ensure the robot does not become trapped, the potential field can be modified [5] or

T. Röfer et al. (Eds.): RoboCup 2011, LNCS 7416, pp. 329–340, 2012.

harmonic functions, for example, streamfunctions from hydrodynamic potential flow theory have been used due to their harmonic properties [10][2][11][6].

The focus of this paper is to create a collision-free path for non-holonomic robots to compete in RoboCup utilising the hydrodynamic APF method. Explicit analytical expressions for the flow field velocity can be derived, and the final trajectory with respect to time can be determined through numerical integration. This is particularly advantageous for time based systems, such as RoboCup mobile robots. Previous research in this area has only considered the avoidance of circular obstacles [11]. Modelling of non-circular obstacles have been achieved by the superposition of circular objects [2] or the panel method [6]. For obstacles with complicated geometry, these methods increase in computational complexity and hence may not be able to be implemented in real time. Also, the robot's velocity is defined within the potential function, and can be difficult to control. In addition the robot's initial orientation has typically not been considered, violating the constraint for non-holonomic robots.

In the proposed approach, elliptical obstacles can be incorporated by transforming circular objects through the Joukowsky transform. This direct approach is significantly more efficient and simple compared to the superposition of circular obstacles. The ability to model elliptical obstacles directly is extremely beneficial in RoboCup, for example, field and goal area boundaries can be represented as a single plate object. In addition, the non-holonomic constraint of the mobile robot is shown to be satisfied by the inclusion of a source object that follows behind the robot. Another advantage of this addition is its ability to constrain the curvature of the resulting path. The inherent issue of uncontrollable speed is addressed through normalising the robot's velocity components. The proposed path planning approach is simulated for RoboCup gameplay scenarios, demonstrating that a C_∞ continuous collision-free path about multiple elliptical obstacles is achieved.

The remainder of this paper will be presented as follows: Section 2 will introduce the fundamental hydrodynamic potential functions. The proposed path planning model is described in Section 3. Section 4 presents simulation results for RoboCup using the proposed method and Section 5 will conclude this paper and present areas of future work.

2   Fundamental Hydrodynamic Functions

One class of the hydrodynamic potential function is the complex velocity potential (CVP) [11][7], defined as:

$$\omega(z) = \phi + i\psi \tag{1}$$

where ϕ and ψ are the velocity potential and streamfunction, respectively. The CVP is a function of the position in the complex plane, $z = x + iy$. The velocity of a point in the complex plane can be determined by differentiating the CVP with respect to z [7]:

$$\frac{dw}{dz} = u - iv \tag{2}$$

where u and v are the velocity components in the x and y directions, respectively.

2.1   Modeling Cylindrical Obstacles

A modified circle theorem is used in this paper to position the obstacle at (b_x, b_y) [11] resulting in:

$$\omega_c(z) = \omega(z) + \overline{\omega}\left(\frac{r^2}{z - b} + \overline{b}\right) \tag{3}$$

where $b = b_x + ib_y$ and $\overline{\omega}$ represents the complex conjugate of ω.

2.2   Joukowsky Transform

The Joukowsky transform is commonly used in fluid dynamics to model flow around aerofoils. This technique transforms the flow around a cylinder of radius r centered at the origin in the complex plane C into a flow around an elliptical obstacle centered at the origin in the complex plane E, while preserving angles between small vectors. The tranformation, $j : C \rightarrow E$, is defined as:

$$z_e = z_c + \frac{\lambda^2}{z_c} \tag{4}$$

where $z_e \in E$ and $z_c \in C$. The transformation constant, λ, governs the ratio between the major and minor axes. Three cases for values of λ can be considered:

Case 1: $0 < \lambda < r$. The flow around the original cylinder shown in Fig. 1(a), is transformed to a flow around an ellipse, as shown in Fig. 1(b). The major and minor axes, m and n, respectively, are defined as:

$$m = r + \frac{\lambda^2}{r}$$

$$n = r - \frac{\lambda^2}{r} \tag{5}$$

Case 2: $\lambda = 0$. The flow remains untransformed after the application of (4) as shown in Fig. 1(a).

Case 3: $\lambda = r$. The flow is transformed to a flow around a flat plate of length $4r$, as shown in Fig. 1(c).

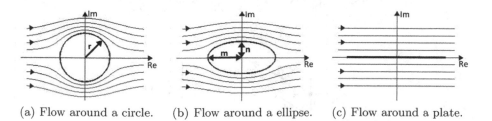

(a) Flow around a circle.   (b) Flow around a ellipse.   (c) Flow around a plate.

Fig. 1. The flow around obstacles modelled using the Joukowsky transform

For the purpose of path planning it is more convenient to express the transformation with respect to velocities. This provides the potential to control path planning at the velocity level, motivated by the inclusion of multiple obstacles and speed regulation. The velocity of z_e can be determined by taking the derivative of (4):

$$\dot{z}_e = \dot{z}_c - \frac{\dot{z}_c \lambda^2}{z_c^2} \tag{6}$$

where $\dot{z}_e \in E$ and $\dot{z}_c \in C$ are velocities in their respective planes.

Inverse Joukowsky Transform. To map coordinates from the complex plane E back to the complex plane C, the inverse transform, $j^* : E \to C$, can be determined from (4):

$$z_c = \frac{z_e}{2} \pm \frac{\sqrt{z_e^2 - 4\lambda^2}}{2}, \quad \text{where} \quad \left| \frac{z_e}{2} \pm \frac{\sqrt{z_e^2 - 4\lambda^2}}{2} \right| \geq r \tag{7}$$

3   Proposed Path Planning Strategy for Non-holonomic Robots

In the proposed model, the destination is represented by a sink at coordinate $e = e_x + i e_y$ [2][11], and a source is included, positioned at $s = s_x + i s_y$, to follow behind the robot. The addition of the source results in a C_∞ trajectory at all times, $t \geq 0$, which is desired for non-holonomic robots. Defining the robot's position and orientation with respect to the positive x axis as $z = x + iy$ and θ, respectively, as shown in Fig. 2. The CVP for a source or a sink [7] is:

$$\omega(z - c) = \frac{Q}{2\pi} log_e(z - c) \tag{8}$$

where $Q > 0$ represents the strength for a source and $Q < 0$ for a sink, and c denotes the source's or sink's position.

Applying the circle theorem from (3) on (8), the CVP becomes:

$$\omega_c(z, c, b, Q) = \frac{Q}{2\pi} \log_e (z - c) + \frac{Q}{2\pi} \log_e \left(\frac{r^2}{z - b} + \bar{b} - \bar{c} \right) \tag{9}$$

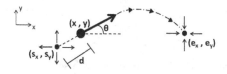

Fig. 2. Defining the robot's position, orientation and the following source

where b and r represent the position and radius of the obstacle, respectively. The velocities for the flow defined in (9) can be determined using (2):

$$u_c(z, c, b, Q) = \frac{Q(x - c_x)}{2\pi((x - c_x)^2 + (y - c_y)^2)} \tag{10}$$

$$- \frac{Qr^2(r^2(x - b_x) + (b_x - c_x)(r_n) + 2(x - b_x)(y - b_y)(b_y - c_y))}{2\pi(r_o^2(r^4 + 2r^2((x - b_x)(b_x - c_x) + (y - b_y)(b_y - c_y))) + r_o^2(r_m))}$$

$$v_c(z, c, b, Q) = \frac{Q(y - c_y)}{2\pi((x - c_x)^2 + (y - c_y)^2)} \tag{11}$$

$$- \frac{-Qr^2(r^2(y - b_y) - (b_y - c_y)(r_n) + 2(x - b_x)(y - b_y)(b_x - c_x))}{2\pi(r_o^2(r^4 + 2r^2((x - b_x)(b_x - c_x) + (y - b_y)(b_y - c_y))) + r_o^2(r_m))}$$

where $r_o = (x - b_x)^2 + (y - b_y)^2$, $r_n = (x - b_x)^2 - (y - b_y)^2$ and $r_m = (b_x - c_x)^2 + (b_y - c_y)^2$. The x and y velocity components for the proposed model containing both a source and a sink are:

$$u_c = u_c(z, s, b, Q_s > 0) + u_c(z, e, b, Q_e < 0)$$
$$v_c = v_c(z, s, b, Q_s > 0) + v_c(z, e, b, Q_e < 0) \tag{12}$$

where Q_s, Q_e, s, e and b are the strengths of the source and sink, and the coordinates for the source, target and obstacle, respectively. From (6), the velocities can be determined for a flow field around an elliptical obstacle: $\dot{z}_e = u_e + iv_e$, where u_e and v_e are the path's velocities around a single elliptical obstacle. The posture of the elliptical obstacle requires the definition of it's angular orientation, this can be considered through the application of rotations on the flow field.

3.1   Source Location and Strength

To ensure that the non-holonomic constraint is satisfied, the source, s, must be positioned directly behind the robot at a distance d as shown in Fig. 2:

$$s = (x - d\cos\theta) + i(y - d\sin\theta) \tag{13}$$

The source's influence can be increased by increasing Q_s or decreasing d, resulting in paths with larger curvature. The ability to influence the curvature can be useful for applications such as RoboCup, for example, while dribbling the ball.

3.2   Multiple Obstacles

A collision-free path around multiple obstacles can be generated through the superposition of the p individual paths for each obstacle [11]:

$$u = \sum_i^p \alpha_i u_i$$

$$v = \sum_i^p \alpha_i v_i \tag{14}$$

where u_i and v_i are the velocities of the path from (12) for obstacle i. The individual paths are weighted by constants, $\alpha_i = f(\mathbf{d})$, where $\mathbf{d}$ represents the distances between the obstacles and the robot's current position. It should be noted that the resulting path, from (14), preserves the C_∞ property of the individual paths. The weighting function is required to be a monotonically decreasing continuous function with a range of $[0, 1]$. A modified weighting function from [11] was utilized in this paper to increase the effect of closer obstacles:

$$\alpha_i = \prod_{j \neq i}^{p} \frac{d_j^4}{d_i^4 + d_j^4} \tag{15}$$

3.3 Speed Regulation

From the model proposed in (12), it is apparent that the robot will accelerate as it approaches the sink. Defining the desired speed of the robot as: $V^2 = u_{\mathrm{reg}}^2 + v_{\mathrm{reg}}^2$, where u_{reg} and v_{reg} are the components of the final robot's velocity. These can be determined by normalising against the current robot speed:

$$u_{reg} = \frac{u \times V}{\sqrt{u^2 + v^2}}$$
$$v_{reg} = \frac{v \times V}{\sqrt{u^2 + v^2}} \tag{16}$$

The resulting motion preserves the C_∞ property from (16), and maintains the robot's initial orientation.

For clarity, Algorithm 1 presents a pseudocode for the proposed approach.

4   Simulation Results and Discussion

From the components of the velocity from (16), the resulting trajectory was determined through numerical integration. The Newton-Euler method with a time step of 128Hz was used.

4.1 Single Obstacle

The path for a single elliptical obstacle without the following source or velocity regulation is shown in Fig. 3(a), with an initial angle of $200°$, where the circle and star symbols denote the start and target locations, respectively, and the sink strength of $Q_e = 1$. It is apparent that the non-holonomic constraint is violated at $t = 0$, requiring the robot to instantaneously turn, as shown in Fig. 4(a). For the remainder of the path, $t > 0$, C_∞ can be observed.

The resulting path with a following source, as defined in Sect. 3.1, is shown in Fig. 3(b). In this scenario, it can be observed that the non-holonomic constraint is satisfied at $t = 0$, as shown in Fig. 4(c). It should be noted that the time required to reach the target has decreased due to the influence of the source.

Algorithm 1. Proposed Path Planning Algorithm

Require: x_n, y_n and θ_n : Current robot pose
Require: x_e, y_e and V_d : Target destination and speed
Require: O : Set of obstacle locations and sizes
Ensure: $x_{n+1}, y_{n+1}, \theta_{n+1}, u_{n+1}$ and v_{n+1} : Pose at next time step for C_∞ motion

   for all $o_i \in O$ **do**
      $s \Leftarrow$ transformed source location in C using (7) on (13)
      $b \Leftarrow$ transformed obstacle location in C using (7) on original obstacle location
      $e \Leftarrow$ transformed sink location in C using (7) on x_e, y_e
      $u_{c_i}, v_{c_i} \Leftarrow$ velocity components in C using (12) for s, b and e
      $u_{e_i}, v_{e_i} \Leftarrow$ transformed velocity components in E using (6) on u_{c_i}, v_{c_i}
      $\alpha_i \Leftarrow$ weighting for obstacle using (15)
   end for
   $u, v \Leftarrow$ velocity of combined path from (14) using the set of u_{e_i}, v_{e_i} and α_i
   $u_{n+1}, v_{n+1} \Leftarrow$ speed regulated velocities from (16) using u, v and V_d
   $x_{n+1} \Leftarrow x_n + hu_{n+1}$, Newton-Euler integration for x position for step-size h
   $y_{n+1} \Leftarrow y_n + hv_{n+1}$, Newton-Euler integration for y position for step-size h
   $\theta_{n+1} \Leftarrow \tan^{-1}\left(\frac{v_{n+1}}{u_{n+1}}\right)$, robot's orientation at next time step
   return $x_{n+1}, y_{n+1}, \theta_{n+1}, u_{n+1}$ and v_{n+1}

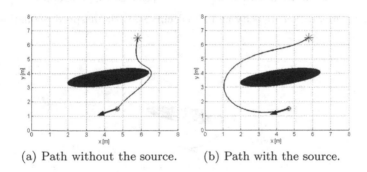

(a) Path without the source.   (b) Path with the source.

Fig. 3. Path without speed regulation, and $Q_s = 0.15$ and $d = 0.03$m

Incorporating speed control, as described in Sect. 3.3, the resulting path for a desired speed of 3m/s is shown in Fig. 5. As the speed regulation has an affect on the robot's position in time, it is expected that the resulting trajectory shown in Fig. 5 would be different in comparison to the path shown in Fig. 3(b). Comparing the speed profiles in Fig. 6(b) against Fig. 4(d), it can be observed that the desired speed is maintained for the entire trajectory.

For dynamic environments, such as RoboCup soccer games, the destination can rapidly change, creating a challenge to generate a C_∞ path, as demonstrated in Fig. 7(a). The incorporation of a following source resolves the discontinuities in velocities at checkpoints 1 and 2, as shown in Fig. 7(b), guaranteeing a C_∞ path at all times, $t \geq 0$.

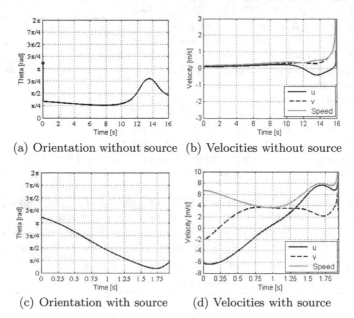

(a) Orientation without source (b) Velocities without source

(c) Orientation with source (d) Velocities with source

Fig. 4. Orientation, velocity and speed curves, without speed regulation

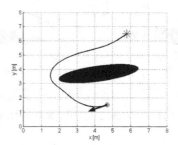

Fig. 5. Path including speed regulation, where $Q_s = 0.15$, $d = 0.03$m and $V = 3$m/s

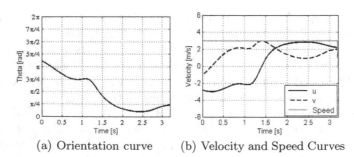

(a) Orientation curve (b) Velocity and Speed Curves

Fig. 6. Orientation, velocity and speed curves for the path including velocity regulation

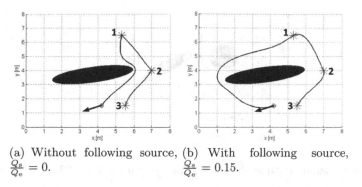

(a) Without following source, (b) With following source,
$\frac{Q_s}{Q_e} = 0$. $\frac{Q_s}{Q_e} = 0.15$.

Fig. 7. Robot path for reaching multiple targets, where $V = 3\text{m/s}$ and $d = 0.03\text{m}$

Effects of the Path Parameters. Parameters that influence the trajectory include Q_s, Q_e, d, and V. Assuming a constant value of d and a desired speed V, the resulting path and minimum allowable curvature, κ_{min}, can be governed by the ratio $\frac{Q_s}{Q_e}$. The resulting path for different values of $\frac{Q_s}{Q_e}$ have been simulated, and are shown in Fig. 8. For a lower value of κ_{min}, as shown in Fig. 8(a), the resulting trajectory is formed below the obstacle due to the allowance of the robot to turn more rapidly. In comparison, a higher κ_{min} forces the resulting trajectory to travel around the obstacle, as shown in Fig. 8(b), which is preferred for a robot to maintain possession of the soccer ball.

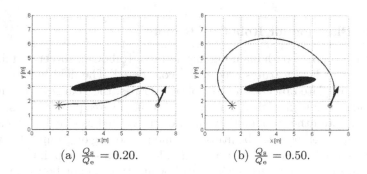

(a) $\frac{Q_s}{Q_e} = 0.20$.   (b) $\frac{Q_s}{Q_e} = 0.50$.

Fig. 8. Paths for different values of $\frac{Q_s}{Q_e}$, with $d = 0.03\text{m}$, $V = 3\text{m/s}$

4.2   Multiple Obstacles

The resulting path for multiple obstacle avoidance, as described in Sect. 3.2 is shown in Fig. 9. Ellipses of different ratios have been included in this example, and it can be observed that the C_∞ property is maintained.

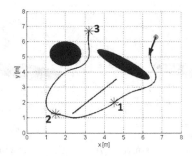

Fig. 9. Path around multiple obstacles using $\frac{Q_s}{Q_e} = 0.15$, $d = 0.03$m, and $V = 3$m/s

Table 1 provides a comparison between the proposed path planning approach and another hydrodynamic APF path planning method.

Table 1. Comparison between the proposed and hydrodynamic approaches

Property	Hydrodynamic method	Proposed method
Contiuity	C_∞ violations at path's start C_∞ path at all other times	C_∞ path at all times
Obstacle type	Circular obstacles	Elliptical and circular obstacles
Speed control	Unable to regulate speed	Can regulate robot's speed
Curvature control	Unable to control curvature	Potential to control path curvature
Multiple obstacles	Can model multiple obstacles	Robot can avoid multiple obstacles

4.3   Path Planning for RoboCup Robots

To apply the proposed path planning method to a RoboCup soccer match, modelling of different field features need to be considered. The field boundaries can be modelled by plate objects. In the RoboCup rules, the goal square cannot be entered, and so can be modelled as a superposition of three plate objects. To prevent collision into other robots, they can be represented as circular objects. For greater flexibility, ellipses can be utilised to model moving or groups of obstacles.

In the proposed method, the robot is modelled as a single point with no radius, for real-life applications, such as RoboCup, this cannot be assumed. The robot's radius, denoted by R, can be incorporated through the remodelling of the obstacle's size or location. The boundary lines can be shifted inwards by a distance R, and the initial goal square dimensions can be increased by a magnitude R as demonstrated in Fig. 10(a). For elliptical obstacles, the major and minor axes can be increased by R, as shown in Fig. 10(b). The result of these modifications is a collision-free path for a robot of radius R. The avoidance of the goal area is demonstrated in Fig. 11, where the solid and dashed paths represent the trajectories with and without the goal area constraint, respectively.

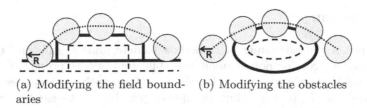

(a) Modifying the field bound-   (b) Modifying the obstacles
aries

Fig. 10. Modifying the field and obstacles to accommodate for the robot's radius. The
solid and dashed lines represent the modified and original boundaries, respectively.

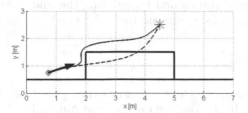

Fig. 11. The different paths for a robot to travel around a goal square

Applying the proposed path planning approach to a typical RoboCup soccer
game setup, the resulting trajectory is shown in Fig. 12. It can be observed that a
non-holonomic robot can travel smoothly while not violating any RoboCup game
rules. Although the proposed approach does not guarantee an optimal path, a
realistic C_∞ path can be efficiently generated for a large number of obstacles. In
addition, the ability to apply curvature constraints on the resulting trajectory
allows the artificial intelligence agent to have greater control on robot motion.

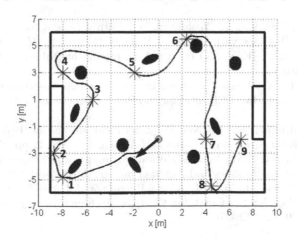

Fig. 12. The robot's path in a RoboCup soccer game setup

5   Conclusion

In this paper, a hydrodynamic APF method to determine a collision-free path to a specified target for non-holonomic mobile robots was presented. It was demonstrated that utilising the Joukowsky transform with the proposed source and sink model, a C_∞ path avoiding multiple elliptical obstacles can be efficiently generated. The adjustment of source and sink strength ratio is shown to have the potential to apply a minimum curvature constraint on the resulting path. The results for the proposed model, incorporating the features and game rules for RoboCup soccer matches have been provided. Future work will focus on incorporating moving obstacles and formulating the mathematical relationship between the curvature constraints and the source and sink ratios.

References

1. Borenstein, J., Koren, Y.: The Vector Field Histogram-Fast Obstacle Avoidance for Mobile Robots. IEEE Trans. on Robotics and Automation 7(3), 278–288 (1991)
2. Daily, R., Bevly, D.M.: Harmonic Potential Field Path Planning for High Speed Vehicles. In: American Control Conference, pp. 4609–4614 (2008)
3. Dracopoulos, D.C.: Robot Path Planning for Maze Navigation. IEEE Int. Joint Conf. on Neural Networks 3, 2081–2085 (1998)
4. Khatib, O.: Real-time Obstacle Avoidance for Manipulators and Mobile Robots. In: IEEE Int. Conf. on Robotics and Automation, pp. 500–505 (1985)
5. Khosla, P., Volpe, R.: Superquadratic Arificial Potentials for Obstacle Avoidance and Approach. In: IEEE Int. Conf. on Robotics and Automation, pp. 1778–1784 (1988)
6. Kim, J., Khosla, P.K.: Real-Time Obstacle Avoidance Using Harmonic Potential Functions. IEEE Trans. on Robotics and Automation 8(3), 338–349 (1992)
7. Schobeiri, M.T.: Fluid Mechanics for Engineers: A Graduate Textbook. Springer, Berlin (2010)
8. Sugiyama, S., Yamada, J., Yoshikawa, T.: Path Planning of a Mobile Robot for Avoiding Obstacles with Improved Velocity Control by Using the Hydrodynamic Potential. In: IEEE/RSJ Int. Conf. on Intelligent Robots and Systems, pp. 1421–1426 (2010)
9. Tilove, R.B.: Local Obsacle Avoidance for Mobile robots based on the Method of Artificial Potentials. In: IEEE Int. Conf. on Robotics and Automation, pp. 566–571 (1990)
10. Wang, Y., Chirikjian, G.S.: A New Potential Field Method for Robot Path Planning. In: IEEE Int. Conf. on Robotics and Automation, vol. 2, pp. 977–982 (2000)
11. Waydo, S., Murray, R.M.: Vehicle Motion Planning using Stream Functions. In: IEEE Int. Conf. on Robotics and Automation, vol. 2, pp. 2484–2491 (2003)
12. Yun, L., Liu, Z., Sun, H., Yuan, J.: A Path Planner of Mobile Robot Based on Multi-Grid Circuit Map. In: The fourth Int. Conf. on Machine Learning and Cybernetics, vol. 2, pp. 1279–1284 (2005)

NaOISIS*: A 3-D Behavioural Simulator for the NAO Humanoid Robot

Aris Valtazanos and Subramanian Ramamoorthy

School of Informatics, University of Edinburgh
Edinburgh EH8 9AB, United Kingdom
a.valtazanos@sms.ed.ac.uk, s.ramamoorthy@ed.ac.uk

Abstract. We present NaOISIS, a three-dimensional behavioural simulator for the NAO humanoid robot, aimed at designing and testing physically plausible strategic behaviours for multi-agent soccer teams. NaOISIS brings together features from both physical three-dimensional simulators that model robot dynamics and interactions, and two-dimensional environments that are used to design sophisticated team coordination strategies, which are however difficult to implement in practice. To this end, the focus of our design has been on the accurate modeling of the simulated agents' perceptual limitations and their compatibility with the corresponding capabilities of the real NAO robot. The simulator features presented in this paper suggest that NaOISIS can be used as a rapid prototyping tool for implementing behavioural algorithms for the NAO, and testing them in the context of matches between simulated agents.

Keywords: Robot Simulators, Robotic Soccer, Strategic Behaviours.

1 Introduction

Software simulators are essential testing and debugging tools for large-scale robotics applications, such as those encountered in the RoboCup domain. The establishment of the NAO humanoid robot as the hardware platform for the Standard Platform League (SPL) has been followed by the development of several such environments, each focusing on different aspects of the complex robotic soccer problem. Most simulators can be classified in one of two extreme categories; on the one hand are the implementations that attempt to generate a full, three-dimensional, physical representation of the robot and its dynamics, aimed at improving low-level sensorimotor skills. On the other hand lie the two-dimensional, kinematics-only simulators, which are normally used in the development of sophisticated strategic algorithms for multi-agent systems.

While the merit of each of these two extremes is indisputable, there has been little work in developing intermediate-type, three-dimensional behavioural simulators, which would bring together features from both categories. In broad terms, a 3-D behavioural implementation would abstract away the complex, low-level

---

* NaOISIS is derived from "NAO" and the ancient Greek word "Noisis" (*Νόησις*), which means cognition.

T. Röfer et al. (Eds.): RoboCup 2011, LNCS 7416, pp. 341–352, 2012.
© Springer-Verlag Berlin Heidelberg 2012

dynamics found in a physical simulator, while still restricting robots to a realistic, partially observable, perspective view of their environment that impacts their decision-making capabilities - the latter feature cannot be modeled accurately in two dimensions.

In this paper, we present NaOISIS, a three-dimensional behavioural simulator for the NAO humanoid robot developed in the MATLAB environment. As outlined above, the simulator builds on the notion of decoupling *motion dynamics* from the implementation of behavioural algorithms, while maintaining basic physical features and properties (such as collisions between the robots, the ball and the goalposts, and kicks of varying speed and directions). Furthermore, our implementation focuses on the accurate modeling of NAO's *sensing* capabilities, by providing simulated cameras and sonar sensors that closely match the real robot's hardware specifications. Thus, it is possible to recreate a realistic *perception-action* cycle, where the simulated agents must make decisions facing the same perceptual constraints as a physical NAO robot.

In the context of developing physically realisable strategic behaviours, one may argue that the abstraction of low-level motion dynamics is too strong a concession to make. However, we claim that this may not necessarily be the case for two reasons.

On the one hand, any experienced physical simulator user can testify that it is extremely difficult to achieve a reliable one-to-one correspondence between simulated motions and motions in the real world - any complex, non-trivial motion must undergo significant transformations before being ported from a simulator to the robot, and vice versa. This complication is magnified when considering closed-loop motions (e.g. walking), where the interplay between the robot's sensors and the environment becomes even harder to simulate accurately.

On the other hand, this abstraction is supported by recent advances in the SPL community, which have seen the development of robust walking engines for the NAO robot; because of the standard platform nature of the competition, it is possible to adopt these engines directly. For example, the latest version of the Aldebaran NAO SDK [2] features a closed-loop implementation that accepts commands of the form $(dx, dy, d\theta)$ and plans a path to move the robot to this desired location. Through such interfaces, the decoupling of motion dynamics from decision making and path planning arises naturally, so it is possible to focus on strategic interactions instead.

In light of the above considerations, our proposed NaOISIS simulator serves as a useful middle ground for developing new behavioural algorithms for the NAO humanoid robot. The selection of MATLAB as a programming environment facilitates experimentation with several sophisticated machine learning, optimal control, reinforcement learning and path planning techniques, implementations of which are not available or are difficult to implement in other programming languages. This choice does come at the expense of computational speed (compared to other candidates such as C or C++), although this is not too restrictive. In the following sections of this paper, we first briefly review other simulators available for the NAO, and we then present the salient features of NaOISIS. Finally, we illustrate how NaOISIS can be used as both a debugging tool and a simulator for full soccer matches, by presenting representative screen shots from sample runs.

2 Related Work

SimSpark [11] is a general-purpose, open-source simulator, designed to accommodate a variety of robotic systems. A specific implementation tailored to the NAO robots was adopted as the official simulator for the RoboCup 3-D Simulation League [9] in 2009. SimSpark features a physical model of the NAO humanoid dynamics, while also modeling the limited field of view of the robot and its perspective view of the environment. The simulator architecture is generic enough so that different behavioural frameworks can plug-in and interact in the context of a simulated match. However, despite being adequate for a simulated match, the underlying dynamics do not yet faithfully resemble the dynamics of a physical robot, as can be seen in videos from recent competitions [6][5]. Other examples of general-purpose simulators used in the RoboCup domain include USARSim [12], which has been mostly used in the context of the Rescue League [8].

A similar NAO implementation exists for the popular proprietary Webots environment [4], providing similar features as SimSpark. The Webots simulator provides an interface for developing controllers in various programming languages (C, C++, Java, MATLAB). A controller plugging directly to the Aldebaran NAO SDK (NaoQi) has also been developed, allowing the sharing of function calls. However, similar problems with the correspondence between real and simulated dynamics exist in this simulator.

B-Human [1], the 2009 and 2010 winners of the RoboCup Standard Platform League, have also released a custom-made simulation environment. This simulator forms part of their software development kit (available for download online) and is therefore mostly suited to their own custom cognitive architecture.

Perhaps the most widely used simulator in the RoboCup domain is the official simulator of the 2-D Simulation League [7]. Although not designed specifically for the NAO robot, it can be used to simulate more complex, multi-agent strategic algorithms, a subset of which could be potentially implemented on real humanoids. However, the lack of a realistic sensor model makes it difficult to achieve a close correspondence between real and simulated strategies.

3 The NaOISIS Simulator

In this section we present NaOISIS, a 3-D kinematic simulator aimed at creating strategic behaviours for robot soccer teams. NaOISIS is written in MATLAB, and is therefore suitable for use with a number of machine learning, path planning, reinforcement learning and optimal control toolboxes. The simulator can also serve as a debugging tool for *state estimation* algorithms, by visualising and comparing the robots' egocentric beliefs to ground truth data. The NaOISIS source code[1] is available for download from [3].

---

[1] NaOISIS is released under a GPL (http://www.gnu.org/licenses/gpl.html) license.

3.1   Environment and Sensors

Field and Robots. The main window of the NaOISIS simulator visualises the soccer field where the robots interact (Figure 1). The dimensions of the field are the same as those used of the RoboCup pitches (6x4m), with the goal posts, central circle, penalty boxes, and penalty kick cross spots similarly scaled (see [10] for a detailed specification). The ball is an orange sphere of a 3cm radius. As in RoboCup, the two goal mouths are colour-coded yellow and blue to assist in vision-based localisation.

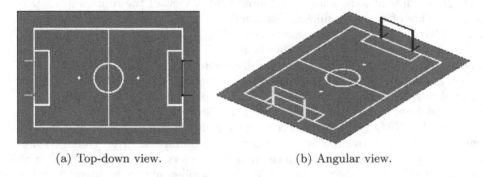

(a) Top-down view.   (b) Angular view.

Fig. 1. Soccer field

The current version of NaOISIS supports two different robot versions. Figures 2(a)-2(b) shows the design which is visually closer to the real NAOs. Unfortunately, this version takes longer to render[2], so a simpler robot design, approximating the NAO through square patches, is often preferred (Figures 2(c)-2(d)). However, it is worth noting that both versions are scaled to match the dimensions of the real humanoid (see [2] for specification), so the difference between them is mainly aesthetic.

As in the RoboCup competition, robots wear either a blue or a pink waistband, depending on the team they are playing for. The waistbands are also included in both robot designs, and serve as an additional cue for visual robot detection and distance estimation.

Cameras. Given that our simulator focuses on developing plausible strategic behaviours for humanoids, it is essential to endow the simulated robots with a realistic sense of their environment. The simulated vision system of NaOISIS was modeled on the cameras of the real NAO robot. Each simulated agent has access to two cameras (Table 1(a)[3]) - a top one, allowing a more complete coverage of the soccer environment, and a bottom one with an additional rotational offset

---

[2] We use the built-in OpenGL libraries provided in MATLAB for scene rendering, and the camera toolbox to set the perspective view for the robots.

[3] The origin is at the midpoint of the two feet, x-axis points to the front, y-axis left, and z-axis up.

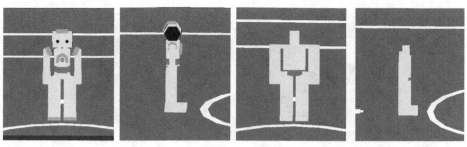

(a) Simulated NAO robot.

(b) Side view of Figure 2(a).

(c) Simplified design using square patches.

(d) Side view of Figure 2(c).

Fig. 2. Soccer field

Table 1. Simulated camera specification

(a) Camera locations.

Camera	x(cm)	y (cm)	z(cm)
Bottom	4.88	0.0	47.09
Top	5.39	0.0	53.72

(b) Head angle limits.

Rotation type	Minimum	Maximum
Yaw	$-35°$	$+30°$
Pitch	$-120°$	$+120°$

of 40° about the y-axis to allow tracking of nearer objects (e.g. the ball when kicking). Additional constraints have been placed on the head yaw and pitch angles of the robots (Table 1(b)). The cameras have a diagonal of 58°, and return 320x240 images in RGB format.

The simulated vision system is illustrated in Figure 3. Figure 3(a) shows an example scene with two robots and a ball. Figures 3(b)-3(i) plot the perspective view as seen by the bottom camera of the robot located on the halfway line, for various head pitch and yaw combinations.

Sonar Sensors. Whereas vision is an important tool for ball detection and localisation, sonar sensing is essential for more interactive tasks such as robot avoidance. NaOISIS features a pair of simulated sonar sensors, also modeled on the corresponding devices of the NAO humanoid. Each of these sensors has a range from 0.15 to 0.75m (these values are modifiable). Moreover, there is an option for adding varying levels of Gaussian - or other - noise to the sensor readings, so as to reflect irregularities that would occur in a physical setting. Figure 4 shows a visualisation of the joint range of a robot's sonar sensors, together with the technical specification.

3.2 Physical Interactions

Despite not modeling motion dynamics explicitly, NaOISIS offers a coarse simulation of physical interactions between robots and their environment. The focus

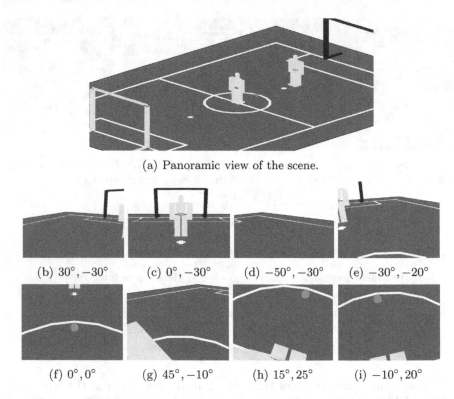

(a) Panoramic view of the scene.

(b) $30°, -30°$ (c) $0°, -30°$ (d) $-50°, -30°$ (e) $-30°, -20°$

(f) $0°, 0°$ (g) $45°, -10°$ (h) $15°, 25°$ (i) $-10°, 20°$

Fig. 3. Field of view for various head angle combinations (yaw,pitch)

is on modeling motions that form an essential part of a humanoid robot's strategy, such as translational/rotational movements and kicking actions. Collisions between robots, balls, and goalposts are also considered. The current version of the simulator does not account for more complex movements, such as getting up from a fall or dives for goalkeepers.

Moving. Each simulated agent has access to a simple motion engine which allows it to navigate around the soccer field. The engine accepts commands of the form:

$$(dx, dy, d\theta) \tag{1}$$

which correspond to the desired linear and angular displacement with respect to the robot's current coordinate frame. To ensure that the executed commands are plausible, the following are also defined before the simulator is run:

– The maximum linear and angular velocities of each robot
– The spread of the error added to each command (which may vary along the three dimensions of the motion)

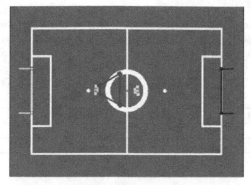

(a) Trapezoidal approximation of the joint range of the left robot's two sonar sensors. Only objects within the range of the trapezium are detected.

Sensor	Angle	Height	Angular range
Left	$+19.48°$	40cm	18°
Right	$-19.48°$	40cm	18°

(b) Specification.

Fig. 4. Sonar sensor range visualisation and specification

Thus, if a robot requests to move by an amount that exceeds its capabilities, this command will be scaled down and/or perturbed accordingly by the simulator.

Kicking. Kicking actions are more difficult to model than translational and rotational movements because they involve interactions between the robots and the ball. Thus, a coarse approximation to the dynamics of these interactions is required to achieve physically plausible kicks. NaOISIS currently supports four different types of (non-directional) kicks:

- Left/right-footed straight kicks
- Left/right-footed side kicks, where the robot first extends the kicking foot to the front and then performs a side motion.

At the start of the simulation, the *maximum* traveling distance for each kicking type is defined. Then, when a robot wishes to kick the ball, it utilises a command with the following information:

- The type of the kick
- The desired speed of the kick, as a number between 0.0 and 1.0, where 1.0 corresponds to maximum allowed speed for the desired type of kick.

The precise trajectory of the ball after a kicking command depends on its *position* and *angle* relative to the kicking foot. A ball will be affected by a kick only if it lies within a rectangle defined with respect to the robot's coordinate frame (Table 2). If the ball lies in the appropriate rectangle, the direction of its trajectory will

Table 2. Admissible rectangles for each kicking type, defined in terms of the robot's coordinate frame (x-axis points to the front, y-axis to the left). For each kicking command, the ball will move only if it lies within the corresponding rectangle. The tips of the left and right feet are at positions (10,5) and (10, -5) respectively.

Kick type	Min x (cm)	Max x (cm)	Min y (cm)	Max y (cm)
Left straight	10	20	0	10
Right straight	10	20	-10	0
Left side	10	22	-15	0
Right side	10	22	0	15

be the angle formed by the line connecting the ball and the tip of the robot's foot (see caption of Table 2), and the line which is normal to the foot on the robot's coordinate frame. Some special kicking cases are also accounted for; for example, if the robot attempts a side kick from the "wrong" side, it will cause a straight kick of reduced speed. Figure 5 shows several examples of kicking commands and the trajectories they incur depending on the ball position.

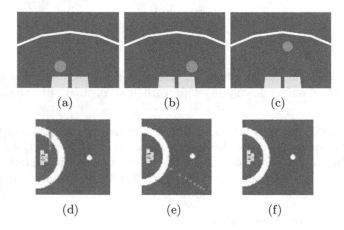

(a)   (b)   (c)

(d)   (e)   (f)

Fig. 5. Robot's perspective (top) and resulting ball trajectories (bottom) for various ball positions and kicking types. (a)-(d): right side kick. (b)-(e): right straight kick with the ball positioned at an angle (but within the required rectangle). (c)-(e): right straight kick - the ball is outside the kicking range so the resulting trajectory is null.

Collision between Robots and Objects. Collisions between robots and the various objects in the soccer field are also coarsely modeled in the NaOISIS simulator. The following types of collision are taken into account:

– Collision between two moving robots
– Collision between a moving and a static robot
– Collision between a moving robot and a goalpost
– Collision between a moving robot and a moving ball (as the result of a kick)
– (Accidental) collision between a moving robot and a static ball

For each robot, two different sets of rectangular bounds are considered: one for its base and one for its torso. For the first three types of collision, the simulator checks for intersections between torso bounds and/or goalpost cylinders; for the latter two, the torso bounds are replaced with the robot's base bounds. Whenever a ball is involved in a collision, the simulator also computes its adjusted trajectory (Figure 6).

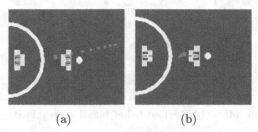

(a)   (b)

Fig. 6. Ball trajectory before (a) and after (b) collision with robot

3.3 Information Exchange

Agents are also provided with a simple message-passing capability, which allows them to communicate messages to their teammates. As NaOISIS is a discrete-step simulator, message exchange occurs once every decision cycle. Each robot updates a *shared data structure* with a list of relevant data, such as its self-localisation belief and the observed position of the ball and/or the other robots. The robots then communicate this structure to their teammates, who may use it to adjust their decisions or to form cooperative strategies. As with physical interactions, there is an option to add a non-zero probability of messages not being delivered successfully.

3.4 Experiments and Testing

The simulator components described in the previous sections can be used to design and test physically plausible behaviours and algorithms. Depending on the user's requirements, NaOISIS may be used to either debug such behaviours and test their properties under varying conditions, or to simulate full soccer matches between teams of autonomous agents.

Developing and Debugging Algorithms. Naoisis provides experimental support for a wide range of algorithmic domains, such as:

- Autonomous decision making under uncertainty
- Path planning and dynamic obstacle avoidance
- Reinforcement learning
- Belief estimation, information filtering and sensor fusion
- Vision-based localisation
- Multi-agent coordination

In most of these domains, it is important to be able to compare the agent's egocentric beliefs and estimates to ground-truth data. NaOISIS provides a set of debugging tools that can be used to visualise this comparison. Figure 7 provides a simple illustration of this functionality in the context of observation-based estimation of the opponent's location.

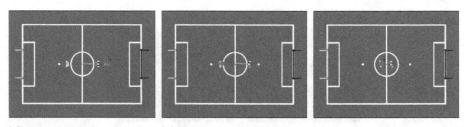

(a) Initial belief - no information available.   (b) Revised belief based on visual estimate.   (c) Further revision using sonar estimate.

Fig. 7. Egocentric beliefs of the right-hand side robot on the location of its opponent. The belief at each stage is indicated by the blue circle.

Simulating Full Matches. The soccer match mode of NaOISIS provides additional functionalities that help simulate full soccer matches between teams of autonomous agents. Most of these are concerned with the position of the ball at the various stages of the game, so as to determine if a goal has been scored or if the ball has crossed the field bounds. In each of these cases, the ball and the robots are repositioned and the score is updated accordingly.

The collision modeling functionality discussed in Section 3.2 is used to determine whether a robot should be penalised for a certain period of time. Whenever a robot is deemed "responsible" for a collision by the simulator, it is placed on the sideline for an interval of time before it is allowed to re-enter the game. Responsibility is judged based on the relative speeds of the colliding robots, and the location of the ball. Furthermore, if a robot does not move for a long period of time while being close to the ball (and thus preventing other robots from reaching it), it is also penalised. Since robot falls are not modeled in the current version of the simulator, penalty shootouts would be highly disfavourable to goalkeepers, and as such they have also been omitted. This will be implemented as an extension in future versions of NaOISIS.

The debug and match modes of NaOISIS are not mutually exclusive; it is possible to plot robot beliefs and other relevant information in the match version as well. Moreover, since the match is carried out at discrete perception-decision-action states, it is possible to record snapshots of the soccer field at various stages. These snapshots can then be synthesised as a movie or used otherwise for evaluation. The frames from the robots' camera feeds can also be recorded and used to debug information processing and state estimation algorithms. Figure 8 shows screenshots from a soccer match betwen two robots.

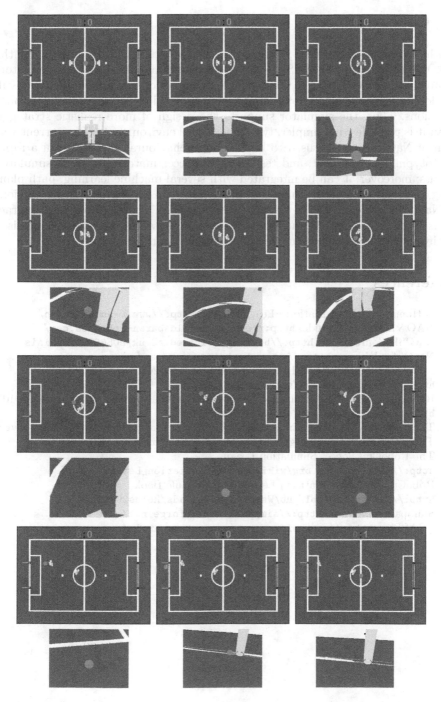

Fig. 8. Simulated soccer game. Top to bottom, left to right - *Odd rows:* field screenshots showing the positions of the robots and the ball at various stages. *Even rows:* the corresponding field of view of the pink robot (initially on the right hand side).

4   Conclusion

We have presented NaOISIS, a three-dimensional behavioural simulator for the NAO humanoid robot. NaOISIS differs from traditional physics-based simulators in abstracting away most of the low-level dynamics of the humanoid, while still modeling salient physical interactions such as kicking actions and inter-robot collisions. Thus, the simulator supports the design of more realistic strategies than it is possible with simpler, two-dimensional environments. The current version of NaOISIS can be used to both debug behavioural algorithms in a rapid prototyping environment, and test them in soccer matches between simulated agents; moreover, it can be integrated with several machine learning, path planning and reinforcement learning that are available in MATLAB. Future versions of NaOISIS will concentrate on the coarse modeling of further physical interactions, such as robot falls and dives, while also providing a more standardised environment for implementing decision making algorithms.

References

1. B-Human Standard Platform League Team, http://www.b-human.de/en/
2. NAO v1.10 User's Guide, http://academics.aldebaran-robotics.com/
3. NaOISIS Source Code, http://homepages.inf.ed.ac.uk/s0566900/naoisis.html
4. NaoQi for Webots, http://www.cyberbotics.com/nao/
5. RoboCup, 3D Simulation League - Final first half (2010),
   http://www.youtube.com/watch?v=EmMt_mlpvWE
6. RoboCup 2010 3D Simulation League - Semifinal second extra half (2010),
   http://www.youtube.com/watch?v=MNCObp1TvNE
7. The RoboCup 2D Soccer Simulator, http://sourceforge.net/projects/sserver/
8. The RoboCup Rescue League, http://www.robocuprescue.org/
9. The RoboCup Soccer Simulation League,
   http://wiki.robocup.org/wiki/Soccer_Simulation_League
10. RoboCup Standard Platform League Official Rule Book,
    http://www.tzi.de/spl/pub/Website/Downloads/Rules2010.pdf
11. SimSpark simulator, http://simspark.sourceforge.net/
12. USARSim, http://sourceforge.net/projects/usarsim/

Facial Expression Recognition for Domestic Service Robots

Geovanny Giorgana and Paul G. Ploeger

Bonn-Rhein-Sieg University of Applied Sciences,
Grantham-Allee 20 53757 Sankt Augustin, Germany
geovanny.giorgana@smail.inf, paul.ploeger@h-brs.de

Abstract. We present a system to automatically recognize facial expressions from static images. Our approach consists of extracting particular Gabor features from normalized face images and mapping them into three of the six basic emotions: joy, surprise and sadness, plus neutrality. Selection of the Gabor features is performed via the AdaBoost algorithm. We evaluated two learning machines (AdaBoost and Support Vector Machines), two multi-classification strategies (Error-Correcting Output Codes and One-vs-One) and two face image sizes (48 x 48 and 96 x 96). Images of the Cohn-Kanade AU-Coded Facial Expression Database were used as test bed for our research. Best results (87.14% recognition rate) were obtained using Support Vector Machines in combination with Error-Correcting Output Codes and normalized face images of 96 x 96.

Keywords: Facial expression recognition, Gabor features, AdaBoost, Support Vector Machines, Error-correcting output codes, One-vs-One multi-classification, Face normalization.

1 Introduction

Facial expression recognition (FER) offers domestic service robots (DSR) a natural way to interact with humans. This channel of information can be used by robots in order to receive feedback on their executed actions as well as to convey empathy.

In general, changes in a face that are due to the execution of a facial expression can be observed in an image as changes in the face texture. Two-dimensional (2D) Gabor filters have proven success for the representation of the instantaneous appearance of a face via texture analysis. One advantage of such filters is the fact that they can provide a spatially localized frequency analysis of the images. Another strong motivation for its use is its similarity to certain mammal's visual cortical cells [7].

Among the most relevant works studying the performance of Gabor filters for FER we find [2], where authors use local Gabor filter banks and PCA plus LDA for dimensionality reduction. Furthermore, [6] presented a local approach where Gabor features are extracted at the location of eighteen facial fiducial points. A very extensive study is presented in [1], where a comparison of different image

T. Röfer et al. (Eds.): RoboCup 2011, LNCS 7416, pp. 353–364, 2012.
© Springer-Verlag Berlin Heidelberg 2012

sizes, feature selectors, classifiers and methods to extend them for the multi-class problem is presented.

In this paper, a fully automatic facial expression recognizer that maps the perceived expressions into one of the following three basic emotions: joy, surprise and sadness, plus neutrality is presented. Our approach finds its cornerstone in the normalization of the input images and the extraction of Gabor features. Since the number of dimensions of the initial feature space is very high, AdaBoost has been used for feature selection. Furthermore, we compare and report the performance of the system when normalized images of different sizes, and different binary learning machines in combination with different multi-classification strategies are employed.

The rest of the report is organized as follows: section 2 explains the system and the employed methods, the experiments and the obtained results are reported and analyzed in section 3, finally, section 4 conclude our work.

2   Methods

2.1   System Overview

Training. of the whole expression recognition system is accomplished after training N binary classifiers whose individual output will be combined in the testing part by different multi-classification strategies. Figure 1 illustrates the six basic stages of the training process. After the detection of faces and eyes in all images of the training set, the located faces are normalized and a pool of Gabor features is extracted from all of them. The subsequent three steps are performed repeatedly in a series of $t = 1, 2, ..., T$ rounds. For each iteration AdaBoost selects the best feature and the whole system is evaluated against another set of images in order to find out how many features to preserve. The amount of features for which the trained classifier presented better results are preserved.

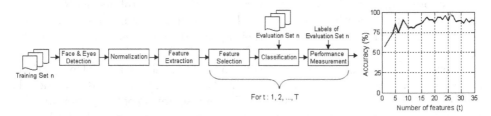

Fig. 1. Diagram of a single binary classifier trainer

Testing. involves similar stages as the training phase (see Fig. 2). After the normalization of the detected faces, only the features selected during training are extracted and used for classification. Classification here is performed by a set of N binary classifiers whose output are later on integrated by one of the analyzed multi-classification strategies.

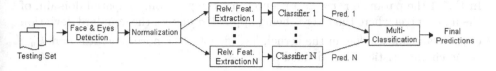

Fig. 2. Diagram of the facial expression recognition system

2.2 Face Detection and Face Normalization

Face and eyes detection is performed with a commercial software of the company L-1 Identity Solutions, Inc. The system has the feature of fast and robust detection in unconstrained environments. However, the face boundaries are not provided by the software. For this reason all face boundaries were post-generated using a geometric face model that employs the distance between the eyes as primary element [9]. Figure 3a illustrates the characteristics of the model and Fig. 3b shows the face regions obtained after cropping some images of the dataset. As the size of the extracted face regions is not unique, we have scaled all images after cropping. Scaling eases feature-feature comparisons since the center of the two eyes are located at a fixed position, and reduces the number of posterior computations without a drastic loss of accuracy. Finally, the histogram of all scaled images was equalized in order to reduce the effects of illumination. In this work, scaled images of size 48 x 48 and 96 x 96 have been investigated.

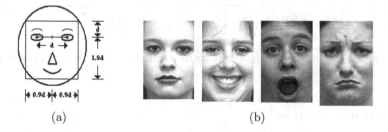

(a)   (b)

Fig. 3. a) Geometric face model, b) Cropped faces with the model

2.3 Gabor Feature Extraction

A Gabor filter results from the modulation of a complex sinusoid with a Gaussian envelope. The mathematical form of a complex Gabor function in spatial domain is given in Eq. 1.

$$\psi_{u,v} = \frac{\|\vec{k}_{u,v}\|^2}{\sigma^2} \exp\left(-\frac{\|\vec{k}_{u,v}\|^2\|\vec{x}\|^2}{2\sigma^2}\right) \left[\exp\left(j\,\vec{k}_{u,v}\vec{x}\right) - \exp\left(\frac{\sigma^2}{2}\right)\right] . \tag{1}$$

In Eq. 1 the parameter $\vec{x} = (x, y)$ represents the position, in spatial domain, of the image pixel where the filter is being applied, σ defines the standard deviation of the Gaussian window in the kernel, $\| \cdot \|$ denotes the norm operator and $\vec{k}_{u,v}$ is the characteristic vector defined as

$$\vec{k}_{u,v} = \begin{pmatrix} k_{(u,v)x} \\ k_{(u,v)y} \end{pmatrix} = \begin{pmatrix} k_v \cos \theta_u \\ k_v \sin \theta_u \end{pmatrix} . \tag{2}$$

From Eq. 2, $k_v = \frac{k_{max}}{f^v}$ corresponds to the scale of the filter and $\theta_u = u \cdot \frac{\pi}{8}$ specifies the orientation of the filter.

In principle, the images can be analyzed into a detailed local description by convolving them with a very large number of Gabor filters at different spatial frequencies and orientations. In our work, we created a bank with 40 Gabor filters at 5 different scales and 8 different orientations. Such Gabor filters were tuned with the following parameters: $\sigma = 2\pi, k_{\max} = \frac{\pi}{2}, f = \sqrt{2}, v = \{0, \cdots, 4\}, u = \{0, \cdots, 7\}$.

After creating the bank, the scaled images are convolved with all filters in the bank to obtain their Gabor representation. Eq. 3 shows the mathematical definition of the convolution of an image $I(\vec{x})$ and a Gabor kernel $\psi_{u,v}(\vec{x})$.

$$O_{u,v}(\vec{x}) = I(\vec{x}) * \psi_{u,v}(\vec{x}) . \tag{3}$$

The magnitude of the complex function $O_{u,v}(\vec{x})$ is then computed to obtain the Gabor features as follows

$$\|O_{u,v}(\vec{x})\| = \sqrt{\Re^2 \left\{ O_{u,v}(\vec{x}) \right\} + \Im^2 \left\{ O_{u,v}(\vec{x}) \right\}} . \tag{4}$$

Finally, all extracted Gabor features from an image are concatenated together to form a feature vector.

2.4   Feature Selection

The dimensionality of the feature vectors is forty times higher than the size of the normalized images. Even small images would bring about very large feature vectors that would either slow down the system or result intractable for most of the existing classification algorithms. Because of that fact, we have employed the AdaBoost algorithm to keep the most useful Gabor features.

AdaBoost is a machine learning algorithm formulated in 1995 by Yoav Freund and Robert Schapire [4]. The learning strategy of this algorithm is based on the Condorcet jury theorem that holds the belief that a group will make better decisions than individuals, given that individuals have a reasonable competence. AdaBoost aims to build a complex, non-linear "strong" classifier H_T by linearly combining T "weak" classifiers $h_t \in -1, +1$. The algorithm runs in a series of rounds $t = 1, 2, ..., T$, and in each round it chooses the weak classifier that achieved the lowest error in a given training set. A benefcal characteristic of the algorithm is the use of two kinds of weights, namely α_t and w_n, where n indicates to what training example w belongs to. On the one hand, the α's weight the h_t's so that the better a weak classifier is the more impact it will have in the last decision; on the other hand, the w's weight all training examples so that in each round the weak classifiers that correctly classified the previously misclassified training examples are favored.

For sake of simplicity, but also due to their proven well performing, decision trees of a single level, also known as decision stumps, have been used as weak classifiers. A decision stump here is defined as a function that evaluates if a Gabor feature $O_m = \|O_{u,v}(\vec{x})\|$ is above or below a certain threshold λ_m. Eq. 5 shows the mathematical definition of a decision stump, whereas Eq. 6 illustrates how the threshold for each dimension of the feature vector was computed [8]. In Eq. 5, $p_m \in +1, -1$ is a parity to indicate the direction of the inequality and λ_m is the threshold value for dimension m. In Eq. 6, P and Z are the number of positive and negative examples, respectively, whereas $\sum_{p=1}^{P} O_{m|y=1}$ and $\sum_{z=1}^{Z} O_{m|y=-1}$ denote the sum of all Gabor features of the dimension m that belong to positive and negative training examples, respectively. In other words, λ_m corresponds to the mean value of the mean of all Gabor features of dimension m belonging to the positive class and all Gabor features of dimension m belonging to the negative class.

$$h_m = \begin{cases} 1 & \text{if } p_m O_m < p_m \lambda_m \\ -1 & \text{otherwise} \end{cases} \tag{5}$$

$$\lambda_m = \frac{1}{2} \left(\frac{1}{P} \sum_{p=1}^{P} O_{m|y=1} + \frac{1}{Z} \sum_{z=1}^{Z} O_{m|y=-1} \right). \tag{6}$$

2.5   Classification

If AdaBoost were used for classification, the T chosen weak learners together with the T α's would be used as in Eq. 7. However, other classifiers can also be used in combination with the selected Gabor features. We also used Support Vector Machines (SVM) with an RBF kernel to create a hyperplane that optimally separates the different classes of data. In order to tune the parameters of the kernel, we performed Leave-One-Subject-Out Cross Validation (LOSOCV) during training because this process gives us some guarantee for generalization to new subjects and helps us to avoid overfitting.

$$H_T(x_i) = \text{sign} \left(\sum_{t=1}^{T} \alpha_t h_t(x_i) \right). \tag{7}$$

2.6   Multi-classification

One-vs-One. (1-vs-1) is a strategy to convert binary concept learning algorithms such as AdaBoost and SVM into multi-classification algorithms. For C

classes, this strategy creates $k = \sum_{i=1}^{C-1} i$ binary classifiers, each comparing two different classes. When an instance is input to the system for classification, all the 1-vs-1 binary classifiers indicate their belief $y_i \in -1, +1$ and then the class with more votes is declared as the winner.

Error-Correcting Output Codes. (ECOC), unlike the 1-vs-1 strategy, offers the possibility to recover from errors made by the individual learners. The first step, as proposed by [3], consists of creating a matrix A whose rows contain unique n-bit code words of zeros and ones. In this case the number of classes C is equal to 4, hence the *exhaustive method* has been used to build A. This method guarantees high inter-row Hamming distance and neither repetition nor complementarity of columns, three necessary conditions to successfully correct errors. The exhaustive method consists of filling row 1 with ones, whereas the remaining rows are filled with alternating runs of 2^{C-i} zeros and ones (here $i = 2, 3, 4$ is the row number). Each word in the matrix represents an expression and has a length $n = 2^{C-1} - 1$. The number of errors this scheme can recover from is $\left\lceil \frac{\Delta_{min}(A)-1}{2} \right\rceil$, where $\Delta_{min}(A)$ is the minimum Hamming distance between any pair of codes in A. When an instance is input to the system for classification, the n ECOC binary classifiers indicate their belief $y_i \in 0, +1$ to create a code word that is then compared to the set of words in A according to Hamming distance. Table 1 shows the coding matrix employed for our 4-class problem, whereas Table 2 shows the distribution of the seven created ECOC classifiers.

Table 1. Coding Matrix A

Expression	f_0	f_1	f_2	f_3	f_4	f_5	f_6
Neutral	1	1	1	1	1	1	1
Joy	0	0	0	0	1	1	1
Surprise	0	0	1	1	0	0	1
Sadness	0	1	0	1	0	1	0

Table 2. Distribution of ECOC classifiers

Classifier	Classifier distribution
f_0	neutral vs rest
f_1	neutral-sadness vs rest
f_2	neutral-surprise vs rest
f_3	joy vs rest
f_4	neutral-joy vs rest
f_5	surprise vs rest
f_6	sadness vs rest

3   Experiments and Results

The eight combinations that can be formed with the two normalized face images sizes, the two learning machines and the two multi-classification strategies described in Sect. 2 were tested against images from the Cohn-Kanade AU-Coded Facial Expression Database [5].

In general, the first and the last frame of the sequences of 96 subjects showing expressions of joy, surprise, sadness and neutrality were selected for training and testing. The test set was collected from 46 subjects and was made up of 37 images of each expression. The training set contained images of the remaining

50 subjects; however, the exact number of training images depended on the multi-classification strategy since we took care that each binary classifier were as balanced as possible. The evaluation set contained the same subjects used for training, but the second frame and one frame before the last were used instead. For each case, the maximum number of features selected per classifier was set to 35 in the training phase (i.e., $T = 35$ according to Sect. 2.1).

Figure 4 illustrates the accuracy rate (black segments), the error rate (white segments) and the tie rate (gray segments) accomplished by all evaluated combinations. In the figure we notice the following:

– The highest accuracy rate (81.76%) was achieved by the two combinations using SVM and ECOC.
– The use of SVM resulted in higher accuracy, except for the case when 1-vs-1 and 96 x 96 normalized images were used together.
– Using ECOC always gave better accuracy than using 1-vs-1, when using the same learning machine.
– When using the same learning machine and multi-classification strategy, 48 x 48 provided better or equal accuracy than 96 x 96, unless AdaBoost and ECOC were combined.
– The use of ECOC turned out into a lower percentage of errors and a higher percentage of ties.

The fact that all combinations using ECOC resulted in a lower error rate and a higher tie rate motivated us to solve the cases of uncertainty with the trained 1-vs-1 classifiers. We applied such classifiers only when the combinations using

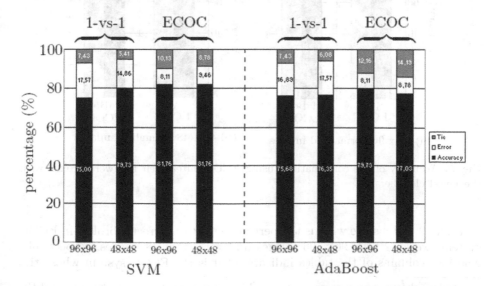

Fig. 4. Accuracy rate, error rate and tie rate of the 8 evaluated combinations

ECOC finished in a tie. We aimed with this to have a tiebreaker that behaves better than random[1].

Figure 5a shows the new accuracy of the combinations using ECOC and 96 x 96 images, while Fig. 5b shows the new accuracy of the combinations using ECOC and 48 x 48 images. Both figures are divided in two parts, each containing two bars. The left part corresponds to combinations using SVM and ECOC, and the right one to combinations using AdaBoost and ECOC. The left bar of each part (black bars) depicts results after breaking ties with 1-vs-1 classifiers that were trained with SVM, and the right bar (gray bars) the results after breaking ties with 1-vs-1 classifiers trained with AdaBoost.

We notice from the figures that:

- All combinations using 96 x 96 images gave better results than those using 48 x 48.
- SVM ECOC outperformed AdaBoost ECOC for both fashions 96 x 96 and 48 x 48.
- The combination that attained the highest accuracy (87.84%) was the one using the SVM algorithm, the ECOC strategy and normalized images of size 96 x 96, and whose ties were broken with 1-vs-1 SVM classifiers.

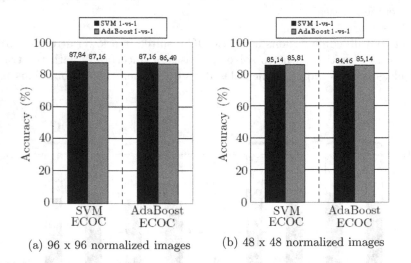

(a) 96 x 96 normalized images   (b) 48 x 48 normalized images

Fig. 5. Accuracy of the combinations using ECOC, after their ties were broken with the 1-vs-1 classifiers

Tables 3 and 4 state what is the percentage of ties correctly broken in ECOC systems working with 96 x 96 images and with 48 x 48 images, respectively. The four columns of the tables indicate what is the ECOC system whose ties

---

[1] Random choices were not necessary in this experiment, but would have been if the 1-vs-1 classifiers had not been able to break all the ties.

were broken, the system used to break the ties, what was the total percentage of ties and what percentage of this was correctly classified, respectively. On the one hand, Table 3 reveals that using SVM 1-vs-1 classifiers when the normalized image is of size 96 x 96 results in a higher amount of ties correctly classified than using AdaBoost 1-vs-1 classifiers. On the other hand, Table 4 shows that AdaBoost 1-vs-1 classifiers are more convenient than SVM 1-vs-1 classifiers when the normalized image is of size 48 x 48.

Table 3. Percentage of ties correctly classified in ECOC systems working with 96 x 96 images

System to be improved	System used to improve	Ties (%)	Ties correctly broken (%)
SVM ECOC 96 x 96	SVM 1-vs-1 96 x 96	10.13	**60.02**
AdaBoost ECOC 96 x 96		12.16	**61.10**
SVM ECOC 96 x 96	AdaBoost 1-vs-1 96 x 96	10.13	53.30
AdaBoost ECOC 96 x 96		12.16	55.59

Table 4. Percentage of ties correctly classified in ECOC systems working with 48 x 48 images

System to be improved	System used to improve	Ties (%)	Ties correctly broken (%)
SVM ECOC 48 x 48	SVM 1-vs-1 48 x 48	8.78	38.50
AdaBoost ECOC 48 x 48		14.19	52.36
SVM ECOC 48 x 48	AdaBoost 1-vs-1 48 x 48	8.78	**46.13**
AdaBoost ECOC 48 x 48		14.19	**57.15**

Table 5 presents the confusion matrix of the system with the highest accuracy. We see from the matrix that all neutral expressions were correctly classified, 81% of the faces posing an expression of joy were correctly classified and 19% were incorrectly classified as neutral, 97% of the faces corresponding to surprise were correctly classified and 3% were mislabeled as sadness, 73% of the sad faces were correctly classified, while 27% were classified as neutral. From the table we can see that neutral and surprise are easier to recognize than joy and sadness. Besides, we observe that the majority of the misclassified expressions were confused with a neutral expression.

Figure 6 presents four correctly classified images, whereas Fig. 7 shows three incorrectly classified. These particular misclassified images have in common that they show not very pronounced expressions, which can be the reason for mis-classification.

The features selected for each classifier of the ECOC system are shown in Fig. 8. In the figures, most of the features are nearby the mouth. This latter might be because in the database this is the feature that varies the most from expression to expression. We also see that for some images, the depicted features match reasonable facial regions. For example, image 8c shows the features that separate surprise from the other expressions; most of the features in this image match the

Table 5. Confusion matrix

		Actual			
		Neutral	Joy	Surprise	Sadness
Predicted	Neutral	100	19	0	27
	Joy	0	81	0	0
	Surprise	0	0	97	0
	Sadness	0	0	3	73

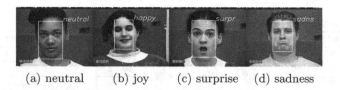

(a) neutral   (b) joy   (c) surprise   (d) sadness

Fig. 6. Four images correctly classified by the system with no ties and highest accuracy

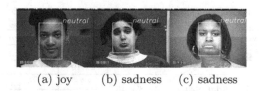

(a) joy   (b) sadness   (c) sadness

Fig. 7. Three images incorrectly classified by the system with no ties and highest accuracy. The actual label of (a) is joy, while the actual label of (b) and (c) is sadness.

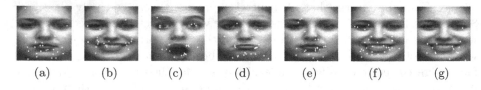

(a)   (b)   (c)   (d)   (e)   (f)   (g)

Fig. 8. Features chosen by AdaBoost for the 7 ECOC classifiers of the system with no ties and highest accuracy. a) neutral vs rest, b) joy vs rest, c) surprise vs rest, d) sadness vs rest, e) neutral-joy vs rest, f) neutral-surprise vs rest, g) neutral-sadness vs rest.

circumference of the open mouth. In image 8d we find features matching the lip corners, which is sensible to separate sadness from the rest.

In total, 170 Gabor features out of 386,640 were used for recognition of the 4 emotions, which represents a 2000-fold reduction (see Table 6).

Table 6. Number of features selected by AdaBoost for each ECOC classifier of the system with no ties and highest accuracy

Classifiers	# chosen features
neutral vs rest	24
joy vs rest	24
surprise vs rest	23
sadness vs rest	27
neutral-joy vs rest	26
neutral-surprise vs rest	30
neutral-sadness vs rest	16

4 Conclusion

This study has focused on exploring the performance of SVM and AdaBoost in combination with ECOC and 1-vs-1 for the problem of facial expression recognition from static images. Two different normalized image sizes have been considered in this analysis, namely, 96 x 96 and 48 x 48. Besides, the use of the 1-vs-1 classifiers to solve the uncertainties generated when the ECOC-based systems cannot find a single winner has been investigated. All evaluated combinations utilized features extracted by convolving Gabor filters at specific positions in the images, and with particular scale and orientation. The experiments revealed that if the normalized images are of size 96 x 96, the SVM 1-vs-1 classifiers convert more ties into correct classifications than the AdaBoost 1-vs-1 ones. On the contrary, if the normalized images are of size 48 x 48, the AdaBoost 1-vs-1 classifiers are more convenient. Another important result is the fact that after tie breaking, 96 x 96 normalized images turned out in about 2% more accuracy than 48 x 48 ones, and that the use of SVM gave better accuracy than the use of AdaBoost. It is also worth mentioning that before tie breaking, ECOC-based systems resulted in better accuracy. AdaBoost as feature selector allowed us to achieve over 2000-fold reduction. The results also show that all combinations have the characteristic of being person-independent and can perform automatically from end to end with the help of an accurate eye detector as the one provided by the L-1 Identity Solutions, Inc.

References

1. Bartlett, M.S., Littlewort, G., Frank, M., Lainscsek, C., Fasel, I., Movellan, J.: Recognizing Facial Expression: Machine Learning and Application to Spontaneous Behavior. In: IEEE International Conference on Computer Vision and Pattern Recognition, pp. 568–573 (2005)
2. Deng, H.B., Jin, L.W., Zhen, L.X., Huang, J.C.: A New Facial Expression Recognition Method Based on Local Gabor Filter Bank and PCA plus LDA. International Journal of Information Technology 11, 86–96 (2005)

3. Dietterich, T.G., Bakiri, G.: Solving Multiclass Learning Problems via Error-Correcting Output Codes. Journal of Artificial Intelligence Research 2, 263–286 (1995)
4. Freund, Y., Schapire, R.E.: A Decision-Theoretic Generalization of On-Line Learning and an Application to Boosting. In: Proceedings of the Second European Conference on Computational Learning Theory, pp. 23–37 (1995)
5. Kanade, T., Cohn, J.F., Tian, Y.: Comprehensive Database for Facial Expression Analysis. In: Proceedings of the Fourth IEEE International Conference on Automatic Face and Gesture Recognition (FG 2000), pp. 46–53 (2000)
6. Koutlas, A., Fotiadis, D.I.: An Automatic Region Based Methodology for Facial Expression Recognition. In: IEEE International Conference on Systems, Man and Cybernetics, pp. 662–666 (2008)
7. Kulikowski, J.J., Marčelja, S., Bishop, P.O.: Theory of Spatial Position and Spatial Frequency Relations in the Receptive Fields of Simple Cells in the Visual Cortex. Biological Cybernetics 43, 187–198 (1982)
8. Shen, L., Bai, L.: Adaboost Gabor Feature Selection for Classification. In: Proc. of Image and Vision Computing, Akaroa, New Zealand, pp. 77–83 (2004)
9. Shih, F.Y., Chuang, C.F.: Automatic Extraction of Head and Face Boundaries and Facial Features. Information Sciences 158, 117–130 (2004)

Effective Semi-autonomous Telepresence

Brian Coltin[1], Joydeep Biswas[1], Dean Pomerleau[2], and Manuela Veloso[1]

[1] School of Computer Science, Carnegie Mellon University, Pittsburgh, PA, USA
{bcoltin, joydeepb, mmv}@cs.cmu.edu
[2] Intel Research, Pittsburgh, PA, USA
dean.a.pomerleau@intel.com

Abstract. We investigate mobile telepresence robots to address the lack of mobility in traditional videoconferencing. To operate these robots, intuitive and powerful interfaces are needed. We present CoBot-2, an indoor mobile telepresence robot with autonomous capabilities, and a browser-based interface to control it. CoBot-2 and its web interface have been used extensively to remotely attend meetings and to guide local visitors to destinations in the building. From the web interface, users can control CoBot-2's camera, and drive with either directional commands, by clicking on a point on the floor of the camera image, or by clicking on a point in a map. We conduct a user study in which we examine preferences among the three control interfaces for novice users. The results suggest that the three control interfaces together cover well the control preferences of different users, and that users often prefer to use a combination of control interfaces. CoBot-2 also serves as a tour guide robot, and has been demonstrated to safely navigate through dense crowds in a long-term trial.

Keywords: telepresence, mobile robots.

1 Introduction

With the advent of the Internet and videoconferencing software, people have become more and more connected. This is particularly evident in the office, where work groups span countries and continents, but still keep in touch. However, these teleconferencing solutions leave much to be desired due to a lack of mobility— the user can only see what the camera is pointing at.

A mobile telepresence platform is much more powerful than these static videoconferencing solutions. Employees can move down the hall to visit their coworker's office from their own home on another continent, engineers can inspect overseas factories remotely, and collaboration between distant coworkers is enhanced. To this end, mobile telepresence robots have been developed which allow users to physically interact with and move through their environment, including the Willow Garage Texai and the Anybots QB [7]. We have developed CoBot-2, a mobile telepresence robot which is controllable through a web browser. CoBot-2 is unique in the extent of its autonomy: although given high-level instructions by a human, it localizes and navigates through the building entirely autonomously.

T. Röfer et al. (Eds.): RoboCup 2011, LNCS 7416, pp. 365–376, 2012.
© Springer-Verlag Berlin Heidelberg 2012

In spite of the ever increasing capabilities and robustness of autonomous robots, they are still incapable of performing some tasks on their own. We postulate that even as autonomous robots become increasingly capable, unexpected situations will remain where the assistance of a human is required. We envision this relationship between humans and robots as being *symbiotic* rather than exploitative, such that both the human and robot benefit from the relationship [13]. In the case of CoBot-2, the robot enables the human to visit remote locations, localizing and navigating autonomously to the greatest extent possible. On the other side, the human observes CoBot-2's motion through video to ensure that CoBot-2 doesn't collide with any unseen obstacles, and may reset CoBot-2's localization if the robot becomes lost.

For both telepresence and other forms of assisted autonomy, the effectiveness of the robot directly depends on the ability of humans to interact with the robot via its user interface. This user interface should empower people with little to no training to fully grasp the state of the robot, and to control the robot to perform complex tasks intuitively and quickly. CoBot-2's user interface is designed to be used from a web browser, and the robot's state is displayed in real-time. The interface allows the user to drive CoBot-2 in three ways: a joystick interface, with directional keys to turn and move forward (with buttons or the keyboard), by clicking on a destination point on the floor in the image from CoBot-2's camera, or by clicking on a destination point on the map of the building. A brief user study was conducted to determine the preferences of users between the three control interfaces. Each of the three interfaces were the preferred control input for some subset of the users, so we conjecture that all of the control modalities should be included in the user interface.

First, we will discuss related work in teleoperation interfaces, and then the hardware and autonomous aspects of CoBot-2's software. Next, we will describe the web interface in detail, followed by the user study procedure and results. Finally, we will discuss CoBot-2's debut as a tour guide in a day-long trial.

2   Related Work

Vehicles have been teleoperated since at least 1941, when pilotless drone aircraft were used to train pilots [5]. Remote operation has since become even more feasible and practical with the invention of the Internet. The first robot capable of manipulating the environment from the web was most likely the Mercury Project, a robotic arm which could explore a sandbox, deployed by Goldberg et. al. in 1994 [6]. This was followed by a mobile robot, Xavier, in late 1995, which could be sent to points throughout a building via a web browser [9]. In 1998, RHINO was developed to guide visitors at a museum. It combined local interactions with museum visitors with a web interface to observe the robot remotely [3]. These early web interfaces required the user to click on a link or refresh the page to update the robot status due to the limitations of HTML at the time. Recently robots teleoperated over the web have become more common, such as [14]. See [12] for a discussion of several other robots teleoperated over the web. CoBot-2's web interface uses AJAX and HTML5 to display live updates of the robot's state in a web browser without the installation of any browser plugins. Currently, at least eight companies have developed commercial telepresence robots; some of them with web browser interfaces [7].

Teleoperation interfaces are used in space missions. On the NASA Mars Exploration rovers, the robots autonomous behaviors are complemented by controllers on Earth. Due to the long communication delay between the Earth and Mars, teleoperation in this environment is particularly challenging [11]. Teleoperation interfaces have also been tested in a competitive environment with the RoboCup rescue competitions. Teams have focused on fusing information with the video display [1], and have also used principles learned from video games [8] to make their interfaces easy to learn and use. These robots operate in difficult to traverse rescue environments, and the operators still require extensive training to perform most effectively. Similarly to telepresence, tele-immersion has enabled people to interact with each other in virtual environments [10].

The human-robot interaction community has extensively studied robot teleoperation. Some of the challenges include a limited field of view, determining the orientation of the robot and camera, depth perception, and time delays in communication with the robot [4]. CoBot-2's teleoperation interface attempts to address each of these problems.

3   CoBot-2's Autonomous Capabilities

CoBot-2 is a new model of the original CoBot presented in [2] and [13], with new capabilities relating to telepresence. In terms of hardware, CoBot-2 uses the same base as the original CoBot (with omnidirectional motion and LIDAR) with the addition of a touch screen tablet, a StarGazer sensor, and a pan/tilt/zoom camera (see Fig. 1)[1]. CoBot-2 is able to localize itself and navigate through the building autonomously.

From CoBot-2's web interface, the user may command CoBot-2 to move to a global position on a map or a position relative to the robot. Although initiated by the user, these commands rely on the autonomous functionality of CoBot-2, particularly localization, navigation and obstacle avoidance.

3.1   Localization

CoBot-2 uses a particle filter on a graph-based map for localization. For sensor feedback, it uses odometry and StarGazer estimates. StarGazer is an off-the-shelf localization system, which uses regularly placed fiducials on the ceiling. Each StarGazer fiducial consists of a pattern of retroreflective dots arranged on a grid, and is detected by a ceiling facing infrared camera mounted on the robot.

The location estimate of the robot is chosen as the mean of the cluster of particles with the largest weight. Localization is used for autonomous navigation

---

[1] Special thanks to Mike Licitra for designing and constructing CoBot-2, and to John Kozar for building the tablet and camera mount.'

Fig. 1. CoBot-2 is equipped with 1) an omnidirectional base; 2) a 4 hour onboard battery supply; 3) a Hokuyo LIDAR sensor to detect obstacles; 4) a touch-screen tablet with wireless for onboard processing; 5) a StarGazer sensor; and 6) a pan/tilt/zoom camera

when moving to a global position. In the rare event that localization fails and CoBot-2 is unaware of its position, it may ask a human for help, who will click a map to set the localization position [13].

3.2   Navigation

CoBot-2 navigates autonomously to move to a global position on the map. The map of the building is represented as a graph of edges connecting vertices. Each edge is associated with a width indicating the distance the robot is permitted deviate from the edge (for obstacle avoidance). This distance is generally the width of the space the robot is navigating. To navigate between two locations, CoBot-2 finds the nearest points on the graph to these two locations and travels on the shortest path between the two points on the graph. CoBot-2 travels in a straight line while entering or exiting the edges of the graph.

3.3   Obstacle Avoidance

Every motion primitive is executed while avoiding obstacles. CoBot-2 avoids obstacles (as detected by its LIDAR sensor) by side-stepping around them. This is possible with the omnidirectional drive, which permits decoupled translation and rotation of the robot. While side-stepping around obstacles, the obstacle avoidance algorithm limits the extent to which CoBot-2 deviates from the center of the edge on the map. If no free path is found to avoid the obstacles, the robot stops and waits until its path is cleared. Once clear of obstacles, the robot moves back to the center of the edge.

Fig. 2. CoBot-2 successfully navigated through dense crowds autonomously at the open house for an entire day, guiding visitors to exhibits

3.4   Open House Guide

In addition to its telepresence capabilities, CoBot-2's autonomous navigation has been used for a tour guide robot. To demonstrate and test these capabilities, CoBot-2 served as a tour guide at an all day open-house for a total of six hours. CoBot-2 successfully guided visitors through dense crowds at the open house for the entire day (see Figure 2).

CoBot-2 was supplied beforehand with the locations of the exhibits, which were marked on the web interface's map and displayed on the touch screen. When the user selected an exhibit, CoBot-2 guided visitors to the exhibit, explaining along the way and avoiding obstacles, including people. If CoBot-2's path was blocked, it said "Please excuse me" until it could move again.

In spite of the dense crowds, which included people of all ages and sizes, and dangerous obstacles such as chairs, table overhangs and posters, which CoBot-2's lasers could not detect, CoBot-2 did not have a single collision or accident for the entire day, while still moving at a reasonable human walking speed. During the open house, CoBot-2 was not closely monitored by humans— at some points, we did not even know where the robot was and had to go search for it.

One interesting aspect of the open house was seeing people's responses to the robot. They quickly became accustomed to its presence, and went from seeing it as a surprising novelty to moving comfortably and naturally without being bothered by the robot's presence. CoBot-2's interactions at the open house showcased the exciting potential of robots which can interact naturally and safely with humans in unconstrained indoor environments, as telepresence robots must do.[2].

4   Web Interface

Users interact with CoBot-2 through a web-based browser interface. With the web-based interface, no special software needs to be installed to interact with

---

[2] A video of CoBot-2's interactions with visitors at the open house is available at http://www.youtube.com/watch?v=0NawEFNXZkE.

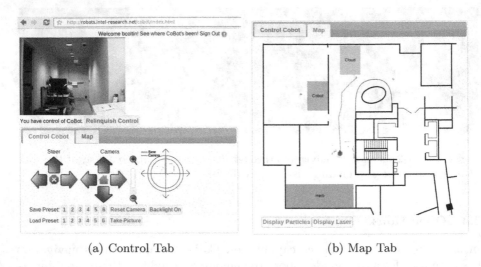

(a) Control Tab   (b) Map Tab

Fig. 3. CoBot-2's web interface, with the Control and Map tabs visible. In the map tab, CoBot-2's current position, path and LIDAR readings are shown. Users may click on the map to set CoBot-2's localization position or travel to a point.

CoBot-2. Furthermore, the software runs on multiple devices— desktop computers, mobile devices such as smartphones, and CoBot-2's own touch screen. The web client repeatedly requests robot state information ten times per second, and requests new images from the robot's camera five times per second.

A username and password are required to control the robot. If a user is not logged in, they may still view images from the robot and its status but are unable to issue commands to the robot. If multiple users are logged in to the server, only one may control the robot at a time to prevent confusion and contention.

Next, we will examine each component of the web interface. The user may switch between two tabs in the web interface: the Control tab, which contains buttons to control the camera and drive the robot, and the Map tab, which shows the map of the environment and the robot's position. The image from the robot's camera is always displayed above the two tabs in the remote interface.

4.1   Camera Controls

From the control tab, users can control the robot's camera. There are arrow keys to move the camera, surrounding a "home" button which returns the camera to a forward facing position suitable for driving. Next to these arrows is a slider with which the user can set the zoom level of the camera. The rate of the camera's motion when using the arrow keys depends on the level of the camera's zoom— at higher zoom levels the camera moves more slowly to allow more precise control. All of these commands have visual icons representing their function (such as magnifying glasses with "+" and "-" for zooming, a picture of a house for the home button) and are associated with keyboard shortcuts (see Fig. 3a). When the user clicks on an arrow or presses the associated keyboard shortcut, the arrow button on the screen becomes pressed, providing visual feedback.

The Control tab also allows the user to configure six preset camera states with a "save" and "load" button for each preset state, which is convenient for attending meetings with CoBot-2 where there are fixed camera positions for looking at slides, the speaker, or other meeting attendees. Adjacent to the preset buttons is a button to turn the backlight on or off. Toggling the backlight enables the user to optimize image quality for different lighting conditions.

4.2   Steering Arrows

Three arrows for steering CoBot-2 are displayed directly to the left of the camera arrows (See Fig. 3a). Although CoBot-2 is capable of omnidirectional motion, these arrows only allow turning in place and moving directly forwards. This is the type of interface we believed would be most familiar to users, and which would also cause the least confusion in conjunction with the movable camera. In the center of the arrows is an emergency stop button. The robot can also be moved with the arrow keys on the keyboard, and can be stopped by pressing space. CoBot-2 autonomously performs obstacle avoidance while controlled with the arrow keys, adjusting its velocity to avoid obstacles.

4.3   Compass

On the right side of the Control tab is a compass-like object which displays the relative orientation of CoBot-2 and its camera, each with a colored "compass needle". The needle representing CoBot-2's camera always points "north" on the compass, and the needle representing the orientation of the robot's base moves relative to this based on the current value of the camera's pan angle. The LIDAR readings are displayed on the compass so that the user can visualize objects in the immediate vicinity of CoBot-2 (see Figure 3a).

The compass is intended to provide situational awareness to a user who has moved the camera, so that they can tell which way the camera is facing, which is essential when driving with the arrow keys. It also allows an experience user to predict which obstacles CoBot-2 will avoid automatically via the LIDAR readings.

CoBot-2 can also be turned precisely using the compass. With the arrow keys, commands are sent to turn a set, small distance when pressed, and to stop turning when the keys are released. However, there is network latency sending the command to stop turning, which will often cause CoBot-2 to overturn. With the compass, CoBot-2 is told to rotate a set angle, so network latency does not cause overturning.

4.4   Camera Image: Visualization and Control

A 320x240 image from CoBot-2's camera is displayed at the top of the display, and refreshed every fifth of a second. Network latency is manageable enough that overseas users can control CoBot-2 without difficulty.

Fig. 4. At left, the robot views a scene at the default zoom level. At right, the user zooms in to the boxed area at the maximum level of 18X. The powerful zoom functionality of the robot allows users to inspect small or distant objects, and even read text remotely.

In addition to displaying the environment, the image is used for control. When a user clicks on the image, the camera moves to center on the selected point, providing more precise control than the arrow keys. Additionally, by scrolling the mouse wheel with the cursor over the image, the camera zooms in or out (see Fig. 4).

By clicking on the ground plane in the image while holding the shift key, the user can move CoBot-2 to a specific spot on the floor. The robot computes the clicked spot by considering the height of the robot and the pan, tilt and zoom levels of the camera. We discard clicks above the horizon, and limit the maximum distance travelled by a single click to reduce the consequences of mistaken clicks. With this control scheme, it is irrelevant which direction the camera is facing, and lack of awareness of the robot's camera and orientation is not an issue. Furthermore, moving the robot by clicking on the image does not suffer from the latency problems of the arrow keys.

4.5   Map

The map is shown as a separate tab so that the user need not scroll the browser window to view the entire interface. It displays a visualization of CoBot-2's environment, position and orientation. While CoBot-2 moves to a point chosen by the user, the path to its' destination is displayed.

Users set a destination for CoBot-2 to travel to by clicking on the map, then dragging and releasing the mouse button to set the orientation. Additionally, by holding the shift key while choosing a location, the user can set CoBot-2's localization position from the Map tab. This feature is useful in the rare event that CoBot-2's localization becomes lost.

In the Map tab, users may still access much of the functionality from the Control tab, such as driving, emergency stop, and moving the camera, through keyboard shortcuts. The image from the robot's camera remains visible.

4.6   Touch Screen and Mobile Interface

One of the goals behind using a web-based interface is to make CoBot-2 accessible from its own touch screen and from mobile devices. However, there are several shortcomings preventing the interface from working as is on touch screens and mobile devices. First, the shift key must be held down to move CoBot-2 by clicking on either the image or the map, but the shift key is not available on touch screens. To resolve this, on touch screens we provide a radio toggle button enabling the user to change modes for both the image and the map. The user then performs a normal click to activate the selected function. On the image, the radio buttons toggle between looking at a point with the camera or moving to a point on the floor, and for the map, the radio buttons switch between setting CoBot-2's localization position and moving to a point on the map.

Second, to set the orientation of CoBot-2 in the map, the user must press the mouse button, drag the mouse and release to set the orientation. This is not possible on certain mobile devices, so for the touch interface, one click sets the position, and a second click sets the orientation.

A third shortcoming is that touch screens are smaller, so the entire interface does not fit on CoBot-2's tablet. To avoid the need to scroll the page, CoBot-2's camera image is only displayed in the Control tab for cell phones and tablets.

5   User Study

To determine the effectiveness of the various user interfaces for controlling CoBot-2, we conducted a small user study. The participants were seven students or recent students in robotics and/or computer science, none of whom had interacted with CoBot-2 before. Their experience with robots varied widely, from people who had never used robots before to people who had worked with them extensively for years. The participants were in different buildings than the robot. The purpose of the study was to demonstrate the telepresence interface's effectiveness and to determine users' preferences for control interfaces among the directional controls (arrow keys), compass, clicking on the image, and clicking on the map. Our hypothesis was that users would prefer to use higher level interfaces: first the map, then clicking on the image, and using the arrow keys least of all.

5.1   Setup

The participants in the study were sent instructions by email, and asked to visit CoBot-2's website and login remotely. They then had to complete a set of tasks via the web interface, each using only one of CoBot-2's control interfaces. For the final task, the participants were told to use whichever control scheme they preferred. See Figure 5 for a detailed explanation of the tasks.

After completing the tasks in the instructions, participants were asked to fill out a survey. The survey asked questions to confirm that the participants had completed the assigned tasks, and to rank from 1 to 10 how much they used

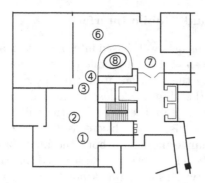

Fig. 5. In the user study, the participants: 1) Begin with CoBot-2 at this position; 2) Read a nametag at position 2 by moving the camera with the arrows; 3) Drive CoBot-2 to point 3 with the arrow keys; 4) Read a secret message on a post-it note at 4; 5) Rotate CoBot-2 with the compass at 3; 6) Navigate to read a poster at 6 by clicking on the image; 7) Navigate to position 7 with the map; 8) Identify a food item in the kitchen area at 8, using any control scheme; and 9) Return to the original position at 1. Not shown on the map are numerous cubicles, chairs and desks arranged in the lab.

each interface in the final task. The participants were also asked open-ended questions to describe the most frustrating and most enjoyable aspects of CoBot-2's interface, and for any other comments they may have.

5.2   Results

All of the participants completed the tasks in approximately fifteen to thirty minutes. The results of the survey did not show a preference for higher level interfaces. We believed that users would greatly prefer driving by clicking on the image or map. However, in the final task, all but one of the subjects used the arrow keys to a significant extent. Furthermore, each of the interfaces was used to the greatest extent by some subset of the participants— two used the map the most, one clicked on the image the most, two used the arrow keys the most, and two used multiple interfaces an equal amount. This indicates that each of the three control schemes is valuable, and the inclusion of multiple control interfaces increases the usability of the web interface. See Table 1 for the complete results.

The impetus for providing multiple interfaces is strengthened by the fact that only two participants used a single interface to complete the final task— subject 4, who used only the arrow keys, and subject 5, who greatly preferred clicking on the image. The remaining participants used a combination of all three control interfaces. Different control methods are appropriate for different situations. We have found the arrow keys to be appropriate for driving small distances and maneuvering in the vicinity of obstacles such as tables, chairs, and posters on easels which cannot be detected by the ground level LIDAR sensor. Clicking on the camera image works very well for line of sight movement, but requires multiple clicks for large distances, and is not feasible for small distances where

Table 1. The responses of the seven participants to the question, "To what extent (from 1 to 10, 10 is the greatest extent) did you use each of the following interfaces when driving CoBot-2 home?"

Subject	Arrows	Click on Image	Map
1	7	5	10
2	8	4	7
3	8	7	8
4	9	1	1
5	1	10	1
6	5	5	5
7	5	8	9

the robot cannot see the floor immediately around its base. Similarly, the map is excellent for traversing long distances, but is not very precise. We suggest that user interface designers provide multiple control interfaces to accommodate both user preferences and the requirements of diverse use scenarios.

Another interesting factor observed in the user study was the amount of trust users placed in the robot. The majority of users were willing to drive by clicking on the map or image and expected CoBot to avoid obstacles. However, subjects 4 and 7 commented in the survey that they did not trust the robot to avoid obstacles while moving autonomously. For subject 4, this lack of trust excluded clicking on the map or image at all in the final task. When asked to use the map interface to go to a point, Subject 7 accomplished this by setting individual waypoints for the robot a small distance apart to prevent collisions, also due to a lack of trust. With more familiarity with the robot and a better understanding of the robot's capabilities, these users would feel more comfortable allowing the robot to move autonomously with the map interface.

6   Conclusion

CoBot-2, a mobile telepresence robot, enables visitors to interact with an environment remotely through its browser-based teleoperation interface. CoBot-2 has been shown to perform effectively and safely in indoor environments with humans over a long term. Control of CoBot-2 is symbiotic with a variable level of autonomy, in which the user controls CoBot-2 through one of three control interfaces. Control interface preference varied between users, with many users taking advantage of multiple interfaces. CoBot-2's semi-autonomous control schemes enabled easy and safe navigation with little user effort.

Acknowledgements. This research was funded in part by the NSF and by Intel. The views expressed in this paper are those of the authors only.

The authors would like to thank Mike Licitra for designing and building CoBot-2, and John Kozar for designing and building the mount for the tablet and camera. Special thanks also goes to the participants in the user study and the guests who interacted with CoBot-2 at the open house.

References

1. Baker, M., Keyes, B., Yanco, H.A.: Improved interfaces for human-robot interaction in urban search and rescue. In: Proc. of the IEEE Conf. on Systems, Man and Cybernetics (2004)
2. Biswas, J., Veloso, M.: Wifi localization and navigation for autonomous indoor mobile robots. In: IEEE International Conference on Robotics and Automation, pp. 4379–4384 (May 2010)
3. Burgard, W., Cremers, A., Fox, D., Hähnel, D., Lakemeyer, G., Schulz, D., Steiner, W., Thrun, S.: The interactive museum tou-guide robot. In: Proc. of the National Conference on Artificial Intelligence, AAAI (1998)
4. Chen, J.Y.C., Haas, E.C., Barnes, M.J.: Human performance issues and user interface design for teleoperated robots. IEEE Transactions on Systems, Man and Cybernetics 37(6), 1231–1245 (2007)
5. Fong, T., Thorpe, C.: Vehicle teleoperation interfaces. Autonomous Robots 11 (2001)
6. Goldberg, K., Mascha, M., Genter, S., Rothenberg, N., Sutter, C., Wiegley, J.: Desktop teleoperation via the world wide web. In: Proc. of IEEE Int. Conf. on Robotics and Automation (1995)
7. Guizzo, E.: When my avatar went to work. IEEE Spectrum, 26–31, 48–50 (September 2010)
8. Kadous, M.W., Sheh, R.K.M., Sammut, C.: Effective user interface design for rescue robotics. In: Proc. of ACM/IEEE Int. Conf. on Human-Robot Interaction (2006)
9. Koenig, S., Simmons, R.: Xavier: A Robot Navigation Architecture Based on Partially Observable Markov Decision Processes, pp. 91–122 (1998)
10. Lien, J.-M., Kurillo, G., Bajcsy, R.: Skeleton-Based Data Compression for Multi-Camera Tele-Immersion System. In: Bebis, G., Boyle, R., Parvin, B., Koracin, D., Paragios, N., Tanveer, S.-M., Ju, T., Liu, Z., Coquillart, S., Cruz-Neira, C., Müller, T., Malzbender, T. (eds.) ISVC 2007, Part I. LNCS, vol. 4841, pp. 714–723. Springer, Heidelberg (2007)
11. Maimone, M., Biesiadecki, J., Tunstel, E., Chen, Y., Leger, C.: Surface Navigation and Mobility Intelligence on the Mars Exploration Rovers, pp. 45–69 (March 2006)
12. Martìn, R., Sanz, P., Nebot, P., Wirz, R.: A multimodal interface to control a robot arm via the web: A case study on remote programming. IEEE Transactions on Industrial Electronics 52 (2005)
13. Rosenthal, S., Biswas, J., Veloso, M.: An effective personal mobile robot agent through symbiotic human-robot interaction. In: Proc. of Int. Conference on Autonomous Agents and Multi-Agent Systems (May 2010)
14. Sim, K., Byun, K., Harashima, F.: Internet-based teleoperation of an intelligent robot with optimal two layer fuzzy controller. IEEE Transactions on Industrial Electronics 53 (2006)

Perceiving Forces, Bumps, and Touches from Proprioceptive Expectations

Christopher Stanton[1], Edward Ratanasena[1],
Sajjad Haider[2], and Mary-Anne Williams[1]

[1] Innovation and Enterprise Research Laboratory, University of Technology, Sydney
cstanton@it.uts.edu.au, edward.ratanasena@student.uts.edu.au,
mary-anne.williams@uts.edu.au
[2] AI Lab, Institute of Business Administration, Karachi
sajjad.haider@khi.iba.edu.pk

Abstract. We present a method for enabling an Aldebaran Nao humanoid robot to perceive bumps and touches caused by physical contact forces. Dedicated touch, tactile or force sensors are not used. Instead, our approach involves the robot learning from experience to generate a proprioceptive motor sensory expectation from recent motor position commands. Training involves collecting data from the robot characterised by the absence of the impacts we wish to detect, to establish an expectation of "normal" motor sensory experience. After learning, the perception of any unexpected force is achieved by the comparison of predicted motor sensor values with sensed motor values for each DOF on the robot. We demonstrate our approach allows the robot to reliably detect small (and also large) impacts upon the robot (at each individual joint servo motor) with high, but also varying, degrees of sensitivity for different parts of the body. We discuss current and possible applications for robots that can develop and exploit proprioceptive expectations during physical interaction with the world.

Keywords: motor learning, Nao, robot soccer, anticipation, collision detection.

1 Introduction

Most animals have a rich sense of touch to provide feedback during interaction with the world. By fusing touch, proprioception, and other sensations they are able to perceive collisions and forces upon their body. Robots however often have hard protective shells, and lack equivalent tactile sensors. For robots to interact meaningfully with the environment, they will need to be capable of detecting expected and unexpected collisions with other objects, people, and themselves. Our aim is to model perceptions related to contact forces on robots that do not possess a dense array of dedicated touch, tactile or force sensors. Examples of such robots include the Sony AIBO or the Aldebaran Nao. We achieve this

T. Röfer et al. (Eds.): RoboCup 2011, LNCS 7416, pp. 377–388, 2012.

by identifying discrepancies in sensed angular motor position values caused by impact forces between the robot's limbs and objects within its environment.

This paper is structured as follows: Section 2 describes the application problem domain - robot soccer, and in particular the Standard Platform League (SPL). Section 3 describes other approaches for detecting physical contact with similar robots (i.e. robots without dedicated touch/tactile sensors). Next, we describe our approach based on proprioceptive motor sensory expectations, and how it differs from other relevant techniques, and the advantages it offers. Our implementation details are described in Section 5. Section 6 describes the results of experiments used to test our approach, and we conclude by discussing the current and future applications of the approach.

2   Problem Domain: The Standard Platform League

Robot soccer, as per human soccer, is a contact sport. It is dynamic, with frequent collisions and subsequent falls (perhaps more so in robot soccer, given the current state of the art). Such collisions can occur between multiple players, between players and referees, between players and the ground, players and the ball, players and the goal posts, and so forth. Players limbs also make (often accidental) impact with other parts of their own body (a self-collision).

In the RoboCup Standard Platform League (SPL), all teams participate with "identical"[1] robots. In the past the league used the Sony Aibo four-legged robot, and today the SPL uses the Aldbaran Nao humanoid robot. In SPL soccer matches, unwanted and undesirable collisions with other robots and obstacles is a common occurrence. With Aibos it could result in "leg-locking" [7], and with Naos it often results in a fall. In both cases, major damage to the robots can occur.

The Aldbaran Nao robots are equipped with sensors such as head mounted colour cameras, front-facing ultrasonic distance sensors, force sensors in the soles of the feet, bump detectors on the toes of each feet, accelerometers, and gyroscope. While the "pushing"[2] rules in the SPL are designed to encourage robots to avoid colliding with each other, collisions still occur with regular frequency. This is due to a number of factors. Both vision-based and sonar-based object avoidance is hampered by the small field of view of both these devices, with many robot-robot collisions occurring when one robot hits another from behind or from the side. Other techniques for detecting collisions involve the use of accelerometers and gyroscopes, but this involves detecting a collision after it happens (often at a point when the robot has already become unstable). Instead, it would be more useful to anticipate collisions, and to react to unexpected collisions almost instantaneously. Thus, a better solution is required.

---

[1] This is the ideal - in practice, teams use similar robots provided by the same manufacturer.

[2] A robot deemed by the referee to be guilty of pushing another robot is removed from the field of play for a period of time.

3   Background and Related Work

Detecting collisions is important for most autonomous robots. For robots to work safely alongside and with humans, they must be capable of detecting physical contact and/or responding appropriately. Many approaches to achieving safe human robot interaction (HRI) focus on building compliant robots through the use of variable stiffness actuators, which can absorb collision energies through compliant mechanisms, while achieving precise joint positioning through variation of stiffness gains. Compliance can be implemented in different ways - mechanically, e.g. through the use of internal springs as a potential energy buffer [9]), or through the use of intelligent software controllers (e.g. that dynamically adjust their commands in response to joint torque sensors [3]). Collision prevention strategies can also be imposed on the controller, e.g. by imposing forbidden workspace area constraints [1].

In the RoboCup domain, approaches to detecting collisions have focused on processing data from accelerometers or joint position sensors. Approaches utilising the accelerometer involve discriminating between patterns of data produced under different gaits and/or environmental conditions. For example, Vail and Veloso [8] demonstrate how the accelerometer sensor of a Sony Aibo can be used to identify different surfaces upon which the robot is walking through a decision tree. In soccer matches, Mericli et al. [6] demonstrate how a Sony Aibo's accelerometer readings can be analysed statistically to probabilisticly discriminate between "normal" walking, and walking in which a collision has occurred.

Another strategy for detecting collisions is to process the joint position data for each degree of freedom on the robot, with the aim of distinguishing normal, collision-free sensor readings, and sensor readings produced when the robot is experiencing a collision. Quinlan et al. [7] describe a fault detection system operating on the Sony Aibo, in which slip and collisions are detected by measuring the normal variation in the robot's motors during normal locomotion tasks, and comparing this with data collected in which collisions and slips occur. After training, a collision or slip is detected if multiple consecutive readings of a sensed motor position value are outside a range of two standard deviations from an expected value. Hoffman and Gohring [5] describe a collision detection process (also implemented on the Sony Aibo) in which they compare command data with sensor data. The difference between command values and sensor values are measured over a period of time (96ms, 12 frames). If the robot's range of motion is impaired because of a collision with another object, the difference between command value and sensor value increases. If the error is above a certain threshold for a particular movement, a collision is detected.

Each of these approaches have their particular strengths and weaknesses. Accelerometers allow the robot to detect collisions independent of the point of contact (useful on a robot without tactile sensors), but require disturbing the robot's body (e.g. if an obstacle hits the robots chest, the accelerometer would detect this, even though there is no motor in the chest). However, using the accelerometer to detect a collision requires the collision to be of sufficient force that it disrupts the stability of the robot (which is a problem on a humanoid robot, such as the Nao as some collisions are detected too late for the robot avoid a fall). Conversely, accelerometers might fail to detect collisions in which a motor's range of freedom is impaired, but the particular motor impedance is not sufficient to effect the stability of the robot. Approaches which rely on detecting collisions through impedances of motors rely on a collision restricting the range of motion of one of the robot's limbs. While this approach may be capable of detecting some gentle collisions, it will fail to detect any collision that does not impede the robot's range of motion. All approaches however, regardless of their input sensory device, rely on discriminating between sensor input during collision-free movement and collision-impaired movement.

4　Our Approach

Our goal was to use motor position sensors to provide the Nao with a sense of "touch" that provided more sensitivity and scope than existing methods. We aimed to build a system to perceive not only the strong, forceful bumps and collisions that occur during robot soccer, but the gentler interactions that might occur when an autonomous robot is interacting with objects and people in its environment - for example, a human gently pushing or impeding a robot's arm or head.

While the previous approaches of [7] and [5] had demonstrated detecting collisions between robots using motor position sensors is possible, these approaches focused mainly on detecting the (sometimes brutal) types of collisions that occur only in robot soccer games[3]. With both approaches, calibration of detection thresholds is done for each category of motion - thus requiring each new motion be calibrated. Also, their detection triggers rely on finding differences that exceed the maximum found in all previous training data for each particular category. Thus, sensitivity and responsiveness can only be improved by creating new categories of motion, and hand-tuning threshold parameters for each particular motion.

4.1　Sensor Command Difference (SCD)

Intuition suggests that previous commands issued to the motor will effect the current sensor reading, and that the difference between the most recent command value and most recent sensor value will not be constant. The motor, being a mechanical device, is subject to physical forces of varying degrees caused by the effects of friction , inertia, momentum, and gravity, as well as the effects of these forces upon the limbs to which it is attached. Impeding forces will likely create a bigger difference between sensor and command had they not occurred, while other forces may push the motor towards a target position, reducing sensor-command difference.

---

[3] Even in robot soccer, it would be advantageous to detect smaller forces earlier, as a soft touch can act as an early warning system to give the robot time to avoid a more serious impact.

We term the difference between the last command value and the current sensor value the "sensor command difference" (SCD). We examined how the SCD varied for each of the Nao's 21 degrees of freedom throughout the course of a robot soccer game. Data was collected for each motor and statistically analysed. As can be seen in Figure 1, predictable periodic patterns can be observed in the SCDs, which probably correspond to foot-ground impacts. Similar patterns were seen for all motors.

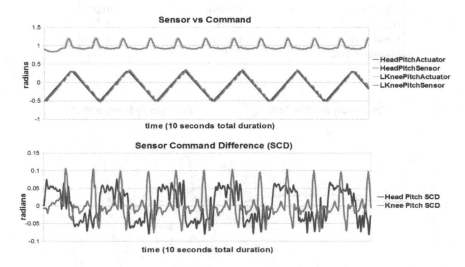

Fig. 1. Top: the sensor and command values for the head pitch and left knee pitch motors are displayed. Below, the corresponding calculated SCDs for head pitch and left knee pitch are displayed. The data was collected while the robot was chasing a soccer ball. Our aim is to predict the SCD every 10ms, and to use discrepancies between SCD estimates and the sensed SCD to infer the experience of an unexpected force/impact.

To further investigate how the SCD values of the different motors of the Nao robot were effected by walking, we instructed the robot to walk forwards (with no rotation and no strafe)[4]. There was no discernible differences between left and right motors. However, as can be seen in Table 1 the magnitude of SCD varied significantly for different parts of the body during the walk. In the legs, the greatest variances and magnitudes of SCD were found in the knees and hip pitch motors, while in the upper body both the shoulder pitch and head pitch experienced large SCD values. The fact that pitch motors (as opposed to yaw and roll) were most effected by walking suggests that impacts between the ground, and also the weight of the robot's body have large effects on SCD.

We speculated that other factors besides walking may also affect SCD values, such as the velocity and acceleration with which a limb is currently moving

---

[4] Using the Aldebaran walk engine, at full forwards velocity, and with motor stiffness set to 80 percent.

Table 1. Variance of SCD for the robot's motors while walking forwards (ordered by greatest variance, top to bottom)

Motor	Max SCD (radians)	Std Dev SCD (radians)
Knee Pitch	0.140	0.040
Shoulder Pitch	0.080	0.050
Hip Pitch	0.060	0.040
Ankle Pitch	0.050	0.020
Head Pitch	0.050	0.020
Hip Roll	0.040	0.020
Hip Yaw	0.020	0.010
Head Yaw	0.020	0.010
Shoulder Roll	0.010	0.005

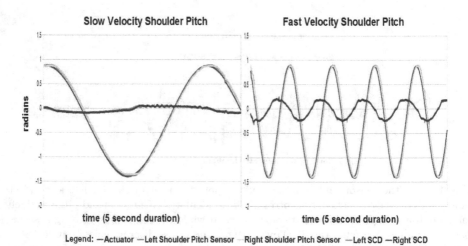

Fig. 2. Slow velocity (left) versus fast velocity (right). The SCD of the high velocity motion is greater in magnitude than the slow velocity motion.

when it is requested to move to a new position. To test the effects of velocity and acceleration, we instructed the robot to move its shoulder pitch and roll motors at varying velocities between two points near the extremities of its range of motion. As can be seen in Figure 2, requesting the motors to travel at higher velocities increases SCD.

5   Implementation

Our aim is to enable the robot to accurately estimate the SCD every 10ms[5]. We then test this estimate as a means of perceiving unexpected forces.

---

[5] On each DCM callback event.

5.1   Design

We train one neural network per degree of freedom to estimate the SCD for that motor every 10ms.

5.2   Data Collection and Processing

We programmed our robots to log all motor position commands and position sensor values every 10ms while performing their normal autonomous duties (e.g. soccer). This data was then processed to calculate instantaneous velocity and acceleration, and also the SCD. Training data is characterised by the absence of the forces we want the robot to perceive. Care is taken that the robot does not receive bumps or pushes from other robots or people. If the robot experiences such events, the training data is discarded. Unavoidable events that may effect SCD, such as foot-ground contact or shifts in the robot's centre of mass are included.

5.3   Learning

Each neural network is trained to approximate a function which predicts the SCD based on the command (c), velocity (v), and acceleration (a), i.e $SCD = f(c,v,a)$. The function is represented by a matrix of values which represents the weights of each node on each other. These weights are optimised to approximate the function (training) as closely as possible using particle swarm optimisation. Once the neural network has been sufficiently trained, the weights are then loaded onto the robots. Using the pre-trained weights, the robot can accurately predict the SCD by calculating its joint command, velocity, and acceleration and feeding it into a neural network with the same configurations as the neural network trained off-line.

5.4   Detection Triggers

We deem an abnormal motor event to have occurred if the predicted SCD differs significantly from the sensed SCD. This comparison is made every 10ms by the robot for every motor of the robot. Since the neural network only approximates the function and considering stochastic errors, our system looks for runs of consecutive discrepancies between prediction and measured SCD to infer the motor is experiencing an unusual force[6,7].

---

[6] To date, we have found 5 to be the best figure.

[7] Initially we modelled the velocity and acceleration over 50ms to reduce the effects of noise. However our neural network is then slow to respond to sudden changes of command.

6   Experiments and Results

Each SCD prediction function learnt by the corresponding neural network was evaluated visually, as per Figures 3 and 4. Neural nets were also evaluated by the maximum discrepancy between their estimate of SCD and the actual SCD.

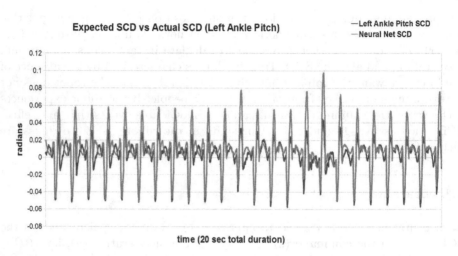

Fig. 3. Left Ankle Pitch: neural net estimate of SCD (displayed in red) versus the actual SCD (blue) experienced by the robot for the left ankle pitch motor for a 20 second duration from a robot playing soccer

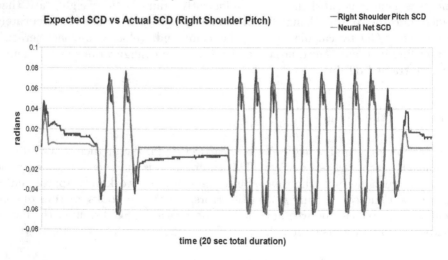

Fig. 4. Right Shoulder Pitch: the neural net estimate of SCD (displayed in red) versus the actual SCD (blue) experienced by the robot for the right shoulder pitch motor for a 20 second duration from a robot playing soccer

We calibrated detection thresholds during a period of experimentation in which we would use our hands to gently grab, push and impede a soccer playing robot. The soccer playing robot used the Aldebaran Walk Engine, and also performed two different kicking motions (both which require the robot to balance on one leg). The human would make contact with the robot at different points of the robot's body. The robot would speak the name of any motors for which it perceived 5 consecutive discrepancies between estimated SCD and the sensed SCD. If more than one motor perceived the impact, the name of the motor with the greatest discrepancy between estimated SCD and sensed SCD would be uttered by the robot.

Our system was calibrated so that it could be used in RoboCup soccer matches - as such false positives can be more damaging (strategically) than false negatives, as it is preferable to chase the ball than avoid a non-existent obstacle. Table 2 displays thresholds which eliminated nearly all false-positives. Unfortunately, we have so far been unable to empirically quantify the magnitude of the force required to trigger a discrepancy. One difficulty we faced in evaluating false-negative error rates is applying forces with a known magnitude. A false-negative can be easily induced with a weak touch, and conversely always avoided with strong touch or bump. The system is very sensitive to touches of the arms and head - only light touches are required (they are gentle enough that that there is little or no disturbance to the robot's movement. The system is less sensitive when it comes receiving bumps to the legs, as these forces need to be of sufficient strength that they disrupt the motion of the leg, and thus the stability of the robot. This may be due to the nature of the motors in these parts of the robots body[8]- it may also be the upper body motors are exposed to less SCD noise caused by walking than the motors in the legs. Also with regards to sensitivity, the further a limb was pressed away from the motor controlling that limb, the less force was required to generate an impact detection, most likely as a result of the leverage provided by the limb against the motor. We attempted to detect the presence or absence of the ball at the point of impact when kicking by instructing the robot to practice kicks without the ball, but the SCD discrepancy caused by ball-to-foot contact with our current kicking action and stiffness settings is so small that this appears impossible[9]. Table 3 describes various types impacts that were used to test our approach, and the motors that would identify the impeding forces.

We are currently using the system in our robot soccer team to detect unexpected impacts. The system is always operating, regardless of higher-level behaviour changes (e.g. a new walking gait). So far the system has proved quite robust to changes in behaviours and walking surfaces. We speculate this is because of the large effect velocity and acceleration have on SCD over a small time-scale (we are calculating velocity and acceleration for the last 20ms), and

---

[8] The motors in the legs appear stronger than the motors in the upper body.

[9] Perhaps with changes to our kicking motion and motor stiffness settings it may be possible to detect foot-ball contact.

Table 2. Threshold values are displayed (left), which were calibrated to remove almost all false positives

Motor	Threshold(radians)
Head Pitch	0.05
Head Yaw	0.05
Shoulder Pitch	0.05
Shoulder Roll	0.03
Hip Yaw	0.04
Hip Pitch	0.05
Knee Pitch	0.04
Ankle Pitch	0.05
Ankle Roll	0.03

Table 3. Impact types and the motors that typically identify them

Description of Forces	Motors typically identified by the robot
Impedance of the head	Head Pitch (vertical impedance), Head Yaw (lateral impedance)
Pushing down on the shoulders of the robot	Hip Pitch, Ankle Pitch, Ankle Roll
Grabbing robot's hand	Shoulder Pitch, Shoulder Roll, Elbow Pitch
Grabbing robot's upper arm	Shoulder Pitch, Shoulder Roll, but requires stronger force than if holding the robot's hand
Robot's foot is impeded while walking (e.g. by a heavy book)	Ankle Pitch, Ankle Roll, Hip Pitch, Hip Roll, Head Pitch, Head Yaw
Robot is unsteady (in danger of falling) but still trying to walk	Ankle Pitch, Ankle Roll, Hip Pitch, Hip Roll, Head Pitch, Head Yaw

that these relationships occur in all types of motions. However, further investigation is required.

Bumps and touches are detected regardless of whether the robot is stationary or moving. Changes to robot configuration are often detected. For example, a robot trained without a shoulder pad would then detect (via the head yaw motor) contact between the head and the shoulder pad when the shoulder pad was replaced. A sensor failure has also been detected via this system[10]. With regards to the collisions that occur in the robot soccer domain, our current results suggest this approach will be very useful for detecting arm-to-arm contact on the Nao robots, and many other collisions that occur in robot soccer matches. When the robot becomes severely unstable, this can be detected by a variety of motors, including head pitch and head yaw. Accelerometer-based approaches [6] would also make an excellent complementary approach.

---

[10] The robot repeatedly uttered "left ankle pitch" - closer investigation revealed the sensor value was a constant value, regardless of the position of the foot.

7   Discussion and Conclusion

We have demonstrated an effective approach to collision detection that relies on "detecting the unexpected". Sensor and command data is collected from a robot in which undesirable impacts do not occur. Machine learning is used to generate a proprioceptive expectation - an estimate of each motor's sensor position, based upon previous commands. Physical contact is inferred when a sensed motor position value differs significantly from a predicted value. Our approach allows the robot to perceive when its limbs physically contact other objects, despite the robot not having any dedicated tactile or force sensors at the point of impact.

Our results suggest interpreting SCDs can provide an estimate of force upon the motors of the robot. Violent actions produce large SCDS; smooth, slow movements small SCDs. We are yet to examine whether this approach can be used to perceive the direction of the force (e.g. is it impeding or pushing?), or whether it can provide a measure of the magnitude of a force.

Future work will extend our approach by fusing motor position values with the sensed electric current to each motor. We also aim to develop appropriate behavioural responses to unexpectedly large SCD stimuli. Currently, our robots only produce a verbal response to unexpectedly large SCD stimuli, and do not adjust or re-plan their motor instructions. As the Nao allows control of the stiffness of the robot's motors (via electric current), a simple approach would be to dynamically adjust motor stiffness in response to, or the anticipation of, large SCDs.

Other extensions to investigate include using neural networks that generate expectations for different types of sensors, such as the robot's accelerometers and the electric current sensors in each of the robot's motors. Also, if accelerometer information is provided to our robots using our current approach, this may assist them in predicting large SCDs, while also detecting collisions in which no range of motion is impeded.

Lastly, in future work we aim to use SCD measures for improving fine motor control. For example, if a walk engine dynamically reduces stiffness in anticipation of large SCDs, this may produce a smoother walk. While unsupervised machine learning has been applied to skills such as walking [2] and kicking [4] in the SPL, feedback for such behaviours is provided through the robot's visual system (e.g. a measure of ball distance in the case of a kick, or recognising land marks in the case of determining the speed at which a robot has walked). We are not aware of any motor learning research in this domain in which proprioceptive expectations are used to provide feedback to improve motor control.

Acknowledgements. We would like to acknowledge Nathan Kirchner and Benjamin Johnston for their useful suggestions.

References

1. Calinon, S., Sardellitti, I., Caldwell, D.: Learning-based control strategy for safe human-robot interaction exploiting task and robot redundancies. In: Intelligent Robots and Systems (IROS), pp. 249–254. IEEE (2010)

2. Chalup, S., Murch, C., Quinlan, M.J.: Machine learning with AIBO robots in the Four-Legged League of RoboCup. IEEE Transactions on Systems, Man, and Cybernetics, Part C: Applications and Reviews 37, 297–310 (2007)
3. Chen, W., Sun, Y., Huang, Y.: A collision detection system for an assistive robotic manipulator. Life System Modeling and Intelligent Computing 97, 117–123 (2010)
4. Hausknecht, M., Stone, P.: Learning Powerful Kicks on the Aibo ERS-7: The Quest for a Striker. In: Ruiz-del-Solar, J., Chown, E., Ploeger, P.G. (eds.) RoboCup 2010. LNCS (LNAI), vol. 6556, pp. 254–265. Springer, Heidelberg (2010)
5. Hoffmann, J., Göhring, D.: Sensor-Actuator-Comparison as a Basis for Collision Detection for a Quadruped Robot. In: Nardi, D., Riedmiller, M., Sammut, C., Santos-Victor, J. (eds.) RoboCup 2004. LNCS (LNAI), vol. 3276, pp. 150–159. Springer, Heidelberg (2005)
6. Meriçli, T., Meriçli, Ç., Akın, H.L.: A Robust Statistical Collision Detection Framework for Quadruped Robots. In: Iocchi, L., Matsubara, H., Weitzenfeld, A., Zhou, C. (eds.) RoboCup 2008. LNCS (LNAI), vol. 5399, pp. 145–156. Springer, Heidelberg (2009)
7. Quinlan, M.J., Murch, C.L., Middleton, R.H., Chalup, S.K.: Traction Monitoring for Collision Detection with Legged Robots. In: Polani, D., Browning, B., Bonarini, A., Yoshida, K. (eds.) RoboCup 2003. LNCS (LNAI), vol. 3020, pp. 374–384. Springer, Heidelberg (2004)
8. Vail, D., Veloso, M.: Learning from accelerometer data on a legged robot. In: Proc. IAV 2004. Springer, Lisbon (2004)
9. Visser, L., Carloni, R., Stramigioli, S.: Energy efficient control of robots with variable stiffness actuators. In: Proceedings of the 8th IFAC Symposium on Nonlinear Control Systems, NOLCOS 2010, Bologna, Italy, pp. 1199–1204. IFAC (2010)

The Ontology Lifecycle in RoboCup:
Population from Text and Execution

Stephan Gspandl, Andreas Hechenblaickner, Michael Reip, Gerald Steinbauer,
Máté Wolfram, and Christoph Zehentner*

Institute for Software Technology, Graz University of Technology, Graz, Austria
`kickofftug@ist.tugraz.at`

Abstract. In RoboCup it is important to build up domain knowledge
for decision-making. Unfortunately, this is a time-consuming and laborious job. At championships easy adaptability of this domain knowledge
can be especially crucial as teams need to be able to change tactics and
adjust to opponent behavior as fast as possible. An intuitive interface to
the agent is therefore necessary.

In this paper, we present a methodology to automatically populate
a domain ontology from natural language text. The resulting populated
ontology can then be deployed in a multi-agent system. This automatic
transformation of text to knowledge for decision-making thus provides
such an intuitive interface to the agents. It is embedded into the broader
(up to now) theoretical context of an ontology lifecycle.

We have created a proof-of-concept implementation in the 2D
RoboCup Simulation League on the base of tactics descriptions from
soccer literature. Experiments show that 71% of tactics are perfectly
transformed and 86% of the actions are executed correctly in terms of
geometric relations.

1 Introduction

In every RoboCup league domain knowledge is necessary for the agents or robots
to perform intelligent tasks. Many teams use expert literature (for example from
books like [14] and [7], or from the HowTo-Wiki[1]) for the purpose of transforming
information how to do things into specific models, plans or source code. Unfortunately, this is a very time-consuming and laborious job which yields many
hybrid and incompatible descriptions of the same concepts. Even worse, as a
rule or knowledge base for robots is generally rather large, its maintenance is
very demanding, too. It commonly takes at least one person to put constant
effort into its manual definition and update. From this point of view we strongly
argue for the automation of the knowledge generation, its incorporation into the
team as well as the maintenance of this knowledge base.

---

* The work has been partly funded by the Austrian Science Fund (FWF) by grant
P22690.
[1] http://www.wikihow.com

T. Röfer et al. (Eds.): RoboCup 2011, LNCS 7416, pp. 389–401, 2012.

Ontologies store data in a structured way under a semantic context and thus provide a suitable basis for knowledge-based agents to operate on. In order to make the definition and adaption of behavior as simple, fast and intuitive as possible, an ontology-based agent system should be able to

- populate a domain ontology based on various sources of information,
- link the ontology classes to actions and
- constantly improve, enrich and verify the ontology during execution.

Thus, descriptions of soccer strategies from books, websites or other knowledge artefacts should be automatically transformed into an ontology the agents can execute. This way, new tactical instructions can be easily incorporated into the agents' knowledge base at championships to exploit specific weaknesses of other teams. Another example would be a logistics robot which has to be trained to deliver parts from the warehouse to a newly installed assembly line. In order to increase the factory's flexibility (and lower costs) a simple way of explaining the desired behavior to the robot in natural language is to be prefered over hardcoding it.

Fig. 1. A sketch accompanying the instructions from [7] to perform a penetration into the opponent field

Therefore, we are constantly working on a methodology for an easy translation of arbitrary, even dynamic (i.e. that also conveys temporal relations) information into behavioral descriptions. Take the following tactics description in natural language from [7] (which is accompanied by the sketch shown in Fig. 1); this will also serve as running example throughout this paper:

* 5 passes the ball to 9 (who has come towards him) and moves
  diagonally to get the return pass;
* 9 passes the ball back to 5;
* 10 sprints deep to the right, criss-crossing with 9 who does the
  same thing from the opposite side;
* 5 can choose to pass the ball to either 9 or 10.

It is easy for humans to understand these sentences and create a mental picture of the described situation. Looking at the according sketch enables them to capture the relations even faster and yields additional information how to interpret it correctly. This is mainly due to the fact, that people are acquainted with soccer and language, so the words can be easily grounded in the soccer domain and (commonly) immediately make sense in the way they have been put together.

The ideas described in this paper follow the same intuition when applied to RoboCup: tactics descriptions should be processed and grounded in the RoboCup soccer domain using a domain ontology and Natural Language Processing (NLP) techniques. The resulting populated ontology, containing the grounded instructions in semantic relation, should then be employed as knowledge base for the agents in the simulations, just like a human remembers during the game what her trainer taught her. Resembling human intuition as well, this ontology should then be verified and improved during gameplay. If a human player notices, for example, that an opposing team adjusts to her tactics, she will try to counter by improving her gameplay or asking her trainer for advice. Similarly an agent could try to vary its behaviour or learn new tactics or variations of the original to win.

The efforts in this paper cover the population of a soccer ontology from text and executing the ontology tactics in the 2D RoboCup Simulation League in the more general scope of an ontology lifecycle. We define this lifecycle as the creation, population and management of an ontology, where the latter consists of the continous evaluation, enrichment and improvement of the populated ontology in the environment. The remainder of this paper is therefore organized as follows: related research concerning the ontology lifecycle is discussed in Section 2. The methodology for automatic population and execution of a domain ontology from text is described in Section 3 in the context of this ontology lifecycle. The experiments and results are covered in Section 4 followed by conclusion and outlook in Section 5.

2   Related Research

Ontologies in software engineering are employed to structure data and store it in a semantic context (accessible to reasoning), or just to conceptualize a certain domain. They define a database of concepts, relationsships and other representational entities for the purpose of performing reasoning, presenting it to and preparing it for users, or to provide a knowledge base for software (knowledge agents).[2]

Naturally, there has been effort to employ ontologies in agent and also multi-agent systems (for example [16], and [19]). For our purposes, an agent is defined as an entity, receiving input from its surroundings and trying to find the best-possible action to perform in the environment on base of a performance measure. In a multi-agent system two or more of such agents have to work together to pursue a mutual goal. It is based on problem-decomposition, trying to reduce the

---

[2] see http://tomgruber.org/writing/ontology-definition-2007.htm

complexity of the given problem by employing many agents, often specialized to certain tasks [24]. The benefits of ontologies (to structure information, to make it accessible to reasoning and to allow for restructuring and presenting the knowledge base in an intuitive format) make them very suitable to control such knowledge-based agents.

Commonly, in most knowledge engineering applications, the users (or knowledge engineers) have to encode the required knowledge or employ large and bulky ontologies not specifically designed for the special domain of application. Thus, one has to accept many drawbacks. In order to keep the knowledge engineering as intuitive, simple, thus minimising time consumption, multiple concepts have to be combined into an integrated process.

A system as proposed which is able to support the complete knowledge engineering process, has to deal with various fields of ontology research: (1) automatic or semi-automatic (i.e. unsupervised or supervised) creation and management of a domain ontology, (2) population of ontologies from various (information or knowledge) sources, (3) grounding the ontology knowledge in a specific agent language, (4) adopting the populated ontology for deployment in the specific (multi-)agent system, as well as (5) refinement, improvement, enrichment and verification of the ontology (further called management of the deployed ontology) in the environment.

A domain ontology can be learned or induced from various sources, most commonly text and websites, but there are also many prepared ontologies for arbitrary topics to start off with. Often called the holy grail of NLP is the extraction of knowledge from completely unstructured text without supervision [23]. The result would be a domain ontology specific to the given problem and certainly well-suited for solving it. Unfortunately, this approach cannot yet provide adequate accuracy. Therefore, in most cases the user has to rely on at least semi-structured or dedicated input or multiple data sets, which have to be maintained separately. An example of a system for automatic learning of domain ontologies is OntoLearn which extracts them from websites and shared documents [20]. Common, though, are systems that supervise and support knowledge engineers in the (cooperative) ontology construction or learning. For example, multiple concepts, tools, data sources and algorithms are incorporated into an integrated system [15].

The main focus of this work comprises the automatic population of ontologies from text. There has been a lot of research into this topic, from semi-automatic to unsupervised concepts. For example, LexOnto provides a way of mapping information to a domain ontology with the help of a lexical ontology [5]. Other researchers use linguistic rules as well as machine learning techniques to populate ontologies and extract information [17]. A semi-automatic system is described in [1] which pre-processes the input texts to generate a conceptual tree and further depends on rules to capture knowledge from it. Focussing on processing the input text for mapping between text and a structured world state, the authors of [13] propose a generative model segmenting text into utterances and mapping these semantically.

The task of deploying the populated ontology in an agent environment is closely related to that of population. This is primarily because an important task in both concepts is grounding. In [12] action ontologies provide natural language interfaces to agents. This paper deals with the problem of representing special action concepts and mapping natural language to suitable actions. Such specific grounding of text onto action concepts is also the topic of [3] and [2], both employing learning techniques to semantically connect commentaries to soccer games and instructions to actions respectively. In the context of grounding and interpreting, it is necessary to identify action structures and intents in the given input texts. [26], and [25] deal with identifying intentions in web search and natural language.

As the generated ontologies will always contains errors and thus have to be continously maintained, management of ontologies is an important issue. Although the same concepts and ideas as shown above can be employed to iteratively improve and refine the ontologies, the feedback of the environment the agents operate in is valuable for this task. Therefore, this should also be captured in a complete knowledge engineering process for (multi-)agent systems. [22] and [6] give an overview over the state of the art in ontology population and enrichment, whereas [28] mainly employs learning techniques for this purpose. Concerning verification [11] computes justifications to explain inconsistencies in ontologies.

The experiments described in this paper are based on the 2D RoboCup Simulation League and a team which has participated in several RoboCup world championships. Detailed information on this league can be found in [4], the according changelogs, on the official website[3] and the championship website[4].

3   The Ontology Lifecycle

Related research yielded an integrated process (shown in Fig. 2) dealing with all necessary steps to automatically generate a self-optimizing RoboCup Simulation League team from soccer knowledge. We call this the Ontology Lifecycle. Before applying the process, relevant knowledge sources have to be identified and evaluated. The books on soccer tactics mentioned in the introduction have provided a good starting point for our purposes. The ideas proposed, though, go even beyond textual input. Proper transformations could also be defined to make use of sketches [9], inspired by research of the MIT Artificial Intelligence Laboratory [10] or the Qualitative Reasoning Group at Northwestern University [8].

In the first stage of the ontology lifecycle, these knowledge artifacts are preprocessed to provide more structured data for the remaining steps. This data will then be interpreted and grounded onto a soccer domain ontology, i.e. the ontology is populated with concrete concepts derived from the data. The domain

---

[3] see http://sourceforge.net/apps/mediawiki/sserver/index.php?title=Main_Page
[4] see http://julia.ist.tugraz.at/robocup2010

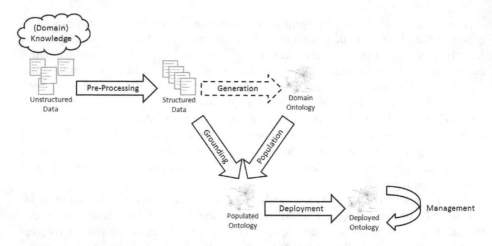

Fig. 2. An integrated process for automatic transformation of knowledge and its deployment in multi-agent systems

ontology may be generated out of suitable data, but is created manually for our tests.

The resulting populated ontology is then deployed in the RoboCup Simulation League to build a virtual soccer team. The process is implemented up to this point and will be concluded by concepts to verify and improve this basic ontology in simulations in future research. In the following sections the ontology population and deployment of a RoboCup team based on the descriptions of [7] (see the running example given above in Section 1) are described in detail.

3.1   Populating a RoboCup Ontology from Text

The automatic population proposed is based on natural language descriptions as found in [7] and a domain ontology. The domain ontology contains general soccer concepts relevant for games in the RoboCup Soccer Simulation League as well as more specific entities for the language processing: this includes players (identified by their uniform number) and player types (e.g. goalkeeper), components of the field (e.g. wing, goal etc.), positions (e.g. center) and the tactic model. For a team to perform a tactic in a game, this model comprises a pre-condition (under which the tactic should be executed), a set of actors (participating in the tactic) and a compound of actions each player has to perform, where actions can be executed in sequence or in parallel.

Pre-processing. Before a description can be actually populated on the base of and into this ontology, it has to be pre-processed to reduce the complexity of correctly grounding the text in the semantics of the ontology. This pre-processing comprises several subtasks, combining state-of-the-art NLP techniques: First, the text is split into atomic actions by a set of stop-words (like 'and') and stop-chars (like ';'). In order to allow for correct identification of terms, the Stanford POS (part-of-speech) tagger [27] is used to assign the lexical category to each word. Now, irrelevant words or irrelevant categories can be removed. On the same basis, pronouns are resolved and the atomic sentence stubs are transformed into a tuple $\{S, P, O\}$ (simply referred to as SPO in the remainder of the paper), with S being the subject, P the predicate and O the object of the sentence. This is done by identifying the verb by POS tag, checking for an actor in the first part and retrieving the object from the rest. After subject and object have been identified, pronouns can then be exchanged by their corresponding entity. For example, after the transformation of the first atomic sentence stub in the running example ('5 passes the ball to 9') into the SPO $\{5, pass, 9\}$, the pronouns 'who' and 'him' in the second ('who has come towards him') are substituted by '9' and '5', respectively.

As the descriptions of the tactics are not linear in time and involve parallel activities as well, temporal adjustments are necessary. By evaluating the tenses and identifying jumps as well as using keywords (for example 'while') in the original sentence stubs, a timestamp can be assigned to each SPO. If an SPO is associated with a timestamp m, it is to be executed before an SPO with timestamp n when $m < n$. The first action(s) have timestamp 0. The second part in the example above would thus be assigned timestamp 0 and the first timestamp 1. Naturally, same timestamps convey that actions have to be performed at the same time.

In the last step of the pre-processing, synonyms are resolved so that different words or phrases with the same meaning can be grounded onto the same semantic concept in the domain ontology. This is done using WordNet [18]: Each word of each SPO is looked-up in the WordNet dictionary (the POS type is transformed into the corresponding SynsetType), reduced to a base form and stored in a synonym dictionary. As in most cases multiple definitions exist for one word, there are two options: either all possible synonyms are saved or the correct is identified by the user (supervised resolution). With a small vocabulary, the first variant is simple and accurate enough, but with increasing size the latter should be preferred as the range of synonym terms might become too large.

Actual Population. Having the SPOs in correct temporal sequence and mapped to unique base forms, the actual population of the ontology can begin. For this purpose, test sequences are generated on the base of the ontology's taxonomy (this concept is easily extended to follow the restrictions as well; please note that formal reasoning will be a topic of future research). That is, the root classes corresponding to each element in the SPO tupel are retrieved from the ontology and specialized test strings are concatenated from simple identifiers (names or specializing qualifiers) associated with each class (as part of the domain modeling) while visiting the subclass relations. For example, the soccer domain ontology contains the following path through the taxonomy tree: *Object - GameEntity - Wing - Left Wing* (with A - B describing that B is a subclass of A). For this simple path, the test strings 'wing' and 'left wing' are created.

This way, each test string is automatically associated with the correct ontology class. The test sequences are then generated as all possible combinations of the subject, predicate and object test strings, yielding for example '5 run left wing'.

Each SPO from the pre-processing can now be compared to each test sequence. As equality measure serves an adapted Levenshtein distance on word level. It derives a similarity factor based on the sequence of words on the one hand as well as the number of equal and the number of words not in both strings. As the synonym resolution is also applied to each word of the test strings, there is no need to compare individual words by characters. The test sequence with the highest equality is saved as target concept.

The SPOs are then traversed to create an action compound for the new tactic in the ontology: for each action a new subclass is derived and the restrictions to subject and object class are set. These classes are put in temporal context by either creating parallel actions or an action sequence depending on the timestamp. The result is the action compound which is associated with the tactic. In order to complete the tactic, it is connected to the set of participating actors and a pre-condition. The latter has to be defined by the user as it is not conveyed in the text.

Post-processing. As the descriptions may contain implicit semantics, the populated ontology can not yet be employed for execution. A set of simple rules is provided to post-process it and fill certain semantic gaps. These rules contain instructions on how to add or adapt ontology actions. After application the ontology is ready for execution.

An example for the need to post-process the populated ontology would be the criss-crossing action in the running example. As actions are generally directed, the criss-crossing of two players is an exception. The populated ontology, thus, contains only one action that player 10 criss-crosses with player 9. In order to resolve this issue, a post-processing rule states that criss-crossing is bi-directed and a criss-crossing action in the opposite direction has to be inserted parallel to the existing.

3.2   Executing the Populated Ontology

After successfully transforming the tactics descriptions into a populated ontology, execution is rather straightforward. Before it can be employed, though, it has to be embedded properly into this execution environment. In order to do so, semantics for execution have to be defined.

Preparing the Populated Ontology. The task of preparing the populated ontology is to provide semantics for the execution. It is thus necessary to define a mapping from ontology action to the method to execute as well as semantics to synchronize the activities correctly. Both have to be defined for the concrete execution environment, in our example a RoboCup simulation league team.

Consider the first two sentences of the running example again: it tells the reader that player 9 runs towards player 5 and then 5 passes the ball to 9. This sounds easy, but two questions arise when it comes to execution. These are (1) what method the agent should call when executing the ontology action associated with 'come towards' (from now on this relation will be represented by angle brackets), and (2) when exactly player 5 should pass the ball. In order to answer these questions we have to (1) define that the ontology action ⟨come towards⟩ maps onto the method *runTo(Position targetPos)* and (2) provide a rule concretizing the behavior of this action, for example that *player1* ⟨come towards⟩ *player2* makes a *player1* move to the point halfways between her and another *player2*.

On the base of these semantics, a condition tupel $\{Pre, Inv, Post\}$ (following the notion of STRIPS [21]) has to be defined for each action of the compound, with *Pre* being the pre-condition, *Inv* the invariant and *Post* the post-condition. The pre-condition adopts the semantics of the action defined above and in a concrete situation thus grounds how the action has to be executed. In the example above, the pre-condition for ⟨come towards⟩ would therefore define the position player 9 runs to (the point halfways between player 9 and player 5) so that the action ⟨runTo⟩ is executed with this position. The invariant is a general model of an actions correct behavior, i.e. the invariant of ⟨come towards⟩ is true as long as player 9 is running towards the described position. Last, the post-condition describes when the action is successfully executed, i.e. that player 9 arrived at this position.

This way, synchronization can be created very naturally: actions committed in sequence or in parallel with other activities have to respect these semantics. This means that actions cannot be executed (stated by pre-condition) as long as actions which have to be performed earlier in the sequence, have been finished (stated by post-condition). Although this is true for most cases, yet experiments have shown that this constraint is too hard in some situations. Again consider our example: if actions would have to be performed strictly consequently, player 5 may pass only when player 9 has already arrived at her destination. Valuable time would be lost with players waiting for other players to finish running or for the ball arriving. Thus, running may be performed semi-consequently: An additional rule is added stating that actions may start after a player has already started running.

The positions the actions arrive at, serve as synchronization points that are expressed in the $\{Pre, Inv, Post\}$ tupels. These positions are progressed on base of the action compound hierarchy using the action semantics. The initial positions a tactic is grounded on, suffice to calculate all further positions. In the example, player 5 knows where player 9 is running towards and can direct her pass to this position. As her next action (please revisit the example) is to move diagonally to intercept, this position can be derived for player 9 (and of course would be for all other players).

Defining these semantics might sound complicated but is not: First, the semantics necessary are rather simple and straight-forward. Second, they can be defined generally for each ontology action (and don't have to be for each appearance in text) and in most cases the definitions are even reusable. Last, in order to perform actions it is necessary for a team to have these ready. It can thus fall back on implementations already available.

Execution. After proper preparation the actual execution is rather simple. Whenever a tactic should be performed (checked by its pre-condition), each participating player (grounded by the tactic's pre-condition) retrieves a linear plan containing only what she has to do. She then executes this plan step-by-step: She starts an action when the pre-condition allows it. The success of the tactic is monitored in each timestep by invariant. That is, each player checks whether all other participating players are still committed to the mutual plan. The invariant is thus a simple conjunction of the individual invariants of all parallel activities. The action the player is performing is executed as long as the post-condition is not fulfilled. If so, this process is repeated with the next action in the linear plan.

4   Experiments

In order to evaluate the concepts developed in this work, we have conducted two series of experiments. On the one hand the quality of the transformation process is measured. 16 tactics descriptions (containing a total of 129 actions) have been chosen randomly and populated into the prepared domain ontology. The ontology tactics are then compared to the original descriptions. The results are that

1. 71% of all actions were identified perfectly
2. an additional 20% of all actions were identified well enough (that is with few flaws which could be corrected by simple rules at execution)
3. the temporal sequence was adjusted correctly in all cases

The resulting populated ontology was therefore well-suited for execution. Most errors left can be easily explained by two factors: on the one hand flaws in POS tagging prevent the correct identification of all SPOs. On the other hand, complex reflexive concepts cannot be interpreted as SPO and require a more complex post-processing. An example would be the phrase 'who does the same thing from the opposite side' from the running example.

Post-processing and execution of the populated ontology were evaluated in simulation. Mapping and semantics as described in the previous setting were created. The quantification in the environment was turned off, so the agent's perception could not influence the simulation results (thus, a single run was sufficient to evaluate each tactic; please note that this was the only aspect that differed from the real championship settings). The remaining errors of the transformation were corrected manually. Then, each tactic was tested in the following setup: a trainer agent positions the number of agents on the field as depicted in the sketch accompanying the description. The agents were provided with the ontology tactic to execute and the simulation was started and logged until all agents had finished their execution.

Each simulation was evaluated whether the geometric relations were as defined by the original tactic description. The results were satisfying as well, yielding that 86% of the tactics were executed correctly.

5 Conclusion and Outlook

We presented a methodology for an automatic population of a domain ontology from text and its deployment as knowledge base for decision-making in multi-agent systems. In order to demonstrate the benefits we have created a proof-of-concept implementation. The results of our experiments in the 2D RoboCup Simulation League show that (1) 71% of actions are perfectly populated into the domain ontology and (2) 86% of actions are executed correctly.

In the future we will prepare the ontology for deployment in tournaments by creating suitable pre-conditions for each tactic. This way, the tactics will be used to enhance the actual gameplay of a team. Extending the methodology by additional sources we will build a team which is solely based on a populated ontology for decision-making.

Furthermore, we will complete the ontology lifecycle in RoboCup: verification and improvement will be based on simulations as well. The defined semantics will be used as model to monitor the correct execution of tactics in the environment. If an interruption is detected, the reason is inferred from this model and the tactic adjusted.

References

1. Amardeilh, F., Laublet, P., Minel, J.L.: Document annotation and ontology population from linguistic extractions. In: Proceedings of the 3rd International Conference on Knowledge Capture (2005)
2. Branavan, S., Chen, H., Zettlemoyer, L., Barzilay, R.: Reinforcement learning for mapping instructions to actions. In: Proce. of the Joint Conf. of the 47th Annual Meeting of the ACL and the 4th International Joint Conf. on Natural Language Processing of the AFNLP, Singapore, pp. 82–90 (2009)
3. Chen, D.L., Mooney, R.J.: Learning to sportscast: A test of grounded language acquisition. In: Proc. of the 25th International Conf. on Machine Learning, Finnland (2008)
4. Chen, M., Foroughi, E., Heintz, F., Huang, Z., Kapetanakis, S., Kostiadis, K., Kummenje, J., Noda, I., Obst, O., Riley, P., Steffens, T., Wang, Y., Yin, X.: RoboCup Soccer Server. In: The RoboCup Federation (Juli 2002)
5. Cimiano, P., Haase, P., Herold, M., Mantel, M., Buitelaar, P.: Lexonto: A model for ontology lexicons for ontology-based nlp (2007)
6. Cimiano, P.: Ontology Learning and Population from Text: Algorithms, Evaluation and Applications, 1st edn. Springer (October 2006)

7. Fascetti, E., Scaia, R.: Soccer Attacking Schemes and Training Execrcises. Reedswain Publishing (1998)

8. Forbus, K.D., Usher, J.: Sketching for knowledge capture: a progress report. In: Proceedings of the 7th International Conference on Intelligent User Interfaces, pp. 71–77. ACM, New York (2002)

9. Gspandl, S., Reip, M., Steinbauer, G., Wotawa, F.: From sketch to plan. In: 24th International Workshop on Qualitative Reasoning (2010)

10. Hammond, T., Davis, R.: LADDER: A language to describe drawing, display, and editing in sketch recognition. In: Proceedings of the 2003 Internaltional Joint Conference on Artificial Intelligence (IJCAI), Acapulco, Mexico, pp. 416–467 (2003)

11. Horridge, M., Parsia, B., Sattler, U.: Explaining Inconsistencies in OWL Ontologies. In: Godo, L., Pugliese, A. (eds.) SUM 2009. LNCS, vol. 5785, pp. 124–137. Springer, Heidelberg (2009)

12. Kemke, C.: An action ontology framework for natural language interfaces to agent systems (2007)

13. Liang, P., Jordan, M., Klein, D.: Learning semantic correspondences with less supervision. In: Proc. of the Joint Conf. of the 47th Annual Meeting of the ACL and the 4th International Joint Conf. on Natural Language Processing of the AFNLP, pp. 91–99. Association for Computational Linguistics, Singapore (2009)

14. Lucchesi, M.: Coaching the 3-4-1-2 and 4-2-3-1. Reedswain Publishing (2002)

15. Maedche, A., Staab, S.: Learning ontologies for the semantic web. In: Semantic Web (2001)

16. Mathieu, P., Routier, J.C., Secq, Y.: Towards a Pragmatic Use of Ontologies in Multi-Agent Platforms. In: Palade, V., Howlett, R.J., Jain, L. (eds.) KES 2003. LNCS (LNAI), vol. 2773, pp. 1395–1402. Springer, Heidelberg (2003)

17. Maynard, D., Li, Y., Peters, W.: Nlp techniques for term extraction and ontology population. In: Proceeding of the 2008 Conference on Ontology Learning and Population: Bridging the Gap between Text and Knowledge (2008)

18. Miller, G.A., Beckwith, R., Fellbaum, C., Gross, D., Miller, K.: Introduction to WordNet: An on-line lexical database (1993)

19. Mousavi, A., Nordin, M.J., Othman, Z.A.: An ontology driven, procedural reasoning system-like agent model, for multi-agent based mobile workforce brokering systems. Journal of Computer Science 6, 557–565 (2010)

20. Navigli, R., Velardi, P.: Learning domain ontologies from document warehouses and dedicated web sites. Computational Linguistics 30 (2004)

21. Nilsson, N.J., Fikes, R.E.: Strips: A new approach to the application of theorem proving to problem solving. Artifical Intelligence 2(3-4), 189–208 (1971)

22. Petasis, G., Karkaletsis, V., Paliouras, G.: Ontology population and enrichment: State of the art. Tech. rep. (2007)

23. Poon, H., Domingos, P.: Unsupervised ontology induction from text. In: Proc. of the 48th Annual Meeting of the Association for Computational Linguistics (2010)

24. Russel, S., Norvig, P.: Artificial Intelligence: A Modern Approach, 2nd edn. Prentice Hall International (2003)

25. Strohmaier, M., Kröll, M.: Analyzing human intentions in natural language text. In: Proceedings of The Fifth International Conference on Knowledge Capture (K-CAP 2009) (September 2009)

26. Strohmaier, M., Lux, M., Granitzer, M., Scheir, P., Liaskos, S., Yu, E.: How do users express goals on the web? - an exploration of intentional structures in web search. In: We. Know 2007 (2007)

27. Toutanova, K., Manning, C.D.: Enriching the knowledge sources used in a maximum entropy part-of-speech tagger. In: Proceedings of the EMNLP/VLC 2000, Hong Kong, pp. 63–70 (2000)
28. Valarakos, A.G., Paliouras, G., Karkaletsis, V., Vouros, G.A.: Enhancing Ontological Knowledge Through Ontology Population and Enrichment. In: Motta, E., Shadbolt, N.R., Stutt, A., Gibbins, N. (eds.) EKAW 2004. LNCS (LNAI), vol. 3257, pp. 144–156. Springer, Heidelberg (2004)

An Overview on Opponent Modeling
in RoboCup Soccer Simulation 2D

Shokoofeh Pourmehr and Chitra Dadkhah

Computer and Electrical Engineering Department of K.N. Toosi University of Technology
shpourmehr@ee.kntu.ac.ir, dadkhah@eetd.kntu.ac.ir

Abstract. This paper reviews the proposed opponent modeling algorithms within the soccer simulation domain. RoboCup soccer simulation 2D is a rich multi agent environment where opponent modeling plays a crucial role. In multi agent systems with adversarial and cooperative agents, team agents should be adapted to the current environment and opponent in order to propose appropriate and effective counteractions. Predicting the opponent's future behaviors during competition allows for more informed decisions. We divide opponent modeling into two categories of individual agent behaviors and team behaviors. Individual behaviors concern modeling the low-level behaviors of individual opponent agents, however in team behaviors, the high-level strategy of the entire team like formation, offensive and defensive system, is recognized. Several methods have been proposed to create different models of opponents to improve the performance of teams in an essential aspect. In this paper, we review the approaches to the problem of opponent modeling published from 2000 to 2010.

Keywords: Opponent Modeling, Soccer Simulation 2D, Robotic Soccer, RoboCup, Multi-agent system.

1 Introduction

The idea of knowledge extraction from other agents' behavior was originally used in the field of the game theory [1]. In order to win a competitive game against an unknown adversary, it is vital to adapt to the dynamics of the environment, mainly caused by opponent's game play. An opponent is an agent that has private strategies and has goals that are conflicting with your own [2]. Opponent modeling predicts and identifies the future behaviors of opponent and proposes an appropriate counteraction [3].

One of the most interesting environments where agent modeling has been used is the robotic soccer domain [4]. The Robot World Cup Initiative (RoboCup) is an ambitious initiative whose ultimate goal is to create, by 2050, a robotic soccer team capable of beating the human soccer world champion. As such RoboCup represents a multidisciplinary area where one of the main domains encompasses the variety of different areas of computational intelligence. Three of the major aspects necessary for

T. R\"{o}fer et al. (Eds.): RoboCup 2011, LNCS 7416, pp. 402–414, 2012.

generating a competitive team are advanced learning, teamwork and opponent modeling concepts [5].

The task of adapting to the environment and opponent is enormously difficult due to the dynamic nature of soccer matches along with the multiple interactions between players. This is where opponent modeling can be used to recognize or predict the behavior of the agents or team such as formation, defense and offense system and pass graphs which allows for more informed decisions. There is a growing body of research on the use of opponent modeling as a common challenge for agents in both RoboCup and general multi-agent systems.

RoboCup soccer simulation provides a good platform for modeling a soccer team in a dynamic and multi-agent domain. The soccer simulation 2D league is based on the publicly available soccer server [45] system enabling 24 client programs (22 players and 2 coaches) to connect through UDP/IP socket [6]. The server simulates the players and the field in a 2D soccer match (shown in Figure 1). The server accepts low-level commands from the players, executes them in an imperfect way and sends (imperfect) perception information to the players [6]. That means, it creates a virtual soccer field and provides all players with local, incomplete and noisy perception information. In RoboCup Soccer Simulation 2D, it is often difficult for players to make a correct decision because of the uncertainty in the field information [7].

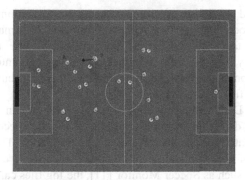

Fig. 1. Soccer simulation 2D's environment as appears in the Soccer Monitor

Each team is allowed to employ a further agent, the coach, which gets a noise-free, global view over the field. It intended to observe the game and to provide additional advice and information to the players. A frequently used opponent modeling approach in RoboCup soccer simulation is to rely on coach agent because it is provided with the complete and noiseless information from the field and can aid in creating agents with adjustable autonomy [8]. In order to focus entirely on opponent modeling, the RoboCup Simulation Coach Competition was held from 2001 to 2006. This competition was situated in the same soccer server, but instead of creating a full soccer team, a single coach agent had to be implemented. RoboCup Coach Competition changed in 2005 in order to emphasize opponent modeling approaches. This challenge calls for research on modeling a team of opponents in a dynamic, multi-agent domain [5]. In this Competition, the teams were directly evaluated based on how its coach agents identify the weaknesses and strengths (patterns) of the opponent from other opponent behaviors without these patterns [9].

Taking a general overview, it can be seen that the task of solving the problem of opponent modeling has been translated in different ways, namely learning of opponent formation, extracting opponent patterns, recognizing opponent plays and forecast the opponent behavior. Thus, various methods have been proposed to create these models. We divide the task of opponent modeling into two classes of individual behaviors and team behaviors. Individual behavior is about modeling the skills of an individual soccer agent including goalie's positioning, intercepting and path planning. Team behavior deals with the high-level strategy of the entire team like formation, offensive and defensive systems.

The goal of this paper is to take a general overview of the opponent modeling concept in soccer simulation 2D. The paper structure is as follows: Section 2 describes the most prominent approaches which are proposed to model team behavior. Section 3 presents the researches on individual behavior of agents. Section 4 concludes the paper.

2   Team Behavior

The classification and prediction of strategies and team formations play an important role in opponent modeling and can represent relevant information to implement counter strategy or tactic to reduce the performance of the opposite team. By analyzing play history, it is possible to gather critical insights into future plays [10]. Prediction of an opponent's behavior requires that their current behaviors can be recognized and classified and recalled later [2].

The techniques of Data Mining for classification problems can be used to solve the problem of behavior detection within the simulated robotic soccer environment. All the matches in simulated robotic soccer are saved in log files. Thus, it is possible to create a repository of data containing the historical records of certain teams. Using specialized software, such as Soccer Monitor [11], the log files can be used to provide a visual perception of the behavior of virtual players in the field. When necessary it can lead to analyze the behavior of the team, forecasting the actions or players positions, using information from previous matches [12]. Team behaviors in soccer domain can involve formation of agents or other strategies and tactics which are called game plays. In this paper, we classify the approaches to team behavior opponent modeling into two parts; formation and game play.

2.1   Game Play

Extensive works have been done on adapting teams' behavior to its opponents. One of the most useful approaches is to do adaptation which relies on classification of the current opponent into predefined models. Opponent modeling process in this approach is comprised of feature extraction, model construction and classification. First of all, in feature extraction part, useful features from the raw sensor data will be abstracted. After that, from the information in the features, opponent models will be constructed which embody the strategic information of how this type of opponent's models behaves. Then, in a matching task, observed opponent will be classified to the predefined models [13].

Riley and Veloso [14] used a windowing approach to extract useful features and a decision tree for classification. They have tried to model high-level adversarial behavior by classifying opponent actions as belonging to one of a set of predefined behavioral classes. Their system could classify fixed duration windows of behavior using a set of sequence-invariant action features. An observation occurred over a fixed length of time (i.e. a window) which affects the accuracy of the classifier and its performance. The classification accuracy for 30-37 classes was around 40% which in order of magnitude is better than guessing (<3%).

Based on this classification algorithm, Riley et al. contributed adaptive setplays which changed and improved throughout a game in response to the opponent team's behavior [15,16]. Their approach named ATAC, standing for Adaptive Team-Adversarial Coaching, used coach agent which was equipped with a repository of predefined hand-written opponent models of behavior. Using a naïve Bayes style algorithm, the coach agent was able to select between different models online. When the game is stopped, e.g., due to an out-of-bound call, the coach rapidly takes advantage of the short available time to create a team setplay plan that is a function of the matched modeled opponent's behavior. For adaptation to modeled opponent the coach agent uses Simple Temporal Network based plan representation and execution algorithm which expresses temporal coordination and monitoring in a distributed fashion.

The downside of this approach is that during execution, the agents do not take advantage of opportunities which may occur. For example if an agent ends up with a good shot on the goal, but the plan is to pass, then it will pass the ball. As they suggested, storing alternative plans and intelligently adding monitors for these plans as in [17] could make the plan execution opportunistic [15, 16]

Iglesias et al. [4] used similar process for recognizing and classifying an observed team behavior. After abstracting useful features from the previous games' log files, they analyzed these features in order to recognize different events. Based on the fact that the actions performed by a soccer team are sequential, they proposed to store events in a trie data structure and use that to obtain useful information. A trie (which is abbreviated from "retrieval") is a kind of search tree similar to the data structure commonly used for page tables in virtual memory systems. Example of this trie data structure is as follow:

"{Pass1to2(R)→Dribble2(R)→Pass2to10(R)→Pass10to11(R)}   and
{Dribble2(R)→Pass2to10(R)→Goal10(R)}"

The advantage of this kind of data structure is that every event is stored in the trie just once. Each event has a number that indicates how many times it has appeared. In addition, they used a statistical dependency test [18] for discovering the significance of sequences and subsequences. To evaluate the relation between an event and its previous events sequence Chi-square test is used [19].

They claimed that this approach works successfully when the pattern followed by a team is related to the players' actions. However, in this research, the different field regions in which the action occurs, has not been represented. As a result, if the pattern followed by the team is related to this aspect, this proposed method would not be viable. Furthermore, if the pattern is related to actions that occur when the player is not in the possession of the ball, this method would not be viable. [4], [20]

They effectively improved their work as represented in [21]. In this work, a soccer agent team behavior is represented as a distribution of its relevant atomic behaviors. In addition, modification of the Chi-Square Test is used in the classification method. In previous method only the expected values are compared and if an observed value is not represented in the expected distribution, it is not considered. In order to solve these problems, the way to compare the two distributions is modified to the sum of the terms which is called (Chi-Square-Obs Test). An important advantage of the proposed test is its speed since only the observed subsequences are evaluated.

Unlike Riley who stored every observation in the opponent models [13], Steffens [22] claimed that a limited number of opponent models can describe a wide range of opponents so his featured-based models contain only a small number (between two to fifteen) of distinct and stable features. Steffens suggested FBDOM method which stands for Feature Based Declarative Opponent Modeling. This method can be used either as an online or offline method for opponent modeling. He demonstrated a feature in RoboCup domain with the following example:

"The opponent often does long pass along the left wing to the forwards."

Then he proposed that rules which map actions to situations are the proper means to express such features. However, a domain-specific language which can formalize situation and action descriptions is necessary. During the match, features in the opponent models which have a probabilistic nature, will be detected by the observations that come in as raw sensor data. In classification part, with a Bayesian classifier, the opponent model with the best value will be chosen. Then a knowledge base will decide which counter-strategy is applicable. Steffens designed several experiments to test if feature-based models are able to represent opponent behaviors, if they generalize to previously unseen teams, and what effect the observation length (i. e. the amount of classification data) has. These experiments showed that the identification accuracy was high for the modeled teams, so the claim that features are a well-suited method to describe opponent behaviors can be supported.

In another work [23], Steffens presented a similarity-based approach to model high-level opponents actions (e.g. shoot towards goal). In its approach, Steffens proposes the use of Case-based reasoning (CBR) in order to predict the opponent's actions from the coach point of view. He increases the classification accuracy by including some derived attributes from imperfect domain theories. Its results showed that similarity-based opponent modeling can benefit from domain knowledge even if it is not known whether the opponent uses the same domain knowledge.

Case-Based Reasoning (CBR) is a powerful and a frequently applied way to solve problems for humans. However, using CBR in highly dynamic environments like soccer simulation 2D results in a large number of cases to be retained, leading to high computational costs for subsequent case management [24].

Ahmadi et al. [24] solved this problem by using an additional layer of cases to an ordinary layer which provides representation, adaptation and similarity measurement parameters for using a case-based architecture where new cases are recognized and stored during the games, but modeling is still done by a coach. In this work, Ahmadi et al. proposed a direct learning strategy, where the agents decision making is directly adapted based on the predicted behavior of opponent agents. This learning approach is appropriate for soccer simulation domain, because the cost of acquiring examples is high, and a high occurrence of samples for a specific situation is not available (required by most learning approaches like Neural Networks and statistical learning approaches). Due to this, every event should be well exploited. They suggested that fuzzification of the cases may also improve the performance of both of the implemented CBR learning systems. Additionally, it provides understandable high-level rules to comprise the game strategies.

In [25], the focus is on the unsupervised autonomous learning of the sequential behaviors of agents from observations of their behavior. Using a hybrid approach, the observations of a complex and continuous multi-variant world state would be translated into a time-series of recognized atomic behaviors. Then this time-series are analyzed in order to find repeating sub-sequences that characterize each team behavior. This system is able to identify different events corresponding to the basic behaviors of the agents using a set of models specialized in recognizing simple and basic behaviors of the agents (e.g., intercept, pass). It should be noted that this approach is based mainly on the tactic of the team without considering the formational behavior of the team.

In [26], a symbolic approach (similar to [27]) for behavior prediction based on association rule mining has been presented. They focused on qualitative representations of information. In their approach a sequential pattern mining algorithm is applied in order to learn frequent patterns in the data. These patterns are then transformed into prediction rules which can be applied to estimate what is likely to happen in the future. One characteristic of the learning approach is high representational power with the potential of learning complex patterns with predicates and variables from relational and temporal data. The drawback of the approach is the high complexity of the learning algorithm. The experiments support the assumption that without limiting the search space during pattern generation the algorithm cannot be used to learn complex patterns on-line due to time and space complexity. They suggested that it is necessary to develop heuristics which allow an efficient learning of patterns without cutting off a large number of potentially good patterns.

Ramos and Ayanegui [28] proposed a model able to manage the constant changes occurring in the game. The solution proposed by these authors is a model that allows the building of topological structures based on triangular planar graphs, able to manage the constant changes in the match. This approach enables multiple relations between the agents. Based on this model tactical behavior patterns had even been discovered in the dynamic conditions.

In another line of work, the champion of the RoboCup Soccer Coach Simulation Competition at 2004 and 2006 used a rule based expert system for modeling the opponent team [3]. Fathzadeh et al. defined the model of the opponent as a collection of multiple identified patterns. In this approach the identification of the opponent's patterns was done by an autonomous coach agent who analyzed the log files of the opponent's past games in offline mode and advises own players. To classify opponent's behaviors, a 3-tier learning architecture was developed. Firstly, sequential events of the game were identified using environmental data. Then the patterns of the opponent were predicted using statistical calculations. Eventually, by comparing the opponent patterns with the rest of team's behavior, a model of the opponent was constructed. In online mode, observing the live game, coach exposes an online model of the opponent and compares it with the stored models in repository. One of the major factors in success of this team was its capability in the handling of the noises and conflicts. In addition, they advised their players to motivate the opponent players to demonstrate the patterns. This trick had assisted them in identifying the opponent behaviors in a simply and fast manner [29-31].

In [32] a relational model to characterize adversary teams based on its behavior has been proposed. In particular, this paper focuses on the tasks of sequence classification using a logic representation for the sequences and the extracted features. The difference of this method compared to previous works is the logical representation language used to model the sequences and the proposal of a distance measure between agent behaviors described as logical sequences.

2.2 Formation

Some of the most important team behaviors are related to tactical plays. In general, tactical plays are planned most of the time and they should occur under the context of formations [28]. Team formation has been defined in different ways. In some works [33], [7], [34] formation is defined as set of player positionings according to the ball position. So the most basic formation would be made out of eleven positionings, ball position information, and the correspondent position of all team players. However, the most popular definition which most of the researchers are using is based on positions of soccer-agents and relevant relations between them. For example, a code 5:2:3 represents a formation composed by five defenders, two midfielders, and three forwards. Goalkeepers are not counted because they are always in a single position [28-31], [12], [35], [36].

A series of computational experiments showed that for this complex task the learning ability of the artificial neural network (ANN) is good [7], [28], [34], [36-37].

For recognizing formations Visser et al. [37] proposed a model based on an artificial neural network. In this model a set of default formations supply information about the opponent team to the online coach. The coach observes the game continually and analyzes the formation of the opponent team at given points in time with an artificial neural network and broadcasts an adequate counter formation to the players during the next interruption. The positions of the players serve as inputs for the ANN, which is trained with the formations most commonly played in test games and the log-files. Whenever the play mode switches to another state than PLAY ON, the coach generates a message for his team. A main consideration about this approach is the lack of any evaluation on counter formations' quality.

Another similar work is proposed in [35] that recognize the formation of the opponent team using a neural networks model. This work feeds the observed player positions into a neural network and tries to classify them into a predefined set of formations. If a classification can be done, the appropriate counter-formation is looked up and transmitted to the players. Like Ramos and Ayanegui, Santiago takes into account multiple relations among defender, midfielder and forward players.

Nakashima et al. in [7] used neural networks to learn opponent's team formation. They employed an off-line learning algorithm where neural networks were used to learn opponent formations from log files that were generated after the matches of target opponent teams.

The idea of using home areas of players to recognize formations comes from the research done by Riley and Veloso in [38]. A home area specifies the region of the field in which the agent should generally be. Thus, they propose that in identifying home areas, the agents can infer a role in the team (defender, midfielder or forward players). A drawback of this approach is that due to dynamic conditions of the world, the player movements can generate such a wide range extending considerably the home areas, which makes the task of determining the role of a player more difficult.

UT Austin Villa, the champion of the RoboCup 2005 coach competition [39], constructed a model of the opponent by characterizing their behavior with a set of features calculated from statistics gathered while observing a game. These features can be characterized as the team formation indicating the general positioning of the agents (e.g. how many players are defenders or attackers) or play-by-play statistics indicating the frequency of game events such as passes, shots, dribbles, etc. Despite its good results, there are many additional potential features that could boost performance within the same framework. In addition, the announcement strategy could be extended to explicitly model the likelihood of correctness relative to the cost of waiting longer to announce the score.

Almeida and colleagues present a Data Mining methodology for the forecast of team formation in [12]. To perform the detection of the formations they used the software Waikato Environment for Knowledge Analysis (WEKA) [40] since it includes a diversity of learning algorithms able to assist in the forecast of the teams' formations and has an easy to use graphical interface. Six learning algorithms were chosen on the basis of popularity and known experimental results: J48, Naive Bayes, K-Nearest Neighbor (IBK), PART, Multi-layer Perceptron and Sequential Minimal Optimization (SMO). Their experiments results show that the learning algorithm that generates the most appropriate model for predicting the formation of a team in simulated robotic soccer matches is the SMO. This study provides a flexible tool for formation forecast which enables the coach of robotic soccer team to be assisted with a decision support system when for instance a given team changes its formation. This tool can be improved by decreasing the percentage of cases incorrectly classified and analyzing the time spent by a team in the transition from one formation to another in order to reduce the incorrectly classified cases.

In a similar work, Faria and colleagues in [36] compared four machine learning techniques in identifying the opponent team and classification of the opponent formations. The conclusions obtained revealed that if a model was trained with certain games, the 3-NN showed better results in predicting formations. However, since the training games are not the same that we wish to predict in a competition, another test set with a different data set of games was applied and the results produced by Support Vector Machines (SVM) were, in terms of accuracy, the best.

3 Individual Behavior

In modeling individual behavior of agents, positioning and interactions between a small numbers of agents are predicted. Stone et al. [41] proposed a technique named "Ideal-Model-Based Behavior Outcome Prediction" (IMBBOP) that uses opponent optimal actions based on an ideal world model to model the opponent's future actions. This work was applied to improve the agent low level skills. For example, it was used to decide when to shoot and when to pass when an agent has the ball close to the opponent's goal. It is also used by agents to determine when the opponents are likely to be able to steal the ball. Stone's work does not directly construct a model.

In [8] a probabilistic adaptive opponent modeling named D-AdHoc (for Dynamic-AdHoc) is proposed. In this approach each agent observes and classifies during the game the encountered opponents into adversary classes which are automatically learned online. Each opponent class predicts the opponent's movements as a positional range where the opponent may be found certain time in the future together with a confidence value. This opponent modeling approach eliminates the assumption of the existence of a coach and the requirement of predefined opponent models.

Ledezma et al. [42] translated the task of acquiring the opponent model into a classification task. It used the logs of opponent team's player to predict its actions using a hierarchical learning schema. In this approach, after learning the agent action numerical parameter of its action will be learnt. The advantage of this trend is that the prediction of the opponent's action is increased since if we do not know the strength of the kick, the agent knows that it is going to kick rather than dashing. In this work Ledezma et al. considered that they had direct access to the opponent's inputs and outputs. In Another work [27] Ledezma and colleagues extend previous approach in the simulated robosoccer domain by removing this assumption. To do so, they have used machine learning to create a module that is able to infer the opponent's actions by means of observation. Next, this module can be used to label opponent's actions and learn a model of the opponent based on their observed input and output behavior, described by low-level actions. In this work, decision trees are learnt to predict the player's action type.

Continuing the previous works, in [43] Ledezma and his colleagues proposed an approach to model low-level behavior of individual opponent agents. In this work, the goalie actions were anticipated by a striker agent using OMBO (Opponent Modeling Based on Observation) so that the striker got as close to the goal as possible and shoot when the goalie was predicted to move. OMBO used machine learning techniques for opponent modeling at three levels. Mapping sensory data into discrete actions build the opponent model, and generate the decision-making algorithm.

The Modeling task itself consisted of two modules; Action Labeling Module (ALM) and Model Builder Module (MBM). In ALM, the last action (and its parameters) performed by any robosoccer opponent will be labeled based on the observations performed by the agent that is going to build the model. The MBM will then label other agent's actions. Once a tagged log of sensory data from agents and their performed actions was ready, the model of the opponent will be created based on data inside of the ALM. In contrast to most activity recognition tasks, where actions are manually tagged by looking at sensory data, Ledezma et al. proposed an automatic way of tagging actions for the robosoccer simulator. Additionally, in this case, there are not "a priori" models of other agent's behaviors.

Similar to this work, Illobre in [44], proposed a method to learn the behavior of a goalkeeper based on the actions of a shooting player. The results show that the method achieves high levels of accuracy with restricted evidence and time. This performance is achieved because of the chosen default assumption of the goalkeeper speed. Another problem is that the movement of the opponent goalkeeper might depend on the position of other opponent players.

4 Conclusion

In this paper, we have explored several techniques of opponent modeling in the competitive domain of RoboCup soccer simulation 2D. From our review it can be inferred that the problem of opponent modeling can be defined in various ways in an application dependent manner. In some applications it is important to model the opponent's formation or game plays, while in some others, the agents' individual behaviors are of interest. Once the opponent's model is defined, these methods usually classify data to fit the defined models using different machine learning methods. The data can vary from low level numerical logs to high level symbolic representations. The so-called coach agent considers this model to make proper counteraction during the decision making process. Although, in theory opponent modeling can be very useful, in practice it is both difficult to accurately do and to effectively use to improve game play. In summary, it was demonstrated that opponent modeling can be definitely applied and enjoy success in improving team or player performance, but it is still an open challenge in adversary games.

References

1. Turocy, T.L., Stengel, B.V.: Game Theory. CDAM Research Report LSE-CDAM (2001)
2. Ball, D., Wyeth, G.: Classifying an Opponent's Behaviour in Robot Soccer. In: Proceedings of the 2003 Australasian Conference on Robotics and Automation, ACRA (2003)
3. Fathzadeh, R., Mokhtari, V., Kangavari, M.R.: Opponent Provocation and Behavior Classification: A Machine Learning Approach. In: Visser, U., Ribeiro, F., Ohashi, T., Dellaert, F. (eds.) RoboCup 2007. LNCS (LNAI), vol. 5001, pp. 540–547. Springer, Heidelberg (2008)

4. Iglesias, J.A., Ledezma, A., Sanchís, A.: A Comparing Method of Two Team Behaviours in the Simulation Coach Competition. In: Torra, V., Narukawa, Y., Valls, A., Domingo-Ferrer, J. (eds.) MDAI 2006. LNCS (LNAI), vol. 3885, pp. 117–128. Springer, Heidelberg (2006)

5. Kitano, H., Tambe, M., Stone, P., Veloso, M., Coradeschi, S., Osawa, E., Matsubara, H., Noda, I., Asada, M.: The RoboCup Synthetic Agent Challenge 1997. In: Proceedings of the Fifteenth International Joint Conference on Artificial Intelligence, pp. 24–29 (1997)

6. Noda, I., Matsubara, H., Hiraki, K., Frank, I.: Soccer server: A tool for research on multi-agent systems. Applied Artificial Intelligence 12(2–3), 233–250 (1998)

7. Nakashima, T., Uenishi, T., Narimoto, Y.: Off-line learning of soccer formations from game logs. In: World Automation Congress (WAC), pp. 1–6 (2010)

8. Marín, C.A., Castillo, L.P., Garrido, L.: Dynamic Adaptive Opponent Modeling: Predicting Opponent Motion while Playing Soccer. In: Eduardo Alonso, E., Guessoum, Z. (eds.) Fifth European Workshop on Adaptive Agents and Multi-agent Systems Proceedings. LIP6, Paris, France (March 2005)

9. Fyfe, C., Tiño, P., Charles, D., García-Osorio, C., Yin, H.: Intelligent Data Engineering and Automated Learning. In: IDEAL 2010. Springer (2010)

10. Laviers, K., Sukthankar, G., Klenk, M., Aha, D.W., Molineaux, M.: Opponent modeling and spatial similarity to retrieve and reuse superior plays. In: Proceedings of the Workshop on Case-Based Reasoning for Computer Games. AAAI Press, California (2009)

11. The RoboCup Soccer Simulator,
http://sourceforge.net/projects/sserver/files/rcssmonitor/

12. Almeida, R., Reis, L.P., Jorge, A.M.: Analysis and Forecast of Team Formation in the Simulated Robotic Soccer Domain. In: Lopes, L.S., Lau, N., Mariano, P., Rocha, L.M. (eds.) EPIA 2009. LNCS (LNAI), vol. 5816, pp. 239–250. Springer, Heidelberg (2009)

13. Riley, P., Veloso, M.: On Behavior Classification in Adversarial Environments. In: Parker, L.E., Bekey, G., Barhen, J. (eds.) Distributed Autonomous Robotic Systems, vol. 4, pp. 371–380. Springer, Heidelberg (2000)

14. Riley, P., Veloso, M.: Recognizing Probabilistic Opponent Movement Models. In: Birk, A., Coradeschi, S., Tadokoro, S. (eds.) RoboCup 2001. LNCS (LNAI), vol. 2377, pp. 453–458. Springer, Heidelberg (2002)

15. Riley, P., Veloso, M.: Coaching a Simulated Soccer Team by Opponent Model Recognition. In: Proceedings of the Fifth International Conference on Autonomous Agents (Agents 2001), pp. 155–156 (2001)

16. Riley, P., Veloso, M.: Planning for distributed execution through use of probabilistic opponent models. In: Proceedings of the IJCAI-2001Workshop PRO-2: Planning under Uncertainty and Incomplete Information, pp. 72–81 (2001)

17. Veloso, M.M., Pollack, M.E., Cox, M.T.: Rationale-Based Monitoring for Planning in Dynamic Environments. In: Proceedings of the Fourth International Conference on Artificial Intelligence Planning Systems (1998)

18. Huang, Z., Yang, Y., Chen, X.: An approach to plan recognition and retrieval for multi-agent systems. In: Proc. of AORC, Sydney, Australia (January 2003)

19. Iglesias, J.A., Ledezma, A., Sanchis, A.: Caos coach 2006 simulation team: An opponent modeling approach. Computing and Informatics 28(1), 57–80 (2009)

20. Iglesias, J.A., Ledezma, A., Sanchis, A.: Comparing behavior in agent modeling task. Structure, 289–296 (2006)

21. Iglesias, J.A., Ledezma, A., Sanchis, A., Kaminka, G.A.: Classifying efficiently the behavior of a soccer team. In: Burgard, W., et al. (eds.) IAS-10, pp. 316–323 (2008)

22. Steffens, T.: Feature-Based Declarative Opponent-Modelling. In: Polani, D., Browning, B., Bonarini, A., Yoshida, K. (eds.) RoboCup 2003. LNCS (LNAI), vol. 3020, pp. 125–136. Springer, Heidelberg (2004)

23. Steffens, T.: Similarity-based opponent modeling using imperfect domain theories. In: CIG (2005)

24. Ahmadi, M., Lamjiri, A.K., Nevisi, M.M., Habibi, J., Badie, K.: Using a two-layered case-based reasoning for prediction in soccer coach. In: Proceedings of the International Conference on Machine Learning; Models, Technologies and Applications, pp. 181–185 (2004)

25. Kaminka, G.A., Fidanboylu, M., Chang, A., Veloso, M.: Learning the sequential coordinated behavior of teams from observations. In: Kaminka, G.A., Lima, P.U., Rojas, R. (eds.) RoboCup 2002. LNCS (LNAI), vol. 2752, pp. 111–125. Springer, Heidelberg (2003)

26. Lattner, A.D., Miene, A., Visser, U., Herzog, O.: Sequential Pattern Mining for Situation and Behavior Prediction in Simulated Robotic Soccer. In: Bredenfeld, A., Jacoff, A., Noda, I., Takahashi, Y. (eds.) RoboCup 2005. LNCS (LNAI), vol. 4020, pp. 118–129. Springer, Heidelberg (2006)

27. Ledezma, A., Aler, R., Sanchís, A., Borrajo, D.: Predicting Opponent Actions by Observation. In: Nardi, D., Riedmiller, M., Sammut, C., Santos-Victor, J. (eds.) RoboCup 2004. LNCS (LNAI), vol. 3276, pp. 286–296. Springer, Heidelberg (2005)

28. Ramos, F., Ayanegui, H.: Discovering Tactical Behavior Patterns Supported by Topological Structures in Soccer Agent Domains. In: International Conference on Autonomous Agents, Proceedings of the 7th International Joint Conference on Autonomous Agents and Multiagent Systems, Estoril, vol. 3, pp. 1421–1424 (2008)

29. Fathzadeh, R., Mokhtari, V., Mousakhani, M., Mahmoudi, F.: Mining Opponent Behavior: A Champion of RoboCup Coach Competition. In: IEEE 3rd Latin American Robotics Symposium, pp. 80–83 (2006)

30. Fathzadeh, R., Mokhtari, V., Mousakhani, M., Shahri, A.M.: Coaching with Expert System Towards RoboCup Soccer Coach Simulation. In: Bredenfeld, A., Jacoff, A., Noda, I., Takahashi, Y. (eds.) RoboCup 2005. LNCS (LNAI), vol. 4020, pp. 488–495. Springer, Heidelberg (2006)

31. Fathzadeh, R., Mokhtari, V., Haghighat, A.T., Mousakhani, M.: Using expert system in robocup soccer coach simulation: An opponent modeling approach. In: Proceedings Second IEEE Latin-American Robotics Symposium, Sao luis-Maranhao, Brazil (2005)

32. Bombini, G., Di Mauro, N., Ferilli, S., Esposito, F.: Classifying Agent Behaviour through Relational Sequential Patterns. In: Jędrzejowicz, P., Nguyen, N.T., Howlet, R.J., Jain, L.C. (eds.) KES-AMSTA 2010. LNCS, vol. 6070, pp. 273–282. Springer, Heidelberg (2010)

33. Reis, L.P., Lopes, R., Mota, L., Lau, N.: Playmaker: Graphical Definition of Formations and Setplays. In: Information Systems and Technologies (CISTI), pp. 1–6 (2010)

34. Uenishi, T., Nakashima, T.: Team Description of opuCI 2D for RoboCup (2009)

35. Ayanegui-Santiago, H.: Recognizing Team Formations in Multi-agent Systems: Applications in Robotic Soccer. In: Computational Collective Intelligence. Semantic Web, Social Networks and Multiagent Systems, pp. 163–173 (2009)

36. Faria, B.M., Reis, L.P., Lau, N., Castillo, G.: Machine Learning Algorithms applied to the Classification of Robotic Soccer Formations and Opponent Team. In: Proceedings of the 2010 IEEE Conference on Cybernetics and Intelligent Systems (CIS) and Robotics, Automation and Mechatronics (RAM), Singapore, pp. 344–349 (2010)

37. Visser, U., Drücker, C., Hübner, S., Schmidt, E., Weland, H.-G.: Recognizing Formations in Opponent Teams. In: Stone, P., Balch, T., Kraetzschmar, G.K. (eds.) RoboCup 2000. LNCS (LNAI), vol. 2019, pp. 391–396. Springer, Heidelberg (2001)
38. Riley, P., Veloso, M., Kaminka, G.: An empirical study of coaching. In: Asama, H., Arai, T., Fukuda, T., Hasegawa, T. (eds.) Distributed Autonomous Robotic Systems 5, pp. 215–224. Springer (2002)
39. Kuhlmann, G., Stone, P., Lallinger, J.: The UT Austin Villa 2003 Champion Simulator Coach: A Machine Learning Approach. In: Nardi, D., Riedmiller, M., Sammut, C., Santos-Victor, J. (eds.) RoboCup 2004. LNCS (LNAI), vol. 3276, pp. 636–644. Springer, Heidelberg (2005)
40. Weka. Weka Machine Learning Project,
    http://www.cs.waikato.ac.nz/~ml/index.html (acessed: October 04, 2008)
41. Stone, P., Riley, P., Veloso, M.: Defining and Using Ideal Teammate and Opponent Agent Models. In: Proceedings of the Twelfth Annual Conference on Innovative Applications of Artificial Intelligence (2000)
42. Ledezma, A., Aler, R., Sanchis, A., Borrajo, D.: Predicting opponent actions in the RoboSoccer. In: IEEE International Conference on Systems, Man and Cybernetics, p. 5 (2002)
43. Ledezma, A., Aler, R., Sanchis, A., Borrajo, D.: OMBO: An opponent modeling approach. AI Communications 22, 21–35 (2009)
44. Illobre, A., Gonzalez, J., Otero, R., Santos, J.: Learning action descriptions of opponent behavior in the Robocup 2D simulation environment. ILP (2010)
45. Chen, M., Foroughi, E., Heintz, S., Kapetanakis, S., Kostiadis, K., Kummeneje, J., Noda, I., Obst, O., Riley, P., Steffens, T., Wang, Y., Yin, X.: RoboCup Soccer Server manual for Soccer Server version 7.07 or Latest.,
    http://sourceforge.net/projects/sserver (accessed on: October 01, (2003)

Multi Body Kalman Filtering with Articulation Constraints for Humanoid Robot Pose and Motion Estimation

Daniel Hauschildt, Sören Kerner, Stefan Tasse, and Oliver Urbann

Robotics Research Institute
Section Information Technology
TU Dortmund University
44221 Dortmund, Germany

Abstract. In this paper, a concept for articulated rigid body state estimation is proposed. The articulated body, for instance a humanoid robot, is modeled in a maximal coordinate formulation and the articulations between the rigid bodies as nonlinear position and linear motion constraints. At first, the individual state of each particular rigid body is estimated with a Kalman filter, which leads to an unconstrained state estimate. Subsequently, the correct state estimate for the articulated rigid body is derived by projecting the unconstrained estimate onto the constraint surface.

1 Introduction

Nowadays robots are primarily used in assembly line production. The deployed robots are mostly fixed-based manipulators. However, researchers all over the world are trying to extend the robots task spectrum into areas requiring more mobility, such as search and rescue services, military operations or carrying out assistant tasks in everyday life. In order to fulfill those tasks, robots must be capable of navigating in various environments of which some are inhabited and designed for human beings. Especially the latter environments are particularly challenging for conventional wheeled robots as stairs or even small objects can become insurmountable obstacles. Thus legged and especially humanoid robots are becoming more and more focus of research. Since humanoid robots try to imitate human appearance in respect of their body design, they have a good foundation to navigate in the proposed environments and therefore are believed to have high potential for future applications. Despite this, currently available humanoid robots have a lack of mobility due to the fact that the generation of stable biped motion is still a major problem, which is not yet solved completely. To control the motion of a robot it is required to determine its current state, including among other things its ego-pose and motion. Exact orientation and velocity information is especially useful for balancing walking motions while measurements of translational movement can also complement the odometry information for improved global localization. Humans determine their state by

T. Röfer et al. (Eds.): RoboCup 2011, LNCS 7416, pp. 415–426, 2012.
© Springer-Verlag Berlin Heidelberg 2012

proprioception - the unconscious perception of movement and spatial orientation arising from stimuli within the body itself - which a robot simply does not have, complicating the process of ensuring motion stability. Proprioception is therefore imitated by estimating the body state with the help of proprioceptive sensors, among them accelerometers, gyroscopes or a compass. Since high-grade sensors are, if available, expensive, low-grade sensors are commonly used in commercially availably humanoid robot designs. Since their sensor reading are mostly not very accurate they in turn lead to erroneous state estimation results further influencing the motion stability of the robot. To overcome this flaw a novel approach to multi body state estimation is proposed in this paper, reducing the estimation error with the help of rigid body simulation. Furthermore this algorithm is capable to combine different sensor readings, allowing to further improve the state estimation by using a multitude of low-cost sensors at different parts of the robot.

After presenting a short overview of current research on this topic in section 2, the mathematical background is formulated in section 4. The derivation of the proposed algorithm is explained in sections 5 and 6 and finally evaluated in section 7.

2   Related Work

Various solutions have been proposed to solve the problem of ego-state estimation of a legged robot. Most of them use some sort of Bayesian filtering technique such as Kalman-filters or one of its variations. The vast majority of those approaches use a greatly simplified robot model in combination with stochastic information about the reliability of the sensor measurement to estimate the state with the highest probability. But only little work towards using a more sophisticated model of the robot has been done (see [1]) so far.

Latter approaches have in common that they try to model the robot as a chain of rigid bodies, the limbs, which are connected via various types of joints. In contrast to the approach, proposed in this paper, so-called reduced coordinate formulations are employed. Indeed this has the advantage that the number of such coordinates only equals the number of the degrees of freedom of the system, which are removed by the joints, and therefore yield in a more compact and intuitive robot model. Nevertheless, this also generally leads to a highly nonlinear model, which makes the utilization of well-known state estimation approaches, such as Kalman-filters difficult or not applicable at all.

3   Proposed Approach

Commonly the estimation of a humanoid robot's position change and orientation in space is estimated using standard Kalman approaches modeling the robot as a single rigid body and using inertia measurements as sensor information. For biped robots however the inertia measurements originating from the periodic walking motions dominate those correlated to actually changing speed or heading. Common approaches simply handle those former signal components as noise, thus having to cope with a significant noise to signal ratio. To be able to use those walking motion signal components as well as the joint sensor information it is necessary to model the complete kinematic structure of the robot to be used for estimation.

To do this and to improve the state estimation of a rigid multi body model a novel approach is presented in this paper, using a maximal coordinate formulation scheme. In this approach the robot is modeled using a maximal coordinate formulation scheme, so that sophisticated prediction methods, such as Kalman-filters, can be applied to improve the pose and motion estimation. In contrast to reduced coordinate formulations, no reduction of the number of required parameters is performed. Instead the connections between the rigid bodies - the joints - are modeled as constraints, which actually introduce additional parameters. While this at first appears to have a counter-productive effect, the modeling allows subdividing the articulated rigid body state estimation problem into a set of smaller single rigid body state estimation problems and therefore can be handled more efficiently.

Furthermore, this approach not only allows the use of a more sophisticated model and therefore a better estimate for the robot's state, but also yields additional benefits. By treating every limb of the robot separately, additional sensor informations can be integrated in the filter. For instance, inertia sensors need not be attached to the torso of the robot only, but could also be mounted to any limb, to improve the overall grade of the estimated state. Furthermore using the robots structural information it is possible to include the knowledge about the current environment the robot is located in. Hence, for example, ground contact or collision information can be modeled as a special type of joints, resulting in contact and collision constraints, which in turn can be included in the state estimation with ease.

4 Definitions

A humanoid robot can be modeled as a collection of n rigid bodies connected by joints. These joints impose a number of $n_{c,p}$ position constraints and $n_{c,m}$ motion constraints that restrict the possible motion of each rigid body. The state x_i of a rigid body can be described as

$$x_i = \left(p_i, q_i, v_i, \omega_i\right)^T \tag{1}$$

whereby p_i and q_i are the position vector and orientation unit quaternion of a rigid body's center of mass. Furthermore, v_i and w_i are the linear and angular velocities. When not mentioned differently all coordinates are relative to a common world coordinate system (WCS). In addition, the state of a rigid body is integrated over time Δt by the following transfer function

$$x_i^{k+1} = \begin{pmatrix} p_i^k + \Delta t \ v_i^k + \frac{1}{2} \Delta t^2 M_i^{-1} f_i^k \\ q_i \left(\Delta t \omega_i^k + \frac{1}{2} \Delta t^2 \ I_i^{-1} \ \tau_i^k \right) * q_i^k \\ v_i^k + \Delta t \ M_i^{-1} \ f_i^k i \\ \omega_i^k + \Delta t \ I_i^{-1} \ \tau_i^k \end{pmatrix} = f\left(x_i^k, u_i^k\right) \tag{2}$$

with $u_i = (f_i, \tau_i)^T$ being the vector of external forces f_i and torques τ_i. Thereby, M_i and I_i are the mass matrix and inertia matrix which describe the first and second order mass distribution of the rigid body in the WCS. While M_i is independ of the current state, I_i depends on the current orientation $R_i = f(q_i)$. Consequently, the angular mass matrix is derived by $I_i = R_i D_i R_i^T$ with D_i being the angular mass matrix in the body local coordinate system (BCS).

Starting with equations 1 and 2 the state of n independent rigid bodies can be comprised as

$$x^k = \left(x_0^k, \ldots, x_{n-1}^k\right)^T$$
$$u^k = \left(u_0^k, \ldots, u_{n-1}^k\right)^T \tag{3}$$
$$f\left(x^k, u^k\right) = \left(f\left(x_0^k, u_0^k\right), \ldots, f\left(x_{n-1}^k, u_{n-1}^k\right)\right)^T.$$

For now, the joints between the rigid bodies are not considered. However, joints can easily be formulated as constraints in a maximal coordinate formulation. Thereby, it has to be distinguished between pure position constraints $C_p(p, q)$ and pure motion constraints $C_m(v, w)$. Other types of constraints are not considered, however, most common joint types can be modeled in this way. As a result, the motion of n articulated rigid bodies can be formulated as

$$x^{k+1} = f\left(x^k, u^k\right) \tag{4}$$

subject to

$$C_p(p, q) - d_p = 0 \tag{5}$$
$$C_m(v, \omega) - d_m = 0 \tag{6}$$

where d_p and d_m are constant offsets, for example the fixed distance between two joints.

5   Unconstrained State Estimation

Consider the following nonlinear time invariant system given by equation 3 and sensor measurements

$$z^k = \left(z_0^k, \ldots, z_{n-1}^k\right)^T \tag{7}$$

whereby

$$z_i^k = g(x_i^k) \tag{8}$$

measures a sub state of the i-th rigid body at time $k\,\Delta t$. Measurements regarding multiple rigid bodies are not considered and thereby are neglected. However, they can be modeled as constraints if the measurement noise is small and can be neglected, as for instance is true for accurate joint state measurements provided by most servo motors.

Furthermore, it is assumed that the control input u_i and the measurements z_i are either unknown or disturbed by noise and therefore modeled as Gaussian random variables. Consequently, the states x_i of the rigid bodies become random variables with unknown probability distributions. Considering this, the first order central moments are

$$E\left(a\right) = \left(E\left(a_0\right), \ldots, E\left(a_{n+1}\right)\right)^T = \widehat{a} \tag{9}$$

and the second order moments are

$$E\left((a - \widehat{a})^2\right) = \begin{pmatrix} \Sigma_{a_0,a_0} & \cdots & \Sigma_{a_0,a_{n-1}} \\ \vdots & \ddots & \vdots \\ \Sigma_{a_{n-1},a_0} & \cdots & \Sigma_{a_{n-1},a_{n-1}} \end{pmatrix} = \Sigma_a \tag{10}$$

with $E(\cdot)$ being the expectation operator, $a \in \{x, u, z\}$ and $a^2 = a\,a^T$.

In this manner, it must be noted that the covariances

$$\Sigma_{x_i,x_j} = \Sigma_{u_i,u_j} = \Sigma_{z_i,z_j} = 0 \ \textit{for} \ i \neq j \tag{11}$$

are assumed zero and therefore will be disregarded. This assumption allows us to tackle the multi body state estimation problem more efficiently as the problem can be decomposed into n single body state estimation problems. Besides, the influence among the rigid bodies is already handled by the constrained projection (see section 6). Thus, well known rigid body state estimation algorithms such as the Extended Kalman Filter [5] and the Unscented Kalman Filter [6] can be applied. Since these types of state estimators have already been discussed in detail in the literature, they are not explained here and further treated as black box algorithms. The completely unconstrained estimation process can be depicted as shown in figure 1.

6   Constrained Projection

In the previous step, an unconstrained state estimate $\widehat{x}^{k+1}$ has been derived. The constrained state estimate $\widehat{x}_c^{k+1}$ can be found by projecting the unconstrained estimate onto the constrained surface. Therefore, a constrained weighed least squares optimization problem can be formulated by

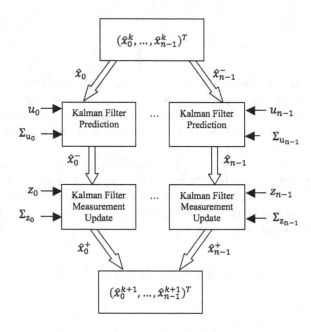

Fig. 1. Unconstrained estimation process

$$\hat{x}_c = \arg\min_x (x-\hat{x})\,W\,(x-\hat{x}) \tag{12}$$

subject to

$$C_p\,(p,q) - d_p = 0 \tag{13}$$
$$C_m\,(v,\omega) - d_m = 0. \tag{14}$$

whereby W is a positive definite weight matrix. However, the position and motion constraints are often contradictory and thus both constraints may not be fulfilled concurrently. As a result equation 12 is solved consecutively for each constraint type.

Nonlinear Position Constraint Projection

For the sake of generality, the position constraints are assumed nonlinear. A solution to equation 12 is found by applying the method of Lagrange-multipliers λ which results in

$$\hat{x}_c = \arg\min_x (x - \hat{x})\,W\,(x - \hat{x}) - \lambda^T\,(C_p\,(p,q) - d_p) \tag{15}$$

subject to

$$C_{p(p,q)} - d_p = 0 \tag{16}$$

Consequently, the optimal $\hat{x}_c$ can be determined by solving

$$W(x - \hat{x}) - J_p^T(\hat{x})\lambda = 0 \tag{17}$$
$$C_p(x) - d_p = 0 \tag{18}$$

with

$$J_p(x) = \frac{\partial C_p(p, q)}{\partial x} \tag{19}$$

being the Jacobian of $C_p(p, q)$.

The solution to the nonlinear problem is derived by linearizing the nonlinear position constraint around the current operating point and iterating until convergence is achieved. The linearization yields

$$C_p(x + \Delta x) \approx C_p(\hat{x}) + J_p(\hat{x})\Delta x - d_p \tag{20}$$
$$\Delta x = (x - \hat{x}).$$

After replacing the nonlinear constraint with its linear counterpart and rearranging equation 17 to Δx, equation

$$\Delta x = W^{-1}J_p^T(\hat{x})\ \lambda \tag{21}$$

is derived. Furthermore, replacing Δx in equation 20 leads to

$$J_p(\hat{x})\ W^{-1}J_p^T(\hat{x})\lambda = d_p - C_p(\hat{x}). \tag{22}$$

Hence, the complexity for solving equation 22 does not depend on the dimension of x but instead it depends on the dimension of λ. As a result, the effort for solving the minimization problem depends solely on the number of constraints and not on the number of rigid bodies involved.

Regarding to [8] the choice of the weight matrix W determines the meaning of the result. On the one hand, choosing

$$W = diag\left(\begin{pmatrix} M_0 & 0 \\ 0 & I_0 \end{pmatrix}, \ldots, \begin{pmatrix} M_{n-1} & 0 \\ 0 & I_{n-1} \end{pmatrix}\right) \tag{23}$$

results in an almost physically correct position change (see [4]). On the other hand,

$$W = \Sigma_x^{-1} \tag{24}$$

favors a maximum likelihood estimate and is therefore chosen. Exploiting the fact that the partial derivatives are

$$\frac{\partial C_p(x)}{\partial (p,q)^T} = J_p(p,q) \neq 0 \tag{25}$$

$$\frac{\partial C_p(x)}{\partial (v,w)^T} = J_p(v,w) = 0 \tag{26}$$

and that the covariance matrix Σ_x can be subdivided into

$$\Sigma_x = \begin{pmatrix} \Sigma_{(p,q)} & \Sigma_{(p,q),(v,w)} \\ \Sigma_{(v,w),(p,q)} & \Sigma_{(v,w)} \end{pmatrix} \tag{27}$$

with

$$\Sigma_{(a,b),(c,d)} = E\left(\left(\begin{pmatrix} a \\ b \end{pmatrix} - \begin{pmatrix} \hat{a} \\ \hat{b} \end{pmatrix}\right)\left(\begin{pmatrix} c \\ d \end{pmatrix} - \begin{pmatrix} \hat{c} \\ \hat{d} \end{pmatrix}\right)\right) \tag{28}$$

being the covariance of random vectors $(a,b)^T$ and $(c,d)^T$, equation 22 can be further reduced to

$$J_p(p,q)\ \Sigma_{p,q}\ J_p^T(p,q)\ \lambda = d_p - C_p(\hat{x}). \tag{29}$$

Altogether, the estimate $\hat{x}_c^j$ at the j-th iteration becomes

$$\hat{x}_c^0 = \hat{x} \tag{30}$$

$$\hat{x}_c^j = \hat{x}_c^{j-1} + \begin{pmatrix} \Sigma_{(p,q)} \\ \Sigma_{(p,q),(v,w)} \end{pmatrix} J_p^T(p,q)\ \lambda. \tag{31}$$

Linear Motion Constraint Projection

After enforcing the position constraints, the motion constraints $C_m(v,w)$ can be applied. Fortunately, motion constraints are mostly linear constraints, such that $C_m(v,w)$ becomes

$$J_m(v,w)^T - d_m = 0 \tag{32}$$

which is linear in $(v,w)^T$. Besides, in the case that the motion constraints are also nonlinear one needs to proceed as described in section 6. However, for linear constraints equation 12 simplifies to

$$\Sigma_{(v,w)}^{-1}\begin{pmatrix} v - \hat{v} \\ w - \hat{w} \end{pmatrix} - J_m^T\ \lambda_{v,w} = 0 \tag{33}$$

$$J_m\begin{pmatrix} v \\ w \end{pmatrix} - d_m = 0 \tag{34}$$

whereby $\Sigma_{v,w}$ is the motion covariance matrix. Further rearranging equation 33 to $(v, w)^T$ and inserting it into equation 34 yields

$$J_m \Sigma_{v,w} \, J_m^T \, \lambda = d_m - J_m^T \begin{pmatrix} \hat{v} \\ \hat{w} \end{pmatrix} \tag{35}$$

which is similar to equation 22 and can therefore be solved likewise.

7   Evaluation

To proof the concept of the described state estimation of a multi body system the algorithm is exemplarily applied to predict the ego-state of the humanoid robot NAO manufactured by Aldebaran Robotics. This robot model has 21 degrees of freedom and an inertia measurement unit in its chest consisting of a 3-axis accelerometer and a 2-axis gyroscope which can not measure the robot's rotational velocity around the z-axis.

To demonstrate the benefit of the multi body model, the trajectory of the robot's center of mass is predicted using the presented approach and compared to a classic *Unscented-Kalman-Filter* (UKF). Furthermore, the proposed approach will be referred to as *Quaternion MultiBody Unscented-Kalman-Filter* (QMB-UKF) for the remainder of this paper. While the UKF can only take the acceleration sensor and gyroscope informations into account, the QMB-UKF additionally uses the robots structural information and joint angle informations provided by the servo motors as constraints.

Since the purpose of this paper is to provide a proof of concept of the QMB-UKF, only a MATLAB implementation is available so far. Due to the fact that this implementation is non optimal in term of runtime, it can not be run on the Nao's AMD Geode processor in real-time at the moment. But given the fact, that algorithm runs near real-time using common desktop hardware, an application to the real hardware using an optimized implementation in addition to mire efficient constraint solvers.

In turn only simulative studies were carried out and thus reference data for the experiments is generated by the simulator *SimRobot* [7]. While it also uses a time discrete rigid body simulation, the ODE [9], the method applied to ensure position stability, the Baumgarte method [2], varies, resulting in a difference and thereby prediction error between SimRobot and the prediction base of the algorithm. This difference is further increased by the fact that the ODE simulation time step of $h_{SimRobot} = 0.005s$ is chosen to yield results closer to reality. The sensors are modeled to mimic the characteristics of the ones used in the real Nao as close as possible in simulation, emphasizing the noise value to yield non-optimal measurements. A detailed overview on sensor modeling can not be given in this paper.

Experiment 1 - Walking in x-Direction

During the first experiment the robot walks straight ahead in x-direction with a constant walking speed of $v_x = 0.07m/s$ for the duration of 20 seconds. The trajectory of the robots center of mass is predicted by the UKF and with help of the proposed algorithm using the QMB-UKF. The results regarding the position and rotation error of the center of mass can be found in figure 2. The errors are displayed additively to demonstrate their relative influence.

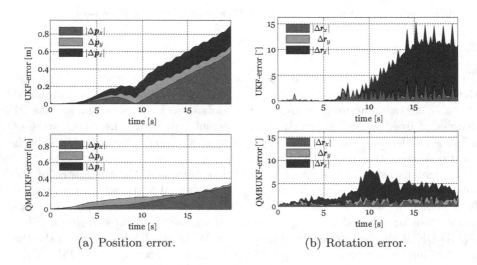

(a) Position error.   (b) Rotation error.

Fig. 2. Experiment 1 - Walking straight in x-direction

The evaluation of figure 2(a) clearly demonstrates the benefit of taking the multi body prediction into account. While the error over time, using the QMB-UKF, still sums up to roundabout 35 cm, this result is obviously superior to the error of over 90 cm using the UKF. The overall rotation error, see figure 2(b), is higher in both cases, with the rotational error around the z-axis having the most influence. Taking into account that the sensors cannot measure the speed or rotation around z-axis of the robot this result is reasonable. Nevertheless evaluating figure 2(b) again demonstrates the benefit of the QMB-UKF.

Experiment 2 - Unforeseen Collision

Since walking with a constant speed represents a favorable use case for a Kalman-filter the second experiment is chosen to evaluate the influence of an unforeseen disturbance. Again the robot walks straight ahead in x-direction with a speed of $v_x = 0.07 m/s$, but collides with an unforeseen object which results in sudden change of both horizontal and angular speed. The results of this experiment are demonstrated in figure 3. Again they are displayed additively to show their relative influence.

As expected both filters have problems tracking the state of the robot since they cannot predict the sudden change of acceleration. But again it is clearly visible that the QMB-UKF highly benefits from taking the multi body model into account and thereby reducing the error in comparison to the UKF.

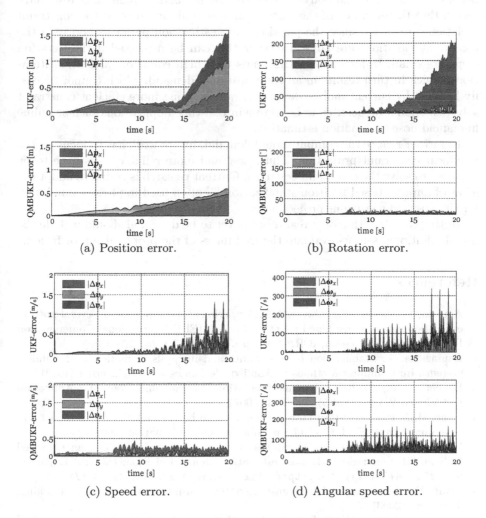

(a) Position error.

(b) Rotation error.

(c) Speed error.

(d) Angular speed error.

Fig. 3. Experiment 2 - Collision with an unforeseen object

8   Conclusion and Future Works

In this paper, an approach towards articulated rigid body pose and motion estimation has been proposed. As a prove of concept, application of the algorithm to a simulated humanoid robot model, has clearly demonstrated an accuracy benefit over conventional body state prediction methods. Preliminary tests have shown that the accuracy of the state estimation only improves if the constraint violations are small, since the position state estimation performs poorly, if the constraint violations are large. However this can be neglected if the iteration steps are small. Unfortunately, this leads to a high count of calculation iterations, which in turn results in an computational demand, which cannot be easily handled by current mobile devices. However, using more efficient constraint solvers it should be possible to be usable in real-time applications such as online humanoid pose and motion estimation.

To further reduce the computational demand, future work has to focus on the nonlinear constraint projection. Thus, new and more reliable ways have to be explored to solve the constraint equation. Current research is trying to adapt the ideas of impulse based position stabilization [3] often used in physics simulation for the position constraint projection.

Testing the algorithm on a real robot is yet to be done and should include the use of additional sensors evaluate the usefulness of the proposed sensor fusion.

References

1. Aoustin, Y., Plestan, F., Lebastard, V.: Experimental comparison of several posture estimation solutions for biped robot rabbit. In: IEEE International Conference on Robotics and Automation, ICRA 2008, pp. 1270–1275 (May 2008)
2. Baumgarte, J.: Stabilization of constraints and integrals of motion in dynamical systems. In: Computer Methods in Applied Mechanics and Engineering (1972)
3. Bender, J.: Impulse-based dynamic simulation in linear time. Computer Animation and Virtual Worlds 18(4-5), 225–233 (2007)
4. Cline, M.B., Pai, D.K.: Post-stabilization for rigid body simulation with contact and constraints. In: IEEE Intl. Conf. on Robotics and Automation (2003)
5. Halvorsen, K., Söderström, T., Stokes, V., Lanshammar, H.: Using an extended Kalman filter for rigid body pose estimation. Journal of Biomechanical Engineering 127(3), 475–483 (2005), http://link.aip.org/link/?JBY/127/475/1
6. Kraft, E.: A quaternion-based unscented Kalman-filter for orientation tracking. Tech. rep. (2003)
7. Laue, T., Spiess, K., Röfer, T.: SimRobot – A General Physical Robot Simulator and Its Application in RoboCup. In: Bredenfeld, A., Jacoff, A., Noda, I., Takahashi, Y. (eds.) RoboCup 2005. LNCS (LNAI), vol. 4020, pp. 173–183. Springer, Heidelberg (2006)
8. Simon, D., Chia, T.L.: Kalman filtering with state equality constraints. IEEE Transactions on Aerospace and Electronic Systems 38(1), 128–136 (2002)
9. Smith, R.: Opensource dynaminc engine (2004), http://www.ode.org

Benchmarks for Robotic Soccer Vision

Ricardo Dodds[1], Luca Iocchi[1], Pablo Guerrero[2], and Javier Ruiz-del-Solar[2]

[1] Dipartimento di Informatica e Sistemistica, Universita "La Sapienza", Rome, Italy
{dodds,iocchi}@dis.uniroma1.it
[2] Departament of Electrical Engineering, Universidad de Chile, Chile
{pguerrer,jruizd}@ing.uchile.cl

Abstract. Robotic soccer vision has been a major research problem in RoboCup and, even though many progresses have been made so that, for example, games now can run without many constraints on the lighting conditions, the problem has not been completely solved and on-site camera calibration is always a major activity for RoboCup soccer teams. While different robotic soccer vision and object perception techniques continue to appear in the RoboCup Soccer League, there is a lack of quantitative evaluation of existing methods.

Since we believe that a quantitative evaluation of soccer vision algorithms will led to significant advances in the performance on perception and on the entire soccer task, in this paper we propose a benchmarking methodology for evaluating robotic soccer vision systems. We discuss the main issues of a successful benchmarking methodology: (i) a large and complete data base or data sets with ground truth; (ii) a public repository with data sets, algorithms and implementations that can be dynamically updated and (iii) evaluation metrics, error functions and comparison results.

Keywords: Benchmarking and Evaluation, Color-based Object Recognition, Robotic Soccer Vision.

1 Introduction

We can generally agree that the camera is the most important sensor to interact with the outer world. The main objective of the perception module is to process the raw image and extract useful information. The inputs for a vision system are images of the camera and sensor readings from internal robot sensors, the outputs are the relative positions of all recognized objects to the robot.

Over all the visual tasks that a computer may perform, recognizing an object present in an image could be the most challenging one. Not only a large number of object categories exist, but also each class can reach a wide number of configurations. In a single image one could find a large number of objects. In the case of video analysis, and tracking several objects in a dynamic environment, things can get worse.

T. Röfer et al. (Eds.): RoboCup 2011, LNCS 7416, pp. 427–439, 2012.

In the RoboCup Soccer League environment, the number of objects of interest is small and there is enough a priori information about their size, color and position (in the case of static landmarks), but still recognizing them remains a complex task. Over the years, in Robocup Soccer competitions, there have been several proposals to solve the problem of object perception. It is possible to find a wide variety of publications about the main stages in object recognition in color coded environment. Since objects of interest have characteristic colors, most of these attempts are based on color vision. Moreover, although code sharing in RoboCup (specially in the Standard Platform League) allows in principle to use and test approaches made by other research groups, this did not occur as desirable.

Indeed, regardless the large amount of work in the area and the fact that a quantitative evaluation is essential to ensure progress, very little has been done on performing a systematic comparison among these approaches. This is mainly due to the lack of a benchmark methodology and of common data sets. Consequently, almost every paper (see next section) provides its own results on its own sequence of images.

Conversely, many other research areas have defined standard benchmarks that are commonly used when introducing new methods. For example, within the SLAM community reference data sets and the most important algorithms are available on-line[1] (often as open-source projects), so that it is easy to directly evaluate a new algorithm with respect to existing ones. Also in the Computer Vision community, many data sets are available for method evaluation: for example, PETS[2] is used for people tracking and human activity recognition.

The main objective of this work is thus to define a benchmark methodology for robotic soccer vision systems, developing both a database and an evaluation methodology. The benchmark can be useful for any RoboCup Soccer League and also for general color-based object recognition. Although at the moment only object recognition and localization is considered, the benchmark methodology can be extended to other relevant issues in robotic soccer vision (like activity recognition, anomaly detection, etc.).

Having a challenging data set helps to track progresses made by current algorithms, stimulates new ideas, improves code sharing and the overall progress of the league. In order to realize such an effective benchmark methodology for soccer robots, the data set must be well organized and sufficiently large (i.e., in use by most/all research groups), the evaluation methodology should take into account different performance metrics, the repository should contain also algorithms (and their implementation), possibly in open-source, with a clear input/output definition, so that can be actually used by the research community.

The paper is organized as follows. In Section 2 we describe the main methods that have been proposed. We give the specifications of the Data Set in Section 3 and introduce the Data Set Description concept. In Section 4 we describe the evaluation methodology for soccer vision systems and its implementation in Section 5. We summarize our work and discuss future directions in Section 6.

---

[1] `radish.sourceforge.net` and `openslam.org`
[2] `www.cvg.rdg.ac.uk/PETS2010/`

2   Overview of Robotic Soccer Vision

Even if the scope of this paper is not to give a full review of soccer vision algorithms, we should describe the most important stages and approaches of a general perception module. Without losing generality, a perception module can be decomposed into two stages: (i) low-level vision, that takes as input a raw image and outputs a set of region candidates and (ii) high-level vision, which uses the regions to determine the 3D relative position of available objects. Despite their differences, many of the proposed techniques can be viewed conceptually in terms of these two stages of processing[3].

In the low-level stage, a bottom-up processing of the image is performed, where image pixels are analyzed in order to extract useful information. This is generally the most computationally expensive task, and at each stage the amount of information that will be processed further is reduced. The first important step is the color segmentation, which uses a color table (look-up table) to map pixels from raw image values to a class of symbolic colors, considerably reducing the amount of information per pixel from 256^3 to the number of classes $|C|$ (normally 8 or 9). Several attempts has been made in order to improve this task. In [3] they use a evolutionary algorithm to convert YUV color space into a easy-to-separate one facilitating the classification task. Following previous work, in [16] colors are not only mapped to unambiguous but also to ambiguous color classes. The ambiguous color classes are resolved based on their unambiguous neighbors in the image. In contrast to other approaches, the neighborhood is determined on the level of regions, not on the level of pixels. Another interesting method is proposed in [6] where they use knowledge of spatial relationship between color classes to calibrate them starting from another classes.

In the early stage of a color-based vision system, the color constancy problem is tackled. As mentioned in [13], variations on the light conditions have a strong effect on the spatial distribution of target color classes. This is known as the color constancy problem. We can agree that manual calibration is a time consuming task and may lead to errors. As lighting will always be different on each playground (even at different times of the day in the same place), teams started developing automated vision calibrations routines. In the literature, there are described two ways for improving color constancy dynamically: (i) adjusting camera parameters or (ii) modifying the color table. These algorithms try to stabilize color recognition, so that the same colors look almost the same under different illuminations.

One of the simplest solutions, presented in [9], is to calculate in each frame the intensity of the image and accumulate an average value in order to detect if there are variations and adjust the gain parameter if necessary. In [5] a genetic algorithm is used to optimize the camera parameters instead of modifying color table.

---

[3] This decomposition assumes that it is possible to define a valid interface between the two levels, general enough to describe all possible configurations of a perception module.

Modifying the color table is the most used approach. In [17] is proposed a simple solution that defines three different lighting conditions: bright, intermediate and dark, and using KL-divergence compares the actual illumination condition and determines which color table to use. The most common method, described in [10] is to adapt the color table using statistics of recognized objects. Similarly, [7] and [1] use geometrical models of field and objects to identify landmarks independently of color classifications. Color information from these landmarks is used build color classes. An adaptive color calibration based on the Bayes Theorem, using chrominance histograms and object shapes to update them is proposed in [4]. Also [11] uses different layers of color representation and updates the color look-up table for each layer using information about recognized objects. A novel approach is presented in [8], where an adaptive transformation of the color distribution of the image is used to adjust static thresholds.

On the other hand the high-level vision module performs a top-down image analysis, using features provided by low-level vision. The main objective is to find objects of interest and estimate their properties. In this stage contextual information and expectations of objects that might be in the image could be used. Usually starting from a list of region candidates of the appropriate color, binary rules are applied to discard false perception. Physical features as size and shape are used to quickly filter wrong candidates. Subsequently deeper analysis are performed. In [14] a context-based vision system is presented. In this work, perception of static objects is improved using a bayesian framework that considers different context-coherence instances. Also, machine learning is a widely used technique in soccer vision. A decision tree learning algorithm, used for compute the pose of visible robot, is described in [18].

So far, we have described algorithms strongly based on a robust color segmentation of the image. Nevertheless there have been proposed different approaches that focus its efforts on high level processing. In [12], assuming that a color core exists independently of the illumination condition, they only use sparse color classification, i.e. name a color blue only if you are sure it is going to be blue for all illumination conditions. Sequently, using edge detection from blobs and counting the number of color pixels inside, the blob color is defined.

Notice that our review and further analysis is only restricted to regular cameras. Although systems running with omnidirectional cameras, share common elements in the perception process, they use techniques that go beyond the scope of this paper.

3 Data Set

The benchmark we are proposing is composed by two main topics: the specification and the gathering of the data sets and the methodology for evaluating and comparing different approaches. The former is discussed in this section, while the latter in the next one. A data set is given by a continuous sequence of data taken with the on-board robot sensors (including the camera) and possibly with additional external sensors that are used to provide ground truth. These data are taken in a specific setting that is described through a Data Set Description.

3.1 Data Set Descriptions

In order to motivate future progresses in robotic soccer vision, well-organized and challenging data sets are required. Data sets must contain both input data to be processed by the algorithms and output data (ground truth) to evaluate the results. Data sets should also consider the many different variabilities that occur in real applications. In this work, we propose to define the characteristics of each data set with a formal *Data Set Description* (DSD) formed by the following items:

- **ID of the robot**, defines from which robot data were taken (e.g., NAO-48),
- **sensor used**, describes which sensors (both on the robot and external sensors) are used to collect data (e.g., robot camera, robot joints, external camera),
- **scene**, describes the environment where the data set has been taken (e.g., Singapore 2010 SPL-Field A),
- **action**, describes the game situation referring to soccer actions (e.g., penalty kick against blue goal),
- **light**, describes the light conditions and whether videos where taken with natural or artificial illumination (e.g., Singapore 2010 venue illumination),
- **occlusions**, describes additional occlusion if any (e.g., other robots or people in the field),
- **shadows**, describes presence of shadows (e.g., due to people around the field),
- **robot dynamics**, specifies if the robot capturing the images is fully, partial (i.e., only the head) or not moving (e.g., still robot tracking the ball with its head),
- **environment dynamics**, describes other elements moving in the field (e.g., ball and other robots moving around).
- other information, such as date and time of capture, duration, author, etc.

Each data set is associated to a DSD that will be useful also for partial analysis: for example, evaluating methods in particular conditions by using only the corresponding data sets. Moreover, DSD are extensible, thus additional features and relevant information can be added in the future, responding to additional uses of the benchmark.

3.2 Data Set Acquisition

Once the characteristics of the data set have been defined, a careful data acquisition procedure must be performed. Since data will be used to evaluate performance of soccer vision methods, three kinds of information must be collected: (i) the images acquired by the robots, (ii) some internal states of the robots (e.g., joints configuration for humanoid robots) in order to determine the pose of the camera with respect to the body), (iii) external sensors used to measure ground truth (in particular the pose of the robot in the environment).

The difficulties in data set acquisition are: 1) acquisition of precise ground truth, 2) data synchronization. Since the field of action of the robots is quite limited, a precise ground truth can be obtained by placing sensors around the field. Both top-view vision cameras and laser range finders can be used for this task (see for example [2]). In case of external vision cameras, we can either place markers on the robot and use an automatic robot pose estimation, or label images by hand. With laser range finders, the use of two or three of them allows for dealing also with multiple robots and occlusions. For data synchronization, internal clocks of the robots can be used to synchronize images and robot states, while if external videos are taken for ground truth, a flashing light pattern, visible from both the robot camera and the external camera, can be used to determine synchronous frames.

In this paper we show some examples of data sets including robot camera videos, robot joints and top-view camera, as shown in Figure 1.

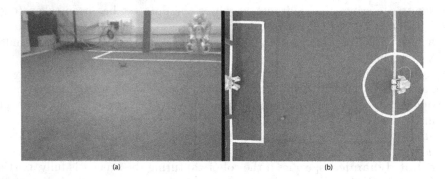

Fig. 1. *Video acquisition* from NAO camera (a) and from a top view camera (b)

3.3   Data Set Labeling

The third step in the Data Set collection consists in providing the ground truth that is needed to evaluate algorithm performance. Since our main objective is the evaluation of performance of soccer vision methods, whose main objective is to recognize field elements (ball, goals, lines, robots,...) and determine their position with respect to the robot, it is necessary to label in each frame all the objects of interest. On the other hand we are not interested in a pixel based recognition, thus we believe it is not necessary to label single pixels in the image.

Therefore, we propose to proceed as follows.

First, we define the elements of interest: *ball, goal post, goal cross-bar, entire goal, field line, field corner, robot*. Then we decide to associate to each of these elements a planar shape. For simplicity, we will use ellipses for the *ball* and quadrilaterals for all the other elements.

Second, a manual annotation of the frames is performed. To this end we developed a tool to simplify this process: it allows the user to move frame by frame within the data set, declare an element type and click over four points that will be used to determine a bounding box (an ellipse in case of the ball) for that element. Examples of such labeling are shown in Figure 2.

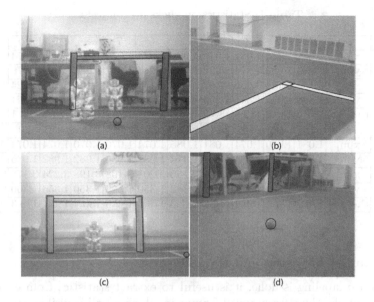

Fig. 2. *Labeling objects of interest.* (a),(c) both entire goals and it parts are labeled using quadrilaterals, as well as lines and corners (b); (d) ellipses are used to label the ball

In case an external camera is used to get videos for the ground truth and automatic detection is not available, also the videos taken from this external camera need to be labelled. For simplicity, we decided to label only the moving objects: the ball and the robots players, including the one that is taken the data set. For the robots we also determine the orientation, although this information will not be very precise. Thus the relative position of the ball with respect to the robot taking the data set can be computed by transforming the ball location from a field reference system to the robot local reference system. While we do not label static elements, thus the distance from the robot taking the data set and the static elements in the field (if needed) may be computed afterwards by knowing the geometry of the field.

Positions and poses extracted by the external camera are provided in metric units within a field reference system. It is also desirable that the external camera introduces as less distortion as possible. In case of high lens distortion, a pre-calibration procedure is suggested, and images in the data set should be provided undistorted.

It is also important to notice that it is unavoidable that this measure introduces small errors when the human operator clicks over an object. In order to estimate this error, we asked several people to repeat the experiment of annotating the ball and a robot within a data set. Results are shown in Table 1.

From these results we can compute standard deviations of human errors in labeling that will be taken into account in the evaluation.

Table 1. Statistics on clicks over objects

	#	1	2	3	4	5	6	7	8	9	10
Ball	x[m]	1.081	1.074	1.081	1.070	1.062	1.074	1.074	1.081	1.085	1.074
	y[m]	1.986	1.982	1.974	1.982	1.982	1.986	1.982	1.982	1.986	1.978
Robot	x[m]	3.155	3.144	3.155	3.152	3.175	3.140	3.163	3.163	3.167	3.224
	y[m]	1.365	1.353	1.361	1.353	1.361	1.361	1.350	1.342	1.353	1.361
	theta[o]	181.2	180.6	180.3	180.9	182.0	181.5	179.8	179.3	181.5	180.1
	#	11	12	13	14	15	16	17	18	19	20	σ
Ball	x[m]	1.081	1.070	1.074	1.081	1.085	1.074	1.070	1.070	1.074	1.074	0.006
	y[m]	1.971	1.982	1.994	1.982	1.974	1.997	1.986	1.982	1.982	1.986	0.006
Robot	x[m]	3.244	3.175	3.190	3.186	3.140	3.201	3.198	3.198	3.224	3.178	0.029
	y[m]	1.342	1.350	1.353	1.342	1.342	1.350	1.353	1.350	1.353	1.350	0.007
	theta[o]	179.7	179.1	180.4	182.2	181.3	179.5	182.0	180.4	180.7	179.3	0.967

3.4   Data Set Statistics and Data Base

Finally, once labeling is done, it is useful to extract statistics from each Data Set: for example, the percentage of frames in which a ball is visible, etc. To this end we have realized a tool that can extract different statistics from a data set.

It is important to notice that this formalization of the data sets allows for creating a real data base of data sets, that can be effectively queried with a simple query language (for example, through a Web interface). In this way it will be possible to retrieve, for example, all the data sets regarding penalty kicks in natural light conditions, with a high percentage of frames containing both the ball and the yellow goal.

4   Evaluation Methodology

In order to evaluate the performance of soccer vision algorithms in a quantitative way, their outputs have to be compared to the ground truth using pre-defined error metrics. Here we propose a *per-image* measure methodology, frequently used in object detection. The evaluation determines the following measures:

- *false positives* and *false negatives* in object detection
- *object detection precision* in the image space
- *object detection precision* in the world space
- *computational efficiency*

We have written an automatic measurement tool that can perform the evaluation of a method, provided that its output is given in a pre-defined format (which is described in the benchmark web site).

In particular, the evaluation compares, for each object of interest that the algorithm declares to detect and for each frame in the data set, the output of the algorithm with the ground truth. For objects, like goal posts and field lines, that may appear multiple times in an image, we evaluate them for each occurrence in the frame. For example, if a frame contains two goal posts and the algorithm returns only one, the missing one will be counted as a false negative.

4.1   False Positives and False Negatives

The first measure we consider is to evaluate the quality of the vision method in detecting the objects of interest in the images.

Let us denote with O the set of objects of interest, and with $o \in O$ a specific object, I_t the frame taken at time t, T the set of all time-stamps, $N = |T|$ the total number of frames, $\delta^o_G(t)$ the number of occurrences of o in I_t in the ground truth, $\delta^o_A(t)$ the number of occurrences of o in I_t in the output of the algorithm, N^o_G the number of frames in which there is at least one occurrence of o in the ground truth, N^o_A the number of frames in which there is at least one occurrence of o in the output of the algorithm.

Many measures can be used to measure the quality of object detection, and while we believe the following ones are the most significant, others can be considered as well in the evaluation methodology.

True positive rate (or detection rate) and *false positive rate* measure accuracy of object detection for a given object o and are defined as

$$TPR^o = \frac{\Sigma_{t\in T}(\min(\delta^o_A(t), \delta^o_G(t)))}{N^o_G}$$

$$FPR^o = \frac{\Sigma_{t\in T, \delta^o_A(t) > \delta^o_G(t)}(\delta^o_A(t) - \delta^o_G(t))}{N}$$

4.2   Object Detection Precision in Image Space

For correctly detected objects, i.e. for all those frames where $\delta^o_G(t) > 0$ and $\delta^o_A(t) > 0$, precision measures will be performed. For objects occurring multiple times a data association problem must be solved in order to match outputs of the method with the ground truth. This must be done even in presence of false positive or false negatives detection. For example, if a frame contains three lines and the algorithm return two of them, one element is counted as a missing detection, but for the other two precision must be computed by comparing them with two of the lines in the ground truth. Data association is solved with a best scoring approach, so the association that returns the best score is chosen. Note that, although this process requires the evaluation of $\delta^o_G(t) > 0 \times \delta^o_A(t) > 0$ combinations, this is in fact a small number, since $\delta^o_G(t)$ is typically limited.

Let us denote with $\Gamma^{o_i}_t$ the bounding box of the i-th occurrence of o in I_t in the ground truth, and $B^{o_i}_t$ the bounding box of the i-th occurrence of o in I_t in the method output. Let $f(i)$ be the association function between the occurrence of an object in the method output and in the ground truth, that returns the best score. Then precision in the detection of the i-th occurrence of o in I_t by the method, which is associated of the $f(i)$-th occurrence of o in I_t in the ground truth, is given by following the PASCAL ([15]) measure

$$\alpha^{o_i}_t = \frac{area(B^{o_i}_t \cap \Gamma^{o_{f(i)}}_t)}{area(B^{o_i}_t \cup \Gamma^{o_{f(i)}}_t)} \tag{1}$$

The average precision in detecting o in all the data set is given by

$$\alpha^o = \frac{\Sigma_{t \in T} \Sigma_{i=1,..,\delta_A^o(t)} \alpha_t^{o_i}}{\Sigma_{t \in T} \delta_A^o(t)}$$

4.3   Object Detection Precision in Field Space

The third evaluation stage is made using the real positions of the object, provided in the data set. Again for each object for which the algorithm returns a position in the world (in robot local ground coordinate frames), the precision of such measure is evaluated by comparing the values in the data set ground truth.

As mentioned in the previous section, we have decided to label only the position of the ball and the poses (position and orientation) of the robots.

Let x_t^{ball} be the relative position of the ball with respect to the robot in the frame I_t computed by the method under evaluation, and χ_t^{ball} the relative position of the ball with respect to the robot in the frame I_t computed with the ground truth.

The evaluation of the location of the ball is based on the Euclidean distance between the two positions x_t^{ball} and χ_t^{ball}. We denote the distance between the tow positions with $||x_t^{ball} - \chi_t^{ball}||$ and the distance of the ball to the robot (i.e., the norm of χ_t^{ball}) as $||\chi_t^{ball}||$.

When this distance is within the standard deviation of the human error in annotating the ground truth σ_d, then this error is fixed to 0.

$$\epsilon_t^{ball} = \begin{cases} 0 & \text{if } ||x_t^{ball} - \chi_t^{ball}|| \leq \sigma_d \\ \left| \frac{||x_t^{ball} - \chi_t^{ball}||}{||\chi_t^{ball}||} \right| & \text{otherwise} \end{cases}$$

A similar measure is defined for poses, by considering both the position and the orientation error. Then, the average error over the entire data set is obtaining by averaging ϵ_t^o over time $t \in T$.

4.4   Computational Efficiency

The last performance measure is computational time of the method. Here it would be important to evaluate it in the actual CPU on which the method will run, thus on the robot CPU. In the case of standard platforms, this is very easy and all the methods can be directly compared. In case of self-built soccer robots, the specification of the CPU, memory and other characteristics of the processing unit on board the robot needs to be provided as well.

5   Benchmark Implementation

The benchmarking methodology described in this paper has been implemented by developing a collection of tools and data sets and by making them available through the web site

http://labrococo.dis.uniroma1.it/RobotSoccerVisionBenchmark/

The web site contains a full description of the components of the data sets, so that it is possible and easy to contribute to it by extending its scope.

At the moment this paper is completed, the benchmark contains 20 data sets, corresponding to about 40 minutes of labeled videos taken with the NAO robots. Additional data sets will be added in the next future, provided also by other teams. Moreover, in order to simplify the use of the benchmarking methodology, we released the following software tools: (i) a tool for data acquisition for the NAO robots, (ii) a tool for labeling sequences of images or videos and (iii) a shared library that facilitates evaluating vision methods.

6   Discussion

It is well known that defining standard benchmarks and common performance metrics is a very important issue in order to improve the general performance of a research topic. In this paper, we propose a standard benchmark and performance metrics for one of the major RoboCup soccer problems: vision and object recognition.

Although we propose a specific methodology and performance metrics that we believe of general interest for evaluating the research in this field, the benchmark can be easily extended in order to capture other specific needs. For example, adding a new performance metric does not affect the collection and labeling of data sets; adding a new object to recognize or additional labels does not affect previous collections.

Moreover, the collection and the use of data sets must not be limited to a single research group, but should be extensively used ideally by all the groups doing research in the field. Therefore the main aim of this work is to involve as many research groups as possible to contribute to the collection of data sets and evaluation of methods, through the benchmark web site that will be constantly updated.

So far we have collected only a few data sets with the methodology presented above in our laboratories. This activity confirmed both the correctness of the proposed approach and of the developed tools and the effectiveness and easy-of-use of the methodology.

Future work is to collect contributions to this benchmark by other research groups, make it widely accepted and use it as a standard evaluation methodology for robotic soccer vision.

References

1. Cameron, D., Barnes, N.: Knowledge-Based Autonomous Dynamic Colour Calibration. In: Polani, D., Browning, B., Bonarini, A., Yoshida, K. (eds.) RoboCup 2003. LNCS (LNAI), vol. 3020, pp. 226–237. Springer, Heidelberg (2004)
2. Ceriani, S., Fontana, G., Giusti, A., Marzorati, D., Matteucci, M., Migliore, D., Rizzi, D., Sorrenti, D.G., Taddei, P.: Rawseeds ground truth collection systems for indoor self-localization and mapping. Autonomous Robots 27(4), 353–371 (2009)

3. Dahm, I., Deutsch, S., Hebbel, M., Osterhues, A.: Robust color classification for robot soccer. RoboCup 2003: Robot World Cup VII, Lecture Notes in Artificial Intelligence (2004)
4. Gönner, C., Rous, M., Kraiss, K.-F.: Real-time Adaptive Colour Segmentation for the Robocup Middle Size League. In: Nardi, D., Riedmiller, M., Sammut, C., Santos-Victor, J. (eds.) RoboCup 2004. LNCS (LNAI), vol. 3276, pp. 402–409. Springer, Heidelberg (2005)
5. Grillo, E., Matteucci, M., Sorrenti, D.G.: Getting the Most from Your Color Camera in a Color-Coded World. In: Nardi, D., Riedmiller, M., Sammut, C., Santos-Victor, J. (eds.) RoboCup 2004. LNCS (LNAI), vol. 3276, pp. 221–235. Springer, Heidelberg (2005)
6. Guerrero, P.E., Ruiz-del-Solar, J., Fredes, J., Palma-Amestoy, R.: Automatic On-Line Color Calibration Using Class-Relative Color Spaces. In: Visser, U., Ribeiro, F., Ohashi, T., Dellaert, F. (eds.) RoboCup 2007. LNCS (LNAI), vol. 5001, pp. 246–253. Springer, Heidelberg (2008)
7. Gunnarsson, K., Wiesel, F., Rojas, R.: The Color and the Shape: Automatic On-Line Color Calibration for Autonomous Robots. In: Bredenfeld, A., Jacoff, A., Noda, I., Takahashi, Y. (eds.) RoboCup 2005. LNCS (LNAI), vol. 4020, pp. 347–358. Springer, Heidelberg (2006)
8. Iocchi, L.: Robust Color Segmentation Through Adaptive Color Distribution Transformation. In: Lakemeyer, G., Sklar, E., Sorrenti, D.G., Takahashi, T. (eds.) RoboCup 2006. LNCS (LNAI), vol. 4434, pp. 287–295. Springer, Heidelberg (2007)
9. Jamzad, M., Lamjiri, A.K.: An Efficient Need-Based Vision System in Variable Illumination Environment of Middle Size RoboCup. In: Polani, D., Browning, B., Bonarini, A., Yoshida, K. (eds.) RoboCup 2003. LNCS (LNAI), vol. 3020, pp. 654–661. Springer, Heidelberg (2004)
10. Jüngel, M., Hoffmann, J., Lötzsch, M.: A Real-Time Auto-Adjusting Vision System for Robotic Soccer. In: Polani, D., Browning, B., Bonarini, A., Yoshida, K. (eds.) RoboCup 2003. LNCS (LNAI), vol. 3020, pp. 214–225. Springer, Heidelberg (2004)
11. Jüngel, M.: Using Layered Color Precision for a Self-Calibrating Vision System. In: Nardi, D., Riedmiller, M., Sammut, C., Santos-Victor, J. (eds.) RoboCup 2004. LNCS (LNAI), vol. 3276, pp. 209–220. Springer, Heidelberg (2005)
12. Lovell, N.: Illumination Independent Object Recognition. In: Bredenfeld, A., Jacoff, A., Noda, I., Takahashi, Y. (eds.) RoboCup 2005. LNCS (LNAI), vol. 4020, pp. 384–395. Springer, Heidelberg (2006)
13. Mayer, G., Utz, H., Kraetzschmar, G.K.: Playing Robot Soccer Under Natural Light: A Case Study. In: Polani, D., Browning, B., Bonarini, A., Yoshida, K. (eds.) RoboCup 2003. LNCS (LNAI), vol. 3020, pp. 238–249. Springer, Heidelberg (2004)
14. Palma-Amestoy, R., Guerrero, P., Ruiz-del-Solar, J., Garretón, C.: Bayesian Spatiotemporal Context Integration Sources in Robot Vision Systems. In: Iocchi, L., Matsubara, H., Weitzenfeld, A., Zhou, C. (eds.) RoboCup 2008. LNCS (LNAI), vol. 5399, pp. 212–224. Springer, Heidelberg (2009)
15. Ponce, J., Berg, T.L., Everingham, M., Forsyth, D., Hebert, M., Lazebnik, S., Marszalek, M., Schmid, C., Russell, B.C., Torralba, A., Williams, C.K.I., Zhang, J., Zisserman, A.: Dataset Issues in Object Recognition. In: Ponce, J., Hebert, M., Schmid, C., Zisserman, A. (eds.) Toward Category-Level Object Recognition. LNCS, vol. 4170, pp. 29–48. Springer, Heidelberg (2006)

16. Röfer, T.: Region-Based Segmentation with Ambiguous Color Classes and 2-D Motion Compensation. In: Visser, U., Ribeiro, F., Ohashi, T., Dellaert, F. (eds.) RoboCup 2007. LNCS (LNAI), vol. 5001, pp. 369–376. Springer, Heidelberg (2008)
17. Sridharan, M., Stone, P.: Towards Illumination Invariance in the Legged League. In: Nardi, D., Riedmiller, M., Sammut, C., Santos-Victor, J. (eds.) RoboCup 2004. LNCS (LNAI), vol. 3276, pp. 196–208. Springer, Heidelberg (2005)
18. Wilking, D., Röfer, T.: Realtime Object Recognition Using Decision Tree Learning. In: Nardi, D., Riedmiller, M., Sammut, C., Santos-Victor, J. (eds.) RoboCup 2004. LNCS (LNAI), vol. 3276, pp. 556–563. Springer, Heidelberg (2005)

Development of an Object Recognition and Location System Using the Microsoft Kinect™ Sensor*

Jose Figueroa, Luis Contreras, Abel Pacheco, and Jesus Savage

Biorobotics Laboratory, Department of Electrical Engineering,
Universidad Nacional Autonoma de Mexico, UNAM

Abstract. This paper presents the development of an object recognition and location system using the Microsoft Kinect™, an off-the-shelf sensor for videogames console Microsoft Xbox 360™ which is formed by a color camera and depth sensor. This sensor is capable of capturing color images and depth information from a scene. This vision system uses a) data fusion of both color camera and depth sensor to segment objects by distance; b) scale-invariant features to characterize and recognize objects; and c) camera's internal parameters combined with depth information to locate objects relative to the camera point of view. The system will be used along with a robotic arm to grab objects.

Keywords: Keywords: Feature extraction, Scale Invariant Feature, Machine vision, Object detection, Pattern recognition.

1 Introduction

Autonomous mobile robots must have several capabilities to perform correctly their tasks: navigation, natural language understanding, object detection and manipulation. The ability for detecting and locating objects in the environment using vision is one the objectives of several research service robots groups. The purpose of our research is to develop a object recognition and location system which takes advantage of the Microsoft Kinect™ sensor, which features both an image sensor and a depth sensor.

2 Depth Vision

The ability of locating objects in three-dimensional space is the most important ability which mobile service robot must have to interact with these objects in the environment. Most of the teams which get to Second Stage in Robocup@Home competition, as well as the teams in the top 5 in the final results have vision systems which combine depth perception and two-dimensional object recognition.

---

* This work was supported by PAPIIT-DGAPA UNAM under Grant IN-107609.

T. Röfer et al. (Eds.): RoboCup 2011, LNCS 7416, pp. 440–449, 2012.

2.1 Kinect

Hardware. The Microsoft Kinect^{TM} sensor for Microsoft Xbox 360^{TM} video-games console, is an input device which allows users to interact with videogames in a controller-less way through gesture and voice recognition. It was release in North America in November 4th 2010 (Fig. 1) with the retail prices of $150.00 USD, in USA, and $2,300.00 M.N., in Mexico.

The vision hardware of Kinect^{TM}is capable of making a three-dimensional reconstruction of the environment by using an infrared camera and an infrared projector, and capturing color images streams. It is based in the PrimeSensor^{TM} Reference Design de la compañía PrimeSense[8], whose vision subsystem is formed by an 640x480@30Hz RGB camera, an 640x480@30Hz infrared camera and an infrared projector.

The PrimeSense's technology for depth information acquisition is based in the Light Coding^{TM}[8] technique, which works by codifying the scene's volume with near-infrared light. The depth vision system utilizes an off-the-shelf CMOS image sensor to read the coded light back from the scene. PrimeSense's SoC chip is connected to the CMOS image sensor, and executes

Fig. 1. Microsoft Kinect mounted on robot^{TM}

a sophisticated parallel computational algorithm to decipher the received light and produce a depth image of the scene. The solution is immune to daylight.

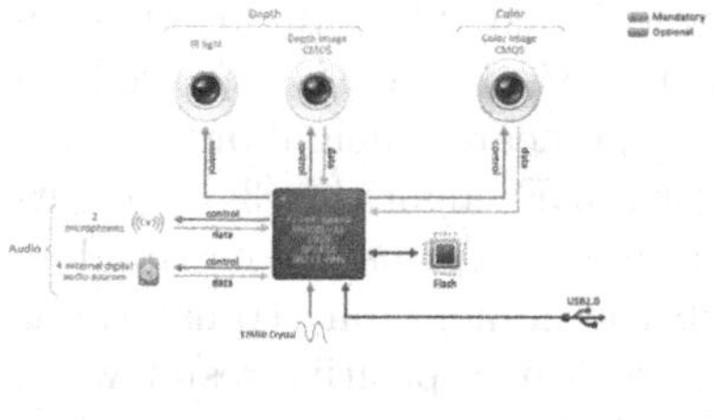

Another capability of the PrimeSensor^{TM} Reference Design hardware is provided by a process called Registration which aligns pixel-by-pixel the elements of the depth map with the color image.

The limitations of Kinect technology are: its effective distance range is from 40 to 625 centimeters, reflective surfaces alter the readings, no depth data can be reconstructed if objects are shadowed from the infrared projector.

Fig. 2. PrimeSensor^{TM} Reference Design

Software. In November 2010, Adafruit Industries offered a bounty to anyone who was able to create an open source driver for Kinect[1], which happened on November 10th, 2010 when a Spanish programmer

called Hector Martín was announced as the winner[2], who created a Linux driver
which allowed using both RGB and depth camera.

The release of this driver and its corresponding forks for Windows and MacOS
started a community for homebrew software which developed applications for
user interfaces, robotics, artistic shows, amongst others. The popularity of Kinect
in the developer community soared to such degree that PrimeSense released
Kinect-enabled versions of their softwares OpenNI and NITE, allowing Kinect
homebrew developers to make body tracking applications.

3 Vision System

Theory. The operations for recognizing and locating an object in a scene are
represented by the first stages of the vision model of David Marr[7]. To begin
with, the primal sketch stage gives the most important information about the
image in two dimensions, such as edges, corners or blobs; in this stage, the
SIFT (Scale-Invariant Feature Transform) algorithm[6] is used to characterize
and recognize objects in a scene. Finally, the 2.5D sketch provides the depth
information of the image; for this stage, the depth sensor of Microsoft Kinect^{TM}
is used to extract the depth information of an object in a scene.

Object Recognition. The objects which are going to be recognized by the vision system are represented by interest points computed by the SIFT algorithm.
The training stage uses a procedure where images of an object are stored in
a list. Once several images are captured, the next step consists in creating the
pattern set, by getting the kd-tree[4] containing the SIFT features of each image
stored in the list, this process is performed using a thread pool, which contains
a loop which checks for any thread which has finished its processing and assigns
a new image to the free thread. Each object is stored in the vision system as an
structure which contains a set of kd-trees, a set of feature matchers and a name
for the object and the set of these structures conforms the objects database.

The recognition process's preparations starts during the training stage, where
a matcher is created for each SIFT feature's kd-tree. Then, when an object is going to to be recognized in a scene, a depth segmented RGB image is captured, its
SIFT features are extracted they are searched for the nearest neighbors on each
kd-tree of the pattern set, using the Best Bin First (BBF) algorithm[3]. The outliers in the nearest neighbor matched features are eliminated using the Random
Sample Consensus algorithm (RANSAC)[5] with the homography transform as
criteria for finding outliers; the matching process returns a positive result when
matches are bigger or equal than four, because the homography matrix is calculated with at least four points.

In the case where an unknown object must be recognized, the recognition
process is performed by matching unknown object's features against the pattern
sets of each object stored in the objects database with BBF, and the greatest
amount of matches is stored during the process. At the end of the matching process of each object database entry, the matches amount is stored in a histogram,

where each bin refers to an object stored in the objects database. Once matching process against all the entries of the object database is finished, the histogram is queried for the entry with the greatest amount, returning an index to an object database entry, which is used for performing an outlier deletion in the matched points using RANSAC. If the number of matches is greater than one, the set of matched points is used on the object location procedure.

Object Location. The set of matched points is used to make a bounding box in the scene image, where the recognized object is located, which is used to make a region of interest for the depth map. The object's location can be computed by using the centroid of the region of interest and get the depth value in that place or by finding the centroid of the three-dimensional point cloud generated by the depth information within the region of interest.

By taking advantage of PrimeSensor^{TM} Registration process, the only Kinect's camera which needs calibration is the RGB one, whose intrinsic parameters and distortion coefficients are obtained by using the Zhang's algorithm [9]. The color image and the pixel-aligned depth image are rectified using the RGB camera parameters and the coordinates of each three-dimensional point are computed with the following equations

$$P3D_x = (x_{RGBD} - cx_{RGB}) * \frac{f\left(depth\left(x_{RGBD}, y_{RGBD}\right)\right)}{fx_{RGB}} \tag{1}$$

$$P3D_y = (y_{RGBD} - cy_{RGB}) * \frac{f\left(depth\left(x_{RGBD}, y_{RGBD}\right)\right)}{fy_{RGB}} \tag{2}$$

$$P3D_z = depth\left(x_{RGBD}, y_{RGBD}\right) \tag{3}$$

where RGBD represents a pixel which contains both color and depth information.

4 Tests

The test started by capturing pictures for training and testing sets, using the RGB camera. The training set consisted in the SIFT features of a set of eight objects (Fig. 3a), and the testing set was formed by two objects which were rather similar to some objects of the testing set (Fig. 3b). Two tests were performed with this data: a control test and a false positives test. Input data for feature extraction was the following: depth-segmented RGB colorspace pixels from a 640x480 image of an object set at 40cm from the camera (Kinect's minimum depth value).

Two different tests were applied, all of them with segmented input pictures, so the pattern just contains objects. The objects used in these tests were a disinfectant spray (lysol), a bottle of Bonafont brand water (water), a generic brand coffee jar (coffee), a Nescafe brand coffee jar (nescafe), a cup, a jar of automotive grease (grease), a bottle of chili sauce (sauce) and a milk carton (milk) for the training set(Fig. 3a), and, a bottle of E-Pura brand water (epura_water) and

(a) Training Objects

(b) False Positive Test
Objects

Fig. 3. Test Setup

a Nescafe Classic brand coffee jar (nescafe_class)(Fig. 3b). All the tests were
made with the camera in a fixed viewpoint, positioned in the lower section of
a domestic service robot which was used by UNAM Robocup@Home team for
competition.

The first test consisted in using the pictures of the object training set to
check if they can be recognized with a theoretical accuracy of 100%. Training
data set was generated from the SIFT features of sixteen (16) images of each
object captured at the same distance from the camera and rotated along the
vertical axis to get images from all the sides of the object, and testing data set
consisted of the training data set.

The second test consisted in using the pictures of the object testing set to
check if false positives could be found. Training data set from former test was

used, and testing data set consisted of the SIFT features of eight (8) pictures of each object the testing data set.

5   Results

In both tests, five (5) runs were done because the randomization part of the RANSAC algorithm introduces variations in running time (Tab. 1, 8). Confusion matrices containing the amount of matches between objects, regardless if they are the same or not, were used to get the false positive counts compared to the amount of true positives. The lowest the false positives rate is, the highest is the uniqueness of the features in a data set (Tab. 3, 4, 5, 6, 7, 10, 11, 12, 13, 14).

For the control test, because the position of the camera and objects are the same during training and testing, it was expected that all the descriptors would have 100% of accuracy, and the results proved this hypothesis true (Tab. 2).

For the false positives test, one of the pictures of the E-Pura brand water bottle got a false positive with the generic brand coffee jar, and the Nescafe Classic coffee jar did not get any false positives (Tab. 9).

Table 1. Control Test: Average Running Time in Milliseconds

object	Test 1	Test 2	Test 3	Test 4	Test 5
coffee	2482.38	2598.44	2520.44	2480.44	2456
sauce	1505.44	1552.19	1573.62	1506.38	1593.19
nescafe	2051.38	2127.5	2301	2056.31	2120.62
milk	3764.5	3696.25	3752.81	3703.06	3608.5
water	1874	1857.38	1705.31	1821.31	1833
lysol	3924.38	4166.19	3857.12	4058.94	3775.25
cup	2401.44	2609.12	2300	2321.5	2363.38
grease	2014.38	2002.69	2002.69	2002.62	2048.5

Table 2. Control Test: Accuracy

object	Test 1	Test 2	Test 3	Test 4	Test 5
coffee	16/16	16/16	16/16	16/16	16/16
sauce	16/16	16/16	16/16	16/16	16/16
nescafe	16/16	16/16	16/16	16/16	16/16
milk	16/16	16/16	16/16	16/16	16/16
water	16/16	16/16	16/16	16/16	16/16
lysol	16/16	16/16	16/16	16/16	16/16
cup	16/16	16/16	16/16	16/16	16/16
grease	16/16	16/16	16/16	16/16	16/16

Table 3. Control Test 1 of 5: Confusion Matrix

object	coffee	sauce	nescafe	milk	water	lysol	cup	grease	none
coffee	16	0	0	0	0	0	0	0	0
sauce	0	16	0	0	0	0	0	0	0
nescafe	0	0	16	0	0	0	0	0	0
milk	0	0	0	16	0	0	0	0	0
water	0	0	0	0	16	0	0	0	0
lysol	0	0	0	0	0	16	0	0	0
cup	0	0	0	0	0	0	16	0	0
grease	0	0	0	0	0	0	0	16	0

Table 4. Control Test 2 of 5: Confusion Matrix

object	coffee	sauce	nescafe	milk	water	lysol	cup	grease	none
coffee	16	0	0	0	0	0	0	0	0
sauce	0	16	0	0	0	0	0	0	0
nescafe	0	0	16	0	0	0	0	0	0
milk	0	0	0	16	0	0	0	0	0
water	0	0	0	0	16	0	0	0	0
lysol	0	0	0	0	0	16	0	0	0
cup	0	0	0	0	0	0	16	0	0
grease	0	0	0	0	0	0	0	16	0

Table 5. Control Test 3 of 5: Confusion Matrix

object	coffee	sauce	nescafe	milk	water	lysol	cup	grease	none
coffee	16	0	0	0	0	0	0	0	0
sauce	0	16	0	0	0	0	0	0	0
nescafe	0	0	16	0	0	0	0	0	0
milk	0	0	0	16	0	0	0	0	0
water	0	0	0	0	16	0	0	0	0
lysol	0	0	0	0	0	16	0	0	0
cup	0	0	0	0	0	0	16	0	0
grease	0	0	0	0	0	0	0	16	0

6   Conclusions

The Microsoft Kinect™ Sensor, combined with accurate object recognition algorithms, can be turned into an affordable solution for object recognition and three-dimensional location, and encourage the development of advanced robotics to people or research teams who could not afford other depth vision technologies.

Table 6. Control Test 4 of 5: Confusion Matrix

object	coffee	sauce	nescafe	milk	water	lysol	cup	grease	none
coffee	16	0	0	0	0	0	0	0	0
sauce	0	16	0	0	0	0	0	0	0
nescafe	0	0	16	0	0	0	0	0	0
milk	0	0	0	16	0	0	0	0	0
water	0	0	0	0	16	0	0	0	0
lysol	0	0	0	0	0	16	0	0	0
cup	0	0	0	0	0	0	16	0	0
grease	0	0	0	0	0	0	0	16	0

Table 7. Control Test 5 of 5: Confusion Matrix

object	coffee	sauce	nescafe	milk	water	lysol	cup	grease	none
coffee	16	0	0	0	0	0	0	0	0
sauce	0	16	0	0	0	0	0	0	0
nescafe	0	0	16	0	0	0	0	0	0
milk	0	0	0	16	0	0	0	0	0
water	0	0	0	0	16	0	0	0	0
lysol	0	0	0	0	0	16	0	0	0
cup	0	0	0	0	0	0	16	0	0
grease	0	0	0	0	0	0	0	16	0

Table 8. False Positives Test: Average Running Time in Milliseconds

object	Test 1	Test 2	Test 3	Test 4	Test 5
nescafe_class	2064.88	1989	1998.62	1996.75	1975.25
epura_water	2361.38	2464.88	2379	2353.75	2380.88

Table 9. False Positives Test: Accuracy

object	Test 1	Test 2	Test 3	Test 4	Test 5
nescafe_class	0/8	0/8	0/8	0/8	0/8
epura_water	0/8	0/8	0/8	0/8	0/8

Table 10. False Positives Test 1 of 5: Confusion Matrix

object	coffee	sauce	nescafe	milk	water	lysol	cup	grease	none
nescafe_class	0	0	0	0	0	0	0	0	8
epura_water	1	0	0	0	0	0	0	0	7

Table 11. False Positives Test 2 of 5: Confusion Matrix

object	coffee	sauce	nescafe	milk	water	lysol	cup	grease	none
nescafe_class	0	0	0	0	0	0	0	0	8
epura_water	1	0	0	0	0	0	0	0	7

Table 12. False Positives Test 3 of 5: Confusion Matrix

object	coffee	sauce	nescafe	milk	water	lysol	cup	grease	none
nescafe_class	0	0	0	0	0	0	0	0	8
epura_water	1	0	0	0	0	0	0	0	7

Table 13. False Positives Test 4 of 5: Confusion Matrix

object	coffee	sauce	nescafe	milk	water	lysol	cup	grease	none
nescafe_class	0	0	0	0	0	0	0	0	8
epura_water	1	0	0	0	0	0	0	0	7

Table 14. False Positives Test 5 of 5: Confusion Matrix

object	coffee	sauce	nescafe	milk	water	lysol	cup	grease	none
nescafe_class	0	0	0	0	0	0	0	0	8
epura_water	1	0	0	0	0	0	0	0	7

References

1. Adafruit: The Open Kinect project - the ok prize - get $3,000 bounty for Kinect for Xbox 360 open source drivers (November 2010),
   http://www.adafruit.com/blog/2010/11/04/the-open-kinect-project-the-ok-prize-get-1000-bounty-for-kinect-for-xbox-360-open-source-drivers/
2. Adafruit: We have a winner - Open Kinect driver(s) released - winner will use $3k for more hacking - plus an additional $2k goes to the eff! (November 2010),
   http://www.adafruit.com/blog/2010/11/10/we-have-a-winner-open-kinect-drivers-released-winner-will-use-3k-for-more-hacking-plus-an-additional-2k-goes-to-the-eff/
3. Beis, J.S., Lowe, D.G.: Shape indexing using approximate nearest-neighbour search in high-dimensional spaces. In: Proc. IEEE Conf. Comp. Vision Patt. Recog., pp. 1000–1006 (1997)
4. Bentley, J.L.: Multidimensional binary search trees used for associative searching. Commun. ACM 18, 509–517 (1975),
   http://doi.acm.org/10.1145/361002.361007

5. Fischler, M.A., Bolles, R.C.: Random sample consensus: a paradigm for model fitting with applications to image analysis and automated cartography. In: Readings in Computer Vision: Issues, Problems, Principles, and Paradigms, pp. 726–740. Morgan Kaufmann Publishers Inc., San Francisco (1987), http://portal.acm.org/citation.cfm?id=33517.33575
6. Lowe, D.G.: Distinctive image features from scale-invariant keypoints. International Journal of Computer Vision 60, 91–110 (2004)
7. Marr, D.: Vision: a computational investigation into the human representation and processing of visual information / David Marr. W.H. Freeman, San Francisco (1982)
8. PrimeSense: Primesense, reference design, http://www.primesense.com/?p=514
9. Zhang, R., Tsi, P.S., Cryer, J.E., Shah, M.: Flexible camera calibration by viewing a plane from unknown orientations. In: Proceedings of the 7th International Conference on Computer Vision, pp. 666–673 (September 1999)

grSim – RoboCup Small Size Robot Soccer Simulator

Valiallah Monajjemi[1], Ali Koochakzadeh[1], and Saeed Shiry Ghidary[2]

[1] Parsian Robotic Center, Electrical Engineering Department,
Amirkabir University of Technology
[2] Computer Engineering and Information Technology Department,
Amirkabir University of Technology

Abstract. Realtime simulation of RoboCup small size soccer robots is a challenging task due to the high frame rate of input vision data and complex dynamic model of robots. In this paper we describe a new multi-robot 3D simulator for small size robot soccer domain named 'grSim'. In order to decrease the model complexity and increase simulation speed, a simplified dynamic model for omni wheels is implemented. grSim has a distributed architecture, feature-rich user interface and supports all aspects of a small size robot soccer game, thus it can completely replace all hardware used by teams during software development. grSim can help software/AI developers design smarter SSL robot teams.

Keywords: Multi Robot Simulator, RoboCup Small Size League, Omni-directional robot modeling.

1 Introduction

Developing artificial intelligence software for mobile robots is one of the most challenging tasks in the process of designing an intelligent robot. Robot software development usually requires a full functional real robot, however due to the hardware problems which experimental robots always suffer from, it's hard to design software during hardware development process. In addition, when a full functional real robot exists, constraints like cost, maximum operation time and possible damages slow down the software development process. Robot simulators can overcome such problems.

A robot simulator is used for developing software without depending "physically" on the actual robot, thus saving cost and time. In advanced simulators, robots and objects are modeled as rigid bodies in a virtual world. After receiving commands from clients or controllers, a physics engine simulates actions and sends simulated robot perception data back to the client. The virtual world can also (but not necessarily) be visualized using a two-dimensional or three-dimensional graphics layer.

Small Size robot soccer focuses on the problem of intelligent multi-agent cooperation and control in a highly dynamic environment with a hybrid centralized/distributed system [3]. Small Size League (SSL) is one of the main leagues

T. Röfer et al. (Eds.): RoboCup 2011, LNCS 7416, pp. 450–460, 2012.

of international RoboCup [2] competetions. In a small size robot soccer game, two teams of five autonomous four wheeled omni-directional robots play against each other in a 6.05m x 4.05m field with an orange golf ball. A global vision system (which is also a shared system) with two overhead cameras and an off-field PC calculates localization data and sends it to each team's AI computer (which is also off-field) via network connections. After a series of high level decision making and low level control algorithms, commands are sent to the robots using wireless communication.

SSL robots must fit inside a cylinder with radius of 9cm and height of 15cm. This size constraint, in addition to high speed of the robots (2 m/s and above) and the fact that there must be five robots to form a team, makes the design and development of the robots a complex and time-consuming task. Software development for SSL robots would be a hard and ineffective task without help from realistic multi-agent simulation environments.

SSL software developers always have suffered from lack of a realistic simulator. Existing multi-purpose simulators are not applicable for this field, therefor designing a realistic simulation environment would be a great help to small size soccer robots AI developers. The main problem in order to simulate small size soccer robots are their complex dynamic model caused by their four wheel omni-directional movement structure.

UberSim [6][10] is a vision centric 3D simulator designed for using in dynamic environment like RoboCup domain. It supports limited numbers of sensors and actuators. Robots are described as a set of C++ classes. The communication between clients and simulator are TCP/IP based. Although The main application of this simulator has been defined on small size soccer robots domain, the simulator is not under active development for years and is somehow outdated. Besides, it is not compatible with new SSL shared vision system, it lacks a powerful and easy to use graphical user interface and run-time configuration panel.

SimRobot [15] is another multi purpose simulator which supports more sensors and actuators. Robots are described using XML description files. It has a rich graphical user interface and it is possible to add some user/world interaction to the simulation environment.

Another multi-purpose simulator is Gazebo [13] [14] which is part of Player / Stage [9] [4] project which supports many types of sensors and actuators, as well as some popular pre-made robots. Robots and environment description is based on XML files. Users can develop controllers using a rich Application Programming Interface (API).

Webots [16] [8] is a commercial Simulator with a rich set of sensors, actuators and pre-made robot models. Webots describes the environments and robots in VRML format. Users can develop robot controllers in almost all popular programming languages such as C/C++,Java,Python,Matlab and Urbi [5] [12].

All aforementioned simulators use Open Dynamics Engine (ODE) [19] as rigid body dynamics engine and Open Graphics Library (OpenGL) [1] for visualization.

The ability to support a broad range of sensors, actuators and configurations, adds some level of complexity to all aforementioned simulators, as a result, their performance drops significantly in multi robot environments like small size soccer robot domain with 10 agents inside. Dramatically all those simulators, except Ubersim, have a complex model for omni wheels which slows down the simulation speed for small size soccer robots.

A feature-rich user interface with abilities like run-time configurations in addition to easy robot and ball manipulation can reduce time and cost and increase efficiency during development of multi robot systems' low level and high level algorithms; a feature which is missing (or is very hard to implement) in the those simulators.

In order to overcome such problems and develop a state of the art 3D simulation environment for small size soccer robot domain, we developed a brand new simulator named "grSim" which will be described in detail in rest of this paper.

2   Overview

grSim is a multi-robot simulation environment designed specially for RoboCup small size soccer robot domain. It is able to completely simulate and visualize a robot soccer game with full details. Teams can communicate with the simulator in the same way they communicate with real world, except the commands should be sent to the simulator via network instead of radio connections to the robots. In this way they can use this simulator as a powerful tool to test and develop low level control and high level decision making algorithms without the need for real robots.

grSim is developed in C++ programming language with Qt framework [17]. Qt is a cross-platform application development framework which makes it easy to develop cross-platform GUI applications. The simulator has been developed for GNU/Linux like operating systems, but it can be easily ported to other operating systems as well.

In each cycle, two data packets are received from clients (artificial intelligent softwares developed by teams) via network connections. These packet contain desired control commands for each robot's actuators (wheels, kickers and the spinner [details in section 3.1]). After interpreting the information, the appropriate commands are fed to the physical layer of the program.

The physical layer uses Open Dynamics Engine (ODE) [19] physics engine. This library supports many types of rigid bodies, joints and collision detection functions. It can also simulate friction and bounciness. Each robot consists of some basic objects and actuators connected together with joints.

The visualization layer of grSim uses Open Graphics Library (OpenGL)[1] API. Due to the native support of hardware acceleration in OpenGL, this layer renders the virtual environment in an acceptable frame rate and in an aesthetic way.

At the end of each cycle, localization data (the current state of the objects which were calculated in physical layer) are sent back to the clients. The format

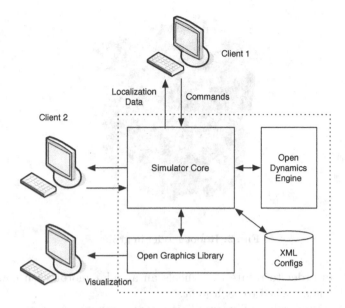

Fig. 1. Overview of the system

used for encoding data is the same format used by small size league's shared vision system, named SSL-Vision. SSL-Vision[21] is the core software of league's shared vision system. As grSim and SSL-Vision use similar data format, clients do not sense any difference between data sent from shared vision system and data sent by grSim, thus no extra effort is needed to convert localization data.

Figure 1 shows an overall view of the simulator's structure.

3   The Simulator

The physical environment consists of 10 robots (5 for each team), a ball, field goals, walls, ground and sky. The dimensions and properties of all objects can be modified in run time. The default values are optimized to comply with latest Robocup SSL rules [7].

3.1   Modeling the Robots in ODE

Each robot consists of four omni wheels, a linear kicker, a chip kicking device and a spin back device. The robot's chassis is a simple cylinder. As the ODE library does not support cylinder with cylinder collision, a dummy sphere is bounded in each cylinder which can only collide with other spheres and cannot collide with any other object. In Figure 2 a small size robot model is depicted.

Wheels. In grSim each robot has four attached wheels with configurable structure. The wheels are solid cylinders attached to the robot's chassis with an ODE's

Fig. 2. Robot model in grSim

Hinge joint. In order to rotate the wheels an angular motor with configurable limited torque is attached to each joint.

Omni directional robots like small size soccer robots, use a special type of wheels called omni (or poly) wheels. In omni wheels there are number of sub-wheels around the circumference which are perpendicular to rolling direction. Sub-wheels help the wheel to side-slide.

Physical modeling of sub-wheels increases the model complexity and is a major cause for simulator deficiency. To overcome this problem we implemented a special model for these wheels inspired by [6]. In this model the wheels have a configurable friction with the ground surface. This friction is different for tangential and perpendicular directions (figure 3). Using this technique, the complexity of robot's physical model is decreased which results in less resource usage and higher simulation speed.

Kicker and Spin-Back Device. Each SSL robot is equipped with a special mechanism to kick the ball. This mechanism will let the robot make a direct or a chip kick. In order to simulate the kicker, a solid cube is attached to each robot's front face. Whenever the kick command is received and the ball is in touch, a proper force will be applied to the ball. The magnitude and type of this force are configurable.

Small size soccer robots use a special actuator in front of their robots in order to manipulate the ball while moving. This device, called spin-back or spinner, helps the robot not to lose the ball's possession while moving. In order to simulate this device, the kicker's solid cube applies a torque to the ball when it touches it. In this way the ball rolls around itself and moves backwards, thus sticks to the robot. In real robots, because of the specific shape of the spinner, ball's side movement during spin back does not happen. In order to avoid ball's side movement, grSim uses the same method discussed in section 3.1 between ball and the ground, i.e. defining different friction coefficients for tangential and

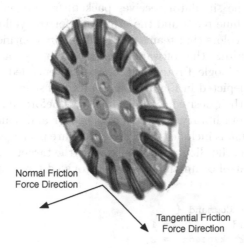

Normal Friction
Force Direction

Tangential Friction
Force Direction

Fig. 3. Omniwheel model in grSim (note that the sub-wheels are textures)

perpendicular directions to the spinner. All of the parameters for the kicker and spinner, like size, maximum torque, maximum force and chip kick angle are configurable.

3.2 Communication

As discussed in section 2 all communications between grSim and clients are network based using User Datagram Protocol (UDP).

Output Packets. The output packet contains the localization data, the position and direction of all ten robots and position of the ball. The packets are generated in the same way that SSL-Vision generates packets. Both SSL-Vision and grSim use Google Protocol Buffer (protobuf) library [11] to encode the data packets. To generate the exact same packet, grSim uses the same protocol configuration file that SSL-Vision uses. The generated packets are sent to the clients with the frequency of 60 times per second, however it can be increased up to 100fps based on simulator rendering frame rate.

To make the simulation output data more realistic, user can specify different kind of noise and disturbances to be added to the localization data. The simulator can apply a two dimensional Gaussian noise to localization data with user specified parameters. It can also add delay to output data to simulate the loop delay which exists in Small Size teams hardware/software architecture. Furthermore, It can add some vanishing effect to output data with user-specified probability. The vanishing effect simulates the temporary loss of an object in vision data. Like other configurations, all these parameters are configurable in run-time.

Input Packets. The simulator receives packets from clients using two UDP sockets, one for the blue team and the other one for the yellow team. Blue and Yellow are standard colors that teams use as the main identifier in SSL matches.

In order to standardize the software's input/output protocols, the commands must be sent using Google Protobuf (section 3.2) encoded data packets. The protocol schema is depicted in Figure 4. The protocol schema describes how the clients must encode their desired control commands before sending data to grSim using Google Protobuf library. The required header and source files to encode data using this schema is included in grSim's software package. According to this schema, in each cycle, the client must first specify the target team, yellow or blue. Next, a series of control commands must be provided for all active robots.

```
message grSim_Robot_Command {
    required uint32 id          = 1;
    required float  kickspeedx  = 2;
    required float  kickspeedz  = 3;
    required float  veltangent  = 4;
    required float  velnormal   = 5;
    required float  velangular  = 6;
    required bool   spinner      = 7;
    required bool   wheelsspeed = 8;
    optional float  wheel1      = 9;
    optional float  wheel2      = 10;
    optional float  wheel3      = 11;
    optional float  wheel4      = 12;
}

message grSim_Commands {
    required double timestamp = 1;
    required bool isteamyellow  = 2;
    repeated grSim_Robot_Command robot_commands = 3;
}
```

Fig. 4. The grSim's Google Protobuf schema for receiving control commands from clients. The protocol documentation and examples can be found in the grSim's documentation and webpage

3.3   User Interface

To ease the software/AI development process, each robot simulator software must be equipped with a good user interface. In order to achieve that, a modern, feature-rich and user friendly Graphical User Interface (GUI) has been developed for grSim. Using the power of Qt framework and its OpenGL integration, the user can modify the objects in the scene, the camera and configurations in an easy way using keyboard or mouse. For example, In order to modify each object's position and orientation, the user can select or move each object by moving the

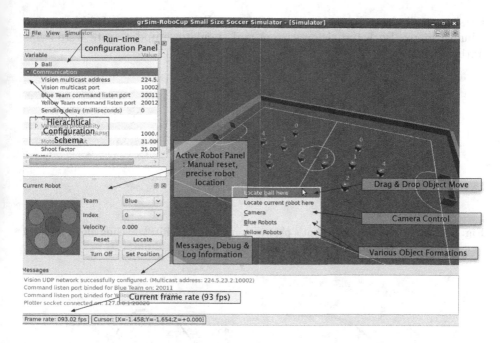

Fig. 5. Features of grSim's Graphical User Interface

mouse and clicking in the scene window. An overview of some UI features are shown in figure 5.

For easy run-time configuration of the simulator parameters, grSim uses Var-Types library[20]. VarTypes is a feature-rich, object-oriented framework for managing variables in C++/Qt4. This library stores all configurable variables and their descriptions in a XML file. It creates a Qt *Widget* for easy modification of variable values during run-time in a thread-safe manner. In this way all of the physical parameters including friction and bounciness of the surfaces, mass of the rigid bodies, actuator parameters and network properties are configurable during program run.

4   grSim in Action

grSim is currently under active development by Amirkabir university's small size soccer robot team named "Parsian"[18]. This simulator plays an important role in the software, control and AI development of aforementioned team.

In order to demonstrate the simulator's accuracy and performance, one robot's simulated motion profile, has been compared to real world data. In the first test, the robot traveled a 4.4m straight line along its local x-axis (its head direction). In the second test, the robot traveled a 3.6m straight line along its local y-axis (perpendicular to its head direction). The position of the robot during the tests are depicted in figures 6, 7 and 8. All the tests were done using 5th generation

Parsian small size robots [18] on a dual core PC with 4GB of RAM running
Ubuntu Linux. Both the vision system and simulator's update rate were 62fps
during the tests.

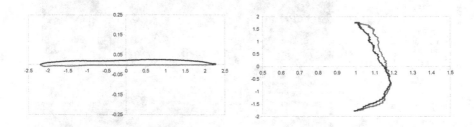

Fig. 6. Comparison between robot's position on the field, while traveling along its head
(left) and perpendicular to its head (right). [Blue is from real data]

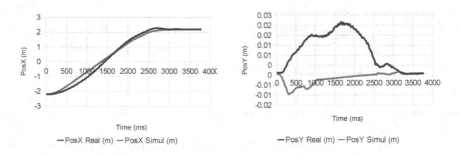

Fig. 7. Comparison between global x and y components of robot's position while traveling along its head direction

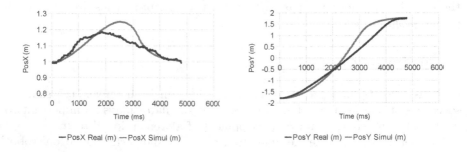

Fig. 8. Comparison between x and y components of robot's position while traveling
perpendicular to its head direction

5 Conclusion

In this paper we described grSim simulator, which can simulate and visualize a RoboCup small size soccer robot game in a realistic way and in real-time. By using a simplified robot model it can reach high simulation speeds (60fps or more). The flexible input protocol and SSL-Vision compatible localization data output make it easy to integrate grSim into any existing SSL software chain. grSim's rich user interface makes it an easy to learn, useful tool in AI development process. Our plan is to release grSim as a free and open-source application to RoboCup small size community in near future. Updates, screenshots and videos are available at http://eew.aut.ac.ir/~parsian/grsim/ .

References

1. OpenGL - The industry standard for high performance graphics (2011), http://www.opengl.org/ (accessed January 2011)
2. RoboCup - official website (2011), http://www.robocup.org/ (accessed January 2011)
3. Small Size Robot League - official website (2011), http://small-size.informatik.uni-bremen.de/ (accessed January 2011)
4. The Player Project - free software tools for robot and sensor applications (2011), http://playerstage.sourceforge.net/ (accessed January 2011)
5. Baillie, J.: Urbi: Towards a universal robotic low-level programming language. In: 2005 IEEE/RSJ International Conference on Intelligent Robots and Systems (IROS 2005), pp. 820–825. IEEE (2005)
6. Browning, B., Tryzelaar, E.: Übersim: A multi-robot simulator for robot soccer. In: Proceedings of the Second International Joint Conference on Autonomous Agents and Multiagent Systems, pp. 948–949 (2003)
7. RoboCup Small Size League Technical Committee: RoboCup Small Size League Rules (2011), http://small-size.informatik.uni-bremen.de/rules:main (accessed January 2011)
8. Cyberbotics: Webots, fast prototyping and simulation of mobile robots (2011), http://www.cyberbotics.com/ (accessed January 2011)
9. Gerkey, B., Vaughan, R., Howard, A.: The Player/Stage project: Tools for multi-robot and distributed sensor systems. In: Proceedings of the 11th International Conference on Advanced Robotics (January 2003)
10. Go, J., Browning, B., Veloso, M.: Accurate and flexible simulation for dynamic, vision-centric robots. In: Proceedings of International Joint Conference on Autonomous Agents and Multi-Agent Systems, AAMAS 2004 (2004)
11. Google Inc.: Protocol buffers - google's data interchange format (2011), http://code.google.com/p/protobuf/ (accessed January 2011)
12. Gostai Ltd.: Urbi - The universal platform (2011), http://www.gostai.com/urbi.php (accessed January 2011)
13. Koenig, N., Howard, A.: Design and use paradigms for Gazebo, an open-source multi-robot simulator. In: Proceedings of 2004 IEEE/RSJ International Conference on Intelligent Robots and Systems (IROS 2004), vol. 3, pp. 2149–2154. IEEE (2005)
14. Koenig, N., Howard, A.: Gazebo - 3D multiple robot simulator with dynamics (2011), http://playerstage.sourceforge.net/gazebo/gazebo.html (accessed January 2011)

15. Laue, T., Spiess, K., Röfer, T.: SimRobot – A General Physical Robot Simulator and its Application in RoboCup. In: Bredenfeld, A., Jacoff, A., Noda, I., Takahashi, Y. (eds.) RoboCup 2005. LNCS (LNAI), vol. 4020, pp. 173–183. Springer, Heidelberg (2006)
16. Michel, O.: Cyberbotics Ltd. Webots TM: Professional mobile robot simulation. International Journal of Advanced Robotic Systems 1(1), 39–42 (2004)
17. Nokia Inc.: Qt - A cross-platform application and UI framework (2011), http://qt.nokia.com/ (accessed January 2011)
18. Poorjandaghi, S., Monajjemi, V., Mehrabi, V., Nabi, M., Koochakzadeh, A., Atashzar, F., Omidi, E., Pahlavani, A., Sheikhi, E., Bahmand, A., Mohaimanian, M., Saeidi, A., Shamipour, S., Karkon, R.: Parsian - Amirkabir university of technology RoboCup small size soccer team. Team Description Paper for RoboCup (February 2011)
19. Smith, R.: ODE - Open Dynamics Engine (2011), http://www.ode.org/ (accessed January 2011)
20. Zickler, S.: Vartypes - A feature-rich, object-oriented framework for managing variables in C++ / QT4 (2011), http://code.google.com/p/protobuf/ (accessed January 2011)
21. Zickler, S., Laue, T., Birbach, O., Wongphati, M., Veloso, M.: SSL-Vision: The Shared Vision System for the RoboCup Small Size League. In: Baltes, J., Lagoudakis, M.G., Naruse, T., Ghidary, S.S. (eds.) RoboCup 2009. LNCS (LNAI), vol. 5949, pp. 425–436. Springer, Heidelberg (2010)

Automated Generation of CPG-Based Locomotion for Robot Nao

Ernesto Torres and Leonardo Garrido

Tecnológico de Monterrey, Campus Monterrey.
Av. Eugenio Garza Sada 2501.
Monterrey, Nuevo León. México
ernesto.torres.d@gmail.com, leonardo.garrido@itesm.mx

Abstract. This paper presents a solution to the biped locomotion problem. The robot used for the experiments is robot Nao by Aldebaran Robotics and it is simulated in Webots mobile robot simulator. Our method of solution does not requires the dynamic model of the robot, thus making this approach usable to other biped robots. For faster results the number of degrees of freedom is kept low, only six are used. The walking gait is generated using Central Pattern Generators with limit-cycle oscillators. For the oscillator connection weights required for synchronization, a genetic algorithm is implemented. Our solution is generated automatically and the best results allow the robot to walk twice as fast as the Aldebaran's webots walk and four times faster than the default walk in Robotstadium.

Keywords: central pattern generator, genetic algorithm, biped robot, robot nao, robot simulation.

1 Introduction

In this paper we present a solution to create a stable and fast walking gait for the robot Nao [1] in the Webots [5] simulator. This proposed solution allows a Nao robot to walk without the requirement of the dynamic model by using a bio-inspired approach. We suggest the use of Central Pattern Generators (CPGs) based on coupled limit-cycle oscillators together with genetic algorithms (GA). The oscillators generate trajectories that are followed by the robot's servomotors. For the oscillator coupling, we use a GA to optimize the synchronization parameters. This work also minimizes the required parameters for the optimization to provide fast results with a relatively low number of evaluations.

We decided to use CPGs [10] to create a controller that is equivalent in the way humans and animals generate locomotion. The CPG is the low level locomotion controller and generates the motion type that a higher level controller requires, such as the interaction between the spinal cord and the brain. We selected genetic algorithms for being very good at finding solutions when the evaluation is simple, in this case, the distance traveled. Both CPG and GA were used to generate the controller in a bio-inspired approach.

T. Röfer et al. (Eds.): RoboCup 2011, LNCS 7416, pp. 461–471, 2012.
© Springer-Verlag Berlin Heidelberg 2012

This paper is divided in six sections. Related Work section gives a brief summary of the state-of-the-art in biped locomotions controllers. In section 3, we describe the CPGs and the oscillator used. In section 4, the Nao robot is introduced together with Webots simulator, also includes the connections proposed for the oscillator coupling and the genome used in the GA. In Section 5, we show the results obtained with this approach, the optimized genome, the oscillation behavior and the walking gait obtained. Finally, Conclusion and Future Works section makes a brief summary on this work and our current and future research projects.

2   Related Work

Designing and implementing locomotion controllers for a robot is a complex task that often requires the knowledge of the dynamic model of the robot in order to generate robust and practical gaits. This is a task far from trivial, and can be approached in many different ways. The most common approach is using control theory. By manipulation of limb trajectories, it creates a desired motion and by using a controller, the balance is achieved.

Several works implement Zero-Movement Point(ZMP) [17] for the generation of trajectories that allows the center of mass to remain inside a stability polygon and thus avoid falling. In [16], ZMP was used to create an engine for an omnidirectional walk. For their dynamic model they use an inverted pendulum with all the mass in the CoM, then a preview controller generates dynamically balanced center of mass trajectories.

Another common approach is the use of heuristic methods to avoid using the dynamic model but still being able to generate a successful motion. Heuristic approaches include: genetic algorithms, reinforcement learning, policy gradient, among others. In [9], genetic algorithms are used together with partial Fouries series and they generate the trajectories based mainly on the CoM to monitor the stability of the gait. In [12], reinforcement learning is used for a robot to learn where to place the swing leg. They also modify the desired walking cycle frequency based on online measurements. Finally, in [3], an extension of the classic Policy Gradient algorithm that takes into account parameter relevance is used. It allows for better solutions when only a few experiments are available.

A third approach exists, a bio-inspired approach. Instead of creating new methods for achieving locomotion, this approach heavily relies on the observation of the living organisms in nature. Works using bio-inspired approaches use Central Pattern Generators (CPGs) as its foundation [6]. In [8], the CPG network consists of the Matsuoka neuron model, and is introduced to realize the locomotion of a bipedal robot, they also use genetic algorithms in several steps to find out large number of parameters according to structure of CPG network. In [7], an oscillator is proposed for a salamander robot to generate the transition between swimming and walking. This same oscillator was used in [2] to generate locomotion for modular robots and in [11] to generate trajectories for the biped robot hoap2. In bio-inspired models, genetic algorithms are often the algorithm of choice for parameter optimization since it is based on evolution.

3   CPG Description

CPGs are neural networks found in vertebrate and invertebrate animals that produce rhythmic outputs without rhythmic inputs. Rhythms in nature are abundant, and are common in animal activities such as the heartbeat, breathing, chewing and digesting.

A CPG can be implemented with oscillators as the basic neural unit and create the connected network by oscillator coupling. An oscillator is a system that executes a periodic behavior. For example, a swinging pendulum executes a periodic behavior once it is released at a certain height and it returns to the same point every cycle. A common oscillator has a characteristic period.

A special form of oscillator often found in nature is a limit-cycle oscillator. A limit-cycle not only has a characteristic period but also a characteristic amplitude. If the limit cycle is affected by some perturbations it is able to return to its original trajectory after a given time. Those characteristics are important for locomotion since it makes the system resistant to small perturbations. A more in-depth information about oscillators and limit-cycles can be found in [15].

The oscillator used for the CPG was proposed by Professor A. Ijspeert from BIRG[7]. It is a non-linear oscillator and has the interesting property of having the limit cycle behave like a sinusoidal signal with an amplitude $\sqrt{E}$ and a period $2\pi\tau$.

$$\tau\dot{v} = -\alpha\frac{x^2 + v^2 - E}{E}v - x \tag{1}$$

$$\tau\dot{x} = v \tag{2}$$

The variable x represents position and v represents velocity. Variables τ, α and E are positive constants. By modifying τ, the period of the oscillator can be manipulated and changing α changes the system speed convergence. The energy of the system is represented as E. For oscillator coupling:

$$\tau\dot{v}_i = -\alpha\frac{x_i^2 + v_i^2 - E}{E}v_i - x_i + \sum a_{ij}x_j + b_{ij}v_j + \sum c_{ij}sj \tag{3}$$

$$\tau\dot{x}_i = v_i \tag{4}$$

Where a_{ij} and b_{ij} represent how oscillator j influences oscillator i. The last summatory allows for sensory input.

A previous work by Mojon [11] suggests a modification on the oscillator to allow the connection strength to be independent from the energy of the oscillators. This is achieved by normalizing the connections:

$$\tau\dot{v}_i = -\alpha\frac{x_i^2 + v_i^2 - E}{E}v_i - x_i + \sum \frac{a_{ij}x_j + b_{ij}v_j}{x_j^2 + v_j^2} \tag{5}$$

$$\tau\dot{x}_i = v_i \tag{6}$$

The last equations represent the final model of the oscillator used in this work. One oscillator per DoF was used.

4   Experimental Setup

The robot used in this paper is the robot Nao from Aldebaran Robotics [1]. This robot is now used in the official Robocup Standard Platform League(SPL) [13,14]. Since the robot requires no hardware modifications is greatly useful for software development. Nao has 21 degrees of freedom(DoF) which makes it highly customizable and has a high number of possible movements. For this paper we tried to reduce the search space all we could to be able to obtain fast results with a relatively low number of experiments. For this reason only 6 DoF were used, 3 from each leg: AnklePitch, KneePitch and HipPitch. These three were selected because the propulsion force for walking is given only in the sagittal plane. Any extra DoF could be used not for propulsion but for stabilization.

The experiments were developed and tested in Webots simulator [5] by Cyberbotics. Inspired by the online robot soccer competition, Robotstadium [4], this software was selected to develop and implement a locomotion controller for Nao robot. The version of the software used is 6.2.4 PRO and the model used is the NaoV3R used for Robotstadium 2009/2010.

Connections subsection includes the DoF used, the oscillators network and the parameters required. Finally, the Genome and Genetic Algorithm section show how we optimized the CPG.

4.1   Connections

Since we use six DoF, six oscillators were used. Each DoF should require 6 or 8 parameters: A(Amplitude), $X0$ for a non-cero centered oscillation, a_{ij} and b_{ij} for a single connection or a_{ij} ,b_{ij}, a_{ik} and b_{ik} for a double connected oscillator, τ for period, and α for convergence speed between connections.

To enable every oscilator to affect and be affected, the links are double connected. With the connections proposed in Fig. 1, we have 4 DoF with 8 parameters and 2 DoF with 6, this makes 44 total parameters to optimize. Forty-four parameters represents a big search space considering each parameter could take real values from $[-1.0, 1.0]$. To reduce the number of free parameters, τ and α are fixed, reducing to 4 or 6 parameter per DoF for a total of 32. We set $\tau \approx 1hz$ and $\alpha = 1$ for maximum convergence speed. Previous work with this oscillator by Mojon [11] gave us some methods for reducing further the search space by obtaining the weight paremeters to force 0, $\frac{\pi}{2}$, π and $\frac{3\pi}{2}$ phase difference. For the first reduction, a certain phase can be forced by controlling a and b parameters. Instead of single evolving every oscillator a simetry is asumed, and only one leg is optimized while the other leg use the same parameters. The trick for this concists in forcing a π phase difference between the left hip and the right hip. With this reduction instead of 32, there are only 16 parameters.

The final model is composed by 3 oscillators. The hip oscillator is connected to the other hip and to its corresponding knee. The knee oscillator is connected to its correspondent hip and ankle. Finally, the ankle is only connected to the knee. Table 1 summarizes the possible values the oscillators are using. Extra adaptations include the maximum range for the amplitude parameter. We allow

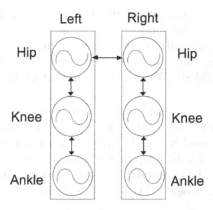

Fig. 1. The Oscillator connections. All the connections are bi-directional

Table 1. Oscillator values. The value range for a variable is displayed in each cell, missing values are values not required.

	$x0$	A	$a1$	$a2$	$b1$	$b2$
Hip	$[-1.0, 1.0]$	$[-1.0, 1.0]$	0	$[-1.0, 1.0]$	$[-1.0, 0.0]$	$[-1.0, 1.0]$
Knee	$[-1.0, 1.0]$	$[-1.0, 1.0]$	$[-1.0, 1.0]$	$[-1.0, 1.0]$	$[-1.0, 1.0]$	$[-1.0, 1.0]$
Ankle	$[-1.0, 1.0]$	$[-1.0, 1.0]$	$[-1.0, 1.0]$		$[-1.0, 1.0]$	

the max amplitude be up to $\frac{\pi}{2}$ for each oscillator. And for the connections strength the maximum allowed is 0.7 to have a smoother synchronization [2]. If the maximum connection strenght is too high the oscillators may not reach synchronization.

4.2   Genome and Genetic Algorithm

The genome used is displayed in Table 2. It corresponds to the values the genetic algorithm optimizes and it is the concatenation of the parameters required by the oscillators. The first five values correspond to the hip, in which $a1$ is always zero, thus ommited from the genome. The next six values corresponds to the knee oscillator and the last four are required by the ankle oscillator. All the parameters can take values from $[-1.0, 1.0]$. The only exception is $b1$ from the hip which can only take values from $[-1.0, 0.0]$.

The $X0$ value indicates the bias for the oscillator to allow for non-zero centered oscillation. A is the amplitude of the oscillation, internally it can only take

Table 2. Genome

X0	A	a2	b1	b2	X0	A	a1	a2	b1	b2	X0	A	a1	a2

positive numbers. The π phase difference between legs is forced by making hip values $a1 = 0$ and $b1 = [-1.0, 0.0]$.

For the Genetic Algorithm(GA) implementation we used GAlib [18]. All the genes are real value numbers and are represented as a float data type. The genetic algorithm is a GASimpleGA which corresponds to the genetic algorithm proposed by Holland.

Tournament selection was used. Two chromosomes from the population are randomly selected and the one with best fitness is selected.

The crossover type used was $GARealOnePointCrossover$, which randomly selects one crossover point to swap genes from the parents. The probability of crossover was 0.90. And for mutation it was used $GARealUniformMutator$, which randomly selects a new value from the alleles $[-1.0, 0.0]$. Mutation probability was set to 0.01.

The population size is 30 individuals, twice the size of the genes. The GA was configured to run until the convergence of the population. A 95% convergence ratio was used as criteria for stopping the GA and it had to be concistent for fifty generations. Elitism was used to preserve the best individual from each generation.

Only the best individual is obtained as a result of each GA execution. Several GAs were executed secuentially and all the best individuals were recorded. The fitness evaluation is the following:

$$Fitness = \begin{cases} 0 & \text{if } Distance < 20 \\ Distance * \frac{StandingSimulationSteps}{TotalSimulationSteps} & \text{otherwise} \end{cases} \quad (7)$$

The $Distance$ was measured in centimeters. $StandingSimulationSteps$ is the number of simulation steps until the robot falls or the 30 second timer expires. $TotalSimulationSteps$ are the total simulation steps in 30 seconds.

5   Experimental Results

First of all, we have three key aspects involved which required careful tunning for optimal performance. These aspects are: the fitness function, the number of oscillators and the phase difference between oscillators.

Several fitness functions were tested, because a bad fitness generates a bad behavior or at least an unwanted behavior. At the beginning, when the fitness only measured the $Distance$, the robot learned to jump to reach for a larger distance instead of generating trajectories that reward the robot later by walking. Another approach used for the fitness function was to allow the robot to walk to anywhere it wants and still obtain a positive fitness. For example, if the robot moved backwards, the distance was multiplied by 0.50 to penalize the wrong direction, but it still receive a non-zero fitness evaluation. Later, it was observed that it was better to give a 0.0 fitness whenever the robot did not move forward. Another important aspect was to encourage the robot to avoid falling. If the reward is too big, the robot learns to reduce the amplitude of the oscillations and it avoids falling, but it does not move forward. The fitness function had to be a balanced equation to give a reward for moving forward but not give too much reward for not falling. To address this, the minimum valid distance required is 20 centimeters.

The number of oscillators and the phase difference between oscillators was fixed to reduce to the maximum the number of free parameters and to allow a faster convergence for the genetic algorithm. Preliminary tests were made with 12 oscillators instead of the 6 presented in this paper. The servos used for each leg were: 3 from each hip, 1 for the knee and 2 from the ankles. The robot was able to walk small distances, but the walking speed was far from good and most of the times the gait of the robot was a weird non human-like movement. For that reason, the number of required oscillators was minimized to generate a walking gait with the minimum required parameters.

The oscillation values are shown in Fig. 2. The servo motors start at 0.0 where the robot is in resting position and later begins locomotion with the synchronization of all the servos. The resulting genome is displayed in Table.3. All the values are within $[-1.0, 1.0]$. As mentioned in the Experimental Setup section, the first five values correspond to the hip, in which $a1$ is always zero, thus ommited from the genome. The next six values corresponds to the knee oscillator and the last four are required by the ankle oscillator

The best results are shown in Fig. 3 with a fitness of almost 350. The average fitness is affected by the mutation and crossover in which some individuals obtain a bad fitness. As it can be appreciated, the best results are obtained aproximately from the 50th generation where it converges. In order to obtain satisfactory results, the genetic algorithm was restarted whenever it converges to increment

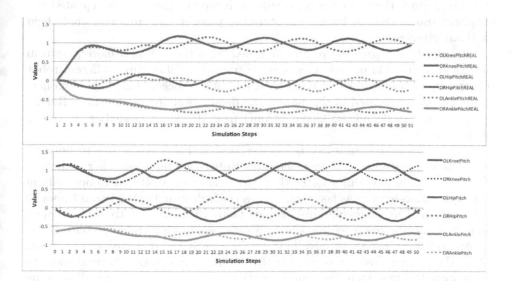

Fig. 2. Comparative between the oscillations generated by the CPG (bottom) and the real oscillations by the servomotors (top)

Table 3. Solution obtained for the CPG (Split in two for display purposes)

X0	A	a2	b1	b2	X0	A	a1
-0.118	0.132	-0.627	0.314	-0.787	0.939	0.119	-0.178

a2	b1	b2	X0	A	a1	a2
0.636	-0.164	0.040	-0.787	-0.049	0.528	0.072

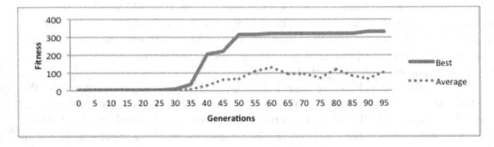

Fig. 3. Fitness per Generation displaying the best individual and the average for each generation

the number of posible solutions. Depending on the population sometimes the convergence can take from several hours to minutes. The Nao learned to walk making small steps to keep balance and avoid falling. It can be appreciated in Fig.4. Its double support phase is very brief, balancing in one feet most of the time. Since the CPG has an oscillator behavior, the cycle is repeated the required times. Once it breaks the standing pose, it can cross the whole field without problem.

In order to compare the results obtained from the GA in a more graphical way, a special world in Webots was programmed to allow three different robot controllers to compete on a race to cross half of the soccer field. The controllers were: the one generated in this paper, the default walking motion from Robotstadium and the default walking controller from the Nao provided as default by Aldebaran Robotics.

The results of the race can be appreciated in Fig.5. From top to bottom: Aldebaran's controller, our controller and Robotstadium's controller. Our controller can walk four times faster than the default controller in Robotstadium and aproximately twice as fast than the Nao default walk by Aldebaran Robotics in naoqi 1.6. The results obtained are satisfactory since our controller is an open loop that may be improved further with reflexes and/or a stability controller.

An additional aspect to have in consideration is the first step of the robot, since the resting position is outside the walking trajectories. Several ways to help with the first step were implemented. The first one was to allow some time for the CPG to synchronize and then connect the servos. It wasn't as useful as we thought and the results were negative. Another variation was to let them synchronize and then when the values were closer to the $X0$ the servos

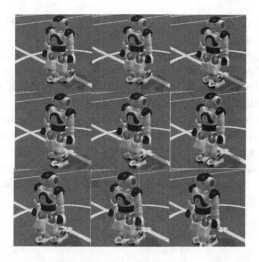

Fig. 4. Nao walking. Can be appreciated from top-left to bottom-right.

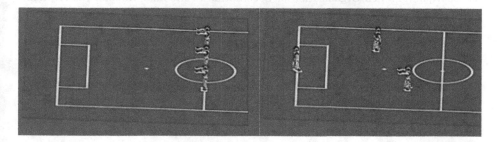

Fig. 5. Race between the controllers. Top blue: Aldebaran's controller. Center black: our controller. Bottom red: Robotstadium's default walk

were connected, but the same negative results were obtained. Finally the last approach was to initiate the servos at the same time of the CPG and allow it to synchronize.

About the time requiered for good individuals, in one 16-hour session good individuals were generated and tested. Our Webots PRO license allows to run in fast mode which gave us 12x the speed of a simulation.

6   Conclusions and Future Work

As the previous section showed, a good locomotion controller can be created by using CPGs. Limit-cycle oscillators provide some resistance to perturbations and allow the robot to generate by itself the first step to break the rest position and walk forward. It was shown that a walking gait can be generated by using only the pitch servomotors: hipPitch, kneePitch and anklePitch. Another advantage of using a low number of oscillators is that the size of the search space is reduced, obtaining faster results.

Research in progress includes the automated generation of side steps. In preliminary experiments we have obtained very good results, but are still not conclusive. Additional experiments will include rotational walk and hopefully an omnidirectional walk with the integration of all the results. At the same time, we are working in optimizing the algorithm in order implement it in the real Nao.Future work will include an implementation in the physical Nao to replace the actual default walk for a hopefully better gait. Additional future work includes the evolution of an adaptive controller simultaneously with the CPG to allow a closed-loop control and achieve faster and more stable gaits.

References

1. Aldebaran Robotics: Nao, official player of the RoboCup since 2008 (2011), http://www.aldebaran-robotics.com/en/node/1168
2. Bourquin, Y.: Self-Organization of Locomotion in Modular Robots. Dissertation, Ecole Polytechnique Federale de Lausanne - EPFL (2004), http://birg.epfl.ch/page53073.html
3. Cherubini, A., Giannone, F., Iocchi, L., Lombardo, M., Oriolo, G.: Policy gradient learning for a humanoid soccer robot. Robotics and Autonomous Systems 57(8), 808–818 (2009), http://www.sciencedirect.com/science/article/B6V16-4W0R0MN-3/2/1de409f7de564e83189ac81bf3a6ca5f , Humanoid Soccer Robots
4. Cyberbotics: Robotstadium: online robot soccer competition (2011), http://robotstadium.org/
5. Cyberbotics: Webots: mobile robot simulation software (2011), http://www.cyberbotics.com
6. Ijspeert, A.J.: Central pattern generators for locomotion control in animals and robots: A review. Neural Networks 21(4), 642–653 (2008), http://www.sciencedirect.com/science/article/B6T08-4SH6B9F-2/2/2e0a2fdad02d315becc218a6602f054d , robotics and Neuroscience
7. Ijspeert, A.J., Cabelguen, J.M.: Gait transition from swimming to walking: investigation of salamander locomotion control using nonlinear oscillators. In: Proceeding of Adaptative Motion in Animals and Machines (2003)
8. Inada, H., Ishii, K.: Bipedal walk using a central pattern generator. International Congress Series 1269, 185–188 (2004), http://www.sciencedirect.com/science/article/B7581-4D338VC-1K/2/ac44f599e008ec39e662ff3e41763cb1 brain-Inspired IT I. Invited papers of the 1st Meeting entitled Brain IT 2004
9. Kong, J.S., Lee, B.H., Kim, J.G.: A study on the gait generation of a humanoid robot using genetic algorithm. In: SICE 2004 Annual Conference, vol. 1, pp. 187–191 (August 2004)
10. MacKay-Lyons, M.: Central Pattern Generation of Locomotion: A Review of the Evidence. Physical Therapy 82(1), 69–83 (2002), http://ptjournal.apta.org/content/82/1/69.abstract

11. Mojon, S.: Using nonlinear oscillators to control the locomotion of a simulated biped robot. Diploma thesis. Ecole Polytechnique Federale de Lausanne - EPFL (2004), http://birg.epfl.ch/page44565.html
12. Morimoto, J., Cheng, G., Atkeson, C., Zeglin, G.: A simple reinforcement learning algorithm for biped walking. In: Proceedings of IEEE International Conference on Robotics and Automation, ICRA 2004, April 1-May, vol. 3, pp. 3030–3035 (2004)
13. RoboCup Organization: Robot world cup initiative (2011), http://www.robocup.org
14. RoboCup Organization: Standard Platform League (2011), http://www.tzi.de/spl/bin/view/Website/WebHome
15. Strogatz, S., Stewart, I.: Coupled oscillators and biological synchronization. Scientific American 269(6), 68 (1993)
16. Strom, J., Slavov, G., Chown, E.: Omnidirectional Walking Using ZMP and Preview Control for the NAO Humanoid Robot. In: Baltes, J., Lagoudakis, M.G., Naruse, T., Ghidary, S.S. (eds.) RoboCup 2009. LNCS (LNAI), vol. 5949, pp. 378–389. Springer, Heidelberg (2010), http://dx.doi.org/10.1007/978-3-642-11876-0_33
17. Vukobratovic, M., Borovac, B., Surdilovic, D.: Zero-Movement Point - Proper Interpretation. Submitted to International Journal of Robotics Research (2004)
18. Wall, M.: Galib: A c++ library of genetic algorithm component (2011), http://lancet.mit.edu/ga/

Application of the "Alliance Algorithm" to Energy Constrained Gait Optimization

Valerio Lattarulo and Sander G. van Dijk

Adaptive Systems Research Group
University of Hertfordshire
Hatfield, UK

Abstract. This paper deals with the problem of energy constrained gait optimization for bipedal walking. We present a solution to this problem obtained by applying a recently introduced heuristic method, the Alliance Algorithm (AA), and compare its performance against a Genetic Algorithm (GA). We show experimentally that the intrinsic ability of the AA to handle hard constraints enables it to find solutions significantly better than the GA. Also with the constraint removed the AA show more reliable optimization results. Finally, we show that the final gait obtained through this method outperforms most solutions to this problem presented in previous works, in terms of walking speed.

1 Introduction

Despite a large amount of literature on the topic, general walking gait generation for bipedal robots is still very much an open problem. It is a hard problem due to the high dimensionality of a typical bipedal robot with many joints, the complexity of the dynamics of the system, and the difficulty of creating an accurate model that could enable supervised learning; given an arbitrary bipedal robot it is not obvious what the specifics of the motion of a good walking gait look like, one can often only formulate requirements that such a walk should meet.

One of the most important of these requirements is *stability*. Physical models of the walker's dynamics can be used to solve this problem, usually done using methods based on stability concepts such as the zero-moment point (ZMP) [13] or zero rate of angular momentum (ZRAM)[3]. But in practice it can be difficult to derive such a model, and gaits generated with these methods generally do not meet another important requirement: *speed*. We apply the gait generation problem to the scenario of competitive robotic football, where speed is a major decider. Finally, in real systems a third requirement is *energy efficiency*; a fast, stable walk is not useful when it depletes the walker's energy source before the end of its task. In this paper we will handle all three of these requirements.

Besides the already mentioned ZMP and ZRAM based methods, a range of other classes of gait generators have been put forward to solve this problem, such as central pattern generators (CPG) [5], and ones based on Fourier series [14], to which the generator used here is related. The methods in these classes

T. Röfer et al. (Eds.): RoboCup 2011, LNCS 7416, pp. 472–483, 2012.

Table 1. Gait generation parameters

Parameter		Range	Unit
T	period	$[0-1]$	(s)
α_{up}	max up/down amp	$[0-40]$	(deg)
α_{fw}	max forward/backward amp	$[0-70]$	(deg)
α_{side}	max sideway amp	$[0-25]$	(deg)
α_{turn}	max turn amp	$[0-40]$	(deg)
α_{lean}	max lean amp	$[0-30]$	(deg)
θ_{vel}	bend velocity factor	$[0-45]$	(deg)
θ_{acc}	bend acceleration factor	$[0-40]$	(deg)
θ_{EMA}	EMA filter coefficient	$[0-1]$	

have different levels of complexity, biological plausibility and required levels of knowledge about the walker's dynamics. However, what they all have in common is that they rely on a set of parameters for which optimal values have to be found. Thus, before deciding upon a gait generating method, it is important to have good optimization methods available. In this paper we will contribute to this by applying the recently introduced Alliance Algorithm (AA)[1] to the problem of finding good parameter settings for a gait generator.

In the following section we will discuss in more detail the bipedal walking problem and the specific gait generator that we use. In section 3 we will describe the Alliance Algorithm. Next, we discuss how this method relates to popular other optimization methods. Section 5 will present the experiments performed to test the performance of the AA, and the results obtained. Finally, we will discuss these findings in section 6.

2 Problem Formulation

The problem that we address here is to find an optimal solution for a walking behavior for a robot in a simulated environment (Spark/RCSSServer3D [9]). Our gait generator is similar to the one described by Morimoto et al [8], in that it works by combining several basic oscillatory movements: 1) move feet up/down, 2) forward/backward, 3) left/right, 4) turn feet, and 5) lean left/right. These basic movements are created by sinusoidal oscillations of the joints, and walking is achieved by selecting the right phase for each and summing the basic motions, weighted relatively depending on the desired walking speed and direction. The phase-locking method to increase lateral stability used by Morimoto et al is not used here; in the scenario considered here frontal stability is a bigger issue. As a basic method to decrease this stability, the torso is bent backward or forward by a certain angle, based on a linear function of the desired walking speed and acceleration. Finally, an exponential moving average (EMA) filter is applied to smoothen changes in direction given by higher level behaviors. This results in the 9 parameters listed in table 1, the collection of which we denote with θ.

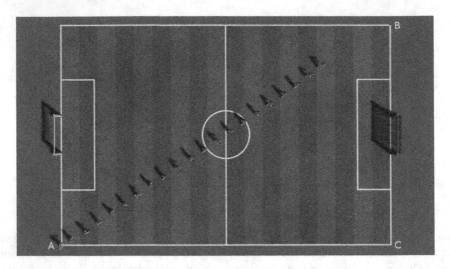

Fig. 1. Example trajectory

The problem then consists of optimizing these parameters in order to maximize the walking speed. As a proxy for this we use an objective function $F(\boldsymbol{\theta})$ that measures the final distance to a target point after an episode of a fixed amount of time, where a smaller distance is better. This choice was made over a function measuring simply the distance traversed in the episode since this could result in a gait that walks fast, but in the wrong direction. The chosen objective function also captures the requirement of stability, since an unstable gait that will make the robot fall will get less far.

In particular, the objective function measures the distance to a target point after an episode of fixed time. Figure 1 shows the trajectory of a sample run. The robot is placed in the bottom left corner of the simulated football field (marked 'A') and aims to walk to the upper right corner (marked 'B'), 21.6m away. It initially faces the *bottom* right corner (marked 'C'), to ensure that stable and fast turning is included in the optimization process. A fixed higher level system determines the walking and turning speeds at each time step, the task of the optimization system is to find the best parameters given this input. A single run consists of 20 seconds of simulation time. Further details of the experimental setup are given in section 5.

As mentioned in the introduction, speed and stability are not the only requirements; another important consideration is energy efficiency. Instead of including the energy use into the objective function, our methods allow us to explicitly pose an a-priori, fixed upper bound. This important benefit removes the necessity of elaborate calibration of the objective function to achieve the correct trade-off.

The energy E_j used by a joint j during time step t for some parameter assignment $\boldsymbol{\theta}$ is determined by:

$$E_{j,t}(\boldsymbol{\theta}) = P_{j,t}(\boldsymbol{\theta})\Delta t = (\tau_{j,t}(\boldsymbol{\theta}) \cdot \omega_{j,t}(\boldsymbol{\theta}))\Delta t, \tag{1}$$

where P is power in Watt, Δt is the length of time step t, τ is torque in Newton meter, and ω is angular velocity in radians per second. These latter three quantities can be obtained from the simulation, and summing the energy values over all joints and all time steps results in the total energy consumption as a function of the chosen parameters: $E_{tot}(\boldsymbol{\theta}) = \sum_{j,t} E_{j,t}(\boldsymbol{\theta})$.

Finally, we can formulate the problem to be solved as:

$$\max_{\boldsymbol{\theta}} F(\boldsymbol{\theta}) \text{ subject to } E_{tot}(\boldsymbol{\theta}) \leq T, \tag{2}$$

where T is a certain fixed threshold. One of the main contributions of the work presented here is the inclusion of a hard constraint, though for completeness we will also handle the unconstrained version.

3   The Alliance Algorithm

The Alliance Algorithm (AA) [1] is based on the metaphoric idea that a certain number of tribes struggle to conquer an environment, which offers resources to survive. Two features characterize each tribe: its strength, and the resources necessary for its survival. In order to increase their strength, the tribes may enter into alliances. This alliance then is characterized by a new strength and amount of necessary resources, dependent on the features of the tribes inside the alliance. The AA proceeds by forming alliances that will be more preferable as their strength is higher than other alliances, but do not consume more than the maximum available amount of resources. The algorithm will end when the strongest alliance is created: it will be able to force out other alliances and conquer the environment, while still meeting its resource constraints. The combination of the tribes that compose the strongest alliance represent the solution of the problem. After convergence it is possible to perform another iteration of the algorithm, starting with tribes that are influenced by the strongest alliance of the previous iteration, to further refine the solution.

Given a certain problem with a certain solution space:

- A single tribe t is composed of the tuple $(\boldsymbol{\theta}_t, s_t, r_t, a_t)$, i.e. a point in solution space $\boldsymbol{\theta}_a$, a strength s_t, a resource demand r_t, and an alliance assignment a_t.
- The set of alliances is a mutually disjoint partition of tribes. Each alliance a again represents a point $\boldsymbol{\theta}_a$ in the solution space, defined by the tribes that compose the alliance and a problem specific joint function.
- The strength s_i of tribe or an alliance is the value obtained with the objective function on the respective solution:

$$s_i = F(\boldsymbol{\theta}_i). \tag{3}$$

- The amount of resources needed by a tribe or an alliance is the value obtained by a constraint function on the respective solution:

$$r_i = E(\boldsymbol{\theta}_i) \tag{4}$$

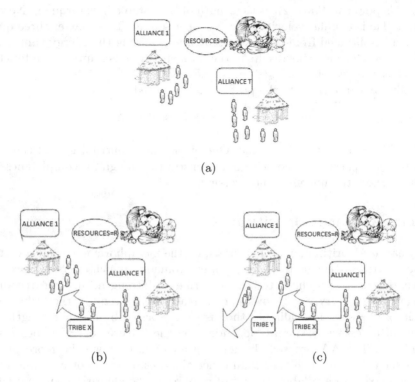

Fig. 2. Dynamics of the Alliance Algorithm. a) Tribes form alliances in an environment with limited resources. b) A tribe can join another alliance. c) By doing so, an inferior tribe can be removed from the alliance.

Initially, N tribes are generated, where the assignments of $\boldsymbol{\theta}$ for each tribe can be taken at random, as done in the experiments presented here, or in a problem specific way when higher level prior knowledge is available. Each is assigned a unique ID $t \in [1-N]$, and its strength s_t and resource demands r_t are determined using $F(\boldsymbol{\theta}_t)$ and $E(\boldsymbol{\theta}_t)$. These properties of the tribes are now fixed; as is shown below, movement through solution space during execution of the algorithm is accomplished by the metaphoric aggregation process, i.e. the formation of the alliances and by doing so the combination of tribes, accomplishes movement in the solution space. To finalize the initialization, each tribe is assigned to its own alliance, such that $a_t = t$. After this point, the ID of an alliance will always be set to the lowest ID of its member tries. This is done to guarantee that each alliance us uniquely identified and to maintain a necessary ordering. Note that an alliance can become empty if the last tribe in that alliance joins an alliance that already contains a tribe with a lower ID. In the remainder of the text we will treat single tribes as 'alliances of one' where applicable.

After initialization, the algorithm starts to iteratively search for the best possible alliance. To do so, at each step first a token is given to an alliance (in our experiments randomly, but again this can be adapted with a different token function), which can now ask a tribe outside of the alliance if it wants to join this alliance. The method of selection a tribe to ask can again be implemented differently for different problems, here random selection is used. However, the alliance receiving the token can choose only tribes that it did not ask before. The request of the alliance to the chosen tribe can have one of several outcomes, dependent on a given decision function.

To exemplify, assume the generic case shown in Fig.2(a), where the tribes of the first alliance are:

$$\left[(s_1, r_1, a_1), \ldots, (s_p, r_p, a_1) \right],$$

and that the tribe that received the request, tribe x, is part of an alliance containing the following tribes:

$$\left[(s_t, r_t, a_t), \ldots, (s_N, r_N, a_t) \right],$$

where 1 is the minimum of the IDs of the tribes in the first alliance, t of the tribes in the second, and $x > t$.

The possible responses (used for our purposes) to the request are:

1. Tribe x can join the first alliance, and by doing so abandon the second alliance, as shown in Fig. 2(b). This results in the two alliances now containing:

$$\left[(s_1, r_1, a_1), \ldots, (s_x, r_x, a_1), \ldots, (s_p, r_p, a_1) \right], \left[(s_t, r_t, a_t), \ldots, (s_N, r_N, a_t) \right]$$

2. Tribe x can join the first, and by doing so abandon the second alliance, and *replaces* an inferior tribe y in this first alliance, as shown in Fig. 2(c). This now results in three alliances:

$$\left[(s_1, r_1, a_1), \ldots, (s_x, r_x, a_1), \ldots, (s_p, r_p, a_1) \right], \left[(s_t, r_t, a_t), \ldots, (s_N, r_N, a_t) \right],$$
$$\left[(s_y, r_y, a_y) \right]$$

3. Tribe x does not join the new alliance when the resulting new alliance is not stronger than the tribe's current alliance, or when the resource consumption of the resulting new alliance is to high.

After that has been decided an action, the effective movement of the tribes and/or alliances and an update of all the data structures is performed, and another cycle is started with the assignment of the token to another alliance. In this way, alliance are formed and broken iteratively, until convergence. This happens when all alliances have asked all other tribes to join without success, and therefore no changes are made. Now, the strengths of all final alliances are compared, and the alliance that has the highest strength without exceeding the available resources wins. The point in solution space θ_a represented by this alliance is the final solution of the problem. Another iteration of the algorithm can be started, seeded with new tribes that can have characteristics similar to the previous strongest alliance, as chosen by the initialization function.

There are several parameter values and functions that have to be chosen when using this algorithm. For the particular problem handled in this paper, a tribe is composed of values for the 9 continuous gait generation variables, within their range as listed in table 1. The initialization function uses uniform random values within this range at in the first iteration, but in subsequent iterations, the values are influenced by the best solution of the previous iteration. Every time that the algorithm restarts, the section of the solution space from which tribes are sampled is centered around the currently known best solution and reduces 5% in size compared to the section used in the previous iteration. A tribe is only allowed to join another alliance only if the passage is convenient for both the entities (in terms of strength and resources) involved in the transition, otherwise the tribe will not join the new alliance. Finally, as mentioned before, the solution of a particular alliance is a function of its tribes, which means that the single tribes are placed in a particular area of the solution place, but when they join together the resulting alliance is placed in another area according to some combination function. For our purposes, this combination is an averaging over the parameter values of the tribes.

4   Related Methods

Similar to a Genetic Algorithm (GA) [4] where individuals can be combined, e.g. through cross-over, the AA moves through solution space by merging solutions. The important difference however, is that in the AA the resulting solutions can be 'broken up' again, when the combination did not turn out to be beneficial in the end. In the AA's pure form, combination is the only search operation, which means that an initial seeding with N tribes defines a set of $\binom{N}{1} + \binom{N}{2} + \cdots + \binom{N}{N}$ possible alliances that the algorithm will search through. However, as we discussed it is possible to perform multiple iterations, where the tribes in a new iteration are generated to be similar to the winning alliance of the last, in order to increase the search space. When doing so, as we have shown, one can also reduce the size of the section of solution space from which new tribes are sampled, to 'home-in' on and refine the best solution, analogous to gradually decreasing the temperature in Simulated Annealing [7].

In the AA, tribes work together to find a better solution, something that is also accomplished, in different ways, by some other optimization algorithms, such as Ant Colony Optimization (ACO) [6] and Particle Swarm Optimization (PSO) [2]. However, in the AA every tribe is essentially selfish and joins with other tribes only if there is an advantage to itself. There is no fixed 'social contract', and every separate tribe and alliance will take any opportunity to conquer the environment. Moreover, in ACO and PSO the individuals move through solution space according to the set of rules of the respective algorithm, whereas in the AA the tribes themselves do not have the tendency towards going near the best solution. Instead, they can join together to form an alliance that will be placed in another area of the solution space, possibly far from the solution represented by the individual tribes.

The selfish behavior of the tribes make the AA also related to greedy search (used in e.g. [11] for the same problem as presented here), but, again, the option of a tribe to leave an alliance directly creates the possibility of backtracking and trying alternative combinations.

Another popular method showing overlap with the AA is the Covariance Matrix Adaptation-Evolution Strategy (CMA-ES). This is an optimization algorithm based on updating the covariance matrix of a distribution over solutions in order to make the sample previously successful search steps more likely. In this algorithm, new samples are combined through weighted averaging to arrive at a new mean of the distribution. This is similar to how we will form alliances by averaging the solutions of their tribes. However, in the AA one can choose any other recombination method, whereas averaging is a fixed integral part of the CMA-ES. Moreover, the mean of all the sampled solutions as used in the CMA-ES is a particular case of the AA in which all the tribes ally together in a single big alliance. This normally however does not happen, as alliances do not accept detrimental solutions, and, in contrast to the CMA-ES, the AA can break up combinations again at a later stage if that turns out to be beneficial.

Finally, the AA aggregates in a way similar to the NEAT algorithm [12], which is a method used for growing the topology of a neural network. The difference with the AA is that there is only one topology that evolves and could increase indefinitely in size, but in the AA there are several alliances that can interact each other and their size is limited by the initial set of tribes.

5   Experiments

In this section we will present the experiments performed to test the performance of AA on the gait generation problem. Besides obtaining pure performance measurements for AA, we will also present a comparison of AA with a GA. This was chosen as the most important comparison method, because it is well understood, and has been used successfully for the purposes presented here, by the authors as well as by others [10]. One of the main benefits of AA is the straightforward way of including hard resource limitations. However, to our knowledge no previous work including such constraints has been performed in the scenario used here, so for completeness and to be able to compare better with related work, we have also performed experiments without energy constraints.

The parameters used for the AA are: number of tribes $N = 10$, and the size of the solution space is reduced 5% after each iteration. For further choices of the different functions of the algorithm, see section section 3.

As mentioned before, a Genetic Algorithm (GA) is used as a reference. Here, each individual consists of a single instantiation of the 9 walking parameters, where each parameter forms a gene. During evolution a population size of 60 is used, with a cross-over rate of 0.1, and a mutation rate on each gene of 0.1. Tournament selection with tournament size 5 is used as the selection method.

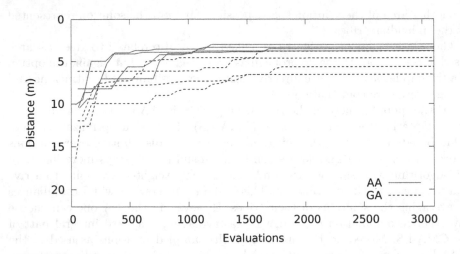

Fig. 3. Best solution found, against cumulative amount of evaluations performed, without energy constraint. AA = Alliance Algorithm, GA = Genetic Algorithm.

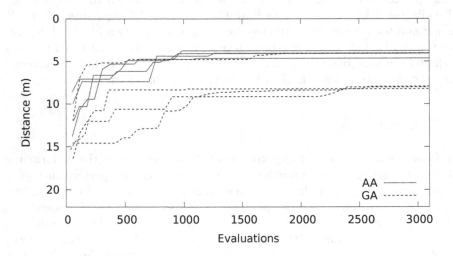

Fig. 4. Best solution found, against cumulative amount of evaluations performed, with energy constraint. AA = Alliance Algorithm, GA = Genetic Algorithm.

Firstly, we have performed 4 unconstrained experiments with the AA and 4 with the GA, and have recorded the best solutions after each iteration/generation. These values are plotted in Fig. 3 against the cumulative number of evaluations of the objective function, i.e. the total number of runs of the scenario of Fig. 1. Note that this number is fixed for each generation of the GA (i.e. 60), but that the number of evaluations for a single iteration of the AA is not fixed, and in our experiments fluctuates around 60.

Secondly, we introduced the energy constraint. After analyzing the solutions found in the first set of experiments, a maximum total energy consumption of 150,000 Joule was chosen as a threshold in order to give a sufficient limitation without constraining the solution space too much. In the AA, this was readily incorporated as the amount of resources available in the environment. Applying such a constraint to a GA, or other similar methods, is less straightforward and can require additional bootstrapping experimentation. In this case, we arrived at including the energy use in the objective function as a negative penalty, but only when this exceeds the maximum of 150,000 Joule. Due to the difference in scale of several orders of magnitude between distance traversed and energy use, this results in a semi-lexicographic selection that preserves a gradient outside of the constraint. However, it is important to note the relative arbitrariness of this choice compared to the AA. The development of the best solutions for these scenarios is shown in Fig. 4.

6   Discussion

The results presented in the previous section offer several insights into the usability of our methods. Firstly we can note that the fastest gait, found by the Alliance Algorithm, achieved an average speed of 0.93 m/s. As a compariosn, Shafii et al [10] have presented to application of PSO with a gait generator similar to, but less restricted than, the one we use, which obtains a speed of 0.77 m/s. Also compared to speeds obtained by some of the high ranking teams in the RoboCup 3D Soccer Simulation competitions, reproduced from [11] in table 2, our best gait scores well. Especially when one notes that the measurements of

Table 2. Comparison of the average speed in different teams

Team	Speed (m/s)
Proposed Approach	0.93
Shafii et al	0.77
BoldHearts	0.68
Wright Eagle	0.67
SEU	1.20
Bats	0.43

the other gaits have been made on a straight walk, whereas in our experiments the agent is presented with the additional handicap of having to first turn into the right direction.

When we look at Fig. 3, we see that the GA is able to find a solution with performance close to that found by the AA. However, this only happened in one out of the 4 performed experiments, where it was lucky to already have a good solution after the first generation, and still after close to twice the amount of evaluations of the objective function than the slowest run of the AA. In the other three experiments all final solutions were inferior. In contrast, each run of the AA was able to find a solution close in performance to the optimal one found, with half of the time doing so within just 500 evaluations.

In the main results of this paper, those of the experiments with energy constraints, this difference is increased further. To our surprise, both methods found solutions with speed not far from that of the best unconstrained solution, even though the solutions in the first set of experiments went well above the energy constraint. However, again the GA only happened on a solution of this quality once; the other three runs resulted in solutions that were much worse. The AA on the other hand again managed to find a good solution in each of the 4 runs. A GA that uses another method to incorporate the energy constraint may perform better. However, such trial-and-error calibration of the algorithm can consume a lot of time, especially in scenarios like the current one where many evaluations of the objective function are very costly. In contrast, the AA naturally accommodates for hard constraints such as the constraint on total power consumption.

Visually it is difficult to discern the qualitative difference between the efficient and inefficient solutions. However, analysis of the solution parameters indicates that in the constrained experiments gaits with lower frequencies are preferred, which results in energy saving thanks to lower angular rates and accelerations.

The results presented in this paper show that the AA can offer a fast and reliable method for finding solutions in complex control problems, such as the gait generation problem for bipedal robots.

Finally, we would like to point out the generality of the algorithm achieved by the possibility of applying any of a wide range of functions to combine and aggregate solutions in a natural way. Moreover, a tribe can represent anything a full solution, but also a partly solution. With this approach it is possible to combine together different entities that together constitute the final solution, such as different behaviors with their separate parameters that combine together into a higher level behavior.

References

1. Calderaro, V., Galdi, V., Lattarulo, V., Siano, P.: A new algorithm for steady state load-shedding strategy. In: 12th International Conference on Optimization of Electrical and Electronic Equipment (OPTIM), pp. 48–53 (2010)
2. Colorni, A., Dorigo, M., Maniezzo, V.: Distributed optimization by ant colonies. In: European Conference on Artificial Life, pp. 134–142 (1991)
3. Goswami, A., Kallem, V.: Rate of change of angular momentum and balance maintenance of biped robots. In: ICRA, pp. 3785–3790 (2004)
4. Holland, J.: Adaptation in Natural and Artificial Systems: An Introductory Analysis with Applications to Biology, Control and Artificial Intelligence. MIT Press, Cambridge (1992)
5. IJspeert, A.J.: Central pattern generators for locomotion control in animals and robots: A review. Neural Networks 21(4), 642–653 (2008)
6. Kennedy, J., Eberhart, R.: Particle swarm optimization. In: Proceedings of IEEE International Conference on Neural Networks, vol. 4, pp. 1942–1948 (August 1995)

7. Kirkpatrick, S., Gelatt, C.D., Vecchi, M.P.: Optimization by simulated annealing. Science 220(4598), 671–680 (1983)
8. Morimoto, J., Endo, G., Nakanishi, J., Cheng, G.: A biologically inspired biped locomotion strategy for humanoid robots: Modulation of sinusoidal patterns by a coupled oscillator model. IEEE Transactions on Robotics 24(1), 185–191 (2008)
9. Obst, O., Rollmann, M.: SPARK – A Generic Simulator for Physical Multiagent Simulations. Computer Systems Science and Engineering 20(5), 347–356 (2005)
10. Shafii, N., Aslani, S., Nezami, O.M., Shiry, S.: Evolution of Biped Walking Using Truncated Fourier Series and Particle Swarm Optimization. In: Baltes, J., Lagoudakis, M.G., Naruse, T., Ghidary, S.S. (eds.) RoboCup 2009. LNCS (LNAI), vol. 5949, pp. 344–354. Springer, Heidelberg (2010)
11. Shafii, N., Reis, L.P., Lau, N.: Biped Walking Using Coronal and Sagittal Movements Based on Truncated Fourier Series. In: Ruiz-del-Solar, J., Chown, E., Ploeger, P.G. (eds.) RoboCup 2010. LNCS (LNAI), vol. 6556, pp. 324–335. Springer, Heidelberg (2010)
12. Stanley, K.O., Miikkulainen, R.: Evolving neural networks through augmenting topologies. Evolutionary Computation 10(2), 99–127 (2002)
13. Vukobratovic, M., Borovac, B.: Zero-moment point-thirty five years of its life. International Journal of Humanoid Robotics 1(1), 157–173 (2004)
14. Yang, L., Chew, C., Poo, A., Zielinska, T.: Adjustable bipedal gait generation using genetic algorithm optimized fourier series formulation. In: IROS, pp. 4435–4440. IEEE (2006)

Robust Algorithm for Safety Region Computation and Its Application to Defense Strategy for RoboCup SSL

Taro Inagaki, Akeru Ishikawa, Kazuhito Murakami, and Tadashi Naruse

School of Information Science and Technology, Aichi Prefectural University,
Nagakute-cho, Aichi, 480-1198 Japan

Abstract. We have proposed a new concept of "safety region" which we use to measure the position of the defense robots[5]. It is defined as a region that the teammate robot(s) can defend the goal when an opponent robot shoots the ball from the inside of the safety region while teammate robots are positioned according to their defense strategy.

Since it is difficult to obtain the accurate safety region in a short time, we need an algorithm that computes an approximate safety region in real time. We proposed such algorithm in the previous paper[5]. However, the safety region obtained by the algorithm is not accurate enough. Therefore, in this paper, we propose an improved algorithm to compute the approximate safety region. We have achieved 95% accuracy and less than 1 msec of computation time, which is adequate for our RoboCup application. We also propose a defense strategy based on the safety region considering the positions of the opponent robots and the pass direction. The achieved results indicate accurate performance for determining the positions of the defense robots.

1 Introduction

In the recent RoboCup Small Size Robot League(SSL), strategies for attacking and defending are growing higher, and the strategy that dynamically changes the number of defense robots depending on the game's situation is often used. In a typical SSL strategies, the potential field[2] and the playbook[3] are used for the action selection and deciding the number of defense robots. Furthermore, cooperative plays such as a direct play [6] are commonly used in recent SSL games. To defend such plays, it is necessary to compute the situation of the game in real time and to determine the positions of the defense robots.

There are some measures for determining a mark robot[4] and deciding a passing robot[7]. We need such a measure for defense robots. We proposed a new concept of "safety region"[5]. It can be used as a measure for determining the positions of the defense robots. We described an algorithm to compute the safety region in real time when defense robots are placed according to the team's defense algorithm, and also we showed an algorithm to place a new (adding) defense robot robot under the measure of safety region[5]. However, the safety region computed by the algorithm is less accurate than the true one. Therefore, improved algorithm is desirable.

T. Röfer et al. (Eds.): RoboCup 2011, LNCS 7416, pp. 484–494, 2012.

In this paper, we propose an improved algorithm to calculate the approximate safety region. It achieves 95% accuracy with respect to true safety region and less than 1 msec of computation time for SSL applications, which is adequate for our purpose. We also propose an improved defense strategy. This is based on the safety region considering the positions of the opponent robots and the pass direction. The results show that it works well for determining the positions of the defense robots.

2 Safety Region

In this section, we define the "safety region" and describe the algorithm to compute the safety region.

2.1 Definition

A concept of "safety region" is simple. It is defined as a region that the teammate robot(s) can defend the goal when an opponent robot shoots the ball from the inside of the region while teammate robots are positioned according to their defense strategy. Remaining region of the field given by removing the safety region is called "unsafety region".

In the following discussion, we do not consider the chip and curved shots.

2.2 Calculation of Safety Region

The calculation of the safety region depends on how the shot action is taken, i.e. a single or assisted shot, and how the team keeps the goal, i.e. defense strategy and the number of defense robots. It takes much time to compute the accurate safety region. Therefore, we describe procedures to compute the approximate safety region. In the following, we discuss the computation model of the safety region for a direct play, which is a play that the first robot kicks the ball to the second robot and the second robot kicks it directly toward the goal. The single shot by the first robot is modeled as well.

Computation Model

Defense robots will move according to their strategy so that we assume the right positions of the defense robots are given at any time. Let $\mathbf{b}$ and $\mathbf{e}$ be the positions of the ball and the shooting robot at time t, respectively, and $\mathbf{r}_i$ be the position of each defense robot i at time t. Let L_r and L_l be the lines connecting $\mathbf{e}$ and the right goalpost and $\mathbf{e}$ and the left goalpost, respectively. (See Fig. 1.) We assume the goalkeeper stands in the defense area and moves along the border line of the defense area while other defense robots stand outside of the defense area and move along the border line. Then, let d_i be the distance between $\mathbf{r}_i$ and border line (, usually equal to the radius of the robot), and let A_i be the curve along with the border line with distance d_i (dotted line in Fig. 1). Let $\mathbf{p}_{r,i}$ and $\mathbf{p}_{l,i}$ be

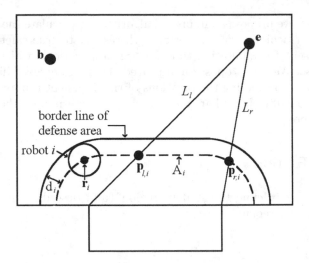

Fig. 1. Definition of symbols

the cross-points between L_r and A_i, and L_l and A_i, respectively. Let $D_{r,i}$ and $D_{l,i}$ be the distance along A_i between $\mathbf{r}_i$ and $\mathbf{p}_{r,i}$, and $\mathbf{r}_i$ and $\mathbf{p}_{l,i}$, respectively.

Assume that the passing robot holds the ball at time t and makes the direct play. Then following equation is obtained for computing the safety region.

Let's define,

$$t_p = \frac{||\mathbf{e} - \mathbf{b}||}{v_p}, \quad t_{j,s} = \frac{||\mathbf{p}_{j,i} - \mathbf{e}||}{v_s} \ (j = r \text{ or } l), \quad t_i = \frac{v_i}{a_i}, \tag{1}$$

where t_p and $t_{j,s}$ are the times for the ball to move from $\mathbf{b}$ to $\mathbf{e}$ with velocity v_p, and from $\mathbf{e}$ to $\mathbf{p}_{j,i}$ with velocity v_s, respectively, and t_i is the time for the robot i to get to its maximal velocity v_i under the condition of maximal acceleration a_i and initial velocity 0.

If $t_i > t_{j,s}$, then compute

$$D_{j,i} < \frac{1}{2}a_i(t_p + t_{j,s})^2 + R_i \quad (j = r, l) \tag{2}$$

otherwise compute

$$D_{j,i} < \frac{1}{2}a_i t_i^2 + v_i(t_p + t_{j,s} - t_i) + R_i \quad (j = r, l) \tag{3}$$

where, R_i is the sum of radii of the defense robot i and the ball.

If Eq. (2) or (3) is satisfied for $j = r$ and l or there is no pass line between $\mathbf{b}$ and $\mathbf{e}$, $\mathbf{e}$ is a point in the safety region. In case there are more than one defense robot, $\mathbf{e}$ is a point in the safety region if at least one of them satisfies Eq. (2) or (3) for $j = r$ and l.

For each point on the field, we compute the above equations and obtain the safety region. However, computation time for this process is high and more optimal solutions is desirable for real time performance.

2.3   Reduction of Computation Time

To reduce the computation time, we use the coarse-to-fine method. The algorithm to compute the safety region using the coarse-to-fine method is shown below. First, assume that the field consists of $W \times H$ grids. We assume W and H are 2's powers here. Prepare an array $F[W][H]$ corresponding to the field. The suffix(i, j) of $F[j][i]$ represents the field position. Let N be a given number $(N = 2^n)$.

Algorithm. Computing the safety region based on the coarse-to-fine method

1. Put a ball on a given point b.(See Fig. 1.)
2. Set $j = 0$, $i = 0$.
3. Set N to 2^n. Compute the algorithm shown in section 2.2 for (j, i) position, where (j, i) is a position e in Fig. 1. Set $F[j][i]$ to 1/0 according to the computation result (safety/unsafety). Also compute the algorithm for $(j, i + N - 1)$, $(j + N - 1, i)$ and $(j + N - 1, i + N - 1)$, respectively, and set the corresponding F.
4. If all four points are in the same region (safety/unsafety), then all the points $F[k][l]$ between $j \leq k < j + N$, $i \leq l < i + N$ are set to the safety/unsafety value. Otherwise, set N to $N/2$ and compute steps 3 and 4 recursively for four divided parts until N comes to 1.
5. Set i to $i + N$. If $i < H$, then goto step 3.
6. Set i to 0 and j to $j + N$. If $j < W$, then goto step 3, otherwise, computation finishes.

3   Experiment

In this section, we show the experimental results of the safety region computation for the direct play[6]. We use the defense strategy of RoboDragons[1] since we know all its details.

3.1   Method of Experiment

The safety region should be calculated analytically. However, it is hard to do the analytical computation, therefore, we calculate it on each of the grid points which are given by dividing the field every 40 mm. We compute the approximate safety region using the model discussed in section 2.2.

To evaluate the correctness or preciseness of the approximate safety region, we need a true safety region. It will be given by the experiment using the real robots. However, it requires (ultra) heavy use of the real robots, causing the breakdown of robots. It is not a good strategy to do such experiments. Instead, we use the simulator of the RoboDragons system which simulates the behavior of the Robots based on the physical law. (Final evaluation should be done in the competition using the system implemented by the proposed algorithm, although has not been conducted in this paper.) The simulation procedure for the direct play is given as followings:

1. Divide the field into n grid points, where $n = 14888$ in this experiment. Put the ball on the initial position **b**, one of the grid points. Also put the attacking robots on the grid points, one around the ball and the other on a given point $\mathbf{e}_i$, a shooting position[1]. Place defense robot(s) on defending position(s) according to the strategy algorithm of the RoboDragons system.
2. At time t, move the ball from **b** to **e** with the passing velocity v_p.
3. Move the ball on one of two shooting lines, L_r or L_l in Fig. 1, which is the farthest line to the defense robots at time t_e, the time that the ball arrives at shooting position **e**.
4. Simulate the movement of the defense robots and judge whether a goal is achieved or not. If the goal is achieved, then the point **e** is a point in the unsafety region, otherwise a point in the safety region.
5. Repeat steps 2 ... 4 for each grid point **e**.

We call a safety region obtained by the above procedure a simulated safety region.

The initial position of the ball used in the simulation is selected from the logged data of the 8 games held in RoboCup Japan Open 2009 and RoboCup 2009. Kick off points, and direct and indirect free kick points are the candidates of the initial position of the ball. The initial position is randomly selected from the candidates. Passing velocity and shooting velocity of the ball are 4.0 m/sec and 8.0 m/sec, respectively, which are the typical values of the robots used in the SSL. The acceleration and velocity of the defense robots are 2.0 m/sec^2 and 0.6 m/sec, respectively, which are the measured values.

3.2   Experimental Results

Coincidence Rate

We compared the approximate safety region with the one obtained by the simulation for the two defense robots under the RoboDragons's defense strategy.

Figs. 2 and 3 show the examples of the experimental results. In the figures, an approximate safety region is placed on a simulated safety region, and the red region is an unsafety region in simulation (simulated safety regin) but not in calculation (approximate safety region), the green an unsafety region in calculation but not in simulation, the blue and white an unsafety region and a safety region, respectively, both in simulation and in calculation. From the view of the fail-safe computation, it is desirable that the red area is 0 and the green area to be smallest. To evaluate these areas, we define the following rates:

$$R_c = \frac{A_b + A_w}{A} , \quad R_a = \frac{A_g}{A} , \quad R_s = \frac{A_r}{A} , \quad A = A_b + A_w + A_g + A_r \quad (4)$$

where A_b, A_w, A_g and A_r are the area of blue, white, green, and red region, respectively. We call R_c the correspondence rate. R_a and R_s are the rate of green and red regions, respectively. Table 1. shows the result of those rates which are

---

[1] We assume that the robot can kick the ball toward any direction.

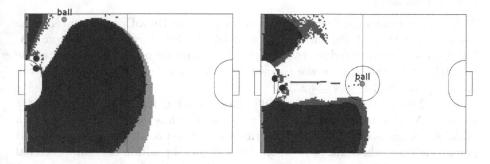

Fig. 2. Safety region: example 1   **Fig. 3.** Safety region: example 2

the average of 10 trials under the condition of 2 defending robots. From Table 1, we think we obtained a considerably accurate algorithm for safety region computation. However, there are still room for improving the algorithm as can be seen in Fig. 3.

Table 1. Rate of the area of blue+white, green and red region

R_c	R_a	R_s
0.951	0.013	0.036

Computation Time

Using the two typical computers, we measured the computation time for the proposed method. Table 2 shows the computation time and the coincidence rate R_c which is the average of 10 trials under the condition of 2 defending robots and using coarse-to-fine algorithm.

Table 2. Computation time and coincidence rate : Coarse-to-fine method

N	computation time(msec)		coincidence rate
	Athron64 X2	Xeon 3.3GHz	R_c
don't use	10.8	6.6	0.950
1	8.5	4.5	0.951
2	3.1	1.4	0.951
4	1.4	0.56	0.951
8	0.94	0.37	0.951
16	0.84	0.32	0.951
32	0.78	0.30	0.951

3.3 Discussion

Judging from our experience to have developed the RoboDragons system, the computation time of the safety region should be less than 1 msec. From Table 2, we can greatly reduce the computation time by using the coarse-to-fine algorithm. We can achieve the time of less than 1 msec without reducing the coincidence rate if $8 \leq N \leq 32$.

Table 2 shows that there are no differences in coincidence rate R_c for various values of N. This is because the initial position of ball is taken from the free kick positions or kick-off positions as described in section 3.1. Note that we should pay attention to the value of N when the initial ball position is around the goal area. Figs. 4 and 5 are examples of resulting safety region. In this case, we cannot get a correct safety region if $N \geq 8$. This fact should be considered when we construct a defense algorithm.

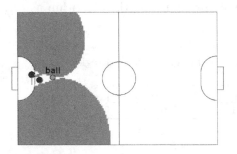
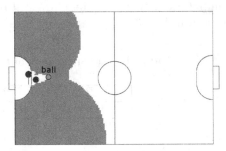

Fig. 4. Safety region obtained by using coarse-to-fine method: N = 2

Fig. 5. Safety region obtained by using coarse-to-fine method: N = 8

4 Defense Algorithms Using the Safety Region

4.1 A Defense Algorithm Considering the Position of the Opponent Robot

A defense algorithm (or an algorithm for positioning the defense robots) is proposed in the reference[5]. In the algorithm the unsafety region is weighted by the position of the opponent robots. Example is shown in Figs. 6 through 8. Fig. 6 shows an unsafety region (gray area) where 3 opponent robots (yellow) and 2 defense robots (blue) are placed. Fig. 7 shows a weighted unsafety region. The darker the area, the higher the weight. Then, we would like to add one defense robot. Where should it be placed? The algorithm proposed in [5] gives its position. As a result, Fig.8 is obtained and the unsafety region is reduced considerably. However, is the proposed algorithm good if the shooting robot is the one at upper-left corner? We must consider the pass direction of the passing robot. In the next section, we define such an algorithm.

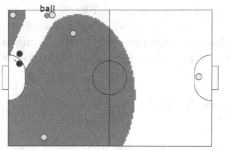

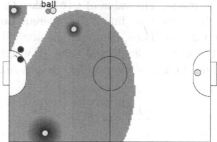

Fig. 6. Safety/Unsafety region   **Fig. 7.** Weighted unsafety region

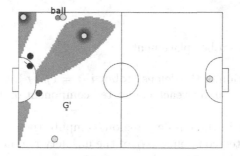

Fig. 8. Safety region: After 3rd defense robot (red circle) is placed

4.2 A Defense Algorithm Considering the Position of the Opponent Robot and the Pass Direction

In a real game, a pass direction is an important factor to evaluate the situation. Therefore, we propose a new algorithm to calculate the safety region considering both the pass direction and a weight function. In the following, we call a line connecting the center of the passing robot and the center of the ball a "pass line".

First, compute safety/unsafety region for already placed defense robots and then compute the weighted unsafety region employing a similar way used in the reference[5]. The weight function $w(\mathbf{e})$ is defined as the following equation for each point $\mathbf{e}$ in the unsafety region.

$$w(\mathbf{e}) = \sum_i max(1, 100 \times (1 - \frac{||\mathbf{r}_i - \mathbf{e}||}{max(M, t_p \times v_i)})) \tag{5}$$

where, $\mathbf{r}_i$ and v_i are a position and a moving speed of the opponent robot i, respectively, and t_p is a time given by Eq.(1). M is a threshold value to keep the weighting area wide when the value of $t_p \times v_i$ is small. We use $M = 270$ in the experiment. The value of $w(\mathbf{e})$ takes the range between 1 and 100.

Next, the pass line is considered. Let L be the line connecting the ball and the point $\mathbf{e}$. Then, compute the angle θ that the pass line and the line L make. If the angle is small, the value of weight function $w(\mathbf{e})$ should be high. So, we define the following modified weight function $w'(\mathbf{e})$.

$$w'(\mathbf{e}) = \begin{cases} w(\mathbf{e}) \times (max(1, 10 \times \cos{(3 \times \theta)})) & (|\theta| \leq \pi/6) \\ w(\mathbf{e}) & (|\theta| > \pi/6) \end{cases} \tag{6}$$

The Eq. 6 shows that the weight is higher if the angle $|\theta|$ is less than $\pi/6$ radian. The weight in Eq. 6 ranges from 1 to 1000. The closer the point $\mathbf{e}$ goes to $\mathbf{r}_e$ or the pass line, the higher weight is gained.

We propose the following algorithm that decides the position of (n+1)th defense robot.

Algorithm. Defense robot placement

1. Let $\mathbf{r}_i$ be a position of the defense robot i ($i = 1 \ldots n$). Calculate safety and unsafety regions. Number each connected component of unsafety region. Let it be $N'_k (k = 1 \ldots)$.
2. For each point $\mathbf{e}$ in the unsafety region, compute the weight $w'(\mathbf{e})$.
3. For each connected component N'_k of the unsafety region, add up all weights of the points in that connected component.
4. Calculate a center of gravity **G'** of the connected component N'_m which has the biggest summation given by step 3.
5. Place the (n+1)th robot at the cross-point r'_{n+1} of the line $L_{g'}$ and the border line of the defense area (a bit outside of the defense area), where $L_{g'}$ is a bisector line of the maximal free angle from **G'** toward the goal.

4.3   Discussion

For the same initial positions of robots and the ball as in Fig. 6, the proposed algorithm in section 4.2 is applied. As a result, the weighted safety region shown in Fig. 9 is obtained. In Fig. 9, the pass line goes toward the robot at the upper-left corner. After placing the 3rd defense robot, which is shown in the red circle in Fig. 10, computation gives a new safety region shown in Fig. 10. The figure shows that the shooting robot is in the safety region.

In Fig. 6, even if the pass line goes toward the other opponent robot, the proposed algorithm provides good placement of the defense robot. Examples are shown in Figs. 11 and 12.

In Fig. 11, the passing robot next to the ball is facing the robot at bottom near **G'**. The defense robot placement algorithm gives the 3rd defense robot position and resulting safety region is shown in Fig. 11. When the passing robot turn to face the robot at upper middle of the field, the algoritnm obtains the

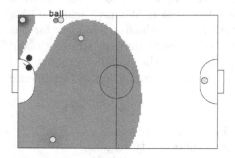

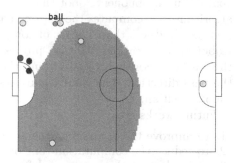

Fig. 9. Weighted unsafety region using the proposed algorithm

Fig. 10. Safety region: After 3rd defense robot (red circle) is placed by the proposed algorithm

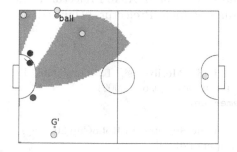

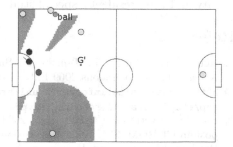

Fig. 11. Safety region: in case the passing robot faces the robot at bottom

Fig. 12. Safety region: in case the passing robot faces the robot at upper middle

3rd defense robot position as shown in Fig. 12. Resulting safety region is also shown in Fig. 12.

It is important to note that this algorithm decide the position of the defense robot depending on the **G'**. For example, assume that, at first, 3rd defense robot (red circle) is placed on the position shown in Fig. 11, and then the passing robot turns toward another robot shown in Fig. 12. In this case, the 3rd defense robot can move to the position in Fig. 12 in a short time since the moving distance is short because of the gravity center comes between two opponent robots in the connected component of the unsafety region. This is a greate advantage of this algorithm.

5   Concluding Remarks

We proposed the safety region as an index for evaluating the situation of the game in the SSL. The safety region is a region that the defense robots kept the goal when an opponent robot shoots in that region. We proposed an improved algorithm that computes the approximate safety region. We have achieved 95% accuracy and less than 1 msec of computation time, which is adequate for our RoboCup application. We also proposed an improved defense strategy using the safety region index considering the positions of the opponent robots and the pass direction. The results indicates accurate determination of the defense robots position.

Future works are

1. to improve the accuracy of the approximate safety region,
2. to reduce the computation time of approximate safety region,
3. to implement the proposed defense strategy into the RoboDragons system and evaluate the algorithm using the real robots.

Acknowledgement. The authors thank Mr. Kentaro Mizushima for his valuable comments and Dr. Salman Valibeik for his proofreading. A part of this work was supported by the chief director's special study fund of Aichi Prefectural University and the president's special study fund of Aichi Prefectural University.

References

1. Achiwa, H., Maeno, J., Tamaki, J., Suzuki, S., Moribayasi, T., Murakami, K., Naruse, T.: RoboDragons 2009 Extended Team Description (2009),
   http://small-size.informatik.uni-bremen.de/
   tdp/etdp2009/small_robodragons.pdf
2. Ball, D., Wyeth, G.: UQ RoboRoos 2004: Getting Smarter. In RoboCup 2004 Symposium CDROM, SSL Team Description Papers (2004)
3. Bowling, M., Browning, B., Chang, A., Veloso, M.: Plays as Team Plans for Coordination and Adaptation. In: Polani, D., Browning, B., Bonarini, A., Yoshida, K. (eds.) RoboCup 2003. LNCS (LNAI), vol. 3020, pp. 686–693. Springer, Heidelberg (2004)
4. Laue, T., Burchardt, A., Fritsch, S., Hinz, S., Huhn, K., Kirilov, T., Martens, A., Miezal, M., Nehmiz, U., Schwarting, M., Seekircher, A.: B-Smart (Bremen Small Multi Agent Robot Team) Extended Team Description for RoboCup (2009),
   http://small-size.informatik.uni-bremen.de/
   tdp/etdp2009/small_b-smart.pdf
5. Maeno, J., Ishikawa, A., Murakami, K., Naruse, T.: Safety region: an index for evaluating the situation of RoboCup Soccer game. JSAI Technical Report. SIG-Challenge-B001-2 (2010),
   http://winnie.kuis.kyoto-u.ac.jp/sig-challenge/
   SIG-Challenge-B001/SIG-Challenge-B001-2.pdf
6. Nakanishi, R., Bruce, J., Murakami, K., Naruse, T., Veloso, M.: Cooperative 3-Robot Passing and Shooting in the RoboCup Small Size League. In: Lakemeyer, G., Sklar, E., Sorrenti, D.G., Takahashi, T. (eds.) RoboCup 2006. LNCS (LNAI), vol. 4434, pp. 418–425. Springer, Heidelberg (2007)
7. Zickler, S., Bruce, J., Biswas, J., Licitra, M., Veloso, M.: CMDragons 2009 Extended Team Description (2009),
   http://small-size.informatik.uni-bremen.de/
   tdp/etdp2009/small_cmdragons.pdf

Local Multiresolution Path Planning in Soccer Games Based on Projected Intentions

Matthias Nieuwenhuisen, Ricarda Steffens, and Sven Behnke

Autonomous Intelligent Systems Group, Institute for Computer Science VI
University of Bonn, Germany

Abstract. Following obstacle free paths towards the ball and avoiding opponents while dribbling are key skills to win soccer games. These tasks are challenging as the robot's environment in soccer games is highly dynamic. Thus, exact plans will likely become invalid in the future and continuous replanning is necessary. The robots of the RoboCup Standard Platform League are equipped with limited computational resources, but have to perform many parallel tasks with real-time requirements. Consequently, path planning algorithms have to be fast.

In this paper, we compare two approaches to reduce the planning time by using a local-multiresolution representation or a log-polar representation of the environment. Both approaches combine a detailed representation of the vicinity of the robot with a reasonably short planning time. We extend the multiresolution approach to the time dimension and we predict the opponents movement by projecting the planning robot's intentions.

1 Introduction

A basic skill for autonomous robots is the ability to plan collision-free paths. In the humanoid soccer domain, a common approach is to determine a gait target vector, which controls the direction and velocity of the robot's motion, by incorporating the target position and the position of obstacles. Hence, perceptions are directly mapped to actions. The mapping may depend on additional factors, like the role assigned to the robot or the game state, but it does not consider the (foreseeable) future. This can lead to inefficient obstacle avoidance, e. g., the robot passes an obstacle on the side closer to its target just to be blocked by the next obstacle in the same direction.

On mobile robot platforms, the computational resources are often limited due to weight and power constraints. Accordingly, exact planning even of relatively small problem instances is not possible in real-time. Moreover, performing time-consuming planning and committing to a long-term plan is no option in highly dynamic domains like soccer. Because of the limited capabilities of the robot's sensors, it is not possible to estimate precise obstacle positions. Therefore, the environment is not only dynamic, but path planning has to deal with uncertainties.

Thus, we propose to use approximate path planning with replanning every time a new state of the environment is perceived. With increasing time since the last perception, the prediction of the world state becomes more uncertain. Consequently, planning steps in the far future should be more approximative than planning steps that have to be executed immediately, as the former will likely be invalid at the time they are executed.

T. Röfer et al. (Eds.): RoboCup 2011, LNCS 7416, pp. 495–506, 2012.

In order to reduce the complexity of the plan representations, we employ multiresolutional approaches, namely, a local multiresolutional grid and a log-polar grid. In both representations, the resolution decreases with increasing distance to the robot.

In arbitrary situations, it is hard to predict the movements of dynamic obstacles without tracking them. Thus, these movements are often modelled as random noise. In contrast, the movements of the opponent in soccer games are aimed at scoring goals. We employ this fact to gain a better estimation of probable obstacle trajectories by projecting the intentions of the planning robot to other field players.

2   Related Work

Our humanoid soccer team uses a hierarchical reactive approach to control the robot's motion [2]. In contrast, our novel approach is based on planning. It considers the foreseeable future to determine obstacle-free robot paths. Continuous replanning allows for always considering the most recent sensory information and for quickly reacting to changes in the environment.

Many planning-based systems exist in the literature. The key challenge is the computational complexity of real-time planning and execution. Kaelbling and Lozano-Pérez reduce the complexity of task planning by top-down hierarchical planning [4]. In their approach, an agent commits to a high-level plan. The refinement of abstract actions is performed at the moment an agent reaches them during plan execution. We follow the same assumption that there is likely a valid refinement at the time an abstract action has to be concretized, and that every plan can be reversed without huge costs in case that there is no such refinement.

A method for resource-saving path planning is the local multiresolution Cartesian grid [1]. It employs multiple robot-centered grids with different resolutions, and nests them hierarchically while connecting them through adjacencies. This representation resembles our approach to approximate plan steps with increasing distance to the robot. In addition, it was designed for soccer robots and, consequently, considers the present circumstances.

Apart from Cartesian occupancy grid maps [11], polar coordinate based grids can be found for egocentric robot motion planning in the literature [3], [5]. In this approach, the environment close to the robot has a high Cartesian resolution that decreases with the distance, due to the fixed angular resolution. Polar grids with hyperbolic distance functions are used to represent infinite radii within a finite number of grid cells. This property is used to plan long-distance paths in outdoor environments [10].

Another kind of polar grids are log-polar grids [6]. Like the local multiresolution grid, this approach emphasizes a more precise path planning in the robot's vicinity. Furthermore, polar grids have the advantage of an easy integration of obstacles perceived by ultrasonic sensors and cameras.

3   Robot Platform

In the RoboCup Standard Platform League (SPL), Nao robots from Aldebaran Robotics are used [7]. The Nao V3+ edition, used in recent RoboCup competitions, has 21 degrees of freedom. The environment is perceived by two cameras, from which only one

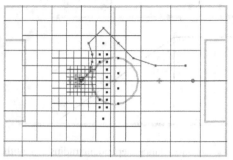 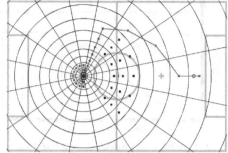

(a) Local multiresolution grid with five levels and 8×8 cells on each level.

(b) Log-polar grid with 16 discrete steps for angle and distance respectively.

Fig. 1. Non-uniform grids allow to cover a given area with multiple orders of magnitude less cells than uniform ones. Depicted are planned paths on a soccer field (red dots and lines) starting from the robot position in the center of the grid to the ball $3\,m$ in front of it. The robot faces towards the positive x-axis. Occupied cells are marked by black dots.

can be used at a time, and two ultrasonic sensors. In our system, we continuously switch between the two cameras, which results in a frame rate of approximately 14 Hz. The ultrasonic sensors are located at the robot's chest covering an angle of approximately $110°$ in front of the robot. The detection range is from $300\,mm$ to $700\,mm$ [1].

The ultrasonic sensors measure the distance towards an obstacle, but not its precise angular position. In contrast, both cameras provide the exact direction towards an obstacle, but no precise distance. This leads to uncertainties which have to be taken into account.

The Nao robot is equipped with a x86 AMD Geode LX 800 CPU running at 500 MHz. It has 256 MB of RAM and 2 GB of persistent flash memory [2]. The built-in processor has the advantage of low power consumption, with the tradeoff of low computational power. Compared to state-of-the-art computer systems, the resources are rather limited. Hence, a low system load is an important requirement for the development of new software components.

Our software is based on the framework of the German SPL-team B-Human [8]. The framework consists of several modules executing different tasks. In addition to modules for, e. g., perception or behavior control, a simulator called SimRobot is provided. For our tests, we integrated a new path planning module into the framework.

4 Path Planning Representations

At RoboCup 2010, we used reactive target selection and obstacle avoidance behaviors. Thus, a gait target vector (v_x, v_y, v_θ), which determines the walking speed in forward, lateral, and rotational direction, respectively, is determined merely by direct perceptions

---

[1] Nao User's Guide Ver 1.6.0, Aldebaran Robotics.
[2] Nao Academics Datasheet, Aldebaran Robotics.

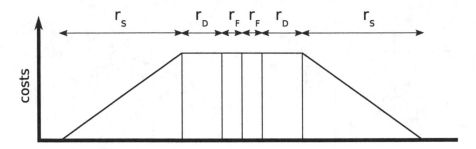

Fig. 2. Obstacles are modeled as a fixed radius r_F and a distance dependant radius r_D with maximum costs and a safety margin r_S with decreasing costs

and active behaviors. Typical behaviors are *go_to_ball* or *avoid_obstacle*. Going to the ball leads to a gait target vector towards the estimated ball position. Obstacles can be seen as repulsive forces affecting the direction of the gait target vector in this notion.

Our approach to planning a path to a target introduces a planning layer between abstract behaviors and motion control. Waypoints provided by this layer are used to determine the needed velocities. The abstract behaviors configure the planner and perceptions are integrated into the planner's world representation.

In our implementation, we use the A*-algorithm with a closed list. We implemented our planner with three different types of representations: a uniform, a local multiresolution and a log-polar grid. All representations have in common that their coordinate system is egocentric, i.e., they are translated and rotated according to the robot's pose. In the following section, we describe our uniform grid representation and detail the two non-uniform grid representations.

4.1   Uniform Grid

A commonly used representation of the environment are uniform grid maps. These discretize the environment into equally-sized cells. Cells marked as occupied correspond to obstacles. Perceived obstacles are initialized with a radius and costs. The radius consists of a static part representing the obstacle radius plus a component that increases linearly with the distance to the grid center. This component represents the uncertainty in distant measurements and the obstacle movements. To keep the overall costs of an obstacle constant, the costs are reduced accordingly. The approximate robot radius and a safety margin with linear decreasing costs are added to the obstacle, before it is inserted into the map. This simplifies the planning problem to finding a path for a robot that is reduced to a point and allows to plan paths through robots standing close to each other, but avoids the vicinity of obstacles if this is possible without huge costs. Our obstacle model is depicted in Fig. 2. The positions of visually perceived obstacles and sonar readings are maintained in different representations outside of the grid and incorporated into it during planning.

The center of a grid cell can be interpreted as a node of a graph with edges to the eight neighboring cell centers. Therefore, it is possible to use standard graph searches as the A*-algorithm (cf. [9]).

Due to the simple structure of the grid, there is no need to explicitly model the cell connectivity. Furthermore, the cost calculation per step is reduced to two cases: Diagonal or straight steps. Because of the simplicity, it can be easily implemented, and arbitrary environments can be represented equally accurate, up to a freely chosen resolution. This makes uniform grids a good choice for many applications. A major disadvantage is the computational complexity, which does not scale well with increasing grid size. Especially in environments with special characteristics, like a soccer field, more efficient representations are possible.

4.2   Local Multiresolution Grid

An efficient path planning algorithm is local multiresolution path planning [1]. Besides the computational advantages of multiresolution, the uncertainty of local sensing and of the own and the opponent's movements are implicitly taken into account with an increasing cell size.

The grid consists of multiple robot-centered grids with different resolutions embedded into each other. With increasing distance to the robot, the grid resolution decreases. This models the uncertainties caused by local sensing with only relative precision and by the dynamic environment. Local multiresolution planning utilizes the fact that the world changes continuously while the agents of the own and the opponent team move. Hence, it is not worthwhile to make detailed plans for the far-future.

More formally, the environment is discretized into a square $M \times M$ grid. Recursively, a grid is embedded into the inner part $\left[\frac{M}{4} : \frac{3M}{4}\right] \times \left[\frac{M}{4} : \frac{3M}{4}\right]$ of the grid at the next coarser level. The cell area of the inner grid is a quarter of the cell area of the outer grid. In order to cover the same area as a uniform $N \times N$ grid, only $(\log_2(N/M) + 1) M^2$ cells are necessary.

Fig. 1a shows an 8×8 local multiresolution grid with a minimum cell size of $10\,cm$ that covers an area of $163.84\,m^2$ with 256 cells using five grid levels. In contrast, a uniform grid covering the same area consists of 16,384 cells.

The neighborhood of an inner grid cell consists of eight neighbors, similar to the neighborhood of the uniform grid. However, the cells at the border to the coarser grid have seven neighbors at the edges and six neighbors at the corner. Likewise, the grid cells at the border to the finer grid have nine neighbors and eight neighbors, respectively. In our implementation, the connections between two neighbors are encoded as edges of a graph.

Incorporating obstacles into the grid is analog to the uniform grid.

For path planning with the A*-algorithm, the costs of each step are calculated by means of the Euclidean distance of the centers of both cells and the added costs of the target cell. The employed heuristic is the distance from the current grid cell to the target.

The advantages of this representation are the low requirements on the robot's memory and on computational power. Moreover, uncertainties occurring in dynamic environments are considered by the increasing cell size with increasing distance from the robot.

4.3   Log-Polar Grid

Although our soccer robots are able to walk omnidirectionally, the best speed can be achieved in forward direction. Furthermore, the robot's sensing capabilities are designed for perceiving the environment in front of the robot. Accordingly, walking in forward direction towards the target, i. e., to the next waypoint of the plan, is preferred for long distances. On a soccer field with only smaller obstacles, it is likely that relatively long straight segments are often part of the plans. Because of our grid representations being egocentric, a target in front of the robot is on the positive x-axis. As this axis is a cell boundary in every resolution, this case is not well supported by the local multiresolution grid. Due to imprecise measurements of the target and inaccurate motion execution, distant targets in a uniform grid presumably change their respective cells, too.

This leads us to a grid representation that fits the robot's motion and sensing characteristics in a more efficient way. In contrast to the Cartesian grid representations, a polar grid representation provides a straight path towards targets in front of the robot if there are no obstacles. Additionally, sensory information relevant to path planning is initially provided in polar coordinates on our robots. Ultrasonic measurements are represented as a distance and an apex angle, and visual perceptions are estimated distances and directions to obstacles. Both can be easily incorporated in a grid representation in polar coordinates.

In polar coordinates, the environment is represented as an angle θ and a distance ρ with regard to the robot's pose, written as a tuple (θ, ρ). In our coordinate system, the robot faces in the direction of the positive x-axis. We discretize the angular component θ into T equally sized partitions. The first partition is chosen in a way that the positive x-axis becomes the angle bisector of the angle represented by this partition. Hence, small angular inaccuracies in the perception or gait execution will not change the grid cell of a waypoint or the target.

To reach the implicit consideration of uncertainties and computational advantages of the multiresolutional grid, the cells of our polar grid grow exponentially with the distance. In order to achieve this, the logarithm of the distance to the robot is partitioned. To define a minimal cell size and still achieve a reasonable growth until the maximum radius, we use a slightly shifted logarithm to avoid the initial strong slope of the logarithm. The calculations to determine the polar coordinates (θ, ρ) of a point (x, y) in Cartesian coordinates, and the corresponding discretized grid cell (r, t) are

$$\rho = \log_b \left(\left(\frac{\sqrt{x^2 + y^2}\,(b-1)}{l} \right) + 1 \right),$$

$$\theta = \arctan \left(\frac{y}{x} \right),$$

$$(r, t) = \left(\lfloor \rho \rfloor, \left\lfloor \left(\frac{T}{2\pi} \right) \theta + 0.5 \right\rfloor \right),$$

where b is the base of the logarithm, l is the minimal cell size and T is the number of angular partitions.

The inverse operation therefore is described by

$$(x, y) = \begin{pmatrix} \cos(\theta) \\ \sin(\theta) \end{pmatrix} \left(\frac{(b^{r+0.5} - 1) * l}{b - 1} \right).$$

In our implementation, we use a base of $b = 1.1789$ and a minimal cell size of $l = 100\,mm$. With 16 cell rings, we reach a maximum radius of $7211\,mm$, which is sufficient to plan paths for any two points within the SPL field boundaries. We use 16 steps for the angular component as well, leading to 256 cells in total, the same number of grid cells used in the local multiresolution grid. The resulting grid is depicted in Fig. 1b.

In the polar grid representation, the obstacles are treated analogously to the local multiresolution grid.

The costs of each step are calculated by means of the Euclidean distance of the centers of both cells, as in the local multiresolution grid. The heuristic is computed likewise. The cell distances are precalculated to decrease the computational complexity. Thus, single node expansions of a planner are not more costly in this representation than in uniform grids.

4.4 Implementation Details

In our 2D planning implementation, we neglect the orientation and velocities of the robot for efficiency reasons. Accordingly, due to the fast replanning, sudden changes of the gait target vector are possible. To avoid this, we introduce a virtual obstacle behind the robot, which represents its starting speed (Fig. 3). The polar grid representation employs a half circle with cost interpolation between a minimum at the edges and a maximum at the midway of the circle segment. In contrast, the Cartesian grid representations use a rectangle having the same characteristic. When the robot is moving, this obstacle is opposed to the gait target vector with costs corresponding to the scalar value of the velocity.

To generate motion commands for a planned path, the planning module sends the next waypoint on the path to the motion control in every execution cycle. Replanning is performed if the robot's movement exceeds a threshold. Between planning calls, the waypoint is adjusted using odometry data. The resulting gait target vector is the weighted vector of position and angle of the waypoint relative to the robot.

5 Evaluation of the Representations

5.1 Planning Time and Node Expansions

The computing time of the three planners was measured in the simulator and on a Nao robot with regard to four different test cases, representing all possible constellations on the soccer field. Those are, 1. that no obstacle is detected, 2. obstacles are detected either in the ultrasonic sensor measurements or 3. through cameras, and 4. obstacles are captured by both, ultrasonic sensors and cameras. The used sensors influence the planning time, because sonar sensors can only perceive close obstacles and have a low angular resolution. This results in only few, but large obstacles in the vicinity of the robot. In every test case, the target was $4000\,mm$ in front of the robot and five other robots were on the field. In order to avoid an influence of noisy sensor data and for a better reproducibility, we set the egocentric obstacle positions manually when testing

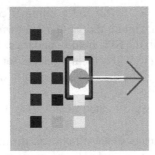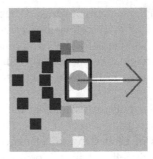

Fig. 3. An artificial obstacle leading to preference of paths towards the robot's current walking direction to avoid sudden directional changes. Cells with higher costs are darker. The red arrow is the gait target vector (left: Cartesian grid, right: polar grid).

Table 1. Planning time (in milliseconds) on the Nao (UG: uniform grid, LMG: local multiresolution grid, LPG: log-polar grid)

test case	UG	LMG	LPG
no obstacles	2.9	0.4	0.3
ultrasonic	12.2	0.8	1.2
camera	18	2.3	3.0
ultrasonic & camera	23	3.0	3.5

on the real robot and used ground truth data in simulation. The measured execution time was averaged over 1000 planner runs.

The results of the tests on the Nao robot are shown in Tab. 1. Overall, the multiresolutional approaches outperform the uniform grid planner clearly. Moreover, there are differences between the local multiresolution and the log-polar grid, as the former has a lower computational complexity.

6   Dynamic Planning with Intention Projection

Most of the obstacles apparent on the soccer field are not static. As described before, we take this into account by smoothing distant obstacles. If the movement of the other robot's can be estimated in advance, it is possible to avoid future positions of these. Naturally, it is not possible to reason about the exact behaviors and targets of the opponent team. Thus, we have to apply situation specific assumptions for probable movements of the opponent. For example, if our robot intends to reach a position where it can get ball possession, this intent is projected to the opponent. Thus, we assume it will move to a position where it can control the ball, too. This assumption is reasonable, as opponents with the same intent as the planning player will more likely intersect the planned path than robots with different intents.

Our dynamic obstacle model extends the static obstacle model by introducing an assumed target relative to the obstacle and a velocity vector v. To be consistent with the multiresolutional representation of space, the obstacle radius is increased and the costs are decreased with increasing distance to the robot. We assume that the opponent may either keep standing still or move to the target with any speed up to a maximum v_{max}. This results in a uniform probability distribution of the opponent's position along a vector $v_{max} \times t$. To represent this the representation of the obstacle grows into the direction of its movement up to the assumed target with advancing time t. The overall costs of the obstacle are kept constant. Hence, the costs of the obstacle decrease.

To represent the resulting different shapes of obstacles at different times, we extend the grid representation to the time dimension. The resulting three-dimensional grid is discrete in time and this discretization may not necessarily equal the space discretization for arbitrary robot speeds. To take this into account, we explicitly calculate the average time the robot needs to follow the planned path up to the current grid cell and discretize it afterwards to determine whether edges connecting nodes within the same time level can be followed.

Figure 4a shows an example where the planned paths with and without dynamic planning are qualitatively different. Our robot aims to reach a position next to the ball to dribble it towards the opponent goal. Therefore it has to walk around the ball. Another player is approaching the ball from the left bottom side of the map. Utilizing a planning algorithm not taking the directed movement of this robot into account, the robot would try to pass the ball at the side the opponent is coming from, possibly having to avoid it later on. The dynamic approach plans the same shortest path, if the opponent is still far away. Otherwise, the ball is passed on the other side, where an intersection with the opponents path is unlike.

Obviously, adding the time dimension makes the planning computationally more involved. For the uniform time discretization we use a 16 s lookahead, discretized into 1 s steps. This results in a local-multiresolution grid with 4096 cells. Thus, we reduce the complexity, following the same ideas of a multiresolutional representation as in the space dimension. The minimal time an opponent robot needs to reach a cell corresponds to the distance of that robot to the cell. Here, we have three qualitatively different cases to consider:

- The robots are close to each other and to the cell. Thus, the cell is represented at a fine spatial resolution and an intersection of the robot's paths may occur soon. A finer temporal resolution is necessary.
- The cell is further away. Hence, it is represented at a coarse spatial resolution and coarse temporal resolution is sufficient, regardless of the position of the opponent.
- The cell is close to the planning robot, but the opponent is further away. In this case a coarse resolution is sufficient to determine, if an intersection of paths in this cell is likely.

Figure 4b depicts snapshots of the ball approaching example using a multiresolutional time representation. The discrete time steps t_m are defined as

$$t_m(i) = \begin{cases} i = 0, 0 \\ i > 0, 2^i \end{cases}$$

The resulting grid contains 1280 cells and covers the same time and space as the uniform grid.

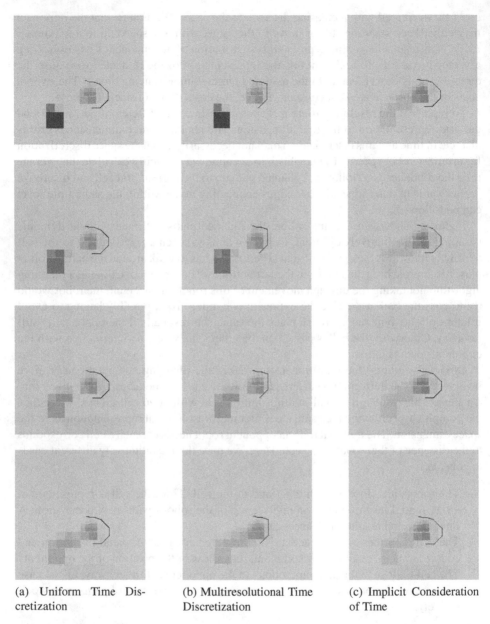

(a) Uniform Time Discretization

(b) Multiresolutional Time Discretization

(c) Implicit Consideration of Time

Fig. 4. The planning robot aims to reach a position behind the ball to dribble it to the goal. An opponent is detected and its position is propagated over the time by elongating the corresponding obstacle shape. At the beginning costs of cell nodes are calculated using a circular obstacle model. With increasing time, the obstacle representation becomes more elongated until it reaches the target, i. e. a position behind the ball opposing our target. The time increases from the top (t=0) to the bottom images (t=8 s). The left image series shows the uniform time discretization, the middle series the multiresolutional time discretization and the right series the implicit consideration of the time.

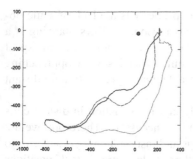 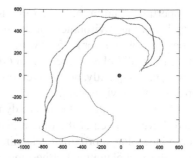

Fig. 5. Comparison of the robot's trajectories while using the planner without time consideration (green/dotted), with uniform time discretization(blue/dashed), and with implicit time incorporation (red/solid). The target is a point behind the ball (circle). An opponent is approaching from the bottom right side. In the example on the left side the opponent is further away than in the example depicted on the right side. Both planners that propagate the opponent's position plan qualitatively similar path. While following the plan without representation of the time, the robot had to replan to avoid a collision with the opponent in the case depicted in the right figure.

As the speed of the planning robot is bound, the time necessary to reach cells increase linearly with their distance to the robot, defining a lower envelope of the reachable cells. Thus, major parts of the three dimensional grid cannot be reached. Especially, the fine grained temporal resolution steps are mostly below this envelope. For the planner it is of primary importance if a path through a cell may possibly cause a collision in the future. Shortest paths often start with long straight segments towards the target. In the case of an apparent obstacle, the segment is split into two segments with an intermediate point that modifies the path to be collision free. The time to follow the first segment can be approximated by a linear function of the distance between robot and endpoint of that segment. This leads us to an implicit incorporation of the time dimension into our obstacle model. We determine the minimal distance to the line segment between the opponent and its assumed target. With the linear time approximation function, we get the approximate time of the probable intersection of the two paths. This approximation is used to estimate the distribution over the possible positions of the obstacle on the field. As time is only implicitly taken into account for planning, the planning problem stays two-dimensional.

We compared the planning time on the Nao robot for the ball approaching example. To plan a path around the ball without considering the opponents movement took on average $1.6\,ms$, with 16 uniform time discretization steps $55.4\,ms$. The multiresolutional discretization with 5 steps took $7.9\,ms$ and the implicit approach took $3.3\,ms$. With every representation of the time component it was possible to use the shortest path, if the opponent was sufficiently far away and to avoid the opponents path if it was closer (see Fig. 5).

7 Conclusion and Future Work

In this paper, we evaluated two approaches to path planning which are applicable to soccer robots with relatively low computational power. Both approaches make use of properties found in the soccer domain. An important property is the lack of static obstacles in the environment of the soccer field. For this reason, it allows planning at a coarse resolution for regions which are far from the robot. Furthermore, one can expect to find a valid plan refinement in order to avoid dynamic obstacles while approaching them. Consequently, our grid representations employ a decreasing resolution for distant parts of the environment.

As virtually all obstacles are dynamic, it is likely that the situation in distant cells will have changed at the time a plan refinement will be necessary. Therefore, we are convinced that approximate planning with continuous and fast replanning is superior to slower exact planning. Furthermore, the estimation of possible future obstacle positions is likely to improve plans with regard to the need of necessary replanning. Thus, we use assumptions about the behavior of soccer playing robots to estimate their trajectories.

The real-robot experiments reveal that the speedup facilitates real-time planning on the Nao.

Acknowledgment. This work was partially funded by the German Research Foundation (DFG), grant BE 2556/2-3.

References

1. Behnke, S.: Local Multiresolution Path Planning. In: Polani, D., Browning, B., Bonarini, A., Yoshida, K. (eds.) RoboCup 2003. LNCS (LNAI), vol. 3020, pp. 332–343. Springer, Heidelberg (2004)
2. Behnke, S., Stückler, J.: Hierarchical Reactive Control for Humanoid Soccer Robots. International Journal of Humanoid Robots (IJHR) 5(3), 375–396 (2008)
3. Borenstein, J., Koren, Y.: The Vector Field Histogram-Fast Obstacle Avoidance for Mobile Robots. IEEE Trans. on Robotics and Automation 7(3), 278–288 (1991)
4. Kaelbling, L., Lozano-Pérez, T.: Hierarchical Task and Motion Planning in the Now. MIT-CSAIL-TR-2010-026 (2010)
5. Lagoudakis, M., Maida, A.: Neural maps for mobile robot navigation. In: IJCNN 1999 (1999)
6. Longega, L., Panzieri, S., Pascucci, F., Ulivi, G.: Indoor robot navigation using log-polar local maps. In: Prep. of 7th Int. IFAC Symp. on Robot Control, pp. 229–234 (2003)
7. RoboCup Technical Commitee: RoboCup Standard Platform League Rule Book (2010)
8. Röfer, T., Laue, T., Müller, J., Burchardt, A., Damrose, E., Fabisch, A., Feldpausch, F., Gillmann, K., Graf, C., de Haas, T.J., Härtl, A., Honsel, D., Kastner, P., Kastner, T., Markowsky, B., Mester, M., Peter, J., Riemann, O.J.L., Ring, M., Sauerland, W., Schreck, A., Sieverdingbeck, I., Wenk, F., Worch, J.H.: B-Human Team Report and Code Release (2010), http://www.b-human.de/file_download/33/bhuman10_coderelease.pdf
9. Russell, S., Norvig, P.: Artificial Intelligence: A Modern Approach. Prentice-Hall (2009)
10. Sermanet, P., Hadsell, R., Scoffier, M., Muller, U., LeCun, Y.: Mapping and planning under uncertainty in mobile robots with long-range perception. In: Proc. of IROS (2008)
11. Thrun, S., Burgard, W., Fox, D.: Probabilistic Robotics (Intelligent Robotics and Autonomous Agents). MIT Press (2001)

Robot Orientation with Histograms on MSL

Fernando Ribeiro, Gil Lopes, Bruno Pereira, João Silva, Paulo Ribeiro,
João Costa, Sérgio Silva, João Rodrigues, and Paulo Trigueiros

Industrial Electronics Department, Univ. of Minho,
Campus de Azurém, 4800-058 Guimarães, Portugal
{fernando,gil}@dei.uminho.pt,
{brunomiguel1987,jsilva86,paulorogeriocp,
joaovazcosta,sergio.fil,pedro50073}@gmail.com,
paulo@trigueiros.com
http://www.minhoteam.org

Abstract. One of the most important tasks on robot soccer is localization. The
team robots should self-localize on the 18 x 12 meters soccer field. Since a few
years ago the soccer field has increased and the corner posts were removed and
that increased the localization task complexity. One important aspect to take
care for a proper localization is to find out the robot orientation. This paper
proposes a new technique to calculate the robot orientation. The proposed
method consists of using a histogram of white-green transitions (to detect the
lines on the field) to know the robot orientation. This technique does not take
much computational time and proves to be very reliable.

Keywords: Robot Orientation, Histogram, Robot Localization, Middle Size
League.

1   Introduction

RoboCup consists of a scientific challenge of Artificial Intelligence, whose main
objective is to build a team of robots to play football against a human team. The
scheduled year for that is 2050. Since 1997, annually new challenges have been added
in order to improve the quality of game on RoboCup. Some of these challenges were
the field size increase and the removal of the corner landmarks.

In order to play soccer each robot uses a vision system to gather information of the
surrounding area, the robot must be able to move, pass and kick the ball, collaborate
with team members, obey to referee commands, etc. Almost all robot decisions are
dependent on its localization. Therefore, one of the main problems in autonomous
football robots can be considered self-localization.

The recent implemented changes on the field are responsible for the blossom of
new ways for self-localization. Before removing the corner landmarks almost all
localization methods relied on those cylinders. The increasing of the soccer field size
also encourages new localization methods, since most robots can only see around
about 4 or 5 meters radius while the field is 18 x 12 meters.

T. Röfer et al. (Eds.): RoboCup 2011, LNCS 7416, pp. 507–514, 2012.

One of the most used algorithms for self-localization is Monte Carlo [8] [1] [5]. Since Monte Carlo is very used, some improvements on it had been proposed [3] [7].

However, rather than proposing a new localization algorithm, this paper proposes a new way of sorting out one step of the localization task, namely the orientation. The proposed approach uses histograms of the transitions between white lines and the green colour on soccer field.

The histogram of the transitions is used to find out the robot orientation. One important aspect to take into account is the time required for the localization task; the better method is dependent on the compromise between the shortest time and the highest rate of correct position.

This paper is organized as follows. In Section 2, the localization state of the art is described. In Section 3, the proposed approach which uses histogram is presented. Section 4 describes the methodology used. Section 5 shows up some results. Section 6 has a discussion about existing localization methods. Finally, Section 7 presents the conclusions and directions for further work.

2   Background

The team robots moving strategy requires localization of all robots of the team, to find the goals to kick the ball. These reasons make the localization one of the main problems on football robots. Before removing the surrounding walls, the landmarks, and the corners and increasing the size of the football field, the localization was not so difficult, but nowadays it is a hard task. In [10] one can read a discussion about the effect on localization after the removal of walls, as it talks about the problem on localization due to field size increase.

At that time, the sensors most used were laser range finder, odometry and vision system. However, the laser ranger was more used before removing the walls and landmarks.

With respect to odometry the problem is the obtained error, since the reading of encoder is not feasible with such slipping wheels like omnidirectional wheels. In order to avoid problems with odometry some works propose a way to fix the reading of the encoders [4]. The sensor still in use, nowadays, is the vision system [7]. The advantage of vision systems is the possibility of observing 360°.

The most used algorithm to self-localization is Monte Carlo [8] [1] [5]. The approach is Sequential Monte Carlo or also called Particle Filter. Since this method was broadly used some improvements had been proposed. This work [3] proposes a method to the number of samples to be adaptive on Monte Carlo. Already on [7] the improvement is to use Genetic Algorithm, the evolution operators to generate points on regions with high post density.

The proposed method in [6] is also very used [9] [2]. It transforms the localization in an optimization problem. The robot position is the minimum error, i.e. the minimum of the function. It uses the Kalman Filter to improve the solution, i.e. to make it smoother. Work [6] has a comparison with Monte Carlo.

3   Histogram Approach

The robot location is composed by [x y θ], being θ the robot orientation. The proposed approach uses histograms to identify the orientation. These histograms count the transitions between white lines and the green colour on the soccer field, and these transitions are found by axial searching lines and by radial searching lines. Histograms are built in two directions; a vertical one and an horizontal one.

Figure 1 shows an example of the system. The image on the centre shows what the robot sees, the green field with the white lines, with axial lines and radial lines, and the transitions represented by dots. The figure on the left shows the transitions represented by white dots, the horizontal histogram (underneath) represented by straight lines and the vertical histogram (on the right side) represented by straight lines, of the grabbed image. The figure on the right represents the same information as on the left part of the image, but with the proper rotation after calculation (transition are now parallel to the X-Y axis).

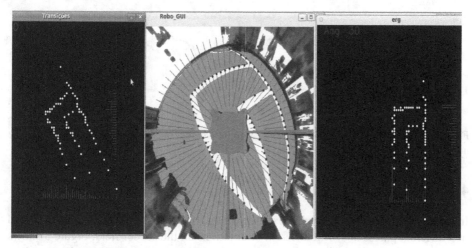

Fig. 1. Screen capture of the real image with the transition white/green dots (middle), the conversion to real distance (left), the rotation and angle displacement finding (right)

The proposed approach in order to calculate the orientation of the robot is: Rotate the histogram (first part) on the interval θ − 40° until θ + 40°, θ is the last known orientation, and verify the histogram maximum value (peak value) for each rotation. Each histogram maximum value, sum of maximum vertical and maximum horizontal, and the corresponding angle should be saved. The angle in which the highest value achieved by summing the horizontal and the vertical histogram values, corresponds to the new orientation of the robot. The 40 was chosen because on one frame the robot can only rotate this value, at most.

On Figure 1 it can be seen that a larger peak value is achieved when the transitions are all aligned (third part), therefore, rotating the histogram (first part) some degrees, in this case 30°, the maximum peak is achieved.

The decision to use the maximum sum of horizontal and vertical histogram values was because in some cases one or two histograms can have more than one maximum, and therefore, more than one solution. The final solution is evaluated with these two histograms to fuse the information but some other approaches could be used, like the Mean of these values or considering only the greater value of both. However, the adopted solution was the maximum sum of them, and when in practice it proves much more reliable. Sometimes there are many similar histogram peak values. For example, on Figure 2, one can see on interval 19 to 23, 29 to 32, and 36 to 40 for vertical histogram. In order to decide the right angle, the sum between horizontal and vertical histograms is used, therefore, removing the ambiguity. In this case the angle is 30°.

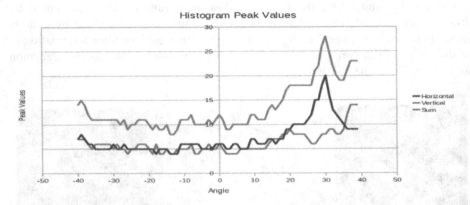

Fig. 2. Histogram peak values of the given example where a maxima was reached at 30° on the horizontal and hence the Sum histogram

4   Experiments

In order to validate the method, 5 experiments were carried out, the robot was placed in 5 different places and with different orientations on the soccer field. Figure 3 shows the locations where the robots were placed to carry out these experiments. The locations chosen were selected in paces with many lines visible, far away from any line, near the corner, near the side line, and near a non-straight line. Due to space restrictions, in the laboratory there is only one half of the soccer field, and therefore, the experiments were carried out only on one side. The size of the half soccer field, in our laboratory, is 8.24 meters width by 6.07 meters length.

The locations and orientations chosen have the aim to validate the method, the transitions occur in higher quantities than in others. The approach in order to compare the angle's results of the histogram used trigonometric methods.

5   Results

The software was developed using C++ language, QtCreator IDE and Ubuntu linux operating system. The computer used was a conventional notebook with a 2.2GHz

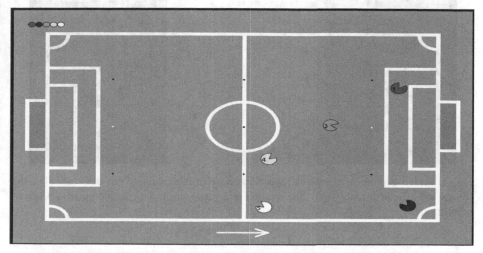

Fig. 3. Conducted experiments with the robot field positions used on the experiments

Core 2 Duo processor. The experiments were: First with 30°, Second with −40°, Third with 102°, Fourth with 48° and Fifth with −51°. In all cases the error obtained was below 2°, which is very acceptable considering that robot moves very fast, and the calculation is carried out every frame. A negative rotation is counter-clockwise (see experiments second and fifth).

Figure 4 shows the result of the third experiment, and Figure 5 shows a graphical representation of the vertical histogram, the horizontal histogram and the sum of these two histograms. In this case, the robot was heading to 120°, and since the search starts 40° before and ends 40°, it started on 80° and ended on 160°. It found the solution on −18°, which corresponds to 120 − 18 or in other words the robot is heading to 102°.

Once again several peaks can be seen on the histograms. On Figure 5, two peaks near each other can be seen on the horizontal histogram, while the vertical is almost constant. However, the peaks are always near.

Figure 6 shows the result of the fifth experiment, and Figure 7 shows the graphical representation of the two histograms and its sum. The robot was heading to angle −20°. Figure 7 shows the maximum value of the vertical histogram was not the right angle, since the right angle was reached because the horizontal overcomes the vertical.

6   Discussion

The proposed method is a new way for robot orientation. It only uses the histograms of transitions between white lines and the green field colour on robotic football fields. These transitions are detected by axial and radial lines. Two histograms are used, one horizontal and one vertical, and to fuse these two histogram a sum of maximum values of each is adopted.

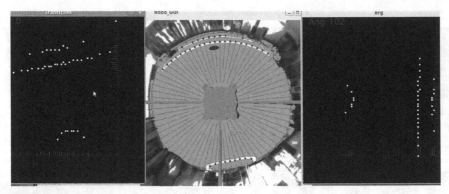

Fig. 4. Screen capture of the real image with the transition white/green dots (middle), the conversion to real distance (left), the rotation and angle displacement finding (right), all of experiment 3

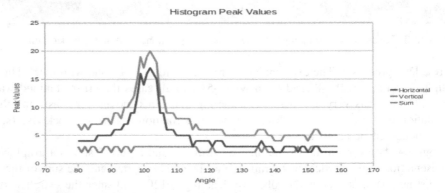

Fig. 5. Histogram peak values of experiment 3 where a maxima was reached at 102° on the horizontal and hence the Sum histogram

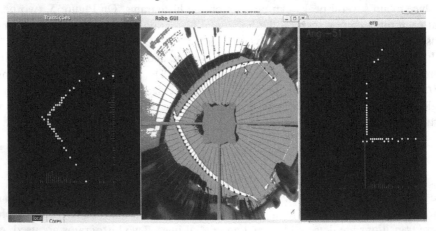

Fig. 6. Screen capture of the real image with the transition white/green dots (middle), the conversion to real distance (left), the rotation and angle dis-placement finding (right), all of experiment 5

The processing time spent for angle detection is only 4 milliseconds. Another advantage is its easiness to use it in other localization methods, because it only needs the transitions, and some methods use these transitions.

A sum of the histograms to remove ambiguity demonstrated to be reliable, because it can calculate the angle when the maximum peaks of the histograms are situated on different intervals.

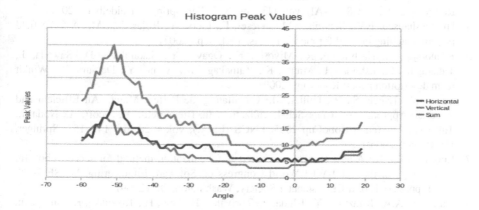

Fig. 7. Histogram peak values of experiment 3 where a maxima was reached at -51° on the horizontal and hence the Sum histogram

7   Conclusions

The localization is an important task on the robotic football environment and one element present on this task is orientation. This paper presents a new way to calculate the robot orientation using only histograms. The approach to build this is the transitions between white lines and the green colour on soccer field. The transitions are found using axial and radial searching lines. There is a vertical and a horizontal histogram.

The proposed method only requires 4 milliseconds to calculate the correct robot orientation. Therefore, it can be used on real football environment without penalizing others computing task, like robot planning and vision systems, allowing the robot's main computer to process between 25 and 30 frames per second.

Acknowledgements. The authors wish to thank all members of the Laboratorio de Automacao e Robotica, at University of Minho, Guimaraes. Also special thanks to the ALGORITMI Research Centre for the opportunity to develop this research.

References

1. Beck, D., Niemueller, T.: Allemaniacs team description paper: Robocup (2009)
2. Hafner, R., Lange, S., Riedmiller, M., Welker, S.: Brainstormers tribots team description paper: RoboCup (2009)
3. Heinemann, P., Haase, J., Zell, A.: A Novel Approach to Efficient Monte-Carlo Localization in RoboCup. In: Lakemeyer, G., Sklar, E., Sorrenti, D.G., Takahashi, T. (eds.) RoboCup 2006. LNCS (LNAI), vol. 4434, pp. 322–329. Springer, Heidelberg (2007)
4. Hundelshausen, F.V., Schreiber, M., Wiesel, F., Liers, A., Rojas, R.,, M.: A force field pattern matching method for mobile robots. Tech. rep. (2003)
5. Komoriya, Y., Hashimoto, R., Miyoshi, K., Ogawa, Y., Sakakibara, D., Sawada, J., Tahara, K., Fukushima, H., Mitani, K., Yamakage, R., Asano, Y., Demura, K.: Winkit team description paper: RoboCup (2009)
6. Lauer, M., Lange, S., Riedmiller, M.: Calculating the Perfect Match: An Efficient and Accurate Approach for Robot Self-Localization. In: Bredenfeld, A., Jacoff, A., Noda, I., Takahashi, Y. (eds.) RoboCup 2005. LNCS (LNAI), vol. 4020, pp. 142–153. Springer, Heidelberg (2006)
7. Luo, R., Min, H.: A new omni-vision based self-localization method for soccer robot. In: Proceedings of the 2009 WRI World Congress on Software Engineering, WCSE 2009, vol. 1, pp. 126–130. IEEE Computer Society, Washington, DC (2009)
8. Nassiraei, A.A., Kitazumi, Y., Ishida, S., Toriyama, H., Ono, H., Takenaka, K., Shinpuku, N., Takaki, M., Fukunaga, Y., Yamada, K., Takemura, Y., Godler, I., Ishii1, K., Miyamoto, H.: Hibikino-musashi team description paper: RoboCup (2010)
9. Su, S., Gu, Z., Chen, X., Dong, L., Huang, Y., Song, X., Wu, L.: Endeavor team description paper: RoboCup (2010)
10. Utz, H., Neubeck, A., Mayer, G., Kraetzschmar, G.K.: Improving Vision-Based Self-Localization. In: Kaminka, G.A., Lima, P.U., Rojas, R. (eds.) RoboCup 2002. LNCS (LNAI), vol. 2752, pp. 25–40. Springer, Heidelberg (2003)

A Low Cost Ground Truth Detection System for RoboCup Using the Kinect

Piyush Khandelwal and Peter Stone

Department of Computer Science, The University of Texas at Austin
{piyushk,pstone}@cs.utexas.edu

Abstract. Ground truth detection systems can be a crucial step in evaluating and improving algorithms for self-localization on mobile robots. Selecting a ground truth system depends on its cost, as well as on the detail and accuracy of the information it provides. In this paper, we present a low cost, portable and real-time solution constructed using the Microsoft Kinect RGB-D Sensor. We use this system to find the location of robots and the orange ball in the Standard Platform League (SPL) environment in the RoboCup competition. This system is fairly easy to calibrate, and does not require any special identifiers on the robots. We also provide a detailed experimental analysis to measure the accuracy of the data provided by this system. Although presented for the SPL, this system can be adapted for use with any indoor structured environment where ground truth information is required.

Keywords: ground truth, robocup, kinect.

1 Introduction

An important prerequisite for most autonomous robot tasks is the ability to self-localize. In order to evaluate self-localization algorithms, the robot's estimate of its location must be compared to the ground truth: its true location. As a result, ground truth data is frequently collected for mobile robots, typically by employing sensors external to the robot. Apart from self-localization, ground truth information can also be used for behavior analysis, by providing the locations of other agents in the environment.

Using a ground truth system requires consideration of a number of factors. These include cost effectiveness, the amount of information that needs to be collected, the accuracy of this information, as well as the ease of use of such a system. These factors typically trade-off against one another. For instance a system constrained by cost may require sacrificing on the amount and accuracy of information. In this paper we present a ground truth system created using the Microsoft Kinect sensor. We apply this system to the domain of robot soccer, namely the Standard Platform League (SPL) from RoboCup. This work was driven with the following constraints in mind.

T. Röfer et al. (Eds.): RoboCup 2011, LNCS 7416, pp. 515–527, 2012.

1. **Low cost** - For obvious reasons, a low cost solution is desirable.
2. **Portability and ease of calibration** - Different space restrictions in different venues dictates that the system should allow for flexible sensor placement. Coupled with a fast calibration time, this allows recording full games between multiple teams at foreign venues.
3. **No special markers on robots** - This is a necessary requirement to record competition games in the SPL, as custom markers are not allowed on the robots.

The Kinect sensor is widely available at a cost of approximately USD 150, and only 2 such sensors are required to span the field completely. This paper describes an open source software infrastructure[1] which will allow RoboCup teams to quickly and easily set up a ground truth system. In this paper, we will also present a detailed experimental analysis of the accuracy of the information that this system provides.

The remainder of the paper is organized as follows. In Section 2, we present a comparison of this system with other related work, and follow with a brief background of the various components of this system in Section 3. We explain our methodology in Section 4 along with an empirical analysis of performance in Section 5. We conclude with an emphasis on directions for future work with a discussion in Section 6.

2   Related Work

Ground truth systems have been used in conjunction with mobile robots both within RoboCup and in other settings. In this section we compare our system primarily with those that have been used within RoboCup. This comparison should also enable the reader to choose the ground truth system best suited to their problem domain.

Our system is similar to the SSL-Vision system [15] in many ways. SSL-Vision is a shared vision system that is used by all teams participating in the RoboCup Small Size League, and is used to estimate the ground truth position of the robots. This system uses 2 RGB cameras to span the entire field, and uses unique identifiers on top of the robots to determine robot locations and orientations. This approach was extended in [13] for the RoboCup Mid Size League. In [11], SSL-Vision was directly used to obtain ground truth positions for the Aldebaran Nao, the robotic platform for the SPL. This was achieved by attaching identifiers on top of the robots' bodies by means of a hardware extension.

An approach requiring the addition of custom hardware to the robot defeats one of the goals of the SPL — to keep a common platform across all teams. Such enhancements are not allowed in competition games, thus limiting such approaches for collecting ground truth information to test environments. Our

---

[1] Instructions for the download and use of this infrastructure are available at http://www.cs.utexas.edu/~AustinVilla/?p=research/kinect

system overcomes the requirement of special markers by utilizing the depth information available from the Kinect. Although we lose orientation information embedded in these markers, we view the gain in the ability to record ground truth information for full games to be advantageous. Additionally, our system is straightforward to implement since it does not require multiple teams to recreate custom extensions to their robots. Note that if orientation information is required, our system can be extended to use special markers as well during testing.

A different approach was used by [9] towards collecting highly accurate ground truth information for a single robot in the SPL. In this approach a motion capture system was used for detecting robot pose by placing a set of 8 identification LEDs on the robot itself. This approach successfully captured a fair amount of information about the robot's pose, including the tilt and roll of the robot body, and the position of the head with respect to the body. The data collected by this approach was made available to the public, which allows teams to run their own localization algorithms on this data-set and test it against the ground truth information. However, to properly evaluate algorithms on the robot, it is necessary to actively collect new ground truth information based on the current behavior and self-localization of the robot. Due to the high cost of a motion capture system, it is impractical for many RoboCup teams to implement such an approach. Additionally, this approach is even more invasive than SSL-Vision in terms of placing additional markers on the robot. In comparison, we believe that our solution has a greater appeal due to its low cost, portability, and ease of use.

3   Background

The Kinect sensor was introduced by Microsoft for the X-Box 360 gaming system. It is is a low cost RGB-D camera, which combines information from a standard CMOS camera with an infrared based depth sensor. It has a horizontal field of view (FOV) of 57 degrees, and a vertical FOV of approximately 43 degrees. The supported depth range from the official specification is 1.2 - 3.5 meters, but in our experiments it has been found to be around 0.7 - 7 meters. The Kinect has already seen some applications to robotics [14], and will be used as a sensor by multiple RoboCup@Home teams for RoboCup 2011 [8] [3]. In our experiments, we have also noted a few drawbacks of this sensor. First, the accuracy of the distance reading is proportional to the distance itself, as the sensitivity of the sensor drops off for larger distances (see Fig. 1b). Secondly, a minimum range of 0.7 meters and an inability to work in direct sunlight makes this sensor unsuitable for some robotics applications.

The software infrastructure of this system builds upon the ROS middle-ware package [10]. We chose to use ROS for two main reasons. First, ROS currently provides 2 drivers for the Kinect, and our system can use either. Second, an implementation of the Point Cloud Library (PCL) [1] is available through ROS. PCL is a relatively new library which provides implementations of a number of

(a) $image_{kinect}$ (b) $pointcloud_{kinect}$

Fig. 1. The image and corresponding pointcloud provided by the Kinect driver in ROS. The colored axes in Fig. 1b give the orientation and location of the coordinate axes for ref_{kinect}

algorithms handling point clouds, some of which have been used in this work. An understanding of ROS may prove beneficial for someone using this system, but it is not entirely necessary.

The Kinect driver from ROS provides information from the Kinect in a number of easily readable formats, and we import this information in 2 formats. The first one is the image that is directly available from the RGB sensor on the Kinect ($image_{kinect}$ in Fig. 1a). The second one is a composite XYZ-RGB point cloud which is created by merging the information from the RGB and depth sensors ($pointcloud_{kinect}$ in Fig. 1b). This point cloud is created by projecting RGB pixels from the camera into 3D space based on the corresponding sensor reading from the depth sensor. The point cloud is provided in a reference frame local to the Kinect (ref_{kinect} in Fig. 1b). ROS also provides sufficient tools for image rectification and geometry to transform a pixel in the RGB image to its corresponding 3D location in the point cloud and vice versa.

4   Methodology

To utilize the information provided by the Kinect, the first necessary step is to obtain the position and orientation of the Kinect sensor with respect to the global coordinate frame of the field (ref_{field}). This global coordinate frame originates at the center of the field, with the $+x$ direction towards the *yellow* goal (see Fig. 3b). Once this information is available, a transformation can be performed on $pointcloud_{kinect}$ to obtain it in the correct reference frame ($pointcloud_{field}$). The calibration procedure to obtain this transformation is explained in Sec. 4.1.[2] To effectively use the color information available in the point cloud, this system also provides a color classification tool, which is briefly explained in Sec. 4.2. Finally, in Sec. 4.3 we explain our methodology for detecting robots and the orange ball.

---

[2] A video demonstrating calibration, color classification and object detection is available at http://www.cs.utexas.edu/~AustinVilla/?p=research/kinect

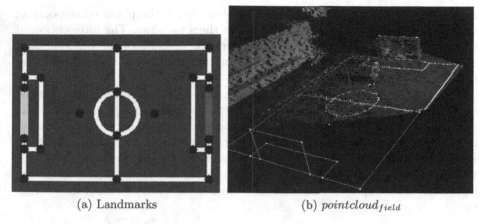

(a) Landmarks   (b) $pointcloud_{field}$

Fig. 2. Obtaining the transformation. Fig. 2a shows the 22 landmarks used for obtaining the transformation, which include all intersections, penalty crosses and goal post bases. Fig. 2b shows the transformed point cloud, along with ref_{field} axes and a wire-frame representation of the field

4.1 Transforming the Point Cloud

The transformation from ref_{kinect} to ref_{field} involves only rotating and translating the coordinate axes, and is therefore a rigid body transformation. If 2 sets of corresponding points are available in both reference frames, a least squares fitting approach can be used to obtain this transformation even in the presence of noise. A survey of such techniques is provided in [6], and we use the approach given by [2], which uses singular value decomposition (SVD) to calculate this transformation. An implementation of this approach is provided in PCL.

To obtain these corresponding set of points, we make use of some *landmarks* on the field. There are a total of 22 such landmarks, all on the ground plane. The position of these landmarks is already known in ref_{field}, and we provide an interface to the user to provide the corresponding points in ref_{kinect}. For a given landmark, the user clicks a pixel in the image, and the system projects this pixel outwards in 3D space to obtain a ray. We gather points in the point cloud which are close to this ray (within $5cm$), and average over these points to obtain the position of the landmark in ref_{kinect}. If a landmark is not visible by the sensor, then we do not obtain a correspondence.

Directly using this method to obtain landmarks has a drawback. Not all pixels in $image_{kinect}$ have a corresponding point in $pointcloud_{kinect}$. This discrepancy occurs because of slight differences in the field of view of the depth sensor and the rgb camera, as well as the failure of the depth sensor to obtain a reading for some pixels. For this reason it may not be possible to collect information about a landmark even when it is present in the image. To counter this problem, we first ask the user to enter any 5 arbitrary points on the ground plane of the field which are visibly present in the point cloud visualization. With these points, we perform least squares plane fitting [5] to obtain an accurate estimate of the ground plane of the field in ref_{kinect}. We then obtain the pixels corresponding to each landmark in the image and project them as a ray. The intersection of the ray with the ground plane estimate gives the position of the the landmark in ref_{kinect}, and we do not require the landmarks to be at their appropriate positions in the point cloud.

We weigh each correspondence by the inverse of the squared Euclidean distance of the point from the camera sensor (squared L2 norm in ref_{kinect}). The advantage of this weighing is that landmarks which are further away are smaller in the image, and incur a greater error in our estimate of their position in ref_{kinect}. We then apply the procedure described in [2] to obtain the transformation $[\mathbf{R}, \mathbf{T}]$, where $\mathbf{R}$ and $\mathbf{T}$ are the rotation and translation components respectively. We can then construct the transformed point cloud $pointcloud_{field}$ as:

$$point_{field} = \mathbf{R}point_{kinect} + \mathbf{T}$$

$\forall\ point_{kinect}$ in $pointcloud_{kinect}$. The transformed point cloud is shown in Fig. 2b.

This calibration procedure of obtaining the transformation takes approximately 5 minutes, and needs to be done only once the Kinect has been placed in the desired location. The ability to calculate this transformation into a global coordinate frame also allows for flexible sensor placement, as long as the requisite portion of the field is visible.

4.2   Color Classification

To utilize the color information also available in $pointcloud_{field}$, It is necessary to classify the raw RGB values into known colors of interest. To this end, we use an approach similar to that of many RoboCup teams for performing color based segmentation [12]. Raw RGB values are translated into a color of interest using a 256 * 256 * 256 color look-up table.

To obtain this look-up table, the user picks a pixel in the image corresponding to a color of interest c. Let us say that the RGB value for this pixel is $value_{rgb}$. Based on a sensitivity parameter adjusted by the user, a neighborhood of raw RGB values around $value_{rgb}$ in the color look-up table are then classified as c. After multiple iterations of entering representative pixels for each color of interest, the color table can be used to easily classify either pixels in $image_{kinect}$ or 3D points in $pointcloud_{field}$.

It should be noted that this method typically needs to be performed only once for a given environment and lighting condition, and does not need to be redone with a change in position of the Kinect. Additionally, since the CMOS camera in the Kinect is fairly consistent in terms of colors, this calibration requires a small number of iterations to complete, and takes only a few minutes.

4.3   Object Detection

The point cloud returned by the Kinect driver was found to be robust to large amounts of noise, i.e. no point in $pointcloud_{field}$ was too far away from its

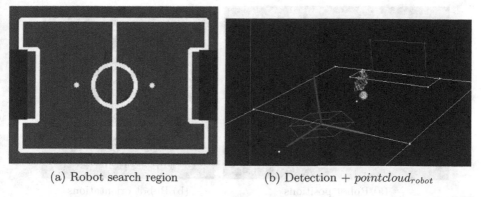

(a) Robot search region   (b) Detection + $pointcloud_{robot}$

Fig. 3. The robot search region in split up between the 2 Kinect sensors, and each system is run independently. Fig. 3b shows the detection given by our system.

true location. This property allows simple heuristics to be used for the object detection task, while at the same time allowing us to obtain the locations of the robots and the orange ball accurately.

Robot Detection: We define a region in ref_{field} in which we expect to find robots. Since $pointcloud_{field}$ is now available in this coordinate frame, we can filter out all points which could possibly belong to a robot. We construct a new $pointcloud_{robot}$ from these filtered points based on the following constraints:

- We select those points which lie in the region marked out by Fig. 3a. We avoid regions near the goal, since noisy points from the netting or goal post may be detected as a robot.
- We select only those points which are above $30cm$ in height.

Since multiple robots may be present in the region of interest, we apply Euclidean clustering on $pointcloud_{robot}$ in an attempt to obtain 1 cluster per robot. We provide a couple of parameters to the clustering algorithm. The first is a *tolerance* of $10cm$, which defines the maximum distance a point in the cluster may have with its neighbors. Secondly, we retain only those clusters with a minimum size of 200 points (min_size). We treat all clusters as possible robot detections, and the x and y coordinates of the cluster centers as the robot's position.

We perform some straightforward error checking for a couple of different cases. First, since it may be required for human referees to step onto the field during a game, we discard any cluster having points above $70cm$ (The nao has a maximum height of $58cm$). Second, if 2 robots on the field are too close, the Euclidean clustering algorithm may return a single cluster for both of them. To avoid a false reading, we also apply some loose size constraints designed to throw out such a cluster. All remaining clusters are reported as robots.

Once a robot has been detected, we attempt to identify its team based on the colored identification marker on its waist. Based on the output of the color

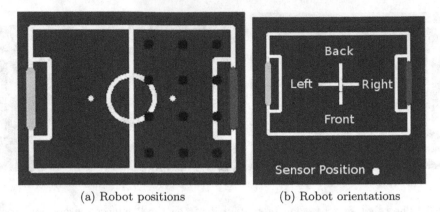

(a) Robot positions   (b) Robot orientations

Fig. 4. Fig. 4a shows the locations where the robot was placed. Fig. 4b shows the orientation for robot placement relative to the field, along with the position of the sensor.

classifier, we count the number of *blue* or *pink* pixels close to the cluster center. With this information we assign a team to each detected robot.

Ball Detection: Ball detection follows a fairly similar approach to robot detection. We construct a new $pointcloud_{ball}$ from $pointcloud_{field}$ based on the following constraints.

- We select points close to the field, and unlike robot detection do consider the regions around the goals.
- We impose a height constraint for selected points. They should lie fairly close to the ground plane, between a height of $15cm$ and $-15cm$.
- We ensure that only points that have been marked as *orange* are selected.

We perform Euclidean cluster extraction on $pointcloud_{ball}$ as well, using the parameters for *tolerance* of $10cm$ and *min_size* of 10. Finally a size restriction is placed on each cluster, based on the true size of the ball. At this point, if more than 1 cluster still remains, we discard all the readings to prevent any false detection of the ball.

Spanning the Entire Field: Up to this point, we have only discussed using information from a single Kinect, which can capture information for at most half the field if placed at a reasonable range. To cover the entire field, we split up the field into separate regions, place a Kinect sensor so that it captures one region, and run a separate instance of our system for each Kinect. The search space depicted in Fig. 3a is then split up according to the placement of these regions and provided to each instance of our system. In practice, we have been able to cover the SPL field with 2 Kinect sensors.

5   Experimental Results

In this section, we present experimental results that give error estimates on the information provided by this system. To calculate the error, we place the robot on some known locations on the field (Fig. 4a), and obtain the position returned by our system. The error is then measured by comparing this value against the known value of the location. To avoid being biased by a particular orientation of the robot, we obtain estimates by placing the robot in 4 different orientations at each position. In these orientations, the robot faces the $+y$, $-x$, $-y$ and $+x$ directions as depicted in Fig. 4b. We name these orientations based on their relative positions with respect to the sensor as *front*, *right*, *back* and *left*, respectively.

We placed the Kinect at a location on the side of the field so that it could sense half the field, and calculated the transformation (see Fig. 4b). Using the transformation, the location of the sensor was recorded at $(-155cm, 397cm, 210cm)$ in the global field coordinates. We then placed the robot in each of the combinations of positions and orientations one by one, and recorded the output of our system. This recorded reading was averaged across 20 frames captured by the system.

Table 1. Average error in the robot's position

Type	Average error (*cm*)
Robot (front)	10.19 ($\pm$5.86)
Robot (right)	10.90 ($\pm$5.87)
Robot (back)	9.72 ($\pm$4.55)
Robot (left)	10.87 ($\pm$6.97)
Robot (overall)	**10.41 ($\pm$5.85)**

The average error for different orientations of the robot is presented in Table 1. During the collection of this data, there were no false positives (defined as any reading more than $50cm$ away from truth point). The robot was detected in 95.64% of the frames recorded by our system. Detection did not occur in all cases because the Kinect driver occasionally returned an incomplete point cloud, in which the robot was not present.

While performing these tests we came across 2 factors that may add to the reported error of our system. First, while marking the locations of the points shown in Fig. 4a, we realized that due to deformities in the field it was impossible to mark all locations *exactly*. Although these deformities crept into our field through years of use, such errors were also recorded by many teams on the newly constructed venues at RoboCup 2010. As a result, in an attempt to provide realistic error estimates we do not correct for this error. The second error was introduced because of manual placement of robots on these markers, and is unavoidable. It is difficult to measure the amount of error added by these 2 factors, but we do not believe it to to be greater than 2–3cm.

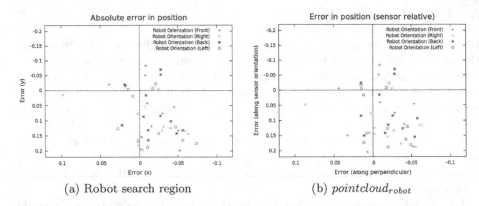

(a) Robot search region   (b) $pointcloud_{robot}$

Fig. 5. The robot search region is split up between the 2 Kinect sensors, and each system is run independently. Fig. 3b shows the detections given by our system.

We also measured the consistency of the calibration procedure for obtaining the transformation. For the given Kinect setup, we entered the set of landmarks 3 times and calculated the transformation separately each time. We then selected the 4 outermost landmarks available in the image, and transformed them in the ref_{field} coordinate frame using the 3 transformations available. We measured the standard deviation of these points across the 3 calibrations, and found it to be $2.02cm$. This is much less than the noise in our system, suggesting that the calibration procedure produces consistent results.

An interesting observation made from Table 1 was that the errors for the *front* and *back* orientations were marginally less than the other orientations. We believe this difference to be a result of a bias introduced by the position of the sensor. Since the points returned by the point cloud are from the surface it hits, these points automatically get moved towards the sensor from the center of the robot. In the *left* and *right* orientations the Kinect sensor faces the side of the robot, which is closer to the sensor in comparison of the front or back of the robot. To further investigate the movement of points towards the sensor, we plot the error of each recorded location with respect to its true location in Fig. 5a. The points are biased in the $+y$ direction, which is roughly the same direction in which the sensor lies. To estimate the amount of this shift, we again plot the error, but rotate the axis for each point in the direction of the sensor (Fig. 5b). The mean error along the direction of the camera is $8.15cm$. Knowledge of this value can help compensate for this systematic shift in points, further reducing the error of our system.

6   Discussion

In this paper, we have presented a new ground truth detection system for application in the Standard Platform League of RoboCup. This ground truth system has many desirable qualities such as a low cost, portability, and ease of use, with a fast calibration time. We have also presented an experimental analysis to indicate the system's accuracy. We aim to use this system for evaluating and improving self-localization algorithms on the robots, and expect to find it useful for other applications within the SPL environment as well.

A RoboCup team can easily apply this system to their own research by acquiring the Microsoft Kinect sensor, and using it along with the open source infrastructure that will be made available with this paper. We also hope that as this system finds greater application in the future, members from multiple teams and research labs may contribute towards its further development.

There are still a number of ways in which this system can be improved. Currently this system is unable to provide the orientation of the robot, which is necessary for a full description of the pose as required for self-localization algorithms. Although the solution of appending markers on the robots to determine orientation is possible, an alternative un-invasive method would be preferable. We believe that it may be possible to perform some further analysis of the data available from the Kinect to make some rough estimates of the robot's orientation. For instance, supervised learning techniques may enable shape modeling for estimating orientation based on features obtained from the image and the pointcloud. We hope to examine such possibilities in the future.

Another area that needs further examination is automating the transformation calculation. Although this procedure is fast and requires little input from the user, while conducting our experiments we have occasionally managed to dislodge the sensor from its position requiring recalibration. This is problematic should it happen while recording a competition game, as a few crucial minutes may pass while the sensor is being recalibrated. By taking the user out of the loop, we may be able to avoid such situations. We believe that automated registration [7] may be possible, by registering the point cloud from the Kinect against an artificially generated point cloud of the field. The only requirement is an approach that can scale to the size of the field and the amount of data provided by the Kinect.

Using automated registration has another advantage. Currently we run separate instances of our system for different regions of the field. Typically there is some overlap between the information produced by these sensors, which we currently discard. We do so because unless the sensors are perfectly aligned with respect to each other, slight differences could allow a single robot to be detected twice. By using a technique based on Iterative Closest Point [4], we may be able to register the point clouds against each other with high accuracy.

We believe that this system can also be extended beyond the SPL. Extending it to other RoboCup soccer leagues such as the humanoid soccer league is a fairly straightforward process. In addition, this system should be suitable for any indoor environment which is structured enough to provide a means of obtaining the transformation into a global reference frame. Given constraints about the robots to be detected, the system's parameters can be adapted to provide ground truth regarding their locations over time.

Acknowledgments. This work has taken place in the Learning Agents Research Group (LARG) at the Artificial Intelligence Laboratory, The University of Texas at Austin. LARG research is supported in part by grants from the National Science Foundation (IIS-0917122), ONR (N00014-09-1-0658), and the Federal Highway Administration (DTFH61-07-H-00030).

References

1. Point cloud library, http://www.ros.org/wiki/pcl
2. Arun, K., Huang, T., Blostein, S.: Least-squares fitting of two 3-D point sets. IEEE Transactions on Pattern Analysis and Machine Intelligence 9(5), 698–700 (1987)
3. Azevedo, J., Cruz, C., Cunha, J., Cunha, M., Lau, N., Martins, C., Neves, A., Pedrosa, E., Pereira, A., Teixeira, A., et al.: Cambada@ home 2011 team description paper (2011),
   http://www.ieeta.pt/atri/cambada/athome/
   docs/CAMBADAHome_TDP2011.pdf
4. Chetverikov, D., Svirko, D., Stepanov, D., Krsek, P.: The trimmed iterative closest point algorithm. Pattern Recognition 3, 30545 (2002)
5. Eberly, D.: Least squares fitting of data. Magic Software, Chapel Hill (2000)
6. Eggert, D., Lorusso, A., Fisher, R.: Estimating 3-D rigid body transformations: a comparison of four major algorithms. Machine Vision and Applications 9(5), 272–290 (1997)
7. Gelfand, N., Mitra, N., Guibas, L., Pottmann, H.: Robust global registration. In: Proceedings of the Third Eurographics Symposium on Geometry Processing, pp. 197–206. Eurographics Association (2005)
8. Jansen, S., Lier, C., Neculoiu, P., Nolte, A., Oost, C., Richthammer, V., Schimbinschi, F., Schutten, M., Shantia, A., Snijders, R., et al.: Borg-the robocup@ home team of the university of groningen team description paper,
   http://www.ai.rug.nl/crl/uploads/Site/BORG_TDP_2011.pdf
9. Niemüller, T., Ferrein, A., Eckel, G., Pirro, D., Podbregar, P., Kellner, T., Rath, C., Steinbauer, G.: Providing Ground-Truth Data for the Nao Robot Platform. In: Ruiz-del-Solar, J., Chown, E., Plöger, P.G. (eds.) RoboCup 2010. LNCS, vol. 6556, pp. 133–144. Springer, Heidelberg (2010)
10. Quigley, M., Gerkey, B., Conley, K., Faust, J., Foote, T., Leibs, J., Berger, E., Wheeler, R., Ng, A.: ROS: an open-source Robot Operating System. In: ICRA Workshop on Open Source Software (2009)
11. Röfer, T., Laue, T., Müller, J., Burchardt, A., Damrose, E., Fabisch, A., Feldpausch, F., Gillmann, K., Graf, C., de Haas, T.J., Härtl, A., Honsel, D., Kastner, P., Kastner, T., Markowsky, B., Mester, M., Peter, J., Riemann, O.J.L., Ring, M., Sauerland, W., Schreck, A., Sieverdingbeck, I., Wenk, F., Worch, J.H.: B-human team report and code release 2010 (2010),
   http://www.b-human.de/file_download/33/bhuman10_coderelease.pdf
12. Sridharan, M., Stone, P.: Real-time vision on a mobile robot platform. In: IEEE/RSJ International Conference on Intelligent Robots and Systems (August 2005)

13. Stulp, F., Gedikli, S., Beetz, M.: Evaluating multi-agent robotic systems using ground truth. In: Proceedings of the Workshop on Methods and Technology for Empirical Evaluation of Multi-agent Systems and Multi-Robot Teams, MTEE (2004)
14. Veltrop, T.: Humanoid robot navigation teleoperation using nao and kinect, http://www.robots-dreams.com/2011/01/ humanoid-robot-navigation-teleoperation-using-nao-and-kinect-video.html
15. Zickler, S., Laue, T., Birbach, O., Wongphati, M., Veloso, M.: SSL-Vision: The Shared Vision System for the RoboCup Small Size League. In: Baltes, J., Lagoudakis, M.G., Naruse, T., Ghidary, S.S. (eds.) RoboCup 2009. LNCS, vol. 5949, pp. 425–436. Springer, Heidelberg (2010)

Adaptivity on the Robot Brain Architecture Level Using Reinforcement Learning

Tijn van der Zant

University of Groningen
Artificial Intelligence dept.

Abstract. The design and implementation of a robot brain often requires making decisions between different modules with similar functionality. Many implementations and components are easy to create or can be downloaded, but it is difficult to assess which combination of modules work well and which does not. This paper discusses a reinforcement learning mechanism where the robot is choosing between the different components using empirical feedback and optimization criteria. With the interval estimation algorithm the robot deselects poorly functioning modules and retains only the best ones. A discount factor ensures that the robot keeps adapting to new circumstances in the real world. This allows the robot to adapt itself continuously on the architecture level and also allows working with large development teams creating several different implementations with similar functionalities to give the robot biggest chance to solve a task. The architecture is tested in the RoboCup@Home setting and can handle failure situations.

Keywords: adaptivity, behavior selection, RoboCup@Home, robot brain development, interval estimation algorithm, reinforcement learning.

1 Introduction

Just a decade ago the main problem with the development of robotic brain was that every group had to implement most algorithms themselves. Many development groups were already happy if the most important modules could be constructed before a deadline such as the RoboCup[6] competitions. In the mean time many groups realized this and groups with common goals started to share code using the Internet. RoboCup assisted in this effort by steering developments through the competition and stimulating teams to share their code. Slowly the problem of developing robotic systems has steered away from *only* building components towards the orchestration of components created by other groups into an overall architecture adding only a few functionalities. This allows the development teams to focus on their own expertises, hopefully also leading to the publication of their code. In the RoboCup@Home competition, for example, it is very common to see groups 'gluing' together all sorts of hardware and software components, focusing only on a few aspects. This is essential since the development of intelligent robots that can operate in the real world requires a huge

T. Röfer et al. (Eds.): RoboCup 2011, LNCS 7416, pp. 528–537, 2012.

effort. It is unlikely that any university-based development group can accomplish a good performance in the RoboCup@Home competitions if they were to build all the modules and components from scratch.

The problem nowadays is how one can choose between all the different implementations and components. The following list represents only a small portion of the possibilities but gives an impression of the overwhelming amount: Carmen[1] [7], Player/Stage[2] [4], MRPT[3], OpenCV with many functions[4], ROS with over 2000 modules[5], OpenSlam with dozens of SLAM algorithms[6], RoboRealm, a commercial robotic vision package[7].

This multitude of modules create the problem of deciding which module to use. How can one determine which SLAM or robot vision algorithm will work for the robot? Which behavior implementation is the best and in what situation is it the best? Which combination of modules are most likely to work correctly together? How can behaviors or other modules automatically be tweaked to improve performance? These are important questions that this article addresses. The main idea is that developers provide implementations for the robot to use, but that the robot decides which implementation to use in a specific situation. This decision process is steered through interaction with the environment and the users interacting with the robot.

A robot can be viewed as a mobile empirical data gathering device. It should therefor be possible that the robot is checking how well it is performing. This is the idea behind many learning algorithms but very few algorithms operate on the architecture level. Researchers are working on approaches to automatically develop components [2] stimulated by the concepts behind Cognitive Developmental Robotics [1]. Also there is focus on the construction of hierarchical behavior systems using reinforcement learning [9,3]. The point of view in this article is different in the sense that it assumes that there is a multitude of modules with similar functionalities and the robot has to choose between them in an automated manner. The components can be manually crafted, learned or a combination of the two. This is useful for the comparison of machine learning algorithms and for large development groups where many people, for example students, create similar behaviors for the robot.

For example, if the robot gets the command to get a certain item in a certain location, how can the robot decide which GoToLocation(location_1), SearchForObject(object), GraspObject(object) and for going back GoToLocation(location_2) to choose from? This is a common problem and can be solved using some strictness on the behavior implementation level. This will be explained in the next section.

---

[1] Carnegie Mellon Robot Navigation Toolkit (http://carmen.sourceforge.net/)

[2] The Player/Stage Project (http://playerstage.sourceforge.net/)

[3] The Mobile Robot Programming Toolkit
(http://babel.isa.uma.es/mrpt/index.php/Main_Page)

[4] The Open Computer Vision Library (http://sourceforge.net/projects/opencv/)

[5] http://www.ros.org/

[6] http://openslam.org/

[7] http://www.roborealm.com/

2   On-Line Adaptation

It happens that the developers arrive with the robot at a certain location and the algorithm that worked well before is not working so well anymore? This is a common tragedy that many development groups have experienced. It happens at places with dynamic environments, for example during competitions such as RoboCup@Home [11] where robots have to operate in an apartment and in a shopping mall. The specifications of the environment are poorly defined and the situations the robot finds itself in are various. Also, in the last years the General Purpose Service Robot test has been introduced where the robot is being put in situations it cannot solve. The mechanisms described in this article solve this problem by keeping track of how well the modules performed in previous, but similar, situations. Using the Interval Estimation [5,12] algorithm the robot can calculate the 95% confidence interval and decide when there is less than 2.5% probability to succeed and revert back to a default action.

This section will explain the adaptive architecture bottom up, starting with the adaptivity on the behavior level. First learning on the level of simple behaviors is explained, gradually building up towards more complex behaviors. Through generalization over the machine learning it is shown how non-behavioral components such as SLAM and robotic vision components can be taken into account.

It is important to keep in mind that we assume that the Markov property holds in the selection of the components. This means that we assume that it does not matter how the robotic system arrived in the state it is in at the moment and that by optimizing the actions in the current state the entire architecture is optimized. It is a research question whether this holds in the real world, but for the time being we explicitly assume it holds.

2.1   Interval Estimation Algorithm

Interval Estimation (IE)-learning [5] is a method for dealing with the exploration/exploitation dilemma in reinforcement learning [8]. The IE algorithm is a method that allows the robot to adapt itself on-line and real-time to the situation at hand. First the algorithm is explained. Then it is shown how the algorithm can be adapted to keep exploring in case of a changing environment, such as changes in user preferences, changes in the physical environment or changes in light conditions.

The problem is that a learning agent wants to select its current best behavior as much as possible (to exploit) and explore to find the optimal behavior at the same time. Since the agent cannot explore and exploit at the same time, there is a dilemma. IE has been successfully applied in [10] for model-based exploration in simulation. The IE-algorithm select the optimal behavior and can be extended to take state information into account.

For any state s the best action a^* has to be chosen. By trying the action, the environment gives feedback about the reward $r_a(t)$ for the action a selected at time t. The optimal action corresponds to:

$$a^* = \arg\max_a E(r_a|a)$$

Where E denotes the expectancy operator. However, we do not know the true expected reward, but only obtain samples around this average. From n samples we can construct sample averages Q_a:

$$Q_a = \frac{\sum_{i=1}^{n} r_a(i)}{n}$$

Purely selecting the action with highest Q_a value does not work well, exploration is necessary [8]. The IE algorithm stores an estimate of the expected reinforcement of an action and some information about how good the estimate is [5]. The IE algorithm estimates the confidence interval of the average of the data obtained when executing actions.

For small amounts of data, typically for a robot, the student T distribution is used to estimate the confidence interval. The upper bound of the confidence interval can be calculated using the following standard statistics, with n as the number of trials a behavior has been selected and $\sum_{i=1}^{n} r_a(i)$ the total reinforcement a behavior a has received. The upper bound of a $100(1-\alpha)\%$ confidence interval for the mean of the distributions is calculated by

$$nub(n, \sum_{i=1}^{n} r_a(i), \sum_{i=1}^{n} r_a(i)^2) = Q(a) + t_{\alpha/2}^{(n-1)} \frac{s}{\sqrt{n}}$$

with $Q(a) = \frac{\sum_{i=1}^{n} r_a(i)}{n}$ as the sample mean, and

$$s = \sqrt{\frac{n \sum_{i=1}^{n} r_a(i)^2 - (\sum_{i=1}^{n} r_a(i))^2}{n(n-1)}}$$

being the standard deviation. $t_{\alpha/2}^{(n-1)}$ is the Student's T function with n-1 degrees of freedom at the $\alpha/2$ confidence level. The IE algorithm selects the actions with the highest or lowest upper bound, depending on maximization or minimization, and is therefore optimistic about the results. If the spread of the data points is high, then the interval is large. As there are more data points collected through time, the interval shrinks because there is more information available.

Initially, the first action will be chosen at random and all actions will be tried out at least once (with no information the upper/lower bounds are: $0 \pm \infty$), but as the bounds tighten, the better the choices become. The IE algorithm balances exploration (a big interval in case of a high uncertainty) with exploitation. The upper bound can be high because it either has little information about the action or because the entire confidence interval is high and the action is good. The DeMoivre-Laplace theorem states that it will converge to its true underlying values in the limit. In practice IE sometimes gets stuck with a suboptimal behavior, because we do not have an unlimited amount of runs on real robots and robots may suffer from "unlucky" experiences.

On a real robot the distributions of the reinforcement values might not be distributed in a way that is favorable for the IE algorithm. While doing experiments on the robot the algorithm got stuck sometimes. To counteract this several solutions exist: one can choose to randomly drop a data point from a randomly chosen action, causing an increment in the size of the interval; or by introducing a discount factor to let actions further in the past have less influence on the calculation of the confidence interval.

With the behavior-based IE algorithm the programmer can give the robot several solutions to solve a particular problem. The developer actively puts the domain knowledge in the system as a hypothesis to be explored. Together with a evaluation criterion (such as time, amount of grasped objects or any other criterion observable for the robot) the robot is able to explore the different behavioral hypotheses and chooses the one that, at that moment in time, it has the highest confidence in that it is the best one. The robot explores and as it gains experiences it will explore less and choose the best action.

It might be the case that the environment changes and that the robot has to adapt itself. For this reason it is useful to have several different behavioral hypotheses ready to use. It is easy to adapt the algorithm to remain adaptive: diminish the influence of a data point related to the amount of time that has past. For example, the amount of data points between the present and when the data point was collected can be used with a discount factor. Both the data point and n can be multiplied with discount factor d using a for the amount of data points that have past, as in d^a. If d is small (< 0.8) then the algorithm will try new hypothesis rapidly, if d is large (> 0.97) then the algorithm is more careful with choosing a new hypothesis.

2.2 Applying the Interval Estimation Algorithm on the Behavioral Level

For every behavior we have several implementation, a postcondition and a criterion for the IE algorithm. This section demonstrates the behavior selection for our grabbing behavior. It is not important how the grabbing behavior is created. In this case two bachelor students created six different implementations using a visual programming environment, since this was the simplest and fastest method to bootstrap the robot. In another case, where the robot has to follow a person, we have two handcrafted implementations and several instantiations using reinforcement learning with different parameter settings.

We used a set of object to train the grabbing behavior on, being a box, a dessert cup (named 'dessertCup' during the tests), a ball, a bottle (named 'bottleP'), a regular cup, a sponge, some tape, a can and a pringles box. The behavior selection was bootstrapped by forcing the robot to use every behavior on every item six times. The results are shown in table 1 and in figure 1.

2.3 Applying the Interval Estimation Algorithm on the Architecture Level

The application of the IE algorithm can be fully automated if one takes into account that the reward function should be inspectable by the robot. This implies that every behavior that is going to be used by the behavior selection mechanism

Fig. 1. Figure 1: Setup of the Nao grabbing an object

Object:	Grab0	Grab1	Grab2	Grab3	Grab4	Grab5
box	100%	100%	100%	0%	0%	0%
dessertCup	83%	17%	17%	0%	0%	0%
ball	67%	50%	67%	0%	0%	0%
bottleP	100%	33%	50%	100%	17%	0%
cup	100%	83%	83%	17%	0%	0%
sponge	100%	17%	50%	17%	33%	0%
tape	17%	83%	100%	33%	0%	17%
can	100%	100%	100%	33%	0%	0%
pringles	100%	100%	100%	0%	0%	0%
AVERAGE	85%	65%	74%	22%	6%	2%

Fig. 2. Table 1: Average success of grabbing behaviors

of the IE algorithm has a postcondition which can be checked upon by the robot. The list below gives some examples that we apply in our RoboCup@Home robot and the postconditions that we use. If the postcondition is satisfied the data is stored, to be used the next time the robot has to execute the behavior. PC means postcondition, IE the criterion to optimize using the IE algorithm using the postcondition.

GoTO(location) Go to a location.
    PC: upon reaching the target location
    IE: minimize time
Search(object) Search strategy for an object near to the robot.
    PC: visual detection of the object
    IE: minimize time

Grab(object) Grab the object, object should be in gripper
  PC: object remains in fixed location in the camera image while the robot moves for at least ten seconds
  IE: maximize success rate (boolean)
Get(object, location) Satisfy   the   following: `GoTo(location)`, `Search(object)` and `Grab(object)`
  PC: `Grab(object) == True`
  IE: none, it is contained in the sub-behaviors

This describes our basic methodology to optimize on the simplest level, only optimizing behavior implementations. While the robot is running, executing behaviors, perhaps executing behaviors with some semi-random variables such as `GoTo( location 1, 1 elemOf listOfLocations)` it collects information about the success of the behaviors. Once the robot has executed a behavior a fair amount of times it can take into account more information. Our research indicates that if, on average, every implementation has to be tested at least six times, preferably a bit more. For example, the previous list of behaviors could then become the following:

GoTO(location, navigationAlgorithm) Go to a location using a specific navigation algorithm.
  PC: upon reaching the target location
  IE: minimize time
Search(object, imageProcessingAlgorithm) Search for an object near to the robot using a specific algorithm for vision.
  PC: visual detection of the object
  IE: minimize time
Grab(object, surface) Grab the object from a specific surface, such as table or floor
  PC: object remains in fixed location in the camera image while the robot moves for at least ten seconds
  IE: maximize success rate (boolean)
Get(object, location) Satisfy   the   following: `GoTo(location)`, `Search(object)` and `Grab(object)`
  PC: `Grab(object) == True`
  IE: none, it is contained in the sub-behaviors

In this second example it is clear that the implementation issues are hidden from the higher level behaviors, in this case `Get(object, location)`. This implies that higher level behavioral script can be generated using, for example, a reasoning mechanism while retaining the dynamics which is needed on the behavior implementation level. It is possible to specify (using natural language, for example) what, according to the user, is important for the robot to pay attention to. When the robot gets more information about the world, it can take more state information into account and react accordingly. We do not have a module at the moment to generate the natural language required for teaching the robot or for interpreting it, but expect to have this ready soon. This should allow us to create scripted behavior which is adaptive just by talking with the robot.

2.4   Automated Analysis of Module Combinations

The last step that can be performed is to test the combinations of states without being connected to a behavior. For example, for the selection of a vision algorithm and navigation method for a specific behavior it is possible to execute `behavior_1( someParameter, visionAlgorithm v, v elemOf listOfVisionAlgorithms , navigationAlgorithm n, n elemOf listOfNavigationAlgorithms)` many times. But if there are other behaviors, as in `behavior_N( anotherParameter, visionAlgorithm v, v elemOf listOfVisionAlgorithms , navigationAlgorithm n, n elemOf listOfNavigationAlgorithms)` then it is possible to calculate the upper and lower bounds of the 95% confidence interval of the combinations of `(v,n)`, `v elemOf listOfVisionAlgorithms`, `n elemOf listOfNavigationAlgorithms`. If there is a new behavior XYZ that uses `(v,n)` as a part of its parameters, it can immediately use the learned confidence interval of the combinations of `(v,n)`. This can be generalized to any amount of modules (parameters) that are being used by the behaviors. This effectively speeds up the the learning and results in the transfer of knowledge from previous behaviors to new behaviors.

2.5   Completely New Situations for the Robot

The robot will, after training in an environment, be utilized in a new location. This does not mean that the learned knowledge is worthless, but it does mean that the uncertainty about the learned knowledge should be increased. Using the IE algorithm this is very easy to do. If one replaces separate data points with the average of those data points, the algorithm will either make the same decisions if the learned knowledge is correct, or it will start exploring using as a start the modules with the highest chance of actually being the best. This manner of a 'soft reset' implies that it is possible to have training schools for robots, and that the knowledge is transferable to new situations and perhaps even to new body types.

For the competition of RoboCup@Home we will reset our learned knowledge and use a discount factor of 0.9. The discount factor is not used with absolute time, with with the amount of times that the modules were executed between that data point and the present. This assures a high rate of adaptability. Because the algorithm is not intended to find the best modules, but to deselect against poorly functioning modules and poorly functioning combinations of modules, we are certain that the robot will behave properly. It is not important to have the best performance, because due to changing circumstances that is a moving target and therefor unattainable. For this reason it is difficult to give results beyond individual behaviors and modules. It is clear how one can select the best image processing and object recognition algorithm for a defined data set. In the real world one cannot measure this because of the rapid changes that occur. A robot that can deal with these changes and one that does not uses poorly functioning modules is much more preferable than a robot that operates perfectly in a constrained but does not know how to adapt itself if the situation changes.

3   Conclusion

This article discusses the use of the Interval Estimation algorithm for the selection of modules on the architecture level of a robot brain. It demonstrates how the algorithm works and how it can be utilized to perform behavior selection, the selection of interpreting modules such as navigation or robot vision, how the learned knowledge can be transfered to new behaviors and how a trained robotic system can get a 'soft reset' to adjust itself fast to a new environment.

The problem with the proposed method is that it is difficult to measure the exact success of it. We think that the success should not be measured in selecting the best combination of modules, but in not selecting poor combinations. Also, in the real world, due to its dynamic properties, there are no optimal solutions. There are good solutions, not so good solutions and bad solutions. The Interval Algorithm calculates this sliding scale and does not need parameterization to decide whether it should explore possible solutions or exploit the gained knowledge.

Acknowledgments. The author would like to thank Christof Oost and Eric Jansen for their work on the testing the grabbing behaviors.

References

1. Asada, M., Hosoda, K., Kuniyoshi, Y., Ishiguro, H., Inui, T., Yoshikawa, Y., Ogino, M., Yoshida, C.: Cognitive developmental robotics: A survey. IEEE Transactions on Autonomous Mental Development 1(1), 12–34 (2009)
2. Bellas, F., Duro, R., Faina, A., Souto, D.: Multilevel darwinist brain (mdb): Artificial evolution in a cognitive architecture for real robots. IEEE Transactions on Autonomous Mental Development 2(4), 340–354 (2010)
3. van Dijk, S.G., Polani, D., Nehaniv, C.L.: Hierarchical Behaviours: Getting the Most Bang for Your Bit. In: Kampis, G., Karsai, I., Szathmáry, E. (eds.) ECAL 2009, Part II. LNCS, vol. 5778, pp. 342–349. Springer, Heidelberg (2011)
4. Gerkey, B.P., Vaughan, R.T., Howard, A.: The player/stage project: Tools for multi-robot and distributed sensor systems. In: Proceedings of the 11th International Conference on Advanced Robotics, pp. 317–323 (2003)
5. Kaelbling, L.P.: Learning in Embedded Systems. MIT Press (1993)
6. Kitano, H., Asada, M., Kuniyoshi, Y., Noda, I., Osawa, E., Matsubara, H.: RoboCup: A Challenge Problem for AI. AI Magazine 18(1), 73–85 (1997)
7. Montemerlo, M., Roy, N., Thrun, S.: Perspectives on standardization in mobile robot programming: The carnegie mellon navigation (carmen) toolkit. In: Proc. of the IEEE/RSJ Int. Conf. on Intelligent Robots and Systems (IROS), pp. 2436–2441 (2003)
8. Sutton, R., Barto, A.: Reinforcement Learning: an Introduction. MIT Press (1998)

9. Vigorito, C., Barto, A.: Intrinsically motivated hierarchical skill learning in structured environments. IEEE Transactions on Autonomous Mental Development 2(2), 132–143 (2010)
10. Wiering, M., Schmidhuber, J.: Efficient model-based exploration. In: Proceedings of the Sixth International Conference on Simulation of Adaptive Behavior: From Animals to Animats 6, pp. 223–228. MIT Press/Bradford Books (1998)
11. Wisspeintner, T., van der Zant, T., Iocchi, L., Schiffer, S.: RoboCupHome: Scientific Competition and Benchmarking for Domestic Service Robots. Interaction Studies 10(3), 392–426 (2009), http://dx.doi.org/10.1075/is.10.3.06wis
12. der Zant, T.V., Wiering, M., Eijck, J.V.: On-line robot learning using the interval estimation algorithm. In: Proceedings of the 7th European Workshop on Reinforcement Learning, pp. 11–12 (2005)

Spatial Correlation of Multi-sensor Features for Autonomous Victim Identification

Timothy Wiley, Matthew McGill, Adam Milstein,
Rudino Salleh, and Claude Sammut

The School of Computer Science and Engineering,
The University of New South Wales
UNSW Sydney NSW 2052
Australia
{timothyw,mmcgill,amilstein,rudinos,claude}@cse.unsw.edu.au

Abstract. Robots are used for Urban Search and Rescue to assist rescue workers. To enable the robots to find victims, they are equipped with various sensors including thermal, video and depth time-of-flight cameras, and laser range-finders. We present a method to enable a robot to perform this task autonomously. Thermal features are detected using a dynamic temperature threshold. By aligning the thermal and time-of-flight camera images, the thermal features are projected into 3D space. Edge detection on laser data is used to locate holes within the environment, which are then spatially correlated to the thermal features. A decision tree uses the correlated features to direct the autonomous policy to explore the environment and locate victims. The method was evaluated in the 2010 RoboCup Rescue Real Robots Competition.

1 Introduction

In urban search and rescue, robots can be deployed to assist rescue workers with the task of exploring damaged buildings and finding victims. To achieve this, robots must be equipped with a variety of sensors, since there is a limit to the information a single sensor can provide. However, data extracted from a variety of sensors must be correlated in order to know more about the environment.

Robots used for urban search and rescue can be teleoperated or autonomous. The latter is the focus of this research. The goal is for the autonomous robot to explore an unknown environment and locate the victims of a disaster.

Several approaches have been developed to solve the problem of autonomously locating victims. Early methods by Birk *et al.* [1] and Pellenz *et al.* [9] used a single a low-cost thermal camera to find victims. If, in any thermal image, a predefined number of pixels is above a manually set threshold, it is assumed that a victim is present. Using a similar approach, both in testing and during the 2009 RoboCup Rescue Thailand Open, we found this method to have a number of problems. The manually set threshold required frequent adjustment as the environment warmed during the day and cooled at night. Spectators,

T. Röfer et al. (Eds.): RoboCup 2011, LNCS 7416, pp. 538–549, 2012.
© Springer-Verlag Berlin Heidelberg 2012

other robots and even warm pockets in the environment can also be detected as victims.

Even when using a temperature range with minimum and maximum thresholds, Markov & Birk [4] still noted false positives. They proposed matching shapes from thermal signatures to projected 2D renders of humans body parts. However, victims that are hidden behind obstacles can produce thermal shapes that do not appear like body parts and are indistinct from false positives. Relying on a single sensor is not sufficient.

To allow features from multiple sensors to be used in conjunction, Nourbakhsh *et al.* [7] proposed the use of a probability measure. The probability indicates the likelihood of a victim being present at a given location. A confidence rating is assigned to each sensor and detection method which produces the probability.

Meyer *et al.* [5] extended this approach by applying an Extended Kalman Filter to the probability measure. This gives temporal information where not all features are observed simultaneously. They also introduce a second Kalman filter to determine the most likely locations of victims.

Both Nourbakhsh and Meyer experimentally showed good performance for these methods. However, they assume features seen in proximity relate to the same victim, even if they are spatially separate. The Extended Kalman Filters are also unintuitive for humans operators. It is not obvious why the robot determines that a given location does or does not contain a victim.

In this paper, we present a set of methods to detect features of a victim and spatially correlate those features. We also present an autonomous policy that performs reliable victim identification. First, we describe our method for thermal blob detection using a dynamically calculated threshold. By aligning the thermal camera with a depth time-of-flight camera, we project the thermal blobs into 3D space (Section 2). We then use edge detection in individual laser scans to perform fast hole detection in 3D point clouds (Section 3). The thermal blobs and holes are spatially correlated (Section 4), and finally, the correlated features are processed and then used by an autonomous policy that controls the robot. (Section 5).

Our methods were tested and evaluated during the 2010 RoboCup Rescue Real Robots competition, where we were awarded the Best-In-Class Autonomy prize. The competition provides a standardised environment to test robotic hardware and software.

2   Camera Sensing

To perform the thermal blob feature detection and projection of the thermal features into 3D space, two cameras are used - a thermal camera capable of reading temperatures in the range of 10-40°C, and a SwissRanger time-of-flight camera that produces an image where each pixel is a measurement of the distance an object in the environment is away from the camera. The feature detection algorithms are applied to the thermal camera images, whereas the range images are used for the 3D projection.

2.1 Thermal Blob Detection with a Dynamic Hot-Pixel Threshold

The method used by Pellenz *et al.* [9] forms the basis of our thermal blob detection, where it is assumed that victims are distinctly hotter compared to the surrounding environment. To overcome issues with a static threshold, we use a dynamic threshold to determine hot-pixels (pixels above the threshold). Hot-pixels are grouped together to form thermal blobs.

The hot-pixel threshold is updated frame by frame as follows.

Let the hot-pixel threshold be hst. There are two cases: either a victim is present in the image, or a victim is not present.

Initially, assume that a victim is present in the image. By this assumption, the victim will produce in the image a region of significantly hotter pixels. A potential value, k, for hst is one that binarizes the image such that only the hot-pixels are above k. Otsu [8] proposed a method to optimally calculate this value, k^*. However, simply setting $hst = k^*$ causes erratic fluctuations. Instead hst is adjusted towards k^*, but only if a victim is present in the image.

The value of k^* is used to determine whether or not a victim is present in the thermal image. If an image contains no victim (or is completely filled by a victim), Otsu's method selects k^* such that a large proportion of the image is above the threshold. Thus, hst does not change, as k^* in these situations is meaningless.

Initially hst is set to the maximum possible value for any pixel in the thermal image. This ensures that the background will not be detected as hot-pixels. When a victim is first encountered, then hst will ramp down to a sufficient value.

The complete algorithm is:

1. Initialise: hst to the maximum pixel value.
2. For the current thermal image, calculate k^*.
3. If the bounding box of the adjoining pixels above k^* (see section 2.2) covers less than cov_{max} pixels, ramp hst by:

$$hst = \eta k^* + (1 - \eta)hst \tag{1}$$

   where η is the ramping rate.
4. Repeat from step 2.

Typically η is set to 0.1 and cov_{max} to 90%. The value of η strikes a balance between sufficiently fast ramping and over ramping. Over ramping can result in the background being detected as a hot-pixels. The value of cov_{max} assumes that a thermal image produced by the background environment is reasonably constant. Therefore, any value of k^* will make the entire image appear to be covered by hot-pixels.

2.2 Projection into Camera Relative Space

Given hst, adjoining groups of pixels above it are grouped together using the '4-connectivity component labelling' technique [10] to form a thermal blob.

The blobs are projected into 3D space in the frame of reference of the thermal camera, so that they can be spatially correlated with other features. This is achieved by calculating the distance the blob is from the camera. The distance is calculated by mapping pixels from the SwissRanger image to the thermal camera image as follows.

A SwissRanger pixel (I_x, I_y) is projected into the 3D space of the SwissRanger using the small angle approximation. It is assumed the x and y Fields of View (FOV) of the camera are evenly distributed across each pixel of the image, as depicted in Fig. 1.

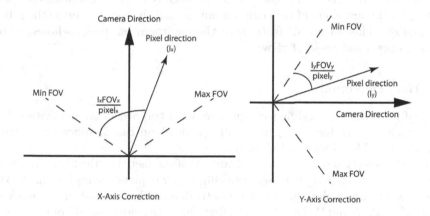

Fig. 1. x and y axis camera correlation

Thus the spherical co-ordinates of the pixel is

$$(d, \theta, \varphi) = \left(d, \frac{\pi}{2} - I_x \frac{FOV_x}{pixels_x} + \frac{FOV_x}{2}, \frac{\pi}{2} + I_y \frac{FOV_y}{pixels_y} - \frac{FOV_y}{2} \right) \quad (2)$$

where is d is the distance value of the pixel (I_x, I_y), $FOV_{x/y}$ is the x and y Field of View of the SwissRanger in radians, and $pixels_{x/y}$ is the pixel dimension of the SwissRanger image. In cartesian co-ordinates the point is

$$(x, y, z) = (d \cos (\theta) \sin (\varphi), d \sin (\theta) \sin (\varphi), d \cos (\varphi)) \quad (3)$$

Given the physical displacement between the SwissRanger and Thermal Camera (T_x, T_y, T_z), the point can be converted into the frame of reference of the thermal camera, where primed values are the corresponding thermal camera properties.

$$(x', y', z') = (x - T_x, y - T_y, z - T_z) \quad (4)$$

Applying the inverse of 3 and 2 gives the corresponding thermal pixel for each mapped distance value.

Finally, by taking an average of the mapped distance values that fall within the blob, the blob's centre can be projected into 3D space (using 2 and 3).

3 Laser Sensing

Robots used in rescue applications are often equipped with a laser range-finder. We use the laser to track the position of the robot in the environment and to produce a 3D point cloud of the environment.

3.1 3D Environment Scanning

Milstein *et al.* [6] proposed a method for 3D position tracking using an occupancy voxel metric. Their method allows us to periodically tilt the mounted laser and produce a 3D point cloud of the environment around the robot, even as the robot is in motion. This allows 3D features of the environment, such as holes, to be detected over a wide field of view.

3.2 Hole Detection

In a real life disaster, surviving victims are often trapped in small cavities. Locating such cavities (or holes) in the 3D point cloud can significantly narrow down the possible locations of victims.

Our hole detection method takes advantage of the fact that the 3D point cloud is generated from aligning 2D laser scans [6]. Fig. 2 depicts a sample scan. Methods for generalised point cloud feature detection, such as found in the work of Gumhold *et al.* [2] and Weber *et al.* [11], first generate nearest neighbours for each point in the point cloud before feature detection is performed. By performing hole detection on each laser scan individually, the nearest neighbour information is obtained for free. Combining data from multiple scans is performed later.

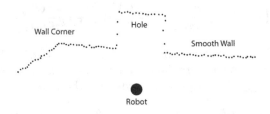

Fig. 2. Sample Laser Scan with typical environment features

Holes are located by initially finding edges (points of sharp discontinuity) in the laser scan. A laser scan is represented as a 1D array, D, of distance values. A value at index i is marked as an edge if

$$\|D_{i-ws} - D_{i+ws}\| \geq \text{threshold} \tag{5}$$

where ws is a window around the value D_i. Typically ws is set to 2. The window compensates for the noisy laser readings typically observed between the front and back walls of the hole, as represented in Fig 2. Each edge is additionally marked as 'positive' or 'negative' based on the sign of the difference in 5.

A hole is 'detected' where a sequence of negative edges is followed (not necessarily immediately) by a sequence of positive edges. By using our 3D position tracker, each 'positive' and 'negative' edge is placed in 3D space. Holes can also be tracked by their 'centre', which is the average of the 'negative' and 'positive' 3D points.

Finally, detected hole centres from all the laser scans are combined. If two laser scans pass through the same hole, their respective centres (in 3D space) will be close. Therefore, to combine the hole centres from multiple scans, we use hierarchal clustering, [3], on the hole centres. The collapsing metric for the clustering is the euclidean distance between hole centres. The maximum value for collapsing clusters is typically 0.2m.

3.3   Hole Detection Post-filtering

The basic hole detection technique will also detect as holes a number of other undesired elements, such as passageways. Post filtering is applied to these false positives to remove any 'holes' that do not meet a given set of properties. Such properties can include the dimension between edges, a hole's height, or location. Useful filtering metrics are described further in section 6.2.

4   Correlating Camera and Laser Data Features

Before correlation can be performed, the frames of reference (such as camera/laser space) of each feature must be aligned. Since the location of sensors on a robot are known, it is straightforward to define kinematic chains, specific to each robot, to perform the alignment.

In conducting feature correlation, care must be taken. Sensor data used by each detection method is captured independently. It is possible that the data are not captured simultaneously. Thus, if the robot has moved some distance between the times at which each sensor captured its data, any correlation is meaningless. Additionally, there may also be minor spatial errors between each sensor.

Two features are said to correlate (ie. relate to the same physical entity) if they have been detected within a small time frame and occur within a small radius of each other.

5   Autonomous Policy for Victim Identification

The autonomous victim identification policy is implemented as a decision tree that considers the blob and hole features before determining what physical action the robot should perform. The decision tree is periodically evaluated, as new features from the environment are detected.

The complete decision tree that calculates and refines various physical settings of the robot (such as motor positions) is described in full detail in [12]. Here, we provide the general rules that govern the decision tree.

The default decision is to explore the environment to find signs of victims. The exploration strategy is unimportant to the victim identification, provided that it covers previously unexplored areas. Once features have been detected, the following choices are made:

1. If both a blob and hole correlate, a victim is assumed to be present at that location. The policy then directs the robot to navigate closer to this location. Once close enough, the human operator is notified that a victim has been located. After the operator has reviewed the sensor data, the policy resumes exploration.
2. If a hole exists uncorrelated to a blob, and is close to the robot (within 2m), the policy centres the thermal camera on the location. If a blob is subsequently found, this can then be correlated to the hole and processed by the next iteration of the decision tree. If no blob is found, this location is discarded. If the hole is not close to the robot, there is no point in the robot trying to find a victim, as the robot must be close to the victim for the operator to verify a victim exists. Such a hole is temporarily ignored until the exploration strategy takes the robot closer to the hole.
3. If a blob exists uncorrelated to a hole, this may indicate a victim, but only if the victim is on the ground. It is possible for victims to be in open corridors, but injured or trapped under rubble, on the ground and therefore not contained within a hole. The policy again navigates the robot towards the victim and notifies the human operator. However, if the blob is above the ground it cannot indicate a victim, and is likely to have been caused by a hot object or sunlight.

Care must also be taken to ensure previous victim locations and locations that were examined but had no victim are not re-examined. A history of the spatial locations of located victims and previously examined locations is maintained [12]. Any features that correlate to these positions are ignored.

6   Results

We evaluated the performance of the dynamic hot-pixel threshold, the hole detection and overall autonomous policy.

The robotic platform used for testing is shown in Fig. 3. The robot has a FLIR ThermoVision A10 Thermal Camera, a SwissRanger SR3000, and an autoleved Hokuyo UTM 30LX laser range-finder that is able to be tilted.

6.1   Dynamic Hot-Pixel Threshold

The ability of the dynamically calculated threshold for hot-pixels in the thermal image was tested using both human subjects and hot objects, such as bottles

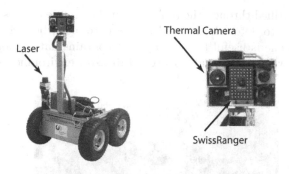

Fig. 3. Robotic Platform

filled with hot water. Fig. 4 shows a low threshold ramping upwards over time to exclude the background. Any pixels below the threshold are set to black. Fig. 5 shows the threshold ramping down, after a bottle with hot water is removed leaving a second bottle containing warm water. In each case, the threshold took 2-3 seconds to ramp to its final value.

Fig. 4. Hot-pixel threshold increasing from left to right with human subject

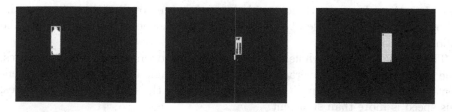

Fig. 5. Hot-pixel threshold decreasing from left to right after the one bottle is removed

6.2   Hole Detection

We tested the hole detection using situations typical of RoboCup Rescue simulation environments. To simulate cavities, victims were placed behind walls or in boxes with holes drilled through them. A standard test case is the scene depicted in Fig 6. The hole to detect is situated in the cardboard box in the middle of the scene. This scene is useful for testing as it contains flat sections, walls and two open passageways, all typical features of disaster environments.

Fig. 6. Sample scene containing a hole

To obtain enough data for reliable hole detection, laser scans over a 5 second period were collected. That is approximately 100 scans collected at a rate of 20 scans per second[1]. Fig. 7 shows a top-down view of the same scene from the data collected by the laser scans.

Fig. 8 shows the results of the hole detection. Negative edges are marked in red, positive edges in yellow, and hole centres in white.

As noted in section 3.3, post-filtering methods are required. For these simulations, we post-filtered the holes by:

- The difference threshold for equation 5 is 0.5m, and a minimum threshold of 0.2m is also used
- Holes must be no wider than 0.5m.
- At least 2 holes centres must cluster together to define a hole.

The algorithm also correctly locates holes if they are not directly in front of the robot. Fig. 9 shows a hole at a height of about 1m being detected where the robot had a side-on view.

The speed at which the hole detection runs is important for real-time robotics. The algorithm takes, on average, 2 seconds for a single iteration. This is dependent upon the number of positive/negative edges in the scene. For our purposes, this speed is more than sufficient.

6.3   Application with Victim Identification

The feature detection and robot behaviours were evaluated in the 2010 RoboCup Rescue Real Robots competition. In various rounds, the robot was placed within

---

[1] The Hokuyo laser produces 30 scans per second. The reduced rate is the rate at which our position tracker can process and align scans

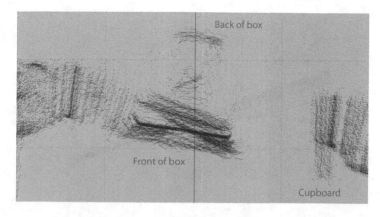

Fig. 7. Top view of laser data of the sample scene

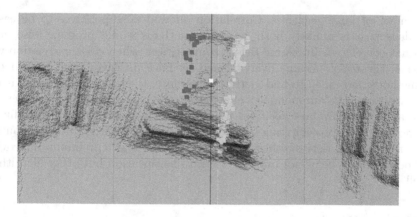

Fig. 8. Top view of holes computed by clustering

an unknown simulated disaster environment that contained 3 to 4 simulated victims. A time frame (between 20-25 mins) was given for the robot to find the victims.

The autonomous policy correctly identified seven out of the eight victims that came within the visual range of the robot. The policy also incorrectly identified one location as containing a victim when no victim was present. This performance was the best among all autonomous robots in the competition, and significantly contributed to our winning the Best-In-Class Autonomy award.

These results demonstrate the effectiveness of our method. Only two feature detection methods are used, and by spatially correlating features the robot was able to solve a complex and challenging task. However, as experienced in past competitions and testing, without correlating these features this would not have been possible.

Both the blob and hole detection produced false positives. In fact, the parameters of both algorithms were tuned so that this occurred, as this reduces

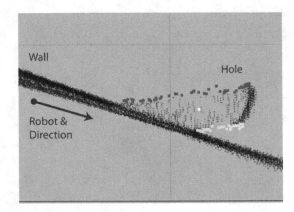

Fig. 9. Hole detected when robot is on an angle to the hole

the number of false negatives. Correlation allowed false positives to be easily discarded, as it is unlikely both detection methods would produce false positives that correlated. In testing before competition, when close correlation was not used, the robot became overwhelmed by false positives from both detection methods and failed to find victims. Therefore, the spatial correlation is the definitive reason the autonomous behaviours could successfully identify victims.

The ability to allow false positives additionally improved efficiency by reducing the amount of pre/post-filtering required. On a platform with limited computing power, running a variety of other systems, CPU time is at a premium. Improving CPU performance helps not only the victim identification but other algorithms as well.

7   Future Work

While effective, the feature detection methods presented are specifically tuned for the layout of the RoboCup Rescue competition. Performing multi-source correlation from a variety of additional camera or other types of sensors will allow the behaviours to be more flexible to variations in victims.

Likewise, the hole detection was very successful for identifying a limited range of hole types. It is possible to vary the types of post-processing filters in order to identify other features such as doorways, wall corners and rounded objects such as barrels.

8   Conclusion

The blob and hole detection algorithms presented in this paper provide a reliable and efficient method of finding features of victims in a disaster environment. Further, by projecting these features into 3D space they can be correlated to provide additional information about the object. The correlation enabled an autonomous robotic behaviour policy to use a limited number of features to reliably locate victims.

References

1. Birk, A., Kenn, H., Carpin, S., Pfingsthorn, M.: Toward Autonomous Rescue Robots. In: Proceedings of the First International Workshop on Synthetic Simulation and Robotics to Mitigate Earthquake Disasters (2003)
2. Gumhold, S., Wang, X., MacLeod, R.: Feature extraction from point clouds. In: Proceedings of the 10th International Meshing Roundtable, vol. 2001, pp. 293–305 (2001)
3. Johnson, S.: Hierarchical clustering schemes. Psychometrika 32(3), 241–254 (1967)
4. Markov, S., Birk, A.: Detecting Humans in 2D Thermal Images by Generating 3D Models. In: Hertzberg, J., Beetz, M., Englert, R. (eds.) KI 2007. LNCS (LNAI), vol. 4667, pp. 293–307. Springer, Heidelberg (2007)
5. Meyer, J., Schnitzspan, P., Kohlbrecher, S., Petersen, K., Andriluka, M., Schwahn, O., Klingauf, U., Roth, S., Schiele, B., von Stryk, O.: A Semantic world Model for Urban Search and Rescue Based on Heterogeneous Sensors. In: Ruiz-del-Solar, J. (ed.) RoboCup 2010. LNCS (LNAI), vol. 6556, pp. 180–193. Springer, Heidelberg (2010)
6. Milstein, A., McGill, M., Wiley, T., Salleh, R., Sammut, C.: Occupancy Vozel Metric Based Iterative Closest Point for Position Tracking in 3D Environments. In: 2011 IEEE International Conference on Robotics and Automation, ICRA (May 2011)
7. Nourbakhsh, I.R., Sycara, K., Koes, M., Yong, M., Lewis, M., Burion, S.: Human-robot teaming for Search and Rescue. IEEE Pervasive Computing 4(1), 72–77 (2005)
8. Otsu, N.: A Threshold Selection Method from Gray-Level Histograms. IEEE Transactions on Systems Science and Cybernetics SMC-9(1), 62–66 (1979)
9. Pellenz, J., Gossow, D., Paulus, D.: Robbie: A Fully Autonomous Robot for RoboCupRescue. Advanced Robotics 23(9), 1159–1177 (2009)
10. Shapiro, L., Stockman, G.: Computer vision, ch. 3. Prentice Hall (2001)
11. Weber, C., Hamann, S., Hagen, H.: Sharp Feature Detection in Point Clouds. In: IEEE International Conference on Shape Modeling and Applications, SMI (2010)
12. Wiley, T.: Autonomous Victim Identification, Honours Thesis, The University of New South Wales (August 2010)

Fast Object Detection by Regression in Robot Soccer

Susana Brandão*, Manuela Veloso, and João Paulo Costeira

Carnegie Mellon University - ECE department, USA
Carnegie Mellon University - CS department, USA
Instituto Superior Técnico - ECE department, Portugal
sbrandao@ece.cmu.edu, mmv@cs.cmu.edu, jpc@isr.ist.utl.pt

Abstract. Visual object detection in robot soccer is fundamental so the robots can act to accomplish their tasks. Current techniques rely on manually highly polished definitions of object models, that lead to accurate detection, but are quite often computationally inefficient. In this work, we contribute an efficient object detection through regression (ODR) method based on offline training. We build upon the observation that objects in robot soccer are of a well defined color and investigate an offline learning approach to model such objects. ODR consists of two main phases: (i) offline training, where the objects are automatically labeled offline by existing techniques, and (ii) online detection, where a given image is efficiently processed in real-time with the learned models. For each image, ODR determines whether the object is present and provides its position if so. We show comparing results with current techniques for precision and computational load.

Keywords: Real-time Perception, Computer Vision.

1 Introduction

In robot soccer, vision plays a crucial role on localization and actuation since both task rely on images to provide ground truth for landmarks and objects localization. One of the biggest challenges faced by robot soccer teams is to provide the robot with adequate models for each class of objects in the field. The current paper presents a highly efficient way of recognizing objects in this environment.

We address the problem of object detection in the RoboCup Standard Platform league that uses the humanoid NAO robots (www.aldebaran.com). In this league, the robots have access to images of high-resolution at a fast acquisition rate and have to process them without totally consuming the limited on-board robot computational resources, which are also needed for all other non-vision task functions. In this context, vision algorithms need to be not only highly reliable, but also computationally efficient and easy to extent to all the objects in the field.

* This work was partially supported by the Carnegie Mellon/Portugal Program managed by ICTI from Fundação para a Ciência e Tecnologia. Susana Brandão holds a fellowship from the Carnegie Mellon/Portugal Program, and she is with the Institute for Systems and Robotics (ISR), Instituto Superior Técnico (IST), Lisbon, Portugal, and with the Department of Electrical and Computer Engineering, Carnegie Mellon University, Pittsburgh, PA. The views and conclusions of this work are those of the authors only.

T. Röfer et al. (Eds.): RoboCup 2011, LNCS 7416, pp. 550–561, 2012.
© Springer-Verlag Berlin Heidelberg 2012

Two widely used approaches for object detection in this domain are: (i) a scan-line based algorithm [8] that effectively reduces the size of the image to samples along vertically spaced-apart scanned lines, but relies on the manual definition of elaborated models of each object; and (ii) a run-length encoding region-based algorithm, CMVision [2], that effectively identifies colored blobs with objects, but is computationally expensive in large images. Other successful more focused approaches include the use of neural networks [6], circular Hough Transforms [4], and circle fitting [5].

Scan-line is a very thorough algorithm that relies mostly on human modeling of the several elements in the field. It creates color segments based on the scanning of just a few columns in the image. To compensate the information lost in the sampling, the algorithm uses human imposed priors on what the segments should be in the robot soccer environment. By requiring the use of a reduced set of lines, the algorithm is very fast. However, the modeling of each object in the field is quite time-consuming and the algorithm is not easily extendable to new objects.

CMVision relies on color segmentation to create blobs that will then be identified as objects. However, since blobs are created based on 4-connectedness, it requires color thresholding of almost all the pixels in the image. Afterward, blobs have to be sorted by color and size, and finally objects are detected based on how well the largest blobs of their respective color fit to a given model, which is again imposed by humans. All this process, albeit quite accurate and easier to extend to new objects than the previous, is computationally expensive.

In this work, we introduce a new object detection approach, ODR, (for Object Detection by Regression), that relies on the offline creation of statistical models for the relation between a whole image and the object position in that image. The use of statistical models enables ODR to use a sampled version of the image, which greatly reduces the online computational cost of the algorithm. Though more complex models could have been used, a further reduction on the computational load is obtained by using linear relations between object positions and images. State of the art offline detection algorithms, e.g. [3], use local statistics of the image to detect an object and thus require exhaustive search at all possible locations and scales, which is quite time consuming. By using the whole image, together with priors provided by the environment such as object colors, we are considering a *global* statistical representation of the image that avoids this exhaustive search. Furthermore, ODR is easy to extend since it relies on the also easy to extend CMVision algorithm to provide the object positions during the offline learning stage.

The paper is organized as follows. Section 2 presents the overall ODR approach with the description of the wide object detection problem to be solved. Section 3 presents the pre-processing required by ODR. Section 4 and Section 5 describe the ODR offline and online algorithms respectively. Section 6 shows experimental results that demonstrate the accuracy and effectiveness in object detection with a rich set of real images taken with a NAO robot, including truncated and occluded objects in noisy situations. We review the contributions of the work and conclude in Section 7.

2 Object Detection by Regression

ODR detects an object position in an image with a robust and computationally efficient algorithm. The algorithm leverages in the repeatability of the RoboCup environment to create a statistical model of the relation between an image and the position of a given object in that image. The statistical model can be learned offline and provides fast online detection by allowing image sampling. Furthermore, it can be extended to other objects without requiring extensive human modeling. As an output to the online stage, the algorithm returns an object position and a confidence on the presence of the object.

To relate images to object positions we use a linear model represented by a matrix W and affine vector b_0. The model is learned offline using a set of training images, represented as a single matrix O. The images contain the object at different positions and this position is know and represented also as a single matrix P. To avoid over fitting to the training data and to provide filtering of noise, the training set is reduced by means of a principal component analysis. The linear model learned relates not the object positions P and the image training set matrix, O, but the object position and the matrix O_r that corresponds to the projection of O on the training set principal components, V. However, since the projection into principal components is a linear operator, the relation between the object position and the image will still be linear and thus efficient to use.

Clearly the relation between images and objects position is more complex than just linear. Typical images are cluttered, objects have different sizes, different colors and suffer from occlusions and pose changes. To allow such a simplification, we consider the existence of a pre-processing stage before ODR. There are no constrains to the type of pre-processing as long as it returns a version of the original image where each pixel is now assigned one of two values: 1 if the pixel is thought to belong to the object, 0 if not. To the set of pixels labeled as 1, we refer to as object hypothesis.

If the object hypotheses was errorless, e.i., all the pixel labeled as 1 belonged to the object and all the object pixels were assigned a value of 1, the object position could be recovered by simple centroid estimation. The centroid estimation is a linear operation over a normalized version of the pre-processed image such that the sum of the value of all the pixels in this image version is 1. This normalization allows to compute the centroid of images with objects with different sizes using the same set of coefficients. Since there is currently no algorithm capable of determining the object hypothesis without error, the pre-processed images will contain noise resulting from robot motion and from illumination.

The objective of ODR is to determine in a robust way the object position in the image, in spite of these errors. To decrease the impact of noise in the detection, we filter the images by projecting them in the principal components of a set of training images. These training images are representative of the set of all possible images containing the object at different positions and are the same as those used to create the statistical model. We also note that ODR could still return a position in images where an object is not present, but there is noise. Interestingly, we then include in ODR a method to estimate a belief in the results of the detection performed using the linear model, inspired by similar work in face classification [7]. During its online processing, ODR projects a given test image into the principal components computed using the training set. It then estimates the belief based on the distance between the image and its projection into the linear subspace generated by the principal components. It detects false positives by thresholding its belief.

ODR results are evaluated using real data acquired by a NAO robot while approaching a ball and a second robot. The data contains all the complexities to be expected from a robot moving. Objects are distorted by blur or/and are occluded and there are images that just contains noise. We present ODR performance using three different metrics: (i) the capacity to detect objects that are in the image, (ii) the capacity to classify an image on whether the object is present or not (iii) the time efficiency of the algorithm.

In Figure 1 we present a diagram for the whole ODR algorithm where we identify each task to be performed in the online and offline stage. Each stage follows three different steps that are closely related across stages. The first step consists in normalizing the images. The normalization accounts for changes in size and fits all images into the same distribution, which is a requirement for the principal component analysis. The second step consists in the projection into the principal components, which are learned in the offline stage and then used for belief estimation in the online stage. The third step differs more significantly between the offline and online stages. In the offline stage, we learn the linear relation between positions and image. In the online stage we use this relation to compute the object position.

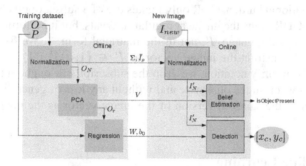

Fig. 1. Online and offline ODR

In the following sections we address the type of pre-processing required by ODR and explain in-depth each of the tasks that compose the offline and online stage.

3 Pre-processing

In the general case of object recognition, the pre-processing required by ODR can be the result of a segmentation or color threshold algorithm. In the case of humanoid robot soccer, color threshold is particularly appealing, since objects in the field are well defined by their color. Color thresholding is also a pre-processing stage in [2] and [8]. When the robot needs to detect the ball in those images, e.g., for kicking it into the goal, the object hypothesis for the ball in each image would be the set of all orange pixels. When the robot wants to identify other robots from a distance, the object hypothesis would be the set of white pixels. In Figure 2, we present an example of a thresholded image and the resulting object hypothesis for the ball and the robot.

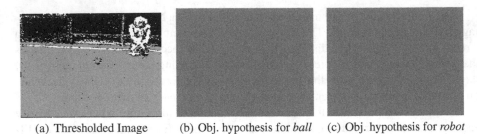

(a) Thresholded Image   (b) Obj. hypothesis for *ball*   (c) Obj. hypothesis for *robot*

Fig. 2. Thresholded images and object hypothesis for *robot* and *ball*

After such pre-processing, an $N \times M$ image I with pixels indexed as $\{i, j\}$, can be represented as a binary vector in a NM space as $I' \in \{1, 0\}^{NM} : I'_k = 1$ if the pixel with indices $\{i, j : k = (i-1)M + j, 0 < j \le M, 0 < i \le N\}$ belongs to the object hypothesis and $I'_k = 0$ if not.

However, the images retrieved by the NAOs humanoid robots have a high-resolution that makes the color threshold of all the pixels a computationally heavy task. To overcome the computational burden, ODR only makes use of a subset of pixels. In a similar approach to [8], ODR scans the image at regular intervals, but it performs the scanning in the vector form of the image, I', not in the matricial form, I. We fix the sampling interval as Δ, and construct the image vector as $I'_s \in \{1, 0\}^D : I'_{s,k} = I'_{\Delta k}$, where $D = NM/\Delta$. To simplify notation, we drop the subscript s throughout the rest of the paper and, except when stated otherwise and without any loss of generality to the algorithm itself, refer to the sampled version of the image vector as the image vector in a $\mathbb{R}^D$ space.

4   ODR Offline Learning

In the offline stage, we have access to a large dataset of labeled, pre-processed and sampled images containing examples of different objects. The images corresponding to a specific object α are collected in an observation matrix O_α and the labels, i.e., the object positions, in a matrix P_α. If the aim was to have the robot identifying n objects, we would have n observation matrices, O_α and n position matrices P_α. Since each object is treated independently, we can again drop the subscript α and when we refer to observations matrix O, it is meant that the matrix only contains observations from a single object.

Each observations matrix, O, is constructed by assigning each pre-processed image to a row in the matrix. For example, if we had a set of L images, $I_1, ..., I_l, ..., I_L$, with N rows and M columns sampled at an interval of Δ pixels, our observation matrix would be defined as:

$$O = \begin{bmatrix} I'_{1,1} & \cdots & I'_{1,1+\Delta} & \cdots & I'_{1,n\Delta+1} & \cdots & I'_{1,NM} \\ I'_{2,1} & \cdots & I'_{2,1+\Delta} & \cdots & I'_{2,n\Delta+1} & \cdots & I'_{2,NM} \\ \vdots & \cdots & \vdots & \vdots & \cdots & \vdots \\ I'_{L,1} & \cdots & I'_{L,1+\Delta} & \cdots & I'_{L,n\Delta+1} & \cdots & I'_{L,NM} \end{bmatrix} \tag{1}$$

where $I'_{l,k} = 1$ if the pixel with coordinates $\{(n,m) \in \mathbb{N}^2 : k = (n-1)M + m\}$ belongs to image l object hypothesis.

The label matrix, $P \in \mathbb{R}^{L \times 2}$, contains the coordinates of the object in terms of its centroid $\mathbf{p}_{l,c} = [x_{l,c}, y_{l,c}]$.

Our training datasets are composed of both synthetic images and real images captured by the robot while it searches and follows a ball. For the synthetic dataset, we simulate images containing a specific object plus random noise. In each image the object is placed in different positions and those position uniformly cover the whole image. The resulting images include random noise and occlusion in edges and corners. For each object, the synthetic dataset is composed of 768 images. Examples can be found in Figure 3 for the ball example. The robot collected data includes balls in different parts of the image, but the sampling is not thorough. The robot is acting according to the ball position and keeps the ball, and the close by objects approximately in the center of the image. The resulting dataset contains fewer examples of ball on the edges of the image. However, real images introduce the variability on the object shape, pose and illumination that the robot will experience during run-time.

From the total of 856 real images we have for Ball and the 204 for Robot, we have occlusion on the image edges (Figure 4(e)) and by other objects (Figure 4(c)). We also have several examples of motion blur (Figure 4(d)) and of random noise (Figure 4(f)) captured by the robot while searching for the ball in the environment. All the real images were labeled automatically using CMVision.

All the offline tasks will be based on the observation matrix O constructed using the training dataset and in the position matrix P. The main output of the online stage is the set of the linear regression coefficients, W and $\mathbf{b}_0$.

Normalization

There are two stages of normalization: the first enable us to deal with objects of different sizes, the second for standardizing the data before performing the PCA. To deal with objects of different sizes, we first need to normalize all the images I': $I'_\rho = I' / \sum_k I'_k$. The observation matrix composed of the normalized images is represented as O_ρ.

For the principal components analysis we compute the observations sample covariance matrix ([1]) and it is a best practice to normalize the values for each pixel so that they follow a unitary zero mean Gaussian distribution. $C(O_\rho) = \Sigma^{-1}(O_\rho - \bar{O}_\rho)^T(O_\rho - \bar{O}_\rho)\Sigma^{-1}$ where O_ρ is the observation matrix corresponding to I_ρ, $\bar{O}_\rho$ is a matrix whose lines are all equal and correspond to the mean of each pixel over all the dataset $\bar{I}_\rho$, and

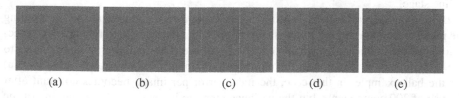

(a)   (b)   (c)   (d)   (e)

Fig. 3. Examples of synthetic images used in training. Images include objects in different positions (Figures 3(a)-3(c) for the ball; Figures 3(d)-3(e) for the robot), occlusion in image borders (Figures 3(a) and 3(d)) and noise (Figures 3(a)-3(e)).

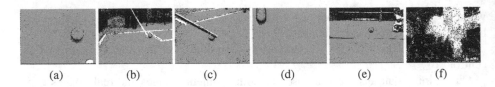

(a)   (b)   (c)   (d)   (e)   (f)

Fig. 4. Examples of real images used for training and testing. Examples include objects of different sizes (Figures 4(a)- 4(e)), motion blur (Figure 4(c)) , occlusions (Figure 4(c) and 4(e)) and random noise (Figure 4(f)).

Σ is a diagonal matrix, whose elements, σ_{ii}, are the standard deviation of the pixel i over O_ρ.

The mean and standard deviation estimated over the training dataset, will be used in the online stage, where each new image is normalized to fit the same distribution.

Principal Components Analysis

The principal components, V, correspond to the sample covariance matrix, $C(O_\rho)$, eigenvectors. The components form an orthogonal set of synthetic images that span the subspace of images with the same object in different positions and with different sizes.

Examples of the principal components obtained using the ball dataset are represented in Figure 5. The examples highlight the hierarchy in resolution of the principal components: the first components, which contain more information, have lower spatial frequency. We can thus reduce the dimensionality on our datasets by projecting the images into the first c components of this new basis, as seen in eq. (2). The effect will be equivalent to filtering in the spatial frequencies domain.

$$O_r = O_\rho V_c^T \tag{2}$$

The number of components used affects the regression results. If ODR uses a large number of components, noise is added to the object model. Furthermore, the number of coefficient to be estimated during regression increases and over-fitting to noise becomes a possibility. If too few components are used, the reduced dataset observation matrix O_ρ may not have enough information to provide good regression results. In particular, all the small examples of the object may be filtered out. Deciding on the number of components to keep for the regression depends on the relative size of the smaller object we want to be able to identify and of the type of noise we expect to find during the online stage.

By training regression models using different number of components and computing the mean detection error per image in an independent dataset which reflects our expectations for the online stage, we can choose a priori the best number of components to use. The impact on precision of the number of components is illustrated in Figure 6(a) for the ball example. In this case, the mean error per image becomes constant after the use of 200 components, but the variance starts to increase. In the remaining of the experiments in this paper, we use only the first 200 components for the ball. For the robot, since it is considerably larger than the ball, we only need to use 15 components to estimate its position.

Regression

After the dataset dimensionality reduction, ODR performs a linear regression between the reduced images $I'_{r,l}$ and the known object positions $\mathbf{p}_l = [x_{c,l}, y_{c,l}]$. The result of the linear regression is the set of coefficients $W_r = (w_{r,x}, w_{r,y})$ and $\mathbf{b}_0 = [b_{0x}, b_{0y}]$, which solve the linear least squares problem in eq. (3) and are given by eq. (4) ([1]).

$$\min_{W_r} \| P - \tilde{O}_r \tilde{W}_r \| \tag{3}$$

$$\tilde{W}_r = (\tilde{O}_r^T \tilde{O}_r)^{-1} \tilde{O}_r^T P, \tag{4}$$

where P is the matrix which row l is the position vector $\mathbf{p}_l$ corresponding to image l, $\tilde{O}_r = (\mathbf{1}, O_r)$ and $\mathbf{1}$ is a column vector with ones that allow us to incorporate the affine bias term in $\tilde{W}_r = [\mathbf{b}_0^T, W]$. The set of coefficients in W and $\mathbf{b}_0$ are the main output of the offline stage.

5   ODR Online Testing

To find the object position in a new image, I_{new}, the robot needs to normalize the image vector following the same steps as in the offline stage. First the image is converted into a probability distribution, $I_{\rho,new}$. Then is normalized so it falls in the unitary zero mean Gaussian distribution estimated in the offline process: $I_{N,new} = (I_{\rho,new} - \bar{I}'_\rho)\Sigma^{-1}$.

Detection corresponds to the application of the linear model learned in the offline stage using equation eq. (5).

$$\mathbf{p}_{new} = \tilde{I}_{N,new} \tilde{W}_r = \mathbf{I}_{N,new} V^T W_r + \mathbf{b}_0 \tag{5}$$

where $\mathbf{p}_{new}$ is the object position in the new image.

We can measure the degree of accuracy of this description by using eq. (6) to project the image in the PCA basis and re-project it back into the images space. The resulting image, I'_{rep}, corresponds to an image with coordinates in the original space of I'_N, but in the subspace generated by the principal components. If I'_N is well described by the components, the I'_N and I'_{rep} should be very similar and the angle θ formed between

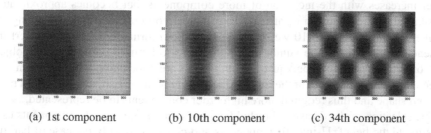

(a) 1st component   (b) 10th component   (c) 34th component

Fig. 5. Examples of principal components corresponding to a dataset composed of the synthetic dataset for the ball

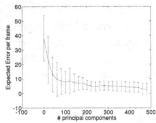
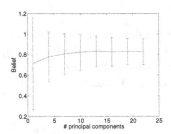

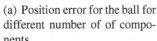

(a) Position error for the ball for different number of of components

(b) Belief estimation for the ball example

Fig. 6. Impact of the number of principal components used in detection and belief estimation in the image case

the two vectors should be very close to 1. We use the cosine of the angle θ, eq. (7), as a proxy for our belief in the existence of the object in the image.

$$I'_{rep} = V^T I_r = V^T V I'_N \tag{6}$$

$$\cos\left(\theta_{object}(I_N)\right) = \frac{I'_N \cdot I'_{rep}}{\|I'_N\|\|I'_{rep}\|} \tag{7}$$

Due to the number of multiplications required, the belief estimation can be very time - consuming. To reduce the computational load, we change equation eq. (6) and choose carefully the number of principal components to be used at this stage.

We change equation eq. (6) by, instead of re-projecting the whole image back into the regular image coordinates, re-projecting only the pixels that belong to the object hypothesis. We are not comparing all the pixels of I'_N and I'_{rep}, but only the fraction that should had been correctly reconstructed.

Furthermore, we note that the number of principal components used at this stage can differ from the number of components used for regression. To determine the minimum number of components required for belief estimation, we compute the mean and standard deviation of belief in the independent dataset previously used for estimating the position error. For the ball example we present the results in Figure 6(b). The mean belief increases with the inclusion of more components, but becomes approximately constant after the inclusion of at least 10 principal components. Thus, by fixing the number of components to 10 we retain most of the information needed to assert the presence of the object. The number of components to be used depends on the object level of detail. While for a ball, just 10 components are good enough for describing the object, the same is not true for the robot. Albeit larger than the ball, the robot has a more detailed shape and thus required more high order components to be represented.

The selection of the decision threshold depends on the number of components used to estimate the belief. Using 10 components and considering only the case of ball detection a threshold of 0.65 takes into account all the examples inside the error bars as exemplified in the plot in Figure 6(b).

6   Experimental Results

In this paper we analyze the results of ODR using elements of the RoboCup environment, such as the ball and other robots. In particular we evaluate the algorithm in three different dimensions. First the capacity of object detection knowing that the object is in the image. Second, the capacity of identifying if the object is in the image or not. Third the time efficiency of the algorithm. We separate the problem of object detection accuracy from the problem of belief estimation in our result presentation. This is motivated by the possibility of changing either one of these parts of the algorithm without affecting the other. Also, since they solve different problems, one provides a position while the other classifies an image, we use different metrics to express results.

The capacity for detecting the object is measured by the percentage of correctly detected objects over the total number of objects that were given to identify. I.e., over all the set of testing images, we only consider those that contained the object. For the ball example, ODR detects correctly 92% of all the ball examples in the dataset. As for to the identification of the second robot, ODR identifies the position of 87.5% of all the robots. These results include changes of pose and occlusion. Examples of detection are provided in Figure 7

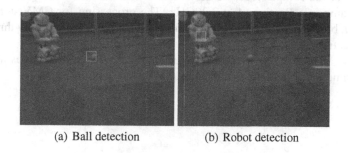

(a) Ball detection   (b) Robot detection

Fig. 7. Examples of ball and robot detection

The capacity for classifying each image according to whether the object is present or not given a belief score is illustrated by the three usual metrics used to evaluate classification algorithms: the precision, the recall and the average precision. The precision evaluates the capacity to differentiate between two classes and is given by the percentage of true positives over the total number of positives. The recall evaluates the capacity of identifying the objects from the desired class and is given by the percentage of true positives over the total number of true examples in the dataset. Both precision and recall metrics depend on the classification criterion. In ODR, the criterion corresponds to a threshold in the belief. Only images with belief higher than the threshold are classified as having the object. The average precision attempts to reconcile precision and recall by considering both metrics at different threshold values. The average precision itself is computed by measuring the area under a precision recall curve, where each point in the curve corresponds to a precision and a recall computed at the same threshold.

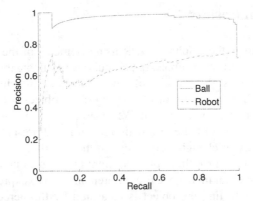

Fig. 8. Precision recall curves for both objects

In the graphic of Figure 8 we present the precision recall curves for the ball and robot. The average precision for the ball is 0.96 while for the robot is 0.67. We compare the performance of both time and accuracy of ODR with respect to other methods. In particular we compare them against CMVision, which was used as the ground truth in training. For the comparisons, we use both methods offline, running each one 1000 times per frame in a Pentium 4 at 3.20GHz. The processing for CMVision includes thresholding, blob formation and ball detection, while ODR includes only thresholding and ball detection.

Results for processing time are presented in Figure 9. ODR achieves an average processing time lower than that obtained by CMVision in both cases.

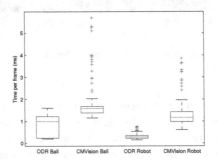

Fig. 9. Comparison of processing times using both ODR and CMVision

7   Conclusions

In this paper, we have contributed a novel object detection by regression approach. Results, which were obtained for the RoboCup case study, show that our learning of offline linear object models for object detection and position provides a fast and robust online performance. ODR is best applicable in general, if the environment and the specific objects to detect and if object presence hypotheses can be pre-computed.

The context of the RoboCup robot soccer is particularly adequate for ODR, as the field and the objects are known ahead, and processing of their color provides a simple prior for the object hypotheses. Our learning models effectively capture the online object images, even given the extreme variations of the images in real game situations. The learned models by ODR are based on a small number of pixels, leading therefore to fast online image processing. Experimental results show that such sampling did not adversely affect position detection precision, which is close to par with the state of the art algorithms when the object is present in the image. Furthermore ODR is significantly more efficient.

ODR is also able to identify if the object is absent in the image based on the learned models. The number of principal components used by ODR affects the ability to reconstruct the image and hence, to identify the absence of the object. The higher the number of principal components, the more precisely the absence of the object is detected, but also the higher the computational cost. In our experiments, we favored a low computational cost, and ODR was still able to successfully identify the absence of the objects with the needed accuracy. In general, setting the tradeoff between the number of principal components used online and the computational cost will depend on the needs of the domain.

References

1. Bishop, C.M.: Pattern Recognition and Machine Learning. Springer (2007)
2. Bruce, J., Balch, T., Veloso, M.: Fast and Inexpensive Color Segmentation for Interactive Robots. In: IROS 2000 (2000)
3. Felzenszwalb, P., Mcallester, D., Ramanan, D.: A discriminatively trained, multiscale, deformable part model. In: IEEE Conference on Computer Vision and Pattern Recognition, CVPR 2008 (2008)
4. Hashemi, E., MRL Team: MRL Team Description Standard Platform League (2010)
5. Hester, T., Quinlan, M., et al.: TT-UT Austin Villa 2009: Naos Across Texas (2009)
6. Nieuwenhuisen, M., Behnke, S.: NimbRo SPL 2010 Team Description (2010)
7. Turk, M.A., Pentland, A.P.: Face recognition using eigenfaces. In: ICCV 1991 (1991)
8. Röfer, T.: B-Human Team: B-Human Team Report and Code Release (2010),
   http://www.b-human.de/en/publications

Real-Time Human-Robot Interactive Coaching System with Full-Body Control Interface

Anton Bogdanovych[1], Christopher Stanton[2],
Xun Wang[2], and Mary-Anne Williams[2]

[1] School of Computing and Mathematics
University of Western Sydney, Australia
A.Bogdanovych@uws.edu.au
[2] Faculty of Engineering and Information Technology
University of Technology Sydney, Australia
{CStanton,XuWang,Mary-Anne}@it.uts.edu.au

Abstract. The ambitious goal being pursued by researchers participating in the RoboCup challenge [8] is to develop a team of autonomous humanoid robots that is capable of winning against a team of human soccer players. An important step in this direction is to actively utilise human coaching to improve the skills of robots at both tactical and strategic levels. In this paper we explore the hypothesis that embedding a human into a robot's body and allowing the robot to learn tactical decisions by imitating the human coach can be more efficient than programming the robot explicitly. To enable this, we have developed a sophisticated HRI system that allows a human to interact with, coach and control an Aldebaran Nao robot through the use of a motion capture suit, portable computing devices (iPhone and iPad), and a head mounted display (which allows the human controller to experience the robot's visual perceptions of the world). This paper describes the HRI-Coaching system we have developed, detailing the underlying technologies and lessons learned from using it to control the robot. The system in its current stages shows high potential for human-robot coaching, but requires further calibration and development to allow a robot to learn by imitating the human coach.

1   Introduction

With autonomous robots becoming increasingly prevalent in society, natural and intuitive methods are required to interact, guide and improve robot behaviour. Learning by demonstration, observation and imitation are approaches to learning in which a teacher (or coach) provides examples of the desired robot behaviour. Examples range from a teleoperated robot recording the actions performed by the teacher, to autonomous robots learning to perform actions by watching a human teacher perform a similar action [2]. Likewise, in the realm of virtual agents, imitation learning has been used to teach autonomous agents in gaming environments to perform complex manoeuvres performed by human experts [4].

   In this paper we consider how to best teach humanoid robots to play soccer in the RoboCup Standard Platform League (SPL). With people, when one person coaches another, the objective of the coach is to use their relevant expert

T. Röfer et al. (Eds.): RoboCup 2011, LNCS 7416, pp. 562–573, 2012.

knowledge and experience to improve the task performance of the person being coached. This knowledge transfer is often verbal, but can be aided by demonstration, pictures, videos, and other forms of communication. However, what is the best way to coach an autonomous robot? In this paper we explore this issue, and present a real-time Human-Robot Interactive Coaching System (HRICS). Our long-term goal is to explore methods of teaching the robot (in real-time) to extend and improve their capabilities without explicit programming. In this paper we focus on one aspect of the system - the teaching of skills by demonstration (imitation learning). To this end, we describe how we use a motion capture suit and portable computing devices (in particular the iPhone) to interact and communicate with the robot.

The remainder of the paper is structured as follows. We begin in Section 2 by describing our specific problem domain of robot soccer. In Section 3 we describe our Human-Robot Interactive Coaching System (HRICS). Section 4 outlines the implementation details of how we connected the motion capture suit to the Aldebaran Nao humanoid robot. In Section 5 we present our initial results and reflections regarding using the prototype system to coach soccer skills. We conclude by discussing the potential of the new approach and future work.

2   Problem Domain: Standard Platform League (SPL)

Robot soccer matches in the SPL involve two teams of autonomous Aldebaran Nao [1] robots competing on an indoor field for two 10-minute halves. Human intervention is not allowed during the matches apart from picking up malfunctioning robots or penalising robots that have violated the rules of the game. The Nao robots need to make a wide range of skills autonomously. Perceptual skills include colour recognition, object recognition and the detection of collisions; motor skills include walking and kicking; and there are strategic decisions to be made (for example, positioning on the field). An enormous effort prior to competition is required for teams to develop and calibrate the software to control such skills and behaviours. The prevailing approach for developing such skills is for a software developer to script the behaviour coarsely, but provide a parameterised interface for modifying the behaviour. Most of these parameters will then be calibrated by hand (a very labour intensive process), while a smaller proportion will be calibrated via an unsupervised learning process on the robot [3]. A challenge facing the developers of autonomous robots is how to best specify, develop and improve the robot's skill-set and decision-making capabilities without tediously hand-crafting and hand-tuning behaviours.

3   Human-Robot Coaching System

Human-robot interactive coaching (HRIC) is a new way of approaching the problem of skill learning and development, and performance improvement. In this section we provide a detailed description of our Human-Robot Interactive Coaching System (HRICS) which has been implemented by integrating a combination of

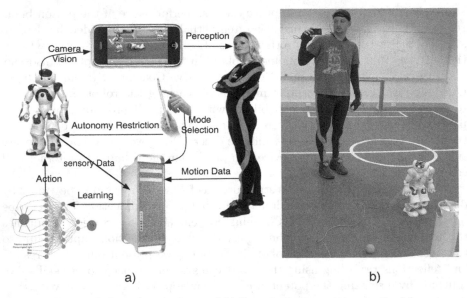

Fig. 1. a) System overview and b) System demonstration in the lab

cutting edge technologies. Figure 1 a) outlines a schematic representation of various components in our system and their interactions; while Figure 1 b) shows a snapshot of the actual system setup that is being tested in our lab.

The focal points of our architecture are the autonomous Nao robot, portable computing devices (iPhone/iPad), a full-body motion capture suit, and data logging for offline processing. The robot is the central part of our architecture it is capable of fully autonomous action, but its autonomy can be restricted by the human coach via the Iphone/Ipad interface. All the data from the robot as well as the user input to the mobile device is stored on the main computer, which processes the data and ensures robot's learning from this data. The perception of the robot is streamed to the mobile device, where it can be annotated for better recognition. The mobile device can also act as a tool for synchronising the perception of the robot and a human coach, which is particularly useful in the imitation mode, where the robot can learn ball searching behaviour, various strategies of approaching the ball or different kicking styles depending on the environment state. While all of the aforementioned components are present in the system and are fully embedded, the main focus of this paper is on integrating its novel element - the full-body motion capture suit. The motion capture suit presents an innovative way of human-robot interaction, where each body part of the human user is involved into controlling the corresponding body part of the robot. The data from the motion capture suit is obtained in real time and is also streamed via WiFi to the main computer, which calculates the necessary transformations to map this data to a robot's motor angles. The technological details of streaming robot vision to a mobile device or collecting robot's sensory data and using supervised machine learning on it [9] are outside of this paper's scope. Next we consider each of the aforementioned system components in detail.

3.1 Aldebaran Nao

The Aldebaran Nao is a humanoid robot with 21 degrees of freedom. The Nao robots are equipped with sensors such as head mounted colour camera, front-facing ultrasonic distance sensors, force sensors in the soles of the feet, bump detectors on the toes of each feet, accelerometers, and gyroscopes.

3.2 iPhone/iPad

We have developed an iPhone/iPad application for interacting with the robot. The application allows the user to interrogate the robot's internal state, stream raw and processed vision in real-time (see Fig. 2), and set operation modes of the robot. The main interface of the application consists of a camera view and a soccer field view which displays raw or processed live images streamed from the robot and the position on the field where the robot believes to be located (Fig. 3). Properly configured robots automatically establish a connection with the portable computer device once they are turned on. An arrowed dot icon is displayed in the soccer field view on the console for every connected robot. The dots on the soccer field correspond to the estimated positions of the robots on the actual field (see Fig. 3).

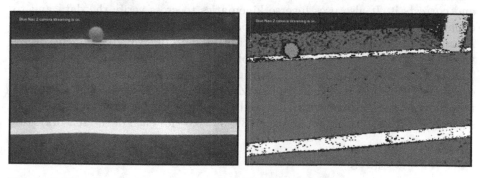

Fig. 2. Raw and processed live images from the robot's camera displayed on an iPhone

For our current experiments (described in the later sections), we provide the human coach live-feeds of raw vision from the robot's camera. Live images captured from the robot camera are transmitted to an iPhone every 200 ms.

3.3 The Motion Capture Suit

As an interface to control and coach the robot we employ a high precision full-body motion capture suit, Xsens MVN[1]. Only recently motion capture suits

---

[1] http://xsens.com/en/general/mvn

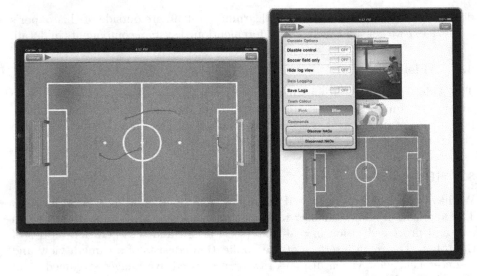

Fig. 3. The main graphical interface of the iPad interactive coaching application

similar to Xsens MVN reached the level of precision when they can correctly capture real-time motion of a human body with no significant data errors. This equipment comes in a form of a lycra suit with 17 embedded motion sensors.

The suit is supplied with MVN Studio software that processes raw sensor data and corrects it. It also uses inverse kinematics to cross-verify the data and to estimate the parameters of additional body joints. As the result, MVN studio is capable of sending real-time motion capture data of 23 body segments using the UDP protocol with the frequency of up to 120 motion frames per second. The key elements of the data being transmitted are absolute (X,Y,Z) position of each segment and its absolute (X,Y,Z) rotation. XSENS MVN is capable of real-time motion capture with very high accuracy. During extensive testing it showed a very small margin of error ($0.8°$, $s = 0.6°$ for each of the sensors) [10].

3.4   Data Logging

Commands from the motion capture suit are translated to robot effector commands (the details of this translation are described in Section 4) and sent to the robot. We programmed our robots to log all effector commands, sensor data, and other internal state variables every 10ms[2]. Data is recorded regardless of whether the robot is operating autonomously or being teleoperated via the motion capture suit. If the robot is being controlled via the motion capture suit, this state information is recorded, as are the walk engine commands (forwards, strafe, and rotation) and head position (pitch and yaw) commands chosen by the human controller. This data enables us to analyse (off-line) the decisions made by the human coach in relation to the robot's perceptual state.

---

[2] On each DCM callback event.

4   Connecting XSENS MVN to a Nao Robot

In our system we directly map each body segment of the human user to the corresponding motor of the robot and adjust those accordingly. The data being transmitted by the suit is measured using the absolute coordinate system of the suit space, where the X axis is aligned with the magnetic north and the origin corresponds to the position at which the suit was turned on. All the rotation data that is being transmitted by the suit comes in the form of quaternions [6].

The robot has a completely different embodiment to a human, so for controlling the robot with a motion capture suit we had to make a number of adjustments. In order to control the movement of its body parts, rather than setting positions and rotations, the robot uses a number of embedded motors, setting the degree of rotation for each of these motors results in the desired movement. To control these motors an SDK allows the programmer to specify the angular rotational position of each motor every 10ms.

When using the motion capture suit for controlling the robot, some of the values we receive for body segments can be directly translated into the appropriate motor rotations on the robot side. For example, there are two motors controlling the head of the robot: pitch motor (headPitch) - responsible for up/down head movement and yaw motor (headYaw) - responsible for side movements. The range of these motors is a bit wider than the corresponding range of the human head motion, but within the range of acceptable human head movement - the angles one must supply to the motors directly map to the Euler angles of the human's head segment.

Fig. 4 provides a graphical explanation for how the data obtained from the motion capture suit is being utilised for controlling the head movement of the robot, its body orientation, as well as forward/backwards/sideways movement of the robot's body. The 3D character shown on the left hand side of the figure corresponds to a reconstructed human model based on the positions and rotations of the body segments received from the motion capture suit. This figure is positioned in the global suit coordinate system, where all coordinates are measured in meters. All the data we receive is in absolute coordinates and angles in relation to this coordinate system.

In order to convert the suit data into the appropriate values for the motor rotations on the robot end, we have to apply the following transformations. First, we must convert the absolute rotation of the head segment into a relative rotation, as the robot operates with angles in the robot space. To do this we have to calculate the relative rotation of the head sensor in relation to the chest segment. The following equation helps to make this translation.

$$Qrot_{relative}(a, b) = \frac{Qrot_b}{Qrot_a} \tag{1}$$

Here $Qrot_{relative}(a, b)$ corresponds to the resulting relative rotation (in the quaternion form) of the head sensor (a) in relation to the chest segment (b). The values of $Qrot_a$ and $Qrot_b$ represent the quaternions defining the absolute rotations of each corresponding body segment.

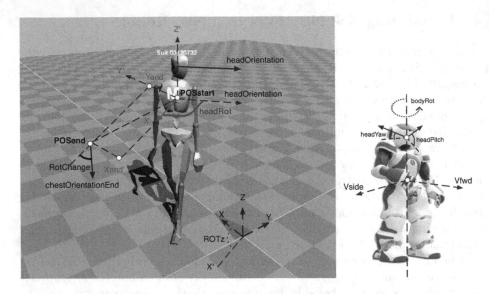

Fig. 4. Translating the MoCap Suit Data into the Robot Motor Angles

The resulting quaternion rotation (that is shown as headRot in the picture) can be represented as a matrix of the following form:

$$Qrot_{relative}(a, b) = [q_0, q_1, q_2, q_3]^T \qquad (2)$$

Once the relative rotation is obtained, we have to transform the quaternion rotation into Euler angles and translate those into correct values for each robot motor. To obtain Euler angles from this matrix - we use the following equation:

$$\begin{vmatrix} rot_X \\ rot_Y \\ rot_Z \end{vmatrix} = \begin{vmatrix} atan2(2(q_0q_1 + q_2q_3), 1 - 2(q_1^2 + q_2^2)) \\ arcsin(2(q_0q_2 - q_3q_1)) \\ atan2(2(q_0q_3 + q_1q_2), 1 - 2(q_2^2 + q_3^2))) \end{vmatrix}$$

Here rot_X represents Euler rotation around the X axis, rot_Y is the Euler rotation around the Y axis and rot_Z corresponds to the Euler rotation around the Z axis. In the case of the head, the yaw motor angle (headYaw) is obtained from rot_Z and the pitch (headPitch) - from rot_X.

Controlling the walking movement of the robot is a bit more complicated than rotating the head or arms. The robot has a completely different joint structure to a human in its legs and a significant difference in its centre of gravity. Direct mapping of human joint rotations onto robot motor angles is a difficult task known in the literature as motion retargeting [7]. To avoid obvious mismatches and prevent the robot from falling over we decided not to move every leg sensor individually, but to play prerecorded standard walking moves of the robot in response to the corresponding move being detected as performed by a human. The robots's SDK provides a function for this purpose, which receives the forward/backward velocity (V_{fwd}), body rotation ($body_{Rot}$) and the side velocity (V_{side}) and results in the desired movement of the robot.

In order to calculate the forward velocity of the robot we first have to compute the forward transformation of the human coach in the direction of the current body orientation. Lets assume that pos_{start} - is the absolute position of the human wearing the motion capture suit in the beginning of forward/backward movement and pos_{end} is the position in the end of this movement.

To obtain the position displacement in the direction of the current body orientation of the human we have to rotate the coordinate system by the angle α, which represents the body rotation of the human (in our case $\alpha = rot_Z$). In order to perform the coordinate rotation we use the following rotation matrix:

$$\Re(\alpha) = \begin{vmatrix} \cos(\alpha) & -\sin(\alpha) \\ \sin(\alpha) & \cos(\alpha) \end{vmatrix}$$

So, after rotating the coordinates using our rotation matrix, the new coordinates ($X_{start}\prime, Y_{start}\prime$) for the point pos_{start} will be:

$$X_{start}\prime = pos_{start}.X * \cos(\alpha) + pos_{start}.Y * \sin(\alpha)$$
$$Y_{start}\prime = -pos_{start}.X * \sin(\alpha) + pos_{start}.Y * \cos(\alpha)$$

And the new coordinates ($X_{end}\prime, Y_{end}\prime$) for the point pos_{end} will be:

$$X_{end}\prime = pos_{end}.X * \cos(\alpha) + pos_{end}.Y * \sin(\alpha)$$
$$Y_{end}\prime = -pos_{end}.X * \sin(\alpha) + pos_{end}.Y * \cos(\alpha)$$

Now we can calculate the forward/backward displacement of the human between those two positions as:

$$\Delta Fwd = X_{end}\prime - X_{start}\prime \tag{3}$$

And the sideways displacement of the robot can be obtained as:

$$\Delta Strafe = Y_{end}\prime - Y_{start}\prime \tag{4}$$

The displacements ΔFwd and $\Delta Strafe$ represent the distances the human body has travelled between two sensor data readings. The data transmission frequency of the suit is manually adjustable and known in advance. Thus to obtain the forward velocity we can use the following equation:

$$V_{fwd} = \frac{\Delta Fwd}{t} \tag{5}$$

And the sideways velocity is obtained as:

$$V_{side} = \frac{\Delta Strafe}{t} \tag{6}$$

In both equations t represents the standard delay between two subsequent suit data transmissions (currently set at 10 ms).

Finally, to make the robot rotate at a desired angle we have to set the value of the corresponding motor. When changing its rotation the robot only deals with relative angles, so it doesn't make any assumptions about its global position and orientation in the world. Similar to obtaining the relative displacements of the sensors we described above, we must compute the change of robot's body rotation in-between two consecutive data measurements of the motion capture suit. To do so, we use the equation (1) and apply it to two measurements of rotations of the chest sensor, so that we obtain the relative rotation ($rotChange = Qrot_{relative}(X', chestOrientationEnd)$) between the chest orientation vector in the first measurement and the chest orientation vector in the last measurement. The $bodyRot$ angle that we send to the robot corresponds to the Z Euler angle of the $rotChange$ variable in the suit space (see Fig. 4). Given that $q_0, .., q_2$ are the dimensions of the $rotChange$ quaternion - we can use the equation (2) to calculate the final argument of our function as:

$$body_{Rot} = atan2(2(q_0q_3 + q_1q_2), 1 - 2(q_2^2 + q_3^2)))$$ (7)

The proportions of the human body and other embodiment characteristics are very different to those of the robot, so no direct mapping can be made between the motion capture data and the data required by the robot. Thus, we had to conduct a series of experiments for obtaining the scaling factors that apply to each of the robot motors.

5   Experimental Evaluation

In order to test the validity of our assumption about the usefulness of the resulting coaching system we conducted a series of experiments where each of the system components was extensively tested[3].

We had a four step plan for evaluating the HRICS. Our first step was calibration of human movements to robot movements via scaling, clipping and the use of minimum movement thresholds. For example, the human controller should not need to run for the robot to walk at top speed; nor should the human fidgeting or other unintentional small movements result in a tiny step from the robot. We calibrated each dimension of the walk engine separately (rotation, strafe and forwards/backwards). Best results were found when small human movements were scaled to larger (proportionally speaking) movements on the robot. For example, for the robot to walk forwards at full speed the human controller was only required to make a small step forward. To remove unintentional human movements from controlling the robot, the human controller was required to stand relatively still. If the robot still moved (due to highly accurate nature of the motion capture suit, even breathing can trigger non-zero values), the minimum value threshold was increased. Head movements were also scaled and offset. As we were using the downwards facing camera of the Nao's two cameras, after

---

[3] The video recording featuring fragments our experimental evaluation can be seen at: http://www.youtube.com/watch?v=XY4nYpEZr5U

head calibration the robot's "eyes" (they are LEDs very display purposes only) would be facing slightly higher than the human's. After conducting extensive testing we came up with the following adjustments for the robot (see Fig. 4). The $body_{Rot}$ variable had to be magnified by 20. Both V_{fwd} and V_{side} require a magnification by 40. The $headYaw$ didn't require any adjustments, as limiting the range of the robot's head to the range of the human didn't seem to have a significant visual impact. Finally, the $headPitch$ had to be magnified by 1.5 and the values beyond the allowed robot's range had to be clipped.

Next, once the suit was calibrated we attempted to play soccer using the human's vision ("super perception"). This involved the human controller walking about the field, with the robot imitating the human's actions[4]. Scoring a goal was not a simple or straightforward task for the robot. Network lag and the nature of the Aldebaran walk engine meant that the robot would imitate a human step direction approximately 1 second after they executed it. The key reason for this delay was the fact that with our currently chosen approach the robot was not able to interrupt its walk half way through the step cycle, but was only able to change the walking direction after the step cycle was finished. Also, our scaling of human movement to robot movement (despite calibration emphasising that small human movements should be converted to large robot movements), meant the human controller was constantly running out of room to move within the 6m by 4m soccer field. Even more difficult was when the human controller would end up in front of the robot and the ball, and thus they had to somehow move themselves behind the robot, but without the robot moving. Despite these difficulties, the human was able to control the robot in a fluent manner after adjusting his behaviour by moving using smaller steps, and reducing the overall movement velocity so that the robot is able to catch up with him.

After experimenting playing soccer with the human's vision, we head-mounted an iPhone feed of the robot's raw camera feed to the human controller[5]. This allowed the human controller to play soccer in a remote location to the robot, but increased the level of difficulty in playing soccer with the robot, mainly due the small field of view provided by the robot's camera. Having no possibility to see the robot made it difficult to adjust the walking behaviour of the experimenter. As the result, it was often required to perform an additional search for the ball as the experimenter's mental representation of the ball and robot's position on the soccer field didn't match the actual positions. Having the vision, though made it possible to quickly detect reference points and update the mental model.

Lastly, we had planned to try and teach the robot skills by logging human command data, together with robot perceptual data, while controlling the robot suit. However, with our current level of fine control requiring more calibration, an autonomous robot soccer player is a more effective goal-scorer than a robot controlled by the motion-capture suit. However, we are optimistic that with further calibration and refinement, there is a great deal of potential with this

---

[4] This was somewhat confronting visually, as the robot's head would not be looking at the ball. Instead it would be looking in the same direction as the human controller.

[5] In the future, we plan to purchase and use wearable video glasses.

approach. A variety of machine learning algorithms operating on robot sensory data are already embedded into our infrastructure [9], so when we reach the desired level of control precision this part of the experiment will be resumed.

6   Discussion

The development of our Human-Robot Interactive Coaching system is inspired by the real-life human soccer coaching and game plays. It is rooted in the view that robot soccer should adopt successful approaches that have been traditionally applied in the human soccer. Programming every facet of the robot behaviour is time-consuming, inflexible, error-prone and limited. It is crucial to understand that our coaching system does not drive the robot's actions directly as an OCU [5] in a soccer game; it is not purely used for remote control but aims to help the robot learn new skills and strategies. The coaching system monitors the autonomous robot's performance in real-time and gives helping instructions to the robots to improve their game playing skills. Our system is used only in robot training and practice matches, and not in real soccer competitions.

The motion capture suit provides the robots with the possibility to learn rich motion dynamics from the human. These capabilities are difficult to implement using the standard computing hardware. Prior to using the motion capture suit we conducted experiments with remotely controlling a team of robots via keyboard input, but it was too difficult to work with so many parameters simultaneously and was hard to achieve the desired level of precision in setting each of the parameters. Correctly updating the value for each of the 21 motors of the robot by pressing the corresponding keyboard button was a difficult task and remembering the keys that correspond to each of those joints was even more challenging. As the result, the autonomous robots prevailed in our experiments and the human participants reported on the high degree of frustration and confusion with controls. In contrast, the use of motion capture suit allowed for a much better and intuitive interface than a keyboard. In particular, control of the robot's head was fluent and natural (as motor commands for the Nao's head were updated very 100ms, as opposed to control of the legs which was updated on every step-cylce). Due to the number of problems we discussed above it was still not feasible to successfully compete against autonomous robots, but the potential of this interface is very high.

The approach we applied for mimicking human's head rotation was successfully tested on moving the arms of the robot. This functionality, however, had to be disabled to make it possible for the user to utilise his hands for interacting with the mobile device rather than for controlling robot's hand movement.

7   Conclusions and Future Work

We presented the progress made in developing a innovative real-time Human-Robot Interactive Coaching system based on an iPhone/iPad interactive coaching console and Xsens MVN full-body motion capture suit for the autonomous robot Nao's motion and behaviour learning. We have shown that our interactive coaching system is a promising solution to effectively improve the performance of Nao robots in autonomous soccer game play through real-time exchange of rich information between the robots in training and the human coach.

The future work will include further fine-tuning of the human-robot motion correspondence, with the aim of allowing a human controller to become an expert coach via teleoperation. We are planning to utilise machine learning to come up with a mapping function that would create correspondences between motion snapshots recorded by the MoCap suit and the resulting posture of the robot. Once this function is in place, it will be possible to design our own walk engine and avoid the lag problems that we discussed earlier. We will also conduct a series of experiments on teaching various tactical and strategical behaviours to the robot from imitating the human user. In particular, we will focus on learning how to map an existing mental model about the state of the world and sensory perception of the robot to an efficient ball searching strategy by learning it directly from a human. Also we will explore using imitation learning for training the robot various kicking styles and selecting an appropriate kicking style depending on the state of the environment.

References

1. Aldebaran: Nao Robot, http://www.aldebaran-robotics.com
2. Argall, B.D., Chernova, S., Veloso, M., Browning, B.: A survey of robot learning from demonstration. Robotics and Autonomous Systems 57, 469–483 (2009), http://portal.acm.org/citation.cfm?id=1523530.1524008
3. Chalup, S., Murch, C., Quinlan, M.J.: Machine Learning With AIBO Robots in the Four-Legged League of RoboCup. IEEE Transactions on Systems, Man, and Cybernetics, Part C: Applications and Reviews 37, 297–310 (2007)
4. Gorman, B., Thurau, C., Bauckhage, C., Humphrys, M.: Believability Testing and Bayesian Imitation in Interactive Computer Games. In: Nolfi, S., Baldassarre, G., Calabretta, R., Hallam, J.C.T., Marocco, D., Meyer, J.-A., Miglino, O., Parisi, D. (eds.) SAB 2006. LNCS (LNAI), vol. 4095, pp. 655–666. Springer, Heidelberg (2006)
5. Gutierrez, R., Craighead, J.: A native iPhone packbot OCU. In: HRI 2009: Proceedings of the 4th ACM/IEEE International Conference on Human Robot Interaction, pp. 193–194. ACM, New York (2009)
6. Kuipers, J.B.: Quaternions and Rotation Sequences. Princeton University Press, Princeton (1999)
7. Nakaoka, S., Nakazawa, A., Yokoi, K., Hirukawa, H., Ikeuchi, K.: Generating whole body motions for a biped humanoid robot from captured human dances. In: IEEE International Conference on Robotics and Automation, pp. 3905–3910 (2003)
8. RoboCup: World Championship and Conference, http://www.robocup.org
9. Stanton, C., Ratanasena, E., Haider, S., Williams, M.A.: Perceiving Forces, Bumps, and Touches from Proprioceptive Expectations. In: Röfer, T., Mayer, N.M., Savage, J., Saranli, U. (eds.) RoboCup 2011. LNCS, vol. 7416, pp. 377–388. Springer, Heidelberg (2012)
10. Supej, M.: 3D measurements of alpine skiing with an inertial sensor motion capture suit and GNSS RTK system. Journal of Sports Sciences 28(7), 759–769 (2010)

A Loose Synchronisation Protocol for Managing RF Ranging in Mobile Ad-Hoc Networks*

Luis Oliveira[1], Luis Almeida[1,3], and Frederico Santos[2,3]

[1] DEEC-FEUP/IT, Universidade do Porto, Portugal
[2] DEE-ISEC, Inst. Politecnico de Coimbra, Portugal
[3] IEETA, Universidade de Aveiro, Portugal

Abstract. Robot motion coordination and cooperative sensing are nowadays two important and inter-related components of multi-robot cooperation. Particularly, when concerning motion coordination, distance information plays a very important role in mobile robotics. In this work, we investigate a new solution based on ad-hoc communication without global knowledge, particularly clock synchronisation, to measure distance between mobile units and to share that information. In order to improve ranging, medium throughput, and application predictability, we propose using a synchronisation protocol that keeps transmissions in the team as separated as possible in time, independently of the topology. Results show around 3.3 times reduction in the number of failed ranges without external interference and an order of magnitude reduction in the asymmetries among the nodes concerning the number of failed ranges when using the proposed synchronisation protocol.

Keywords: TDMA, synchronisation, cooperation, information exchange, relative localisation.

1 Introduction

Mobile autonomous robotics are key elements to many current applications which, similarly to the present trend towards multi/many-core computing platforms, exploit the benefits of parallelism. Using multiple such units can increase the effectiveness of surveillance, improve the rate of coverage in search and rescue, enable the transport of large parts, etc. However, achieving cooperation among multiple robots requires information exchange to enable, for example, cooperative sensing [10][8] and inter-robot motion coordination [11][3]. In addition, global services such as managing formations and sensor fusion typically require localisation services.

In this paper we explore both topics, i.e., relative localisation using an RF ranging method, and information sharing by means of a novel broadcast protocol for wireless ad-hoc multi-hop networks. This protocol enforces loose synchronisation among the units transmissions so that they are periodic but as much

---

* This work was supported by project FCT PTDC/EEA-CRO/100692/2008.

T. Röfer et al. (Eds.): RoboCup 2011, LNCS 7416, pp. 574–585, 2012.
© Springer-Verlag Berlin Heidelberg 2012

separated in time as possible and we show that this feature has a strong positive impact on the effectiveness of the ranging and thus on the quality of the localisation service. The paper is organised as follows. Section 2 discusses related work. Section 3 presents the information sharing approach. Section 4 presents the proposed synchronisation protocol. Section 5 shows the worst-case topologies in terms of information dissemination latency. Section 6 presents an experimental validation of the proposed framework, and finally, Sect. 7 conlcudes the paper.

2 Related Work and Contribution

Different approaches to relative (anchor-free) localisation based on wireless communications have already been explored [6]. One typical approach is to measure the pairwise distances and then share them among the units. With such distances, each node can use an adequate algorithm, such as Multi-Dimensional Scaling (MDS), to compute positions in a coordinate system. Concerning the measurement of pairwise distances, two main possibilities have been explored. One uses the received strength of the RF signal in message exchanges as distance estimation [4][7]. However, the received signal strength, despite its simplicity of use, is affected by many phenomena that hinder its relationship with distance and thus other approaches have been emerging such as using the Time-of-Flight (ToF) [6]. This approach, which we follow in our work, is independent of signal amplitude being thus much more robust. However, it is also more complex to use and takes substantially longer to measure.

Concerning the sharing of the pairwise distances a broadcast protocol must be used. We propose using the one in [2] and extended in [4] where a so-called extended connectivity matrix, filled in with the measured distances, is disseminated across a multi-hop network. This dissemination can be done without [4][6] or with [2][9][1] synchronisation of transmissions to reduce collisions. In the latter case, the synchronisation can be based on a global clock [2] or just relative [9][1] but [9] does not work in ad-hoc networks and [1] considers the network fully linked.

In this context, our contribution is a relative (loose) synchronisation protocol, based on the Recongurable and Adaptive Time Division Multiple Access (RA-TDMA) approach proposed in [9], to support the coordination and dissemination of pairwise distance measurements obtained with ToF. The protocol is fully distributed, works in ad-hoc networks supporting dynamic topologies including the separation and joining of cliques (subnetworks formed by partitions) and does not rely on clock synchronisation.

3 Information Sharing

Sharing information throughout the network is done with a broadcast protocol that disseminates a set of shared variables, each having one single producer and multiple consumers. This protocol makes use of a set of controls that regulate the updating of those variables in order to enforce consistency between the copies at the producer and consumers. They ensure that newer produced data eventually reaches all copies of each variable at the consumers as well as that stale information is detected and removed following a unit crash, departure from the team, or simply a link rupture. These controls are the following:

1. Local time-stamps, indicating the freshness of the data
   - one time-stamp per shared variable
   - time-stamps are reset when their respective information is updated (t_u), allowing to control the age
   - information is removed if not updated after a pre-set variable-dependent validity interval (t_{val})
2. Sequence numbers indicating between copies of the same shared variable which is the one containing fresher information
   - one sequence number per shared variable
   - each sequence number is increased by the producer unit right before it is sent together with the new information
   - larger number corresponds to newer data

Finally, note that each unit cleans up its own variables, i.e., removes (deletes) stale information, every time it broadcasts them, just before transmission. This means removing all variables for which $t_{now} \geq t_u + t_{val}$ (Fig. 1).

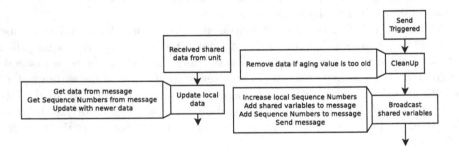

Fig. 1. Broadcast protocol - sending and receiving procedures

3.1   Sharing Distances

To share the pairwise distances obtained with the RF ranging mechanism we create in each unit k an extended connectivity matrix $M_{n \times n}^k$ similarly to [4], whose element (i, j) is the measured distance between nodes i and j and n is the number of units in the team. Each unit i writes in the ith line, only, so that $M^k(i)$ contains the view unit i has of the network, stored in unit k. When two nodes are out of range, a special code Ω is written in the matrix to represent such situation.

Each of these lines will be one shared variable, thus having an associated time-stamp and sequence number (Fig. 2).

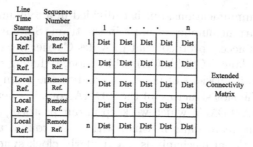

Fig. 2. Connectivity matrix and associated control variables

The distances are obtained using the nanoLoc development kit [5] that measures the ToF (ranging). We configured the ranging to be done in two phases (Fig. 3). The first phase measures $r_1 = V \times (t_1 - t_2)/2$ and the second one measures $r_2 = V \times (t_3 - t_4)/2$, where V is the propagation speed of the RF signal. Finally, r_2 is sent back and the values are averaged, thus the whole ranging procedure returns $Dist = (r_1 + r_2)/2$.

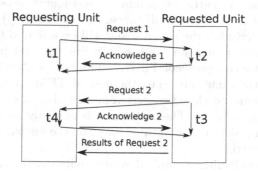

Fig. 3. ToF - Ilustration of the ranging process

One of the problems with this method is the latency resulting from each ranging ($t_{ranging}$), since each unit can only range another one at each time and such latency is variable depending on whether the ranged unit is online or not:

$$t_{ranging} = \begin{cases} 20ms, & \text{if unit is online} \\ [30, 100+]\,ms, & \text{if unit is not online} \end{cases} \quad (1)$$

Therefore, in order to avoid long latencies, we keep track in a vector of the current neighbourhood of each node, i.e., nodes from which messages were received in the previous cycle, and we range one unit from that vector per cycle.

4   Reconfigurable Ad-Hoc Synchronisation Protocol

Our approach follows the RA-TDMA protocol [9] in which the team units transmit in a round. Moreover, the round duration is predetermined, since it sets the reactivity of the communications, but it is divided in a dynamic number of slots according to the current number of units in the team. Similarly, we also use an underlying medium access protocol that provides Carrier Sense Multiple Access with Collision Avoidance (CSMA/CA) arbitration, reducing the collisions with alien traffic, i.e., transmissions of nodes external to the team, and even among team units while the slot structure of the TDMA round is being adjusted. The main purpose of RA-TDMA, which we keep in our protocol, is to separate the transmissions of the team units in time, within the round, as much as possible, without using global mechanisms, particularly clock synchronisation. This is done synchronising on the receptions of the messages sent by the other team units as shown further on.

However, unlike the original proposal, our protocol must cope with ad-hoc networks and dynamic topology, which requires new approaches to the propagation of the information in the network and to the agreement on the slots structure and assignment at each instant. These new features are supported on the extended connectivity matrix $M_{n \times n}^k$ presented before, which combines pairwise distances with topological information. The lines of the matrix present the vision each unit has of the network, and the columns present the vision the network has of each unit. In particular, a unit j is considered to be on-line if the jth column contains at least one valid distance, i.e., $\exists i : M^k(i,j) \neq \Omega$. Otherwise, it is considered to have left the network and will be removed from the team.

The current number of team units n is a fundamental parameter for the proposed synchronisation protocol. Firstly, all update periods for all nodes (t_{up}) are configured to the same value, i.e., the desired TDMA round period. Then, each unit autonomously divides this period in a number of slots equal to n with the duration of $t_{slot} = t_{up}/n$. Then, each slot is uniquely assigned to one unit, which is done with a slot allocation table based on the knowledge retrieved from the connectivity matrix.

Naturally, this mechanism requires all connectivity matrices of connected units to be consistent, which is enforced within a bounded interval (see Sect. 5) by the updating rules shown in Sect. 3. This interval sets a limit on the rate of topology changes that our protocol is capable of handling. Faster rates may prevent the team to reach consistent connectivity matrices.

Given a team of units, we define that all units reached an agreement when all have the same slot allocation table. In order to simplify consistency when updating the table we use the same strategy as in RA-TDMA, based on a unique identifier per unit. Whenever the table is changed we sort the list of on-line units by growing identifiers and assign them to the n slots in order, starting from slot 0.

Figure 4 shows a situation in which units 3 and 4 are connected through unit 0. The matrix transmitted from unit 0 to unit 4 carries the knowledge of a new team unit, 3 in this case, which allows unit 4 to build a consistent table.

Beyond the consistency of the slot allocation table, the units must also agree on the start of their slots. This particular aspect is also handled similarly to RA-TDMA but in a localised fashion in which each unit synchronises in each

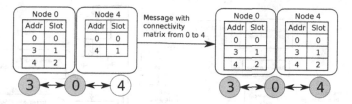

Fig. 4. Updating the slot allocation table

round with the units in range, only, using the messages received from them. This synchronisation propagates to the whole network through any connection path. In the beginning of each slot, each node sets the start of the next slot as one round later ($t_{tx}^{nxt} = t_{tx}^{now} + t_{up}$). Then, upon reception of a message in slot m at t_{rx}^m and duration t_{len}^m, algorithm 1 is executed to possibly adjust the start of the next slot. This causes a phase shift of the whole TDMA round.

Algorithm 1. Re-synchronisation upon reception of message in m^{th} slot

1. $t_{tx}^{nxt'} = t_{tx}^{now} + t_{rx}^m - t_{len}^m + (n - m) \times t_{slot}$
2. $t_{tx}^{nxt} = max(t_{tx}^{nxt'}, t_{tx}^{nxt})$

Figure 5 shows the synchronisation mechanism where the initial slots are marked with dashed lines. A delay in unit 1 is noticed by unit 3 that delays its next slot setting a new timeframe, marked with full lines. Units 2 and 4 are still unaware of this delay and keep their initial slots. Once unit 3 transmits in the adjusted slot, unit 2 is made aware of this adjustment and will synchronise. Finally, unit 4 will also synchronise after receiving a message from unit 2.

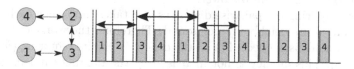

Fig. 5. Propagating slots synchronisation

Figure 6 shows the complete sending-receiving procedures of our ad-hoc broadcast and synchronisation protocol. In each round each node will receive at most once from each of its neighbours, aggregate all received matrices with its own, update its neighbourhood vector, possibly resynchronise the round timeframe, range one of its neighbours, update its matrix and transmit it.

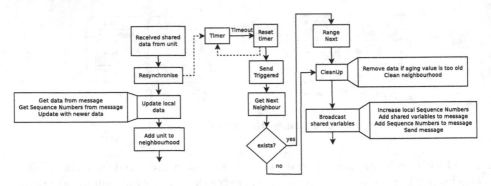

Fig. 6. Complete broadcast protocol - sending and receiving procedures

5   Upper Bounds to Information Propagation

Despite the unreliability of the wireless medium, it is reasonable to consider the medium lossless for the purpose of establishing some baseline properties that are intrinsic to the protocol. Here we will analyse the conditions that maximise the information propagation latency in the absence message losses.

First we analyse the worst case topology for information sharing. This situation is similar to the one reported in [2] and corresponds to the case in which all units form a line but sorted such that the identifiers decrease in the direction of the propagation of the information. For example, Fig. 7a shows the worst-case topology for propagating information from unit 4 to unit 0. In this case we will need one initial round for unit 4 to transmit its new information, such as a new node, which will be received by unit 3 that will transmit it it to unit 2 in the following round until the information gets to unit 1. At that point, one slot is enough to finally transfer the information to unit 0. The total worst-case latency is $(n-2) \times t_{up} + t_{slot}$. If any two units switch position, or if there is any parallelism, the latency will be lower.

However, acquiring the pairwise distances also takes time since each node will range only one of its neighbours per round. Therefore, the worst-case situation occurs when a node is connected to all the other $n-1$ units requiring $(n-1) \times t_{up}$ time (Fig. 7b). This latency is not increased by the ranging carried out by the other nodes since they occur in parallel in different slots.

Considering both information dissemination and ranging, simultaneously, we realise that for each extra ranging one unit has to perform, one less round is needed for disseminating the information. For example, in a line topology (Fig. 7a) unit n ranges unit $n-1$ and then, in the same round, transmits its information taking $n-2$ hops plus 1 slot to get to unit 0. If unit n was also connected to unit $n-2$, then it would take two rounds to range both units but it would take one less hop to get the information to unit 0, thus taking the same time as before. Consequently, the worst-case number of rounds required for all units to be in agreement after a topology change is upper bounded by $n-1$.

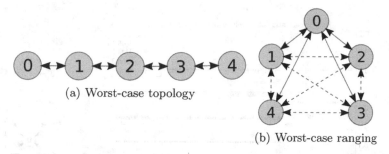

(a) Worst-case topology

(b) Worst-case ranging

Fig. 7. Worst-case data propagation latency scenarios

Finally, in terms of the maximum age that a given information might develop in the network, let us consider $max_{val} = t_{val}/t_{up}$ as the number of rounds in the information validity interval. Suppose now that a given information is max_{val} old when transmitted by unit n in a worst-case line topology. This information will arrive at unit 0 in less than $n-1$ rounds, thus within $max_{val} + (n-1)$ rounds after its generation. Then, unit 0 will keep it for another max_{val} rounds before removing it. Therefore, the maximum number of rounds that a piece of stale information can remain in the network before being removed is upper bounded by $2 \times max_{val} + (n-1)$.

6   Experimental Results

An experimental validation was carried out with a Nanotron's nanoLoc development kit [5]. This kit includes 5 nodes, each using an Atmega128l μC, communicating in the 2.4GHz ISM band according to IEEE 802.15.4 with a chirp modulation, which allows RF ranging using ToF.

We organise the experiments in two sections, firstly showing the synchronisation capabilities of this algorithm in a small room environment and secondly, exhibiting the improvements of using synchronisation in the ranging process. In all cases, $max_{val} = 10$ rounds, the ranges resolution is 1 byte, expressing distance in dm, and the value 255 is used to signal the out-of-range condition (Ω).

6.1   Validating the Synchronisation Protocol

We start by setting $t_{up} = 500ms$ and activating units 1, 3 and 4 which run the protocol. Unit 0 is used for monitoring purposes, only. As shown in Fig. 8, there are two disjoint subnetworks, one with unit 1 and the other with units 3 and 4. Note that the protocol allows each subnetwork to synchronise internally independently of each other (Fig. 8 left plot, up to round 26). Then, at that point, unit 2 is switched on and connects to both subnetworks, joining them, thus allowing the synchronisation to propagate across. After a short transient of

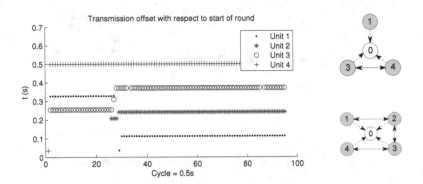

Fig. 8. Synchronising disjoint subnetworks with one unit

2 rounds, all units are synchronised with their transmissions separated as much as possible (125ms).

Figure 9 shows a case with 5 units and $t_{up} = 200ms$, where several consecutive network reconfigurations occur, with nodes joining and leaving. Units 0, 2 and 4 form a network and unit 3 joins at round 24 causing a resynchronisation from 3 to 4 slots. At round 45 unit 1 also joins. Then, at round 60, unit 4 leaves, which causes a resynchronisation later on, after $max_{val} + 1 = 11$ rounds, which is when it is removed by the nodes it was connected to. The same happens when unit 2 and later unit 3 leave the network. All protocol timings were verified.

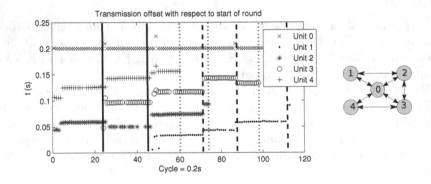

Fig. 9. Synchronisation protocol operating with joining and leaving units

6.2   Improvements in Ranging Success

In the following experiments, we set up a fully linked network with 5 nodes but two physical layouts aiming at analysing the impact the synchronisation has in the ranges performance, both on accuracy and in failure rate.

Concerning accuracy, we used the physical layout of Fig. 10a with a separation of $1m$ between two consecutive units. Unit 0 ranged every other unit and logged the results. We analysed 3500 rounds of operation with, and another 3500 without, synchronisation, with $t_{up} = 200ms$. The ranging results showed that the accuracy was similar in both cases with a negligible difference of the average errors, $abs(mean(D_{error}^{synch}) - mean(D_{error}^{\overline{synch}})) < 0.01m$. The difference in terms of standard deviation of the distance measurements was also small, despite larger, $abs(std(D^{synch}) - std(D^{\overline{synch}})) < 0.1m$.

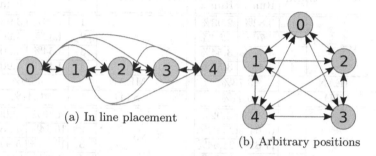

(a) In line placement

(b) Arbitrary positions

Fig. 10. Two fully linked physical layouts

Concerning ranging failure rates, we used the layout shown in Fig. 10b but we used units 1 to 4 only. Two runs of 5000 consecutive rounds were logged using a $t_{up} = 200ms$ and for both cases, with and without synchronisation. The percentage of failed ranges for each unit is shown in Table 1a. Then we repeated these experiments after having switched on unit 0, which was programmed to send a 127B packet every $20ms$ ($64.3kbps$), without synchronisation, just to create interference. The percentages of failed ranges for each unit, are shown in Table 1b.

The results of the synchronised experiments show a residual percentage of range failures that is similar for all units, between 3% and 4.4%. The results with interference show a minor degradation, with percentages of losses from 3.5% to 4.9%. On the other hand, the results without synchronisation show substantial degradation with certain nodes, in one case going up to 25%. Nevertheless, even without synchronisation it is still possible to find units exhibiting range failures similar to the synchronised case. This is easily explained looking at Fig. 11. In fact, without synchronisation some units will end up transmitting almost at the same time, which causes a high number of failures due to collisions, and other units will transmit very far apart, thus similarly to the synchronised case. For example, in Fig. 11b unit 1 has a very high clearance from the other units while units 2 and 3 are transmitting very close to each other. This log corresponds to the baseline Run 2 experiments without synchronisation in Table 1a. With synchronisation the team units do not practically interfere with each other. Consequently, the average range failure rate without synchronisation and without interference is 3.3 times higher than that obtained with synchronisation (12.5% compared to 3.7%), and the standard deviation of the units failures rates is one order of magnitude higher without than with synchronisation (8.48% and 4.62%

Table 1. Experimental results using topology in Fig. 10b

(a) Baseline measurements

$t_{up} = 200ms$
5000 samples
Interference: No

Synchronisation	Unit	Error Rates	
		Run 1	Run 2
Yes	1	3.82%	2.98%
	2	3.70%	3.53%
	3	4.36%	3.54%
	4	4.42%	3.52%
MEAN		3.73%	
STD		0.47%	
No	1	4.20%	3.92%
	2	13.64%	23.94%
	3	11.82%	24.54%
	4	3.62%	14.70%
MEAN		12.55%	
STD		8.48%	

(b) Measurements with noise

$t_{up} = 200ms$
5000 samples
Interference: $127Bytes/20ms$

Synchronisation	Unit	Error Rates	
		Run 1	Run 2
Yes	1	4.16%	3.92%
	2	4.32%	3.96%
	3	3.54%	3.50%
	4	4.46%	4.90%
MEAN		4.09%	
STD		0.47%	
No	1	4.36%	12.00%
	2	16.14%	4.88%
	3	10.82%	5.94%
	4	4.28%	13.02%
MEAN		8.93%	
STD		4.62%	

compared to 0.47%). However, the actual degradation of the situation without synchronisation depends on too many factors, such as starting conditions, and is thus very difficult to characterise accurately.

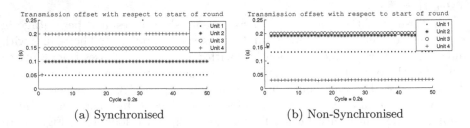

(a) Synchronised   (b) Non-Synchronised

Fig. 11. Periodic dissemination with and without synchronisation

7   Conclusions

Robot motion coordination and cooperative sensing are two areas that benefit from multi-robot cooperation based on wireless communication. In this paper we proposed a novel ad-hoc synchronisation / broadcast protocol that integrates the dissemination of ranging data through the network in an effective way contributing to an improved relative localisation service. This protocol extends a previous one named RA-TDMA which is meant for infrastructured scenarios.

Experimental results with IEEE 802.15.4 nanoLOC nodes validate the properties of the protocol, namely its ability to enforce synchronisation in ad-hoc scenarios even when a single path connects different nodes, its ability to acquire and efficiently disseminate ranging information thorough the network, as well as its effectiveness in reducing the failure rates of the ranging operations. As future work we plan to experiment on the limits of units velocities that the protocol can cope with.

References

1. Crenshaw, T., Tirumala, A., Hoke, S., Caccamo, M.: A robust implicit access protocol for real-time wireless collaboration. In: ECRTS 2005 - 17th Euromicro Conference on Real-Time Systems, pp. 177–186 (July 2005)
2. Facchinetti, T., Buttazzo, G., Almeida, L.: Dynamic resource reservation and connectivity tracking to support real-time communication among mobile units. EURASIP J. Wireless Communications and Networking 2005(5), 712–730 (2005)
3. Fazenda, P., Lima, P.: Non-holonomic robot formations with obstacle compliant geometry. In: IAV - IFAC Symposium on Intelligent Autonomous Vehicles (2007)
4. Li, H., Almeida, L., Carramate, F., Wang, Z., Sun, Y.: Connectivity-aware motion control among autonomous mobile units. In: SIES - International Symposium on Industrial Embedded Systems, pp. 155–162 (June 2008)
5. Nanotron: nanoloc development kit (2010),
   http://www.nanotron.com/EN/PR_nl_dev_kit.php
6. Neuwinger, B., Witkowski, U., Rückert, U.: Ad-hoc communication and localization system for mobile robots. In: Kim, J.-H., Ge, S.S., Vadakkepat, P., Jesse, N., Al Manum, A., Puthusserypady, S.K., Rückert, U., Sitte, J., Witkowski, U., Nakatsu, R., Braunl, T., Baltes, J., Anderson, J., Wong, C.-C., Verner, I., Ahlgren, D. (eds.) Advances in Robotics. LNCS, vol. 5744, pp. 220–229. Springer, Heidelberg (2009)
7. Oliveira, L., Li, H., Almeida, L.: Experiments with navigation based on the rss of wireless communication. In: ROBOTICA 2010 10th Conference on Mobile Robots and Competitions (March 2010)
8. Ryan, A., Tisdale, J., Godwin, M., Coatta, D., Nguyen, D., Spry, S., Sengupta, R., Hedrick, J.: Decentralized control of unmanned aerial vehicle collaborative sensing missions. In: ACC 2007 - American Control Conference, pp. 4672–4677 (July 2007)
9. Santos, F., Almeida, L., Lopes, L.: Self-configuration of an adaptive tdma wireless communication protocol for teams of mobile robots. In: ETFA - IEEE Int. Conference on Emerging Technologies and Factory Automation, pp. 1197–1204 (2008)
10. Song, K.T., Tsai, C.Y., Huang, C.H.C.: Multi-robot cooperative sensing and localization. In: ICAL - IEEE International Conference on Automation and Logistics, pp. 431–436 (September 2008)
11. Wang, Z., Hirata, Y., Kosuge, K.: Checking movable configuration space in c-closure object for object caging and handling. In: ROBIO - IEEE International Conference on Robotics and Biomimetics, pp. 336–341 (2005)

Real-Time 3D Ball Trajectory Estimation for RoboCup Middle Size League Using a Single Camera

Hugo Silva, André Dias, José Almeida, Alfredo Martins, and Eduardo Silva

INESC PORTO/ISEP
ISEP, Instituto Superior de Engenharia do Porto
ESEIG, Escola Superior de Estudos Industriais e de Gestão
Rua Dr Antonio Bernardino de Almeida 431 4200-072 Porto, Portugal
{hsilva,adias,jma,aom,eaps}@lsa.isep.ipp.pt
http://www.lsa.isep.ipp.pt

Abstract. This paper proposes a novel architecture for real-time 3D ball trajectory estimation with a monocular camera in Middle Size League scenario. Our proposed system consists on detecting possible multiple ball candidates in the image, that are filtered in a multi-target data association layer. Validated ball candidates have their 3D trajectory estimated by Maximum Likelihood method (MLM) followed by a recursive refinement obtained with an Extended Kalman Filter (EKF). Our approach was validated in real RoboCup scenario, evaluated recurring to ground truth information obtained by alternative methods allowing overall performance and quality assessment.

Keywords: Monocular Vision, 3D Trajectory Estimation, Maximum Likelihood Method, Extended Kalman Filter, RoboCup Middle Size League.

1 Introduction

This work addresses an important problem in RoboCup Middle Size League (MSL)[5], which is the real-time ball trajectory estimation in 3D space, using only a perspective monocular camera. Our MSL Team ISePorto robots, have as primary mean of perception, three perspective monocular cameras. However, due to a restrictive number of factors mainly related to the lack of camera synchronization together with the currently available cameras field of view, makes stereo vision not our most suitable option.

Considering previous restrictions, 3D ball estimation by a single camera is a demanding task, since current MSL robots are able to kick a soccer ball at very high velocities (more than 10 m/s). This is particularly challenging when the ball is kicked into the air at high velocity and describes a 3D trajectory that must be estimated in real-time.

Our camera system provides a sequence of observations which are non-linearly related to the ball position in 3D. The estimation of parameters from a sequence of non-linear observations is a non-linear filtering problem which can be tackled using recursive algorithms such as the Extended Kalman Filter and the Unscented Kalman filter [2]. These methods are appropriate for real-time applications since they recursively update the estimates every time a new observation is received. However, they require a

T. Röfer et al. (Eds.): RoboCup 2011, LNCS 7416, pp. 586–597, 2012.

Fig. 1. Image sequence of a typical 3D ball trajectory in Robocup Middle Size League Scenario, (game between Cambada and TechUnited)

good initialization. Unfortunately, the available sensor does not provide an accurate estimate of ball position and velocity after a kick. This prevents a widespread use of the previous methods as stand-alone techniques.

To overcome this difficulty we combine two methods: the Maximum Likelihood method (MLM) which is able to provide an initial estimate of the trajectory and the Extended Kalman filter (EKF) which is able to recursively update the trajectory using new information. The MLM uses the first few frames of the ball trajectory to obtain the initial estimate and the EKF tracks the ball in the following frames, both methods are combined in a Multi-Target data association layer, that follows validated ball candidates presented in the retrieved images. This association layer uses image target dynamics in robot world coordinates to distinguish between possible ball candidates and false positive ones.

The 3D ball estimation system is based on the analysis of 3D motion parameters estimated from 2D image observations: image ball coordinates are detected using vision system described in [9]. Physical ball motion model and camera projective model are combined in order to obtain a set of measurements, non-linearly related to the 3D ball position and velocity of the ball at the initial time instant. After, a ML estimator is applied to the ball trajectory, pure parabolic trajectory is assumed. The number of available data measurements influences ML estimator error, so by analysis of a cost function together with the number of available data measurements it's possible to define when the ML estimator error is below a given threshold.

This paper is organized as follows:

- Section 2 presents related work associated to the problem; some of the methods already address the real time concerns.
- Section 3 formulates the problem and the describes the architecture proposed in this paper.
- Section 4 presents experimental results and a comparison with ground truth information.
- Section 5 presents conclusions and discusses future work.

2   Related Work

The 3D ball estimation problem, in Robocup Scenario was first approached by Voigtlander et al.[10], who used a mixture of Perspective and Catadioptric cameras to estimate the 3D ball trajectory. The 3D-ball position was given by basic stereo triangulation extended by a 2D Kalman filter approach to the 3D scenario. Motivated by the same challenge Birbach et al. [2] developed a monocular vision system with aide of a inertial sensor to track a moving ball. The pose of the camera is obtained by integrating gyro and accelerometers measurements, to minimize the error caused by inertial measurements, visual landmarks of the RoboCup field are used to correct the camera pose estimation. The camera provides the direction towards the ball through it's image position, depth information comes from the effect of gravity over time, which directly influences the ball radius in the image. This paper also compares results obtained using a ML estimator approach with an Unscented Kalman filter(UKF), but unlike our approach there is no integration of both methods. There are also other approaches that performs online 3D trajectory estimation using EKF [7], however this technique only works if the camera is on a stationary position, becoming a straight forward solution.

Furthermore, 3D trajectory ball estimation is also being applied in other types of systems, namely the ones related to computer game study of sports. The work realized by Chen et al. [4], estimates 3D ball motion from 2D frames in basketball games. It incorporates physical characteristics of ball motion to reconstruct the 3D trajectories and estimate the player shooting locations. Another 3D ball estimation work related to sport analysis was developed by Yu et al. [11], who conducted ball estimation analysis in soccer videos.

Other approach is presented by Benovoli et al. [1], that uses MLM to help estimate the ballistic target problem. Some of the work proposed by Ribnick et al. [8], also uses MLM in the estimation of 3D projectile positions and velocities using monocular cameras.

In the literature, there are other application scenarios related to different types of sensors namely radar systems, Farina [6], also applies ML estimation techniques to track airborne objects.

3   Implementation of 3D Ball Estimation Method

3.1   3D Ball Estimation Architecture

This work is based on achieving 3D ball estimation with a single camera for use in real time applications, such as Robocup Middle Size League.

The 3D ball estimation is achieved using a robot that has a differential traction wheelbase with an additional rotating upper body housing the onboard computer, wireless communications module, inertial and magnetic sensors and a kicking device. The kicker mechanism has two fixed cameras, one used mainly for close range ball detection and the other for 3D aerial ball estimation. Has mentioned previously, due to having a short baseline, stereo vision is not our most suitable option. On top of the upper body, the robot has a rotating head with a camera for long range target tracking.

Overall system architecture is described in figure 2. The vision system starts by acquiring images at a pre-determinate frame rate (in this case 30 fps 640x480 image resolution).

Afterwards, image segmentation is conducted and relevant object (ball) information is extracted from the image, according to the work described in detail by Silva et al. [9].

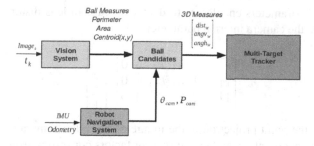

Fig. 2. Ball estimation Architecture

Basically, image ball information is retrieved using image blob extraction techniques together with edge detection that uses weighted least squares circle estimation, in order to obtain a more robust ball estimation. This information, contains all relevant ball information measures such as: ball perimeter area, bounding box, ball centroid and ball radius, all of this in 2D image plane.

Due to the fact that MSL League has a very dynamic background environment, more than one ball candidate may appear in the same image, thus the system must be able to cope with multiple ball target candidates,

Additionally, information is provided by the robot Navigation System, namely: odometry information provided by wheel encoders and gyro information given by the robot Inertial Measurement Unit. This information is fused to obtain the camera position and bearing (P_{cam}, θ_{cam}) in World coordinates, see figure (1,2). The World frame is centered in the middle of the Robocup Field with x axis oriented from our team side to the opponents side.

$$\theta_{cam} = \theta_{kicker} + \theta_{robot} \tag{1}$$

$$P_{cam} = P_{robot} + R(\theta_{cam}) \times P_{camkicker} \tag{2}$$

Having the 2D ball information fused with the robot information, we can obtain 3D information of the ball related to World frame but in other coordinate frame whose center is the robot middle point (at the wheelbase).

The 3D ball candidates information is managed by a Multi-target Tracker framework. This framework starts by initializing a ML estimator for each of the possible ball candidates. After the ML estimator converges, we can initialize a EKF-filter to track the ball. The objective of this framework is twofold: First, we use the tracker to eliminate erroneous measures of balls that appear in the public, by analysing it's dynamic related to the robot motion. Second, we are able to initialize the EKF-filter only when ML estimator has good convergency and analyse it's behavior when we update the ball information in consecutive frames.

3.2 Camera Model

Considering a pinhole camera projection model, image points relate to the camera reference frame by means of the camera intrinsic parameters. To obtain these parameters an offline calibration [3] is performed, by acquiring measures to a known 3D target.

The intrinsic parameters enables us to deal with camera lens distortion factor and allows to obtain the camera internal parameters (3).

$$\begin{bmatrix} x \\ y \\ 1 \end{bmatrix} = \begin{bmatrix} fkx & k_\theta & x_0 & 0 \\ 0 & fky & y_0 & 0 \\ 0 & 0 & 1 & 0 \end{bmatrix} \begin{bmatrix} c_x \\ c_y \\ c_z \\ 1 \end{bmatrix} \tag{3}$$

Where (x, y) is the point projection in the image plane, f the camera focal length and (x_0, y_0) the principal point, kx and ky are scaling factors converting from space metrics to pixels in image and k_θ is an additional skew factor that usually is set to 0.

Lens distortion (such as occurring in wide angular lens) can be corrected prior by correcting the relevant pixel through a pre-calculated look up table.

We can relate the image reference frame with the camera in the kicker coordinated frame by means of some pre-determined (calibrated) camera parameters ($R_{\theta cam}$, $T_{\theta cam}$), (4). These angles are have their initial point in the camera but they are aligned with the World coordinate frame.

$$P_{img} = A \left[R_{\theta cam} | T_{\theta cam} \right] P_{camkicker} \tag{4}$$

After the ball estimation points in the kicker coordinate frame are obtained it's possible to calculate the horizontal and vertical angle ($ang_h w$, $ang_v w$) in polar coordinates by means of simple perspective projective geometry. In figure 3, one can observe the different coordinate frames, from the image plane to world coordinate frame.

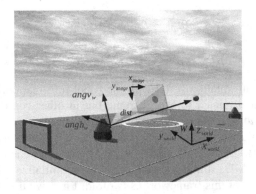

Fig. 3. System Reference Frames

In order to obtain the distance to the ball, we need to acquire bearing measures of two points of the ball. The estimated ball radius and added with the estimated ball center is used to obtain the other ball point. Since the ball radius is previously known and it's equal to a FIFA size 5 ball, the distance to the ball is thus given by, (6).

$$\theta_{ball_{cxy}} = \arctan \left(\frac{\cos(\arctan\left(\frac{y - C_{py}}{F_{cy}}\right)) \tan(ang_{hw})}{\cos(ang_{vw})} \right) \tag{5}$$

$$dist_w \simeq \left(\frac{R_{ball}}{angh_w - angh_{wplimit}} \right) \tag{6}$$

Where $angh_{wplimit}$ is the polar coordinate horizontal angle to a image point of the edge of the ball at the right of the ball center.

3.3 Parabola Model

Ball trajectory is considered to be a parabola in the x, y and z World coordinate frame (see figure 3). The ball trajectory can be formally represented by a dynamical system given by ordinary differential equations as a function of time T that follow classical mechanics such as gravitation.

Trajectory equations are given by (7):

$$\begin{cases} x(T) = x_0 + v_x T \\ y(T) = y_0 + v_y T \\ z(T) = z_0 + v_z T - \frac{1}{2} g T^2 \end{cases} \tag{7}$$

Where g is the gravity acceleration and v_x, v_y and v_z are the velocity components in relation to x, y and z. The trajectory described by the ball belongs to a plane which determines an angle with the x axis having a start point $P_{init}(x_0, y_0, z_0)$ with $(z_0 = 0)$ and $T = 0$, a maximum height $P_{max}(x, y, z)$ and an impact point $P_{end}(x, y, z)$ for $T \neq 0$ at $z(T) = 0$.

$$\frac{dz(T)}{dT} = v_z - gT = 0$$
$$T_{Zmax} = \frac{v_z}{g} \tag{8}$$

Where T_{Zmax} is the ball maximum height.

3.4 Maximum Likelihood Method

In 3D ball estimation, ball trajectory can be identified by the initial point $P_0(x_0, y_0, z_0)$ and by it's initial velocity vector, whose components are v_x, v_y and v_z, if we obtained the kicking instance t_0. Taking in consideration all k images frames, measurements of distance $(dist_w(k))$, vertical angle $(angv_w(k))$ and horizontal angle $(angh_w(k))$, collected into the vector measurement:

$$z(k) = \left[dist_w(k)\; angv_w(k)\; angh_w(k) \right]^T + n_k \tag{9}$$

with a measurement noise which is Gaussian, zero-mean and with the following covariance matrix:

$$R = \begin{bmatrix} \sigma_{dist_w}^2 & 0 & 0 \\ 0 & \sigma_{angv_w}^2 & 0 \\ 0 & 0 & \sigma_{angh_w}^2 \end{bmatrix} \tag{10}$$

$\sigma^2_{dist_w}$ $\sigma^2_{angv_w}$ $\sigma^2_{angh_w}$ are the variances of the ball position measurement errors. The MLM calculates the unknown vector parameter $x = [x_0, y_0, z_0, v_x, v_y, v_z]^T$. The unbiased estimator $\hat{x}$ is obtained by solving the minimization problem $\hat{x} = argmin_x \Lambda(x)$, where $\Lambda(x)$ represents the ML estimator functional and T the camera frame rate.

$$\Lambda(x) = argmin_x \sum_{k=1}^{N} [Z_k - h(x)]^T R^{-1} [Z_k - h(x,k)] \tag{11}$$

$$h(x,k) = \begin{cases} \sqrt{(xo+vxt(k)-Pcx)^2 + (yo+vyt(k)-Pcy)^2 + (zo+vzt(k)-\frac{1}{2}gt(k)^2-Pcz)^2} \\ \\ arcsin \frac{zo+vzt(k)-(1/2)gt(k)^2-Pcz}{(xo+vxt(k)-Pcx)^2+(yo+vyt(k)-Pcy)^2+(zo+vzt(k)-\frac{1}{2}gt(k)^2-Pcz)^2)} \\ \\ arctan \frac{yo+vyt(k)-Pcy}{xo+vxt(k)-Pcx} \end{cases}$$

$$\tag{12}$$

3.5 3D-Extended Kalman Filter

To perform 3D ball estimation, we used an Extended Kalman Filter approach, that was initialized using the ML estimator fit.

This is useful because the EKF employs a model that only possesses an approximate knowledge of the state. Unlike the linear Kalman Filter algorithm, the filter must be accurately initialized at the start of operation to ensure that we obtain a valid model. If this is not done, the estimates computed by the filter will simply be meaningless. So, we initialized $\hat{X}$, using the MLM estimator and thus calculate the ball trajectory, based on those estimations.

We developed two different approaches to the EKF filter, one where the State was the target (ball) current position and velocity, and the other one which is presented in this paper, where the state is the ball initial position and velocity(x0, y0, z0 , vx0, vy. We preferred the initial position approach to the EKF filter, because estimating the initial position of the ball, is much smoother than tracking at each time instant, whose measure are more noisy.

Filter State is:

$$\hat{X} = [\hat{X}_{x_0} \ \hat{X}_{y_0} \ \hat{X}_{z_0} \ \hat{X}_{v_x} \ \hat{X}_{v_y} \ \hat{X}_{v_z}]$$

With the Predict Step:

$$\hat{X}_{k|k-1} = \hat{X}_{k-1}$$

$$\hat{P}_{k|k-1} = \hat{P}_{k-1} + Q$$

and Update Step:

$$\tilde{y}_k = z_k + h(\hat{X}_{k|k-1})$$

$$S_k = H_k P_{k|k-1} H_k^T + R$$

$$K_k = P_{k|k-1} H_k^T S_k^{-1}$$

$$\hat{X}_{k|k} = \hat{X}_{k|k-1} + K_k \tilde{y}_k$$

$$P_{k|k} = (I - K_k H_k) P_{k|k-1}$$

4   Experimental Setup

Experimental setup is divided into two parts: First, the ISePorto robot setup and second the external stereo vision system used to produce the 3D ground-truth analysis.

Robot Setup. Tests were conducted using the ISePorto Robot in a middle size field. The robot was positioned fixed at the goal. Robot vision hardware is composed by three USB cameras, working at 30 FPS with VGA (640x480) resolution connected to a Intel Pentium Dual Core PC 2 GHz and 2GB memory running a RT-Linux operating system. High level software is composed by several modules, being one of this modules the acquisition and image processing system [9], implementing the ball detection algorithm.

Ground-Truth 3D Ball Tracking. In order to evaluate the quality of the 3D ball estimation from the ISePorto robot, an exogenous system of cameras in a stereo baseline was introduced as a ground-truth(GT), figure 4. Cameras were positioned with a baseline of 13 meters and connected via GigEthernet to a Intel Pentium Dual Core PC 2 GHz running a Linux operating system. The cameras used are Jai CB080GE working at 20 FPS with 1024x768 resolution with external sync trigger.

In performed tests, the ball is moving during most of the time. Therefore, first step in ball detection by ground-truth cameras is to extract moving objects by a background frame subtraction operation. The pixel-wise subtraction extracts all the moving objects in the frame obtaining not only a ball but also other type of objects. We eliminate all objects that are not possible ball candidates by performing colour threshold segmentation in HSV space

5   Results

We performed multiple tests to evaluate the ML estimator fit and EKF estimation, in several parabolic kicks. Some of those results are presented in this section. First, we evaluated the results obtained by the robot MLE estimator versus the Ground-truth, using the images acquired by the stereo vision GT system. In figure 4, the red line is ML estimator from the ground truth and in black is the ML estimator from the robot.

In figure 5, the ground-truth error for a parabolic ball trajectory is displayed. One can observe that ground-truth ball does not describe a true parabolic trajectory, but

Fig. 4. MLE fit from Ground-Truth stereo compared to the Robot MLE estimator

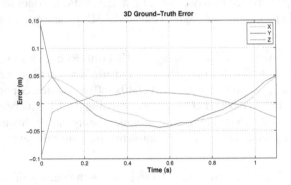

Fig. 5. Ground-truth error

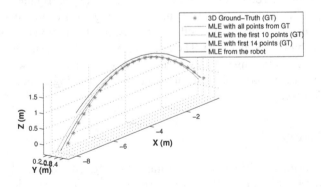

Fig. 6. Ground-Truth and MLE from the Robot with different initialization points

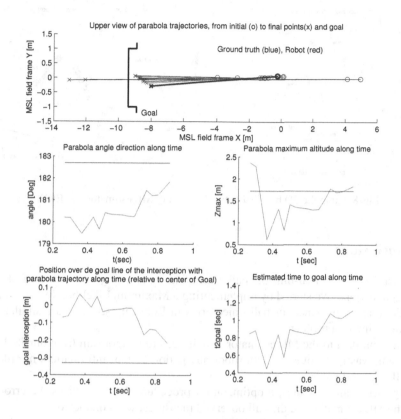

Fig. 7. 3d Ball Estimation Trajectory Analysis

shows some side effects. In figure 6, one can observe results obtained by the robot MLE estimator for different observations measures.

In figure 7, one can observe the 3D ball trajectory estimation analysis. On top, an upper view of the parabola trajectories are displayed in World coordinates. It shows the converges of the 3D ball trajectory has the MLE estimator of the robot gets more ball observation measures. It also displays the convergency between robot and GT MLE estimators. In the bottom left of the figure, one can observe the position over the goal line of the described parabola trajectory. It's possible to observe compared to figure 5, that the ball describes a small trajectory deviation at the beginning of the trajectory.

From figure 8 one can observe the MLE estimator behavior as a initialization mechanism to the EKF filter. This results represents another benchmark test performed with the robot and ground-truth..

In figure 8, results obtained by the ML estimator together with EKF estimation for the same kick are displayed. The ball described two parabolic trajectories in the image. The difference between the ML estimator and the EKF estimation does not exceed 2 cm standard deviation.

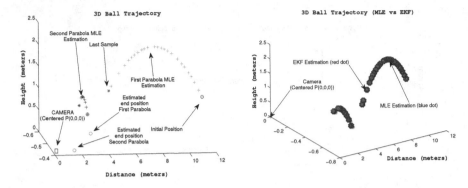

Fig. 8. Parabolic 3D ball Estimation Trajectory ML estimator and EKF

6   Conclusions

This paper describes a real-time 3D ball estimation using a single camera for Robocup league applications. We started by implementing a Maximum Likelihood Estimator for parabolic trajectories, since air ball trajectories in RoboCup scenario are similar to a projectile estimation problem.

As an input data to the ML estimator ,a ball detection algorithm from our ISePorto MSL Team, was used together with robot navigation system information, in order to obtain 3D ball estimates.

Using these ball estimates, a optimization procedure that minimizes the error between the ML estimator and the ball observed measures, was conducted.

However, our ML estimator fit works in batch computation and requires at least 8 measures to have a good estimate of the ball trajectory, this is equivalent to almost 0.5 seconds of our camera frame rate time. In order to try to extend the ML estimator approach to be computed in real-time, we developed an Extended Kalman Filter that used the ML estimator fit has initial condition. Furthermore, we state that the EKF and ML estimator both estimate ball trajectory with very small standard deviation.

All results were evaluated using a ground-truth stereo vision system, proving that results obtained by the robot using only a single camera are similar to the ones obtained by external synchronize stereo vision system.

Acknowledgements. This work was supported by Portuguese FCT under grant SFRH/BD/47468/2008 and PCMMC project with reference PTDC/EEA-CRO/100692/ 2008, and by Porto Politechnic Institute under grants SFRH/PROTEC/49796/2009, SFRH/PROTEC/49795/2009 and SFRH/PROTEC/49716/2009.

References

1. Benavoli, A., Farina, A., Ortenzi, L.: Mle in presence of equality and inequality nonlinear constraints for the ballistic target problem. In: IEEE Radar Conference, RADAR, vol. 2008, pp. 1–6 (23-30, 2008)

2. Birbach, O., Kurlbaum, J., Laue, T., Frese, U.: Tracking of Ball Trajectories with a Free Moving Camera-Inertial Sensor. In: Iocchi, L., Matsubara, H., Weitzenfeld, A., Zhou, C. (eds.) RoboCup 2008. LNCS (LNAI), vol. 5399, pp. 49–60. Springer, Heidelberg (2009)
3. Bouguet, J.Y.: Camera calibration toolbox for Matlab (2008),
   `http://www.vision.caltech.edu/bouguetj/calib_doc/`
4. Chen, H.T., Tien, M.C., Chen, Y.W., Tsai, W.J., Lee, S.Y.: Physics-based ball tracking and 3d trajectory reconstruction with applications to shooting location estimation in basketball video. Journal of Visual Communication and Image Representation 20(3), 204–216 (2009),
   `http://www.sciencedirect.com/science/article/B6WMK-4V42J7T-1/2/ee679c824d68763735882866737dd33d`
5. Dias, A., Almeida, J., Martins, A., Silva, E.: Traction characterization in the robocup middle size league. In: IEEE Robotica 2007 - 7th Conference on Mobile Robots and Competitions, pp. 2007–(2007)
6. Farina, A., Di Lallo, A., Timmoneri, L., Volpi, T., Ristic, B.: Crlb and ML for parametric estimate: new results. Signal Process 86(4), 804–813 (2006)
7. Herrejon, R., Kagami, S., Hashimoto, K.: Online 3-d trajectory estimation of a flying object from a monocular image sequence. In: Proceedings of the 2009 IEEE/RSJ International Conference on Intelligent Robots and Systems, IROS 2009, pp. 2496–2501. IEEE Press, Piscataway (2009),
   `http://portal.acm.org/citation.cfm?id=1733023.1733149`
8. Ribnick, E., Atev, S., Papanikolopoulos, N.P.: Estimating 3d positions and velocities of projectiles from monocular views. IEEE Trans. Pattern Anal. Mach. Intell. 31(5), 938–944 (2009)
9. Silva, H., Almeida, J.M., Lima, L., Martins, A., Silva, E.P.: A Real Time Vision System for Autonomous Systems: Characterization during a Middle Size Match. In: Visser, U., Ribeiro, F., Ohashi, T., Dellaert, F. (eds.) RoboCup 2007. LNCS (LNAI), vol. 5001, pp. 504–511. Springer, Heidelberg (2008)
10. Voigtlander, A., Lange, S., Lauer, M., Riedmiller, M.: Real-time 3d ball recognition using perspective and catadioptric cameras (2005)
11. Yu, J., Tang, Y., Wang, Z., Shi, L.: Playfield and Ball Detection in Soccer Video. In: Bebis, G., Boyle, R., Parvin, B., Koracin, D., Paragios, N., Tanveer, S.-M., Ju, T., Liu, Z., Coquillart, S., Cruz-Neira, C., Müller, T., Malzbender, T. (eds.) ISVC 2007, Part II. LNCS, vol. 4842, pp. 387–396. Springer, Heidelberg (2007),
    `http://portal.acm.org/citation.cfm?id=1779090.1779135`

Author Index